THE PRE-ESTABLISHED PHYSICS AND THE QUANTUM IMPULSE

THE REFUSAL OF THE PHYSICS IN APPLYING ITS OWN SCIENTIFIC METHOD TO COVER UP THEIR OUTDATED FUNDAMENTALS

J. P. HANSEN

Copyright 2019 - J P Hansen
jph.hansen@gmail.com
Independent Publishing - USA
ISBN – Number: 9781794546431

ACKNOWLEDGEMENTS

Thanks to all who participated in this publication. To outstanding CreateSpace Publishing of unparalleled vision and worldwide information. With appreciation to the Superior Officer of Sciences of the Brazilian Navy, Emmanuel Gama de Almeida; To the Director of ONE-Energy Optimization, Dr. Everton Novais and to all Colleagues and Enterprise Contacts, through Scientific and Technological dialogues; And loved ones.

CREDITS

Figures, Cover, Photos, Tables and Documents: By the author and or adapted from Pinterest and Google images, and References mentioned.

CONTENTS

PART 1 – INITIAL CONCEPTS...1

I.	INTRODUCTION………………………………………………1
II.	PRIMORDIUM………………………………………………….1
III.	SPHERE UNIVERSE……………………………………………4
IV.	DOGMAS AND NEW IDEAS ……………………………...13
V.	EARTH AT THE CENTER OF THE UNIVERSE……………15
VI.	SCIENCE AND RESISTANCE………………………………..17
VII.	NEW HYPOTHESES AND NEW RESISTANCES…………..24

PART 2 – EVOLUTION OF BASIC CONCEPTS OF PHYSICS UNTIL THE QUANTUM......................35

VIII.	FORCE, AGENT OF MOTION, VACUUM, SPEED, VELOCITY, IMPETUS, SPECIES OF MATTER, MOMENTUM, INERTIA AND MASS, IMPULSE, ACCELERATION AND ACTION, ENERGY.............................35
IX.	GRAVITY...46
X.	MASS, MOLECULE AND CHEMICAL ELEMENTS...............52
XI.	PRESSURE, VOLUME, TEMPERATURE, MOTION, HEAT, PARTICLES, SHOCK SECTION, PARTICLE FLOW..............57
XII.	PARTICLES AND WAVES OF LIGHT..............................61
XIII.	MAGNETISM, ELECTRICITY AND OPTICS...........................66
XIV.	THE CATHODE RAYS AND THE DISCOVERY OF THE ELECTRON..108
XV.	X-RAYS..114
XVI.	RADIOACTIVITY – ALPHA, BETA AND GAMMA RAYS..116
XVII.	COSMIC RAYS..128
XVIII.	QUANTUM 130

PART 3 – THEORY OF RELATIVITY, RELATIONAL MECHANICS AND QUANTUM MECHANICS...133

IXX.	THEORY OF RELATIVITY...133
XX.	ABSOLUTE SPACE AND FORCE AT A DISTANCE..............142
XXI.	RELATIONAL MECHANICS...144
XXII.	CHARGES, MASS NUMBERS, ATOMIC NUMBER AND ISOTOPES………………………………………………...…..147
XXIII.	QUANTUM ATOM 149
XXIV.	NEUTRINO, POSITRON AND NEUTRON 165
XXV.	PAIR CREATION AND ANNIHILATION 166
XXVI.	ATOMIC NUCLEUS..169
XXVII.	FIELD AND INSTANT COMMUNICATION, THEORETICAL MASS, PHOTON MASS, IMAGINARY MASS, NEGATIVE MASS, SPIN WAVES, PUNCTUAL SPIN WAVES, PUNCTUAL SPIN, CASIMIR EFFECT, STATISTIC AND FIELD...176
XXVIII.	MESONS..179
XXIX.	DIVERGENCES, MATHEMATICAL ABSTRACTIONS, NEGATIVE ENERGIES, MORE MATRICES, STATISTICS AND PROBABILITIES FIELDS, "PHANTOM PARTICLES" WITHOUT MASS AND STRANGE PARTICLES.....................182

PART 4 – QUANTUM IMPULSE 189

XXX.	THE THEORY OF QUANTUM IMPULSE	189
XXXI.	DIRAC QUESTIONS ON FUNDAMENTAL CONSTANTS AND THEORY OF QUANTUM IMPULSE	219
XXXII.	PROTON MODEL BY QUANTUM IMPULSE THEORY	220
XXXIII.	MASER, LASER AND QUANTUM IMPULSE THEORY	223
XXXIV.	NEUTRINO MASS AND QUANTUM IMPULSE	226

PART 5 – FROM PARTICLES MASS TO ZERO MASS IN PARTICLES AND FIELDS, FROM ZERO MASS TO MASS CREATED OF FIELDS AND PARTICLES 229

XXXV.	UNIVERSAL BETA INTERACTION AND NEW PARTICLES	229
XXXVI.	ELECTRIC AND MAGNETIC FIELDS WITHOUT MASS AND WITH MATHEMATICAL ABSTRACTIONS	230
XXXVII.	PARTICLES WITHOUT MASS AND MASSLESS FIELD THAT CREATES MASS BY MATHEMATICS: THE HIGGS FIELD	231
XXXVIII.	ELECTRO WEAK THEORY	232
XXXIX.	THE CREATION OF ELEMENTARY PARTICLES, OCTETS, ACES AND QUARKS, GLUONS OR "PASTE WITHOUT MASS"	233

PART 6 – QUANTUM IMPULSE AND GRAVITATIONAL RADIATION 235

XL.	GRAVITATION AND QUANTUM IMPULSE	235

PART 7 – THE QUANTUM IMPULSION AND THE RETURN OF MEDIEVAL FORCES 245

XLI.	NATIONAL AND WORLD DISSEMINATION OF QUANTUM IMPULSE THEORY	245
XLII.	THE MACKENZIE UNIVERSITY-BRAZIL AND QUANTUM IMPULSE THEORY	255
XLIII.	NEUTRON AND QUANTUM IMPULSE	262
XLIV.	UNKNOWN SOLUTIONS	267
XLV.	NEW EXPERIMENTS	269
XLVI.	TRIBUTE TO ISTVAN ERDELYI – IMMORTAL CREATOR OF QUANTUM IMPULSE THEORY	270
XLVII.	NEW PROPOSITIONS	270
XLVIII.	INTERNATIONAL SYMPOSIUM ON ASTRONOMY DIRECTED BY MYOPICS ASTRONOMICAL!	271
XLIX.	ADVANCEMENT IN TEACHING AND THE NUCLEAR AREA	271
L.	FORMAL REPORT OF CURRENT PHYSICS ON UNIDENTIFIABLE AERIAL OR FLYING OBJECTS	273
LI.	THE QUASARS AND GRAVITATIONAL RADIATIONS OF QUANTUM IMPULSE	274
LII.	FORMAL CONFESSION OF FORMAL PHYSICS	274
LIII.	ASTRONOMY OBSERVATORY AND THE QUANTUM IMPULSE THEORY	275

LIV.	THE BRAZILIAN SOCIETY OF PHYSICS-SBF AND THE QUANTUM IMPULSE THEORY	280
LV.	CRYSTALLIZED PHYSICS AND THE FANTASTIC PROBABILITY OF FREE NEUTRON DISINTEGRATION	282
LVI.	THE TRIAL OF "THE NEW FACE OF THE PHYSICAL WORLD"	283
LVII.	"SPACE COMPASS" INITIAL EXPERIMENTS	286
LVIII.	UPPER LIMIT	287
LIX.	SCIENCE & CULTURE JOURNAL OF SCIENCE PROGRESS BRAZILIAN SOCIETY-SBPC AND THE DEFENSE OF THE INQUISITION SCIENTIFIC BY THE GREATEST EXPONENT OF THE MODERN ANTI-HERESY HANDBOOK	288
LX.	QUANTUM IMPULSE EXPLAINING THE KEY EXPRESSION OF THE THEORY OF RELATIVITY FOR THE PERIHELION MOTION	289
LXI.	AN IRREFUTABLE PROOF BY THE OWN THEORY OF CURRENT PHYSICS, OF THE ELEMENTARY QUANTUM OF ENERGY h' OF QUANTUM IMPULSE THEORY	291
LXII.	INTERNATIONAL COMMITTEE ON SPATIAL RESEARCH	292
LXIII.	THE INSTITUTE OF ENGINEERING OF SÃO PAULO-BRAZIL AND SCIENCE	294
LXIV.	SCIENCE & CULTURE JOURNAL AND THE INQUISITION ACTING	295
LXV.	ELECTRON DIFFRACTION AND QUANTUM IMPULSE THEORY	300
LXVI.	THE REPORT OF THE PRESIDENT OF THE NUCLEAR ENERGY NATIONAL COMMISSION, PROFESSOR ENGINEER AND PHYSICIST OF SÃO PAULO AND CATHOLIC UNIVERSITIES	301
LXVII.	EVIDENCE OF ELECTROMAGNETIC FIELD IN MASSES	304
LXVIII.	THE GRAVITATIONAL INVARIANT	304
LXIX.	DIAMETER OF THE BLACK HOLE BY QUANTUM IMPULSE THEORY WITH THE GRAVITATIONAL RADIATION OF 10^{19} CYCLES PER SECOND	305
LXX.	THE PHYSICS IN THE DECADE OF 70 WITH MORE RENORMALIZATION, QUARKS AND GLUONS WITHOUT MASS, SPECIAL OR "COLORED" CHARGES AND STRING PARTICLES	307
LXXI.	MASSLESS NEUTRINO BY PRE-ESTABLISHEAD PHYSICS	310
LXXII.	THE SWEDISH ACADEMY OF SCIENCES AND THE NOBEL PRIZE IN PHYSICS FOR THE UNIFICATION OF FORCES	312
LXXIII.	"MODERN LESSONS" FROM CURRENT PHYSICS	314
LXXIV.	NEW EXPERIMENT ON EARTH VELOCITY	314
LXXV.	THE SCIENCE PROGRESS BRAZILIAN SOCIETY-SBPC AND FREEDOM OF EXPRESSION	315
LXXVI.	NUCLEAR ENERGY AT THE INSTITUTE OF ENGINEERING OF SÃO PAULO-BRAZIL	316
LXXVII	ATTESTATION OF INVARIANT GRAVITATIONAL FREQUENCY BY MESQUITA AND APPROVAL OF SCIENCE PROGRESS BRAZILIAN SOCIETY-SBPC AGAINST THE OWN SBPC	316
LXXVIII.	A MATTER NEVER CLOSED	317
LXXIX.	DENSITY AND AGE OF THE UNIVERSE	317

LXXX.	THE INSTITUTE OF ENGINEERING OF SÃO PAULO-BRAZIL AND THE NOBEL PRIZE IN PHYSICS	318
LXXXI.	UNIFIED FIELD	319
LXXXII.	NEW PROOF OF GRAVITATIONAL RADIATION BY THE CONICAL MOTION	320
LXXXIII.	SCIENCE AT THE INSTITUTE OF ENGINEERING OF SÃO PAULO-BRAZIL	320
LXXXIV.	SYMMETRY DETECTED	320
LXXXV.	THE FIRST THEORETICAL PHYSICIST IN BRAZIL AND HIS CONCLUSIONS ABOUT THE PHYSICS	321
LXXXVI.	NEW RADIATION DETECTED PROVES QUANTUM IMPULSE THEORY	325
LXXXVII.	NEW GENERATION OF GRAVITATIONAL WAVE DETECTION	325
LXXXVIII.	NEW INFORMATION-SCIENCE AT THE INSTITUTE OF ENGINEERING OF SÃO PAULO-BRAZIL	326
LXXXIX.	CREATION OF NEW IDEAS AND IMMORTAL CREATION	329
XC.	HOW DISCOVERIES OCCUR IN PHYSICS	329
XCI.	THE COMPLEXITY THAT WAS LACKING	330
XCII.	"MASSES, WHAT ARE THEY AND WHERE DO THEY COME FROM"?	330
XCIII.	SUPERSYMMETRY AND SUPERSTRINGS FOR 50 YEARS	331
XCIV.	GENERALIZED FIELD OF ALL PARTICLES, STANDARD MODEL, GRAVITATIONAL FIELD	331
XCV.	FORMAL OMISSION	332
XCVI.	NEUTRINO MASS	336
XCVII.	LEARNING OF THE MILITARY DICTATORSHIP BY SCIENCE PROGRESS BRAZILIAN SOCIETY-SBPC	337
XCVIII.	IN THE NAME OF SCIENCE!	337
XCIX.	"A BRIEF HISTORY OF THE TIME" OF HAWKING AND THE QUANTUM IMPULSE THEORY	338
C.	"MATTER CLOSED"	345
CI.	ASSEMBLY OF PERPETUAL OPPRESSION	345
CII.	INTENSE ACTIVITY OF HERETIC HUNTERS!	346
CIII.	ELECTROMAGNETIC PHENOMENA UNKNOWN BY CURRENT PHYSICS	352
CIV.	SCIENTIFIC AMMUNITION OF THE BRAZILIAN NAVY	352
CV.	NOTHING SHOULD BE ADDED TO SCIENTIFIC KNOWLEDGE	354
CVI.	VOTE OF CONFIDENCE FOR SCIENTIFIC ANTIOPPRESSION	360
CVII.	NUCLEAR FUSION PORTABLE	363
CVIII.	NOTHING NEW!	364
CIX.	SCIENCE PROGRESS BRAZILIAN SOCIETY-SBPC ON TRIAL	364
CX.	SPECIAL MEETING OF THE SBPC WITH BRAKES ON THE HUNTERS OF HERETICS!	365
CXI.	STANDARD MODEL WITHOUT GRAVITY AND UNIFICATION OF THEORIES OF RELATIVITY AND QUANTUM WITH STRINGS OF DIMENSIONS "COILED"	368
CXII.	TRIPLE MEETING	368
CXIII.	THE JOURNAL NATURE AND ARTICLE SUBMITTED OF LASER EXPERIMENT, WHICH PROVED THE FALLACY OF RELATIVITY	370
CXIV.	OTHER INTERNATIONAL PHYSICS JOURNALS	375
CXV.	PHYSICAL REVIEW LETTERS	376

The Pre-Established Physics and the Quantum Impulse

CXVI.	AMERICAN JOURNAL OF PHYSICS	376
CXVII.	FOUNDATIONS OF PHYSICS JOURNAL	376
CXVIII.	IL NUOVO CIMENTO JOURNAL	377
CIXX.	PHYSICS LETTERS A JOURNAL	378
CXX.	ASTRONOMY AND ASTROPHYSICS JOURNAL	379
CXXI.	THE ASTRONOMICAL JOURNAL	379
CXXII.	THE ASTROPHYSICAL JOURNAL	380
CXXIII.	GENERAL RELATIVITY AND GRAVITATION JOURNAL	381

PART 8 – EXPERIMENT PUBLISHED IN INTERNATIONAL JOURNAL OF FUNDAMENTALS OF PHYSICS PROVES BY QUANTUM IMPULSE THE FALLACY OF BASES OF CURRENT PHYSICS..........382

CXXIV.	PHYSICS ESSAYS JOURNAL	382
CXXV.	ONE YEAR AFTER: MARCH 2003	386
CXXVI.	DISCOVERY OF THE MASS OF THE NEUTRINO THAT HAD NOT MASS BY IMPOSED PHYSICS!	387
CXXVII.	QUANTUM MASS	389
CXXVIII.	"SPACE COMPASS": THE VELOCITY INDICATOR	390
CXXIX.	INERTIAL VELOCITY OF THE LASER LIGHT AND CERN-MINOS-FERMILAB, OPERA-ICARUS-CERN	400
CXXX.	SPIN, HELIX, PARTICLE CHARGE AND QUANTUM IMPULSE THEORY	426
CXXXI.	ELECTRON STRUCTURE AND QUANTUM IMPULSE	430
CXXXII.	RADIATION BRAKING ("BREMSSTRAHLUN"), CHERENKOV RADIATION AND THE PHOTON RADIATION OF QUANTUM IMPULSE	433
CXXXIII.	NUCLEAR FORCE, ABSTRACT SPACES, STATISTICS, ELECTROMAGNETIC AND GRAVITATIONAL FIELD, AND QUANTUM IMPULSE THEORY	435
CXXXIV.	THE CASIMIR-POLDER-LIFSHITZ EFFECT AND THE QUANTUM IMPULSE	438
CXXXV.	PLASMA AND QUANTUM IMPULSE	439
CXXXVI.	GLUE WITHOUT MASS, THE GLUONS AND THE QUANTUM IMPULSE	440
CXXXVII.	ENTANGLEMENT, MOZART SYMPHONY AND THE QUANTUM IMPULSE	441
CXXXVIII.	HIGGS FIELD AND QUANTUM IMPULSE THEORY	444

PART 9 – THE UNIVERSE OF THE MOTION STARS...447

| CXXXIX. | MOTION STARS | 447 |
| CXL. | GALACTIC VELOCITY, SPACE AND THE QUANTUM IMPULSE | 477 |

PART 10 – UNIVERSE FORCES UNIFICATION............482

CXLI.	UNIFICATION OF ALL PHYSICAL INTERACTIONS	482
CXLII.	CREATION IN THE QUANTUM VACUUM: MAKE UP THE LIGHT!	483
CXLIII.	THE GENERAL SCIENCE JOURNAL AND PHYSICS ESSAYS	485

CXLIV.	TEAMS, PROJECTS AND ACRONYMS OF CURRENT PHYSICS	487
CXLV.	THE GREATNESS OF THE UNIVERSE	496
CXLVI.	THE FORCE OF THE UNIVERSE	496
CXLVII.	GRAVITATIONAL WAVE DETECTION	497
CXLVIII.	NASA-FERMI-JAMES WEBB SATELLITES	508

PART 11 – ADVANCED CONNECTIONS ... 508

CXLIX.	BASIC CONCEPTS "ADVANCED" – SPACE, TIME, MASS	508
CL.	BIOPHYSICS, FREE WILL AND DIVINITIES	510
CLI.	PLANET OF THE APES	511
CLII.	OTHER CONNECTIONS	516
CLIII.	ANTIGRAVITY AND GRAVITATIONAL-ELECTRIC-MAGNETIC FIELDS ENERGY	524

PART 12 – EPILOGUE ... 531

REFERENCES ... 561

THE AUTHOR ... 586

PART 1 – EARLY CONCEPTS OF SCIENCE

I. INTRODUCTION

In the full millennium of 2000 computer equipment, cellphones, televisions, spacecraft, external and internal imaging processors, transmitting wave radiations from biological organisms and objects, are in the presence of billions of inhabitants of the planet. And the current most recent scientific theories that deal with these radiations of light or radio waves, television, infrared, X-rays, microwave, lasers, gamma rays and gravity, are the relativity and quantum mechanics originated from a quantity of radiation. But different perceptions occur in the study of the constitution of these radiations, leading to the conflicting consequences and unexpected connections to humanity. "*The Physics of Quantum Momentum or Impulse*" deals with the fundamental bases of physical nature, from the dawn of civilization to the present state that involves these radiations, showing troubled facts in the scientific and technological environment, which are unknown to humanity.

II. PRIMORDIUM - 40 MILLION YEARS to 320 BC

Primates – It is admitted by current science, based on geological and fossil studies, that around 40 million years ago the first types of primates appeared. And between 25 and 5 million years were the apes. Four million years ago, the first apes of hominid primates appeared, the ancestors of the current race. And 2 million of years ago it originated the hominid "habilis" and with 1.5 million of years the hominid of erect gait, "erectus". Next around 400000 years the first primitive or archaic hominids would appear, the "homo sapiens". And around 150000 years would have arisen the "homo sapiens sapiens" of characteristics of the present man, in which fossils of these early humans indicated their origin of Africa and following towards the lands of Europe, Asia, Oceania e America.

Gods and the Universe – In the early days of civilization, around 4000 BC, located with center in Mesopotamia (present-day Iraq), like that coming out of nowhere, all the elements of the essence of a modern civilization arose, as the domain of agriculture, irrigation, livestock, ceramics, weaving, metallurgy, perfumery, vessels, weights and measurements, writing, construction with grandiose monuments, medicine, the wheel for transport, music, musical instruments, laws, government and the astronomy for knowledge of the heavens. And interacting with other civilizations around, which arose in the present region of Egypt, Greece and India, the inhabitants identified the "Most High Lords" or later denominated of Gods that would have come from the heavens. However, evidence arose from previous civilizations, not admitted by current science, and monumental works of antiquity throughout the world, built without concrete evidence of transport and preparation of immense amounts of stones up to 100 tonnes, with the highest degree of precision of cuts. And that the current science only comes to explain how built up only by orders of the rulers in civilization that little had to more than chipped stone tools used even by apes!

Numbers, Figures, Space, Infinite and Time – For sizing or quantities were fixed numbers and shapes of the bodies, to relate the facts of the sky and the Earth. Around 3500 BC, numbers were applied to figures such as circumferences, circles, spheres, triangles, pyramids, cones, squares, cubes, rectangles, cobblestones, polygons, polyhedra, determining positions, this is, distances, areas and volumes. And the inhabitants of Sumer, the current Iraq, discovered in the rectangle triangle the relationship that the square built on its larger side was equal to the sum of the squares built on the other two sides, also obtained by Pythagoras of Samos in 570 BC. The circumference was divided into 360 parts, each part to a degree, combining with the days of the year in 360 days that was the value observed by the Sumerians and subsequently in 365 days when the Sun returned to the same position in relation to almost all the stars of the firmament. And the Sumerians determined the relationship between the circumference and radius of the circle, in ~ 3.14 times the diameter ($\sim 2 \times 3.14$ the radius) for the length of the circumference, and the length of the radius in the circumference, the radian = 57.3... degrees and 1 degree = 0.017... radians.

As inscriptions of Antiquity around 580 years BC, Anaximander of Miletus observed space as a notion of infinity or apeíron, the place of positions of bodies, the matter, and its variation of positions, this is, the motion. And time was observed as a period of occurrence and the notion of eternity. For India Kanad around 550 BC space and time were the physical foundation of complet rest and all motions.

Zodiac – The Sumerians observed the groupings of stars and attributed names of animals (zodiac), and figures of humanoids and objects. The motions of the Sun, Moon and planets were determined by "planete", a traveler, for presenting a motion in which their trajectories returned and after they went on again, that is, a retrograde or errant motion.

Numerical Trajectories in the Sky – Around 1000 BC, the Chinese presented the compendiums: "Arithmetic" or "Number Theory"; "Algebraic" or "Reduction and Transfer" and "Geometric or Measurements of the Earth", between numerical expressions in "Mathematics" or "Knowledge in Circular Trajectories in the Sky", with astronomical calculations, properties of triangles and fractions.

Rule of Segments – And also around 1000 BC the Hindus treated astronomy, numerical calculations and geometry, in which rules of segments they were presented, as the basic relationship between sides of triangles, as the opposite side of the larger side, that is, of the sine of angle, and the cosine relating the adjacent side with the larger side. So for any right triangle, knowing the dimension of the side opposite the angle, was obtained the dimension of the larger side.

Ether – In 600 BC, the space where the stars were rotating would be filled with a continuous medium, the ether, a kind of clean air or fluid, according to Anaximenes of Miletus.

Matter – For Tales of Miletus in 600 BC, all concrete bodies occupying the space, with shapes and dimensions, constitutes the matter. And the primordial matter of the universe was water, *"Upon which the Earth floated and was the beginning of all things"*. Already for Anaximenes in 550 BC would be the primary matter air that would be reduced to water by simple compression. In turn, Heraclitus of Ephesus in 520 BC proposed to be the fire that matter. However for Xenophones of Ionia, in 500 BC, was the earth that would constitute the raw material of the universe.

Light in Ether – In addition to the four primordial materials, earth, air, water and fire, one more came to be added: the matter of the sky, this is, the ether, as Aristotle of Stagira in 350 BC. All these subjects would follow their natural place. The bodies with the greatest amount of material on earth would follow their natural place to the same, while the fire upward, and the water and the air between them. And the light would be similar to the sound vibrations, in which a luminous object would have vibration that would be propagated through a transparent medium of the ether.

Atoms – Similar to the primordial materials that constituted everything in the universe, perhaps there could also be minimal particles of all matter. Around 550 BC, Kanad presented that all matter consisted of fundamental particles, the atoms. And between

470 and 420 BC according to Leucippus of Miletus and Democritus de Abdera, if the matter were continuous and divided up to infinity there would only be a single particle, because ahead of infinity would be the point of no dimension. Thus, the matter would be formed by indivisible particles, atoms, extremely subtle as a veil, which constituted the flavours emitted by the substances and would also constitute the light. And based on the hypothesis of the indivisible atoms of Leucippus and Democritus, the empty space, the vacuum, was admitted, so that the atoms could move, according to Epicurus of Samos in 320 BC: *"The Gods are higher than men, and they cannot act in the world, and they are really indifferent to human affairs. The atoms with their motion in the vacuum determine the materials and make up the living beings, but they are not together with the Gods that are far above. Atoms separate the men of the Gods. What results from atoms, by the product of accidental combination, is the world in its infinite variety".*

And according to J. T. Jeans in 1943: *"The world that up to that moment had been a place of recreation, pleasure, became his prison. Until then, it was understood in beauty, sweetness, enthusiasm, being the facts delivered by the gods to men. These facts would be in reality the work of man's imagination. Human perceptions and emotions saltwater in light incidents in the world and indicated the existence of an objective world, external, independent and indifferent to man. There was sweet and bitter, warm and cold, the colours, but in reality what existed were atoms and the emptiness, forming the atoms with their motions all there was in the world and then denying free will.*

III. SPHERE UNIVERSE – 650 BC to 80 BC

Shape of the Earth – The Sumerians and after the Greeks sought the rationalization of what occurs in nature, as the position of the Earth in the universe. To the Sumerians the Earth was spherical. And in 650 BC, as Tales emerges an explanation of the shape and position of the Earth in the universe, admitting as a thick plane floating in the water, and the water going up to the edges of stones at the full extent of it, positioned at the center of the universe. Or, the Earth admitted as a cylinder thick on waves of water, wrapped in a sphere where the stars were rotating around the Earth and water, in which the Sun passed under the Earth and water, occurring at night, according to Anaximander.

Spherical Harmony – In 550 BC, according to Pythagoras, the universe would be governed by harmonious forms of spheres where the planets and stars were, producing inaudible music in their orbital motions of the spheres and following numerical relationships. And Anaximenes recognized by motion the planets differing from the stars.

Spherical Earth – Reports of navigators indicated that according to the beginning of navigation sites, the amount of nights spent was different when moving in a certain passage in latitude. This indicated a spherical shape for the Earth and water. Thus, Pythagoras around 500 BC conceived the Earth as spherical and recognizing that the Morning Star, and the Vesper star of the afternoon, the two were one only, the planet Venus.

Bodies and Earth Turning to Central Fire – Around 500 BC Philolaus de Tarentum presented the hypothesis that the stars, the Sun, the planets, the Moon, and the Earth, would revolve from West to East, around another celestial body, a "Central Fire", distant and not visible by the position that the observers were in the Earth, and that would be the center of the universe, according to the studies of Philolaus and Hicetas of Syracuse around 550 BC. And as Hicetas, Pythagoras and Ecphantus of Syracuse presented the rotation of the Earth on its axis every 24 hours.

Multiworlds – Around 450 BC, Anaxagoras de Lampsaco presented that because there were an enormous number of stars in the firmament, there should be other inhabited worlds. But the Earth would no longer be the center of the universe.

Earth without Motion – But around 400 BC Ecphantus placed the Earth as the center of the universe and motionless.

Earth in the Center of the Circle and Retrograde Motion – And between 380 BC and 310 BC, Plato of Athens, Eudoxus and Calipo of Cnidus, Aristotle, Heraclides of Points presented that the motions of the Earth, Moon, Sun, planets and stars, were the result of a set of spheres centered on the Earth, and it was the Earth that remained motionless or rotating to alternate day and night. And the circular shape would be the most perfect curve for these motions of the stars and the Earth as the center of the circular motion that would be the most natural and therefore inevitable. And the motion of the planets in the firmament, in which their trajectories returned and then went on again, that is, the retrograde or errant motion was to be explained by the imaginary spheres where were the Sun, Moon, stars and planets, which carried complex motions to achieve this effect. There were 27 rotating spheres to explain the effect of the motions of the planets in the firmament. And the Earth would come to be in the center of the spheres, according to these scientists.

Above and Below – According to Aristotle, the universe was made up of two distinct worlds: above the Moon was the abode of God, the creator of the universe, eternal place and splendor in temples and arts; and below the Moon, a corruptible place, with everything transforming, beginning, middle and end. *"The celestial bodies and the stars are all immutable, but on Earth everything becomes"*, as Aristotle.

Unbound Motion – The rotating Earth on its own axis would explain with simplicity and clarity the day and night without the incredible motion of the spheres of the stars around the Earth. *"But if the Earth rotated an object flung vertically upward it would not fall in the same place it had been launched. For, while rising in the air, the ground should move Eastward, and the object would fall West from the starting point. As this fact does not occur, the conclusion is that the Earth motionless. The object thrown into the air would be disconnected from the motion of the Earth. And if the Earth rotated, the object would fall at a point far from where it had been launched"*, according to Aristotle.

Sphere Action – As Aristotle, the spheres where the stars were, were triggered by a driving action creating vortexes through the creation of God. However, in the motion of some celestial bodies in a given sphere, different motions, this is, between slow and fast, were observed. A ratio of their change of position of all the trajectory, the distance by the time, the speed, or the displacement, this is, the length of the shortest trajectory between these positions by time, the velocity. Because these stars are similar in dimensions and matter, there should be the same motion for these bodies in this sphere. Then, these vortexes of Aristotle were questioned, as reported by Theophrastus of Eresis in 330 BC in "Opiones", the first book on the History of Science.

In the words of Jeans in 1948: *"But it prevailed that the Earth was the center of the universe, and the driving force of the outer sphere rotating all the bodies placed men in the grandeur of a great Earth, and not in a miniature fragment of the universe, besides a great Earth being impossible to move in space without any driving force acting, that is, the Earth should be triggered by gods, as was the Physics seen by Aristotle"*.

The Greater Sun – In the inaudible music of the harmony of the universe, in 380 BC Eudoxus came to calculate the diameter of the Moon in relation to the Earth, during the Moon's eclipse. It observed the apparent diameters of the Moon and the Sun, which could be measured in the visual reading plane (around 50 cm). For the Moon and the Sun the apparent diameters are coincidentally of similar values, around 1 degree (equivalent to a segment around 15 cm at this visual distance). Figure 1.

D_{aSun} = apparent diameter of the Sun
D_{aMoon} = apparent diameter of the Moon
D_{Sun} = Sun diameter
D_{Moon} = Moon diameter

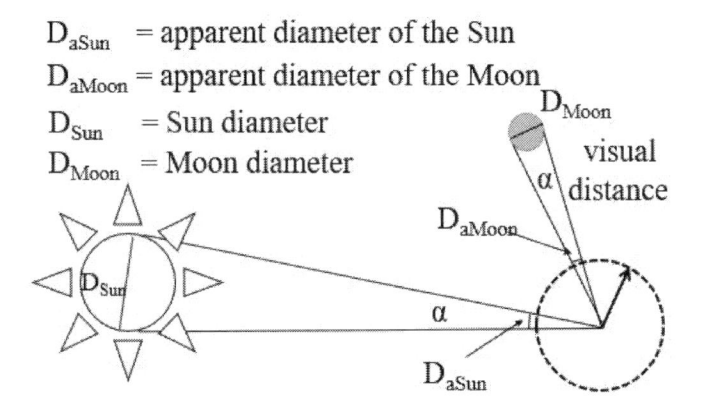

Fig. 1 – Apparent diameters (visuals) of the Sun and the Moon. Adapted from the clipart of *Editora Europa*.

Eudoxus observed the diameter of the Earth being designed by the shadow on the Moon during the lunar eclipse, from the beginning to the end of the eclipse. It measured the apparent diameter of the Earth's diameter shadow designed on the Moon as equal to once the apparent diameter of the Moon.

When the Moon was at a vertex between Earth and the Sun, the angle was very close to 90° between the Earth, Moon and the Sun. By the Figure 2, cos 83.5° = ~ (11/100) = d_{EM} / d_{ES} resulting d_{ES} = ~ 9 x d_{EM}. Despite the same angular diameter (apparent) of the Sun being similar to that of the Moon, the Sun was farther from the Earth.

D_{aS} = apparent diameter of the Sun
D_{aM} = apparent diameter of the Moon
d_{EM} = Earth-Moon distance
d_{SM} = Sun-Moon distance
d_{ES} = Earth-Sun distance

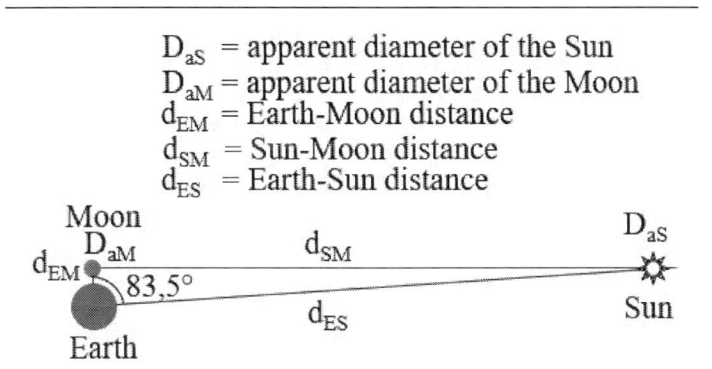

Fig. 2 – Despite the same apparent diameter of the Sun and Moon, the distance from the Sun was greater than that of the Moon. Adapted from the clipart of *Editora Europa*.

Because the apparent diameters of the Moon and the Sun are similar, this is, a same angle seen by the observer, Eudoxus conceived to relate the Earth relative distances to the Moon and Earth to the Sun with its actual diameters. Figure 3.

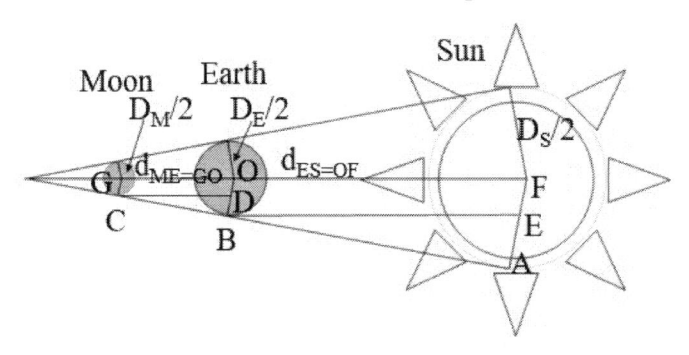

Fig. 3 – Eudoxus observed that the Sun with the same apparent diameter as the Moon has a greater distance proportionality from the Earth. Adapted from the clipart of *Editora Europa*.

With the similar BCD and ABE triangles:
$(D_E/2 - D_M/2)/(D_S/2 - D_E/2) = d_{ME}/d_{ES}$

And being similar to triangles OGD and OFK: $D_{ME}/d_{ES} = (D_S/2)/(D_M/2)$. Figure 4.

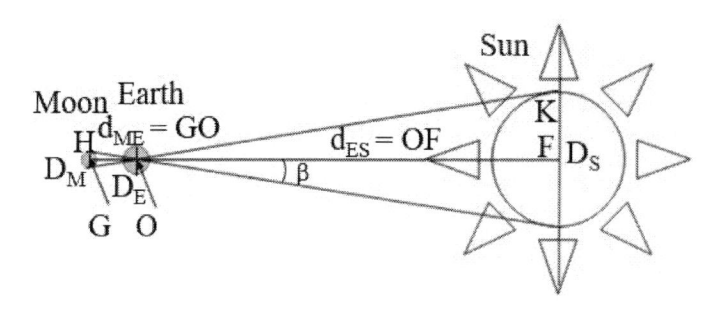

Fig. 4 – In relation to similarity triangles Eudoxus calculated the actual diameter of the Earth, the Moon and the Sun. Adapted from the clipart of *Editora Europa*.

As $d_{ES} = \sim 9\, d_{ME}$, and $d_S/2 = \sim 9\, d_M/2$ by the similarity of triangles of the same angle β maintaining proportionality between the sides, obtained:
$$(d_E/2 - d_M/2)/(\sim 9 d_M/2 - d_E/2) = (\sim 9\, d_M/2)/(d_M/2)$$

And calculating $d_M = (\sim 10/18) d_E$, or $d_E = (\sim 18/10) d_M$, that is, the actual diameter of the Earth being calculated in a little more than one and a half times the actual diameter of the Moon.

Eudoxus estimated the diameter of the Earth D_E in $\sim 103,000$ lengths of course of competition stadium, that is, measure of stadium (equivalent to $\sim 0, 185$ km). That is, D_E equivalent to $\sim 19,000$ km in the metric pattern adopted in 1799. With that diameter D_T estimated, and in the OFK triangle where Eudoxus seems to have measured the apparent angular diameter in 2° with β *equal* 2°/2 = 1°. So: tg 1° = $\sim 17/1000$ = $(D_S/2)/d_{ES}$ = $(\sim 9D_M/2)/d_{ES}$ = $(\sim 9 \times \sim 10/18 \times D_E/2)/d_{ES}$, obtaining: d_{ES} = ~ 15 million stadiums equating to ~ 2.8 million km. E, d_{ME} = ~ 1.6 million stadiums equivalent to 300,000 km; D_M = $\sim 50,000$ stadiums ($\sim 10,000$ km); D_S = $\sim 500,000$ stadiums ($\sim 90,000$ km).

These results, from the diameter of the Sun in ~ 9 times greater than the diameter of the Earth, and from the distance from the Sun to the Earth in ~ 9 times the distance from the Moon to the Earth, were transcribed by Euclid, because Eudoxus's work had disappeared.

The relevant fact is Eudoxus to prove the Sun greater than the Earth. The greatest place in the universe was not the Earth, where man lived with his Gods, but was another star like the Sun. The calculation of Eudoxus was irrefutable, because the small apparent diameter of the Sun did not prevent its true dimension. Any long-distance physical entity is seen as a small dimension, but in reality it can present a considerable dimension.

Sun in the Universe Center – Around 290 BC, Aristarchus of Samos formulated the heliocentric model, in which in the motion around the Sun by the Earth it was observed that the stars remained fixed, for by the immense distance between the stars and the Earth the effect of two sides of vision, that is, the parallax, was undetectable.

And around 270 BC Aristarchus presented new measurements of distances and diameters of the Sun, Earth and Moon.

When the Moon was at a vertex between Earth and the Sun, the angle was very close to a circle quadrant (90°) between the Earth, Moon and the Sun, and Aristarchus measured the equivalent of 87°. As Figure 5, sen $\alpha = (d_{EM}/d_{ES})$ ∴ sen 3° = $\sim (5/90$ to $5/100) = (d_{EM}/d_{ES})$ ∴ $d_{TS} = \sim 19 \, d_{EL}$

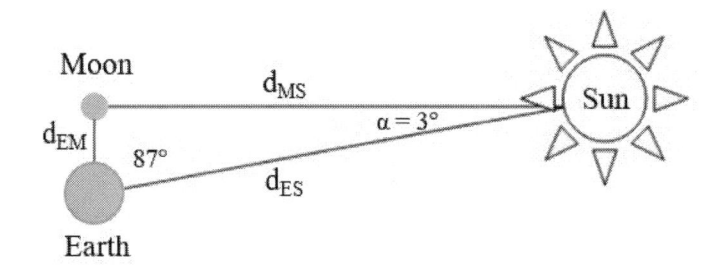

Fig. 5 – Measurement by Aristarchus of distances between Earth, Sun and Moon. Adapted from the clipart of *Editora Europa*.

And relating distances and diameters of the Earth, Moon and Sun, Aristarchus determined related distances between Earth, Moon and Sun. By Figure 6, being similar to the BCD and ABE triangles:

$$(D_E/2 - 2D_M/2)/(D_S/2 - D_E/2) = D_M/D_S$$

And by proportionality between the sides of the rectangle triangles with the same equal angle, is obtained: $D_S = \sim 19\ D_M$, and $D_S/2 =\sim 19\ D_M/2$ and $D_M =\sim (20/57)\ D_E$, or $D_E =\sim (57/20) = D_M$, that is, the actual diameter of the Earth being calculated in a little more than two and a half times the actual diameter of the Moon.

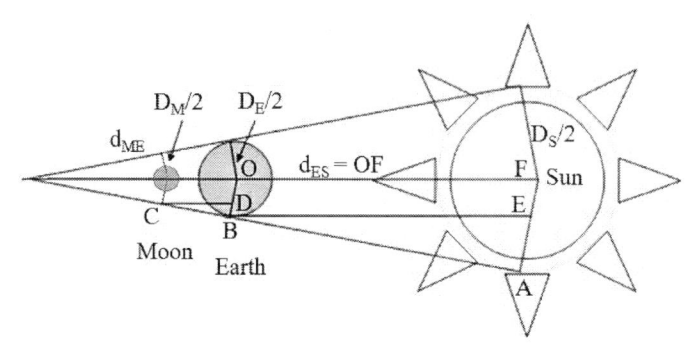

Fig. 6 – Relationship between the diameters of the Moon, Earth and Sun determined by Aristarchus. Adapted from the clipart of *Editora Europa*.

For the apparent angular diameter Aristarchus seems to have measured in $\sim 1°$, and admitting the diameter of the Earth at $\sim$ 103,000 stadiums ($\sim$ 19,000 km) as Eudoxus had admitted, obtained the distance from Earth to the Sun in $\sim$ 43 million stadiums ($\sim$ 8 million km); D_M *equal* $\sim$ 35,000 stadium ($\sim$ 6,600 km); D_S *equal* $\sim$ 650,000 stadiums ($\sim$ 120,000 km). In some revision, Aristarchus came to find for the diameter of the Earth at $\sim$ 14,000 km.

By the greater volume of the Sun and greater amount of matter, it would be much greater by the Sun to have a displacement of matter than a lighter body. Thus, Aristarchus came to conclude that the Earth revolved around the Sun and this as the center of the universe.

But despite these more substantiated measurements of the motions of the celestial bodies, other scholars preferred to be on the side of the Gods. Thus, it was to follow the method of moving spheres for the celestial bodies, with the Earth at the center of the universe, and the drive of the driving action by the outer sphere triggered by the greater God of the Greeks, Zeus. And leaving also atomism, to be very close to the Gods.

Circumference and Slope of the Earth – Reports reported that in the city of Siena, in Egypt, the Sun was in the summer month over a well, without presenting any shade, i.e. vertically focusing. And around 250 BC, Eratosthenes of Salahudeen effected the first determination by calculating the diameter of the Earth, in which at the same time, in the city of Alexandria, it could be obtained the angle between the two cities. Figure 7. And obtaining by geometry, determining the terrestrial circumference in ~ 250,000 stadiums of measurement (around 25,000 miles or 46,250 km) and the diameter of the Earth around 14,700 km.

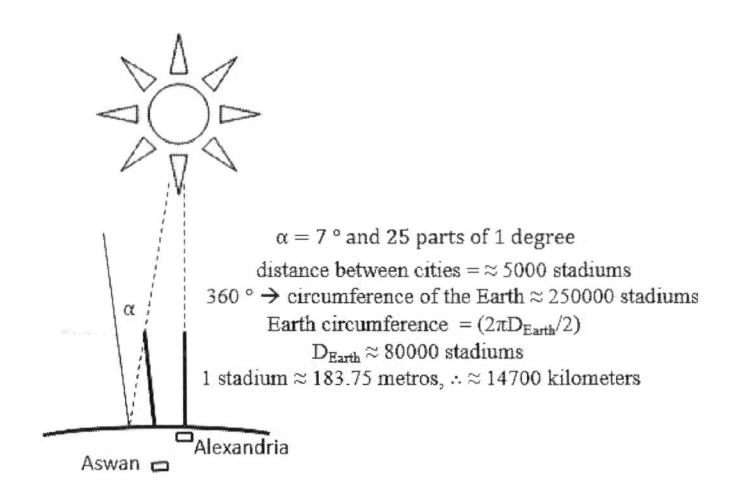

$\alpha = 7°$ and 25 parts of 1 degree
distance between cities = ≈ 5000 stadiums
360° → circumference of the Earth ≈ 250000 stadiums
Earth circumference = $(2\pi D_{Earth}/2)$
$D_{Earth} \approx 80000$ stadiums
1 stadium ≈ 183.75 metros, ∴ ≈ 14700 kilometers

Aswan   Alexandria

Fig. 7 – Determination of the diameter of the Earth by Eratosthenes. Adapted from the clipart of *Editora Europa*.

Similarly, it was measured the difference between the shadow that was designed in summer and winter. When it was considered the Earth as the center of the universe, the Sun was had as slanted, the ecliptic (apparent orbit of the Sun around the Earth), by its sphere at these seasons. And when the Sun was admitted as the center of the universe, the Earth would have the inclination explaining the seasons of the year with longer days in the summer and smaller in the winter, as Eratosthenes around 235 BC.

Epicycles – Another explanation for the retrograde motion of the planets and other motions of the stars, without the spheres in which the stars would be, but keeping the Earth as the center of the universe, was to be the epicycle in conjunction with a circular motion around the Earth. It consisted of the epicycle in which the planets or other stars rotated to a circumference, the deferential, that is, the transporter circle of the planet where in the center of the same was the Earth. Or, the Earth effecting a small, central circular motion would be obtained from the retrograde motion effects to the planets, as Apollonius of Perga around 230 BC.

Around 220 BC, Apollonius explained the retrograde motion of the planets through epicycles and deferential, in which the center of a smaller circle, the epicycle moves along a larger circle, the deferential. Figure 8.

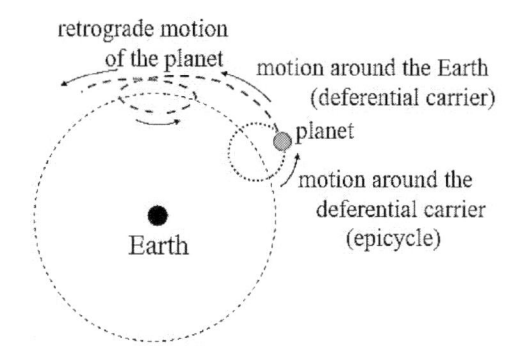

Fig. 8 – Retrograde, deferential and epicycle motions.

Arithmetic, Geometric and Algebraic Advancement – And between 600 BC to 500 AC, Tales, Anaximander, Pythagoras, Hipparchus of Nicaea, Tarentum, Hippocrates of Kos, Theodorus of Cyrene (present-day Shahhat in Greece), Plato, Eudoxus, Cleomedes of Lysimachia, Meneceno of Alupeconnesus, Zenon of Elea, Euclid of Alexandria, Archimedes of Syracuse, Apollonius, Heron of Alexandria, Eudemian of Rhodes, Strabo of Asia, Democritus, Gemino of Rhodes, Conon of Samos , Aristarchus Theon of Alexandria, and Eratóstenes, Philolaus, Xenocrates of Chalcedon, Cnidus's Liposuction, Parmenides of Elea (current Ascea), Hippias of Elis, Nicômano of Gerasa, Cidenas of Babylon and Claudius Ptolemy of Egypt, Pappus of Alexandria, Iamblicus of Apamea, Eutocio of Ascalon had presented new calculations for lines, surfaces and plans. Calculations as geometric relationships between angles and triangle sides, polyhedron figures inscribed in circles, parts of measurements of the main measure. Also, measurements of angles related to the sides of triangles inscribed in a circle were presented; calculation of the circle included for

being considered by them to be the most perfect form of nature; tangent of angle being the relationship between opposite side and adjacent side; sine of angle = opposite side/larger side; cosine = adjacent side/larger side; secant = larger side/adjacent side, cosecant = larger side/opposite side, cotangent = adjacent side/opposite side, for a circle of 360°.

More Distant and Larger Sun – Around 150 BC, Hipparchus determined the distance between the Earth, the Sun and the Moon. With greater precision in measuring the angle formed by the Moon, Earth and Sun when the Moon is at the vertex with the Earth and the Sun, Hipparchus obtained the distance d_{ES} from the Sun to the farthest Earth from the Earth, the distance d_{EM} from the Earth to the Moon, the diameter D_E of the Earth and the diameter D_M of the Moon.

$D_E = \sim 80,000$ stadiums $= \sim 14,700$ km
$d_{ES} = \sim 100$ million stadiums $= \sim 18.3$ million km
$d_{EM} = \sim 2.4$ million Stadiums $= \sim 440,000$ miles
$D_S = \sim 960,000$ stadiums $= \sim 170,000$ km
$D_M = \sim 23,000$ Stadiums $= \sim 4,200$ miles

Smaller Earth and Even More Distant Sun- Around 80 BC the Earth became smaller by the calculations of Poseidonius of Apamea, with diameter D_E of the Earth with $\sim 50,000$ stadiums $= \sim 9,200$ km and distance d_{ES} from the Earth to the Sun even farther away, with ~ 345 million stadiums $= \sim 64$ million km.

IV. DOGMAS AND NEW IDEAS – 6000 BC to 33AC

In 550 BC, as Siddhartha Gotama: *"Make no mistake for what is tradition or rumours, by simple logic and deduction, nor after reflecting on an opinion and approving it, nor because it contains the future. Do not believe in anything because it came from some great authority or of teachers and priests. Believe what you have experienced and discovered according to your studies and ethics".*

Doctors of Laws and Sciences – In the year 30, according to Christ, *"You, the Doctors of the Law, who have been the key to Science, and yourselves not, but impedistes to those who came to enter".* (Bible, New Testament, Luke, Chapter 11, verse 52).

Crucifixion of New Ideas – Resistance to new ideas against the interests and ideas prevailing in the dominant dome, Christ was crucified and dead in the year 33.

Resurrection of New Ideas – In 6000 BC, according to ancient Indian texts, "Mahabbarata", "Ramayana", "Rig Veda", "Vimanika Shastra", they mentioned flying machines: *"The flying* *vehicle moved more easily than a bird in the sky and curved toward the Sun and the Moon, and landed in the Earth with great noise. The flying vehicle was adorned with jewels and faces of demons. Then appeared the flying vehicle that had the speed of thought, and action by propellant. That flying vehicle shone like the Sun. Celestial vehicles that were in the front end, were moving with extraordinary speed, and possessed a fuselage that shone like gold. heavenly beings moved freely in space and on Earth. Groups came in divine aerial vehicles that moved on their own, and the gods appeared in their own flying cars. And the Lord of the Heavens came in a special flying object that could harbor 36 divine beings. The divine vehicle came from heaven and the heavenly Being descended from the apparatus by an elegant staircase made of shiny metal. The gods had come from a faraway place, in heaven, and turned to Earth".*

And in 5000 BC, according to the apocryphal Book of the Secrets of Enoch, written around 1300 BC appeared celestial contacts: *"There appeared two beings who gleaming like the Sun, their eyes as a flame, and from their lips came out a corner and a fire of violet color in appearance; their wings were brighter than gold, his hands whiter than snow. The eternal God has called you, and you shall rise to heaven with us. The angels took me to the first sky and saw the ether and a sea greater than the Earth's sea. I saw the elders and the leaders of the heavenly orders and 200 angels who directed the stars and their functions in the heavens, and flew with their wings and seemed all that sail. I've seen all the rays of Sun and the Moon. I measured and compared their motions and their lights, and saw that the Sun was greater than that of the Moon. I saw a great light and the flaming hosts of the great angels. I saw angels who carried terrible weapons, directed against those who dishonor God, who corrupt by demonic sorcery and enchantments to the deeds of evil. The angels hasten me to go to heaven definitively. And the angels took him to the highest heaven where the Lord lies".* Around 875 BC, according to the Book of the Bible-Kings II – Chapter 2, verses 1 and 11, new appearances of flying machines are reported: *"Here's what happened on the day the Lord snatched Elias, leader of the Israelites' successors, to heaven in a whirl: suddenly a fiery chariot with fire horses separated him from Elisha, a leader who succeeded Elias who ascended to heaven in a whirlwind".*

And around 600 BC, according to the Book of the Bible-Ezekiel, Chapter 1, verses 1, 3 to 7, 13 to 19, 22 to 24; Chapter 3, verses 13 and 14, was the reappearance of the flying machines, by the account of Ezekiel, leader of the Jews who succeeded the Jewish and the Israelites: *"On the banks of the Cobar river, the heavens opened and behold divine visions. In this place came the hand of the Lord upon me. It blew from the north side a fiery wind, a thick cloud with a beam of blazing fire, and in the center came out of the middle fire something that possessed a red glow. The image of four beings that appeared to be human form was distinguished in the center. Each had four faces and four wings. Their legs were upright and the plants of their feet resembled those of the bull and twinkled like polished bronze. In the midst of these beings, something resembled glowing embers like torches that circulated among them; and from that fire that projected a dazzling light, come out lightning. Beings zigzagged like lightning. I have dihered a wheel over the Earth next to each of the four beings. All four wheels resembled each other and looked built into each other. They could move in four directions without returning to their motions. His hoops were of an astonishing height, garrisoned of eyes across the circumference. When living beings moved or rose from the Earth, presentations all and rose with them. Hovering above these beings, there was a vault, limpid as crystal, stretched over his head. Under that vault they stretched their wings until they touch. And I listened to the noise of his wings. I heard the clamor of the flapping of the wings of living beings and the noise of their wheels beside them, a portentous noise. The spirit then transported me".*

And in the year 33, according to the Book of the Bible, New Testament Lucas, Chapter 24, verses 4 and 6, John, Chapter 20, verses 12 to 18, Matthew, chapter 28, verses 5 and 6: *"And two characters appeared with resplendent robes that heralded the resurrection of Christ who ascended him to heaven in a bright light"*. Thus, resulting in the continuity of new ideas against those who controlled the keys of Laws and Sciences.

V. EARTH IN THE CENTER OF THE UNIVERSE – 150 to 1250

Geocentric System – Around the year 150, Ptolemy presented the geocentric system. The Earth admitted by the power of the church as definitively immovable in the center of the universe, needed to bring a more improved system to explain the most precise positions of the planets, Moon, Sun and stars in the firmament. Thus, a new system of circular motions, by mathematical calculations, sought to explain these positions, with the planets and the Sun describing orbits around the Earth in lieu of the rotating spheres that required the same speed for all the stars. In this system, the planets were positioned at distances by their orbit periods in relation to the zodiac, in which the largest period was Saturn in ~30 times, 30 years, more than Earth, Jupiter in ~12 years, Mars ~2 years, Venus ~7 months and Mercury ~two and a half months. For a planet with longer period, greater was considered the distance from the same to the center of the universe

However, it was observed that in addition to the different periods that the planets completed their orbits in relation to the constellations, the zodiac, also the actual motion of the planets presented different speeds according to their positions in their orbits. Ptolemy then introduced the equant point, for equal amount of motion, by which as Figure 9 the angle (α) traveled by the planet at one time was equal in different positions of the orbit (A to B and from C to D), despite different speeds (V_1 and V_2). Thus, the planets would present a circular motion around the center of the deferential circle (transporter), which was the center of the universe near Earth, and a uniform motion around equantes, besides the epicycles with centres at the distances determined of these equantes. Besides the deferentials, the epicycles and the equantes of the planets, also the planets were in spheres that swayed to correspond to what was in the sky observed by the immobile Earth, where all the planets, the Sun and all the stars revolved around them, beside the center of the universe, as Ptolemy in "Almageto, Megiste Sintaxis Mathematica".

Thus it would be maintained the harmonious world of the thought of the Gods, or of the Supreme Creator, with the sphere and the circle with its perfect curve and with uniform motion.

Alfonso X, King of Castile and Lion, the Sage of Astronomy, a partisan of Ptolemy and who was a candidate for the papacy came to mention in 1250 that: *"If God had asked me to advise you, by creating the universe, I would have given you better suggestions"*

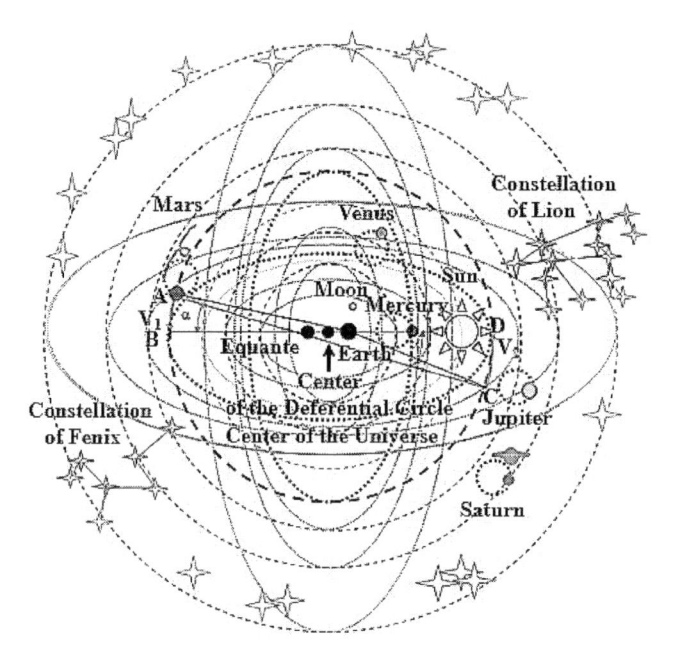

Fig. 9 – Ptolemy's Geocentric System

Ptolemy emphasized Aristotle's argument, adding that if the Earth rotated the bodies in it would present an escape, an escape force, as was the case with the launch of sling at the time; If it turned, it would be harder to shoot an arrow East than Westward, because the Earth turning from West to East would move farther away from the arrow, because it was admitted that the motion of the Earth was untied from the motion of a body out of touch with the soil of the Earth; and a terrible wind would always be occurring, so that a bird that would fly in the direction of that wind would not be able to return to the nest; also, there would be permanently temporal and headaches waves of tides, destroying all the Earth; and if it moved around the Sun, a parallax of the stars should be observed (deviation from the apparent position of the stars due to change in the position of the observer), which was not observed, but also did not occur deviation with the constellations of the stars in which they did not they noticed changes in their forms and in their positions in the firmament.

VI. SCIENCE AND RESISTANCE – 230 to 1558

Not To Analyze, To Believe – Christ had presented new laws and ideas. However, the motto followed by the church and built in the name of Christ was: *"Do not analyze, believe"*, as analyzed by Jeans.

Therefore, this motto of the Total or Catholic Church, contrary to scientific knowledge and respect for the thought of the next, was God the creator of the Earth for man. If Aristotle's science was favorable to the Earth being the center of the universe, and the geocentric system explaining the whole mechanism of the stars in the firmament, so that science if the holy Bible already taught it as the center of God's creation! It was enough to "believe"! And the Greek scientific teaching should not be followed, especially by following other Gods. Reconciling the teachings of the Greeks, as not disagreeing with the biblical teachings, theories such as atomism and others could explain many facts by science and not by direct intervention from God. Reconciling these teachings would be ideas considered heretical, this is, would be received as false ideas. So it was a great fear to deal with scientific knowledge. Around the year 230, Origen of Alexandria received full revulsion by attempting to reconcile the teachings of the Greeks with the Old and New Testaments of the Bible.

Knowledge Temples – Between 285 and 313 was founded the University and the Library of Alexandria, the largest scientific center at the time, by Ptolemy I and Ptolemy II. In 390, the Catholic Church continued in opposition to the scientific teachings of the Greeks. These teachings brought studies about the nature, position of the Earth and other stars, interpreted as contrary to divine studies of the heavens and eternity. According to St. Ambrose, of Treves: *"To discuss the nature and position of the Earth does not come to aid in the hope of eternal life. You lose sight of the sky as you try to measure the Earth"*. And near Alexandria, in the previous Constantinople city of Byzantium of Greek origin alongside the black and Aegean Sea, the opposition to those who did not follow the bigotry of the Catholic Church's Summit, was contrary to the teachings of the ancient Greeks who had their gods themselves. And next to the Lighthouse of Alexandria, one of the wonders of the construction of the time, almost completely erased the knowledge of the Library of Alexandria. Considered as the fundamental basis of the anti-Catholic heretical teaching, the Library of Alexandria, the center of knowledge of the ancient Greeks, came partially destroyed by Teophilus, Bishop of Alexandria. The tyranny of the political and religious power summits was widened against any differences in opinions imposed by them.

In Athens, around 470 in the Scientific Academy of Plato, opinions were presented against the biblical description of the creation of the universe, confronting the Catholic Summit, as reported by Proculus of Constantinople.

Around 480 were published compendiums, with emphasis on links of Christian doctrine with the studies of Plato and Aristotle, by Boethius of Rome that for political reasons came to be arrested and executed.

In Persia around 490 in the School of Jundisharpur, Christian scholars for theological divergences considered heretics, moved from Constantinople to this School where there was ample freedom of opinion. They had been left with copies of the works of the ancient Greeks and translated into other languages.

Scientific Refuge – Plato's Scientific Academy came to be closed and with the partial destruction of the Library of Alexandria by the dogmatics, from 530 scholars headed for Constantinople, where theological divergences between the Christian Church of Constantinople of ties with the Greeks and the Catholic Christian Church of Rome they came to allow a lesser reaction to the scientific knowledge of the ancient Greeks in Constantinople. Thus, it would be a center to avoid the total destruction of all studies carried out, preserving the original ideas presented.

Ideas in the Fire – In 642 came the final destruction of the Library of Alexandria by fire, burning all works written by religious fanatics, followers of religious summits, claiming that if those writings of the ancient Greeks coincided with the Book of God they would be useless, and if they disagreed, they would be pseudo-truths, heretics, false opinion, and they must be destroyed.

Scientific Unification with Theology – Around 700 came the first synthesis of Aristotle's texts with hyped Christianity by Origen, Aurelius Augustinus and Boethius, through John of the city of Damascus.

City and House of Wisdom – In 762, the Arab Empire founded its capital in the city of Baghdad, through the caliph of the Islamists, Harun Al Rashid, coming to support the doctrine that a great value to reason would reach the greatest depths of religiosity. Translations of the works of the ancient Greeks, Chinese and Hindus were sponsored, and these works were developed by the Arabs and recovering the cultivation of Greek science and other peoples, keeping all knowledge prepared by man until that time. And around 820 the caliph successor, Al Mamun, founded the House of Wisdom in Baghdad, for translations of Greek works brought from Constantinople and works of Origins of India and China.

Arab Science – Between 820 to 1450, scholars of the Arab Empire and their legacy developed with this knowledge the Geometry, Trigonometry, Algebra, Astronomy, Optics, Chemistry, Philosophy, by: Thabit Qurra, Al Battani, Al Karismi, Al Kindi, Abu Kindi, Al Jahari, Kamal Al Din, Ibn Shakir, Abul Farghani, Al Farabi, Habach Hasib, Al Abbas, Abu Machar, Sinan Qurra, Al Kujandi, Abu Iraq, Al Karaji, Abul Hassan, Al Nasawi, Abu Bujani, Abu Quhi, Al Maghribi, Ibn Qunfudh, Ibn Ayub, Al Chatir, Al Qalassadi, Abu Yunes, Al Kayami, Abu Zargali, Al Ghazali, Al Asfizari, Abu Biruni, Abul Kazini, Al Tusi, Al Samaw, Al Qazwni (Al Schirazi), Ibn Sina (Avicena), Al Haytham (Alhazen), Abu Sufi (Azophi), Al Bitruji (Alpetragius), Jabir Ibn Aflah (Geber) e Abul Ruchd (Averróis).

Around 900, Al Haytham (Alhazen) in his Treatise of Optics came to explain that the vision of objects was solely resulting from the reflection of light by sight, emphasizing the absurdity of the teaching of the ancient Greeks in which the ray of light came from the bodies that were sent to the eyes and these to objects returning to the eyes. and Abu Sufi (Azophi) came to present details of constellations with the stars in their positions, glitters and colours, identifying them as Altair, Aldebaran, Betelguese, Rigel, of Arabic names.

In 950 was founded the University of Cairo and on 970 in Cordoba in Spain the Caliphs Alhakem II founded the University of Cordoba, coming to extend to all Europe translations of Arab works with all the knowledge of the same and of the ancient Greeks.

Freedom of Thought – Around 1050 the main philosopher of the Roman Empire of the East M. Psellos presented the defense of freedom of thought and critical study, against the obscurantism of the religious power of the Catholic Church.

Understand To Believe and Believe Without Understanding – Around 1100, P. Abelard in France, made the defense of scientific knowledge widely when commenting that: *"It would be necessary to understand to believe"*. To believe in heaven, hell and other worlds, scientific knowledge would be necessary.

However, a row within Islam, a religion founded by a prophet in Arabia around the year 600, came to criticize the excessive value to reason, which would draw attention to the deity with his revelations which should be amply exalted. Was a variant in relation to dogmatics of Christianity represented by the Summit of the Catholic Church.

European Universities – European Universities were built with the teaching of Theology, Physics, Medicine and Law, based on the intellectual activity of the Arabs since the beginning of the formation of the Arab Empire, through the freedom of thought as to the studies existing in this time on the works of the ancient Greeks. Were the universities of: Salerno, 1050; Bologna, 1088; Paris, 1150; Oxford, 1167; Montpellier, 1180; Cambridge, 1209; Salamanca, 1243; Sorbonne, 1253; Coimbra, 1290; Rome, 1303; Prague, 1348; Krakow, 1364; Vienna, 1365; Heidelberg, 1386; Leipzig, 1409; Freiburg im Breisgan, 1457; Munchen, 1472, in which there were around 50 Universities worldwide until this year.

Mathematics and Asian Astronomy – Around 1150 in India, mathematical calculations were developed by Bhaskara from India.

And in China, developments in Astronomy and Mathematics: Around 1250 the "Treatise of Mathematics" by Chin Chiu Shao; "Marine Circle Measurements Mirror" by Li Chih (Li Yoh), and by Yang Hui; And around 1300 were presented "Introduction to Mathematical Studies" and "Precious Mirror of the Four Elements", by Chu Shih Chieh.

Expanded Publications – Around 1200, European scholars transcribed Arabic and Greek publications, disseminating throughout Europe the scientific knowledge of ancient Greeks and developments by the Arabs. By: John Hollywood (Sacrobosco), Domingo Gonzalez, Juan Sevilla, Gerardo (of Cremona), Michel Scot, Adelard of Bath, Robert of Chester, Herman el Manco, Gerbero (Silvestre II), Leonardo (of Pisa), Afonso X and John Peckham.

Resistance to Science in Latin or in Any Language – Robert Grosseteste of the University of Oxford had provided the restoration and translations for the Latin works of Plato and Aristotle, from Arab and Greek versions. But because Aristotle's scientific ideas were broader than Plato's, they were more colliding with the dogmatic ideas of the Catholic Church Summit. In 1210, when Grosseteste attended the University of Paris, a Council of the Catholic Church where this church was a great center of Christian Theology condemned the works of Aristotle. They were works by this Council that put the reason and scientific knowledge above the belief in "God Almighty".

For Aristotle there was a higher science, the wisdom ("sophia") that sought ("philia") the fundamentals of the whole of knowledge, i.e. knowledge through philosophy. And only by it would it be possible to approach the maximum value of wisdom that would be God, which was the first cause of existence and motion, and it was for him the first trigger of all that existed. As Aristotle, because God was the trigger of the fullness of everything, it was not necessary the creation of the universe, so for Aristotle the universe had always existed and eternally moved following the same harmony. Thus, for him there would be no beginning and no end to the universe, for it was as eternal as the trigger itself that moved it. God was then to Aristotle the "Being Perfect", eternal.

E Aristotle focused thought as the most noble and perfect activity among beings, so God should be thought himself and the greatest thought was to think of himself. For, if God came to think less, for example, in the being "homo", it would be a form of degradation, as Aristotle. Thus, for Aristotle, God could not know the world that was moved by Him, and much less that man could establish any bond with God. So Aristotle did not accept religion, that is, the connection between man and God. It would be God just the pawn of the order that sustained the eternal and harmonious motion of the universe, and then religion was for Aristotle only a way of leading the devotees and serving the interests of the Governing Summit of it. For Aristotle, only the ignorance of the people and the exploitation of this ignorance for their own benefit of maintaining their power would explain religion.

Despite the resistance to the disclosure of these translations into Latin, for this "good council" not to disclose the ideas of Aristotle, applied by this Council of the Catholic Church Summit, the University of Paris around 1220 accepted Aristotle's ideas as an object of study.

But it would be evident, by these ideas of this "heretic" Aristotle, that some "coming from above" drive would be introduced into the universe to go against the defenders of this "heretic".

Nature of the Facts – Around 1240 Alberto Magno coming from the University of Padua defended the importance in questioning the nature of facts, knowledge based on every observation, And science could not consist in believing what someone had said without the right to any questioning. He paraphrased all the works of Aristotle, commenting widely in agreement or criticism. Despite criticizing Aristotle's works, it was little to be tolerated by the Catholic Summit.

Barriers to Knowledge – Around 1250 Roger Bacon of Oxford University also advocated scientific knowledge and publications, contrary to the rules for which an express authorisation of superiors would be required for these publications. He also considered reason and philosophy as superior to simple beliefs. He said that physical science would lead to the knowledge of all the facts and also to the knowledge of its creator. As aggravating the confrontation with the Catholic Summit that was contrary to these ideas considered heretics, Roger Bacon was also contrary to Ptolemy's system, by the Earth as the center of the universe. He mentioned several obstacles to knowledge: weak and inept authority; ancient habits; cloaking of ignorance by an apparent wisdom and mentioned that: *"These obstacles are still valid in this season"*.

"Masters of Knowledge and Power" – In 1252, among the obstacles to knowledge, by Summits formed of "knowledge and power owners", Pope Innocent IV approved and decreed torture for the confession of suspected heresy or witchcraft. Heresy and witchcraft entered as pretences for those who disagreed with the opinions of the dominant power.

"Physical World Written by God" – Around 1260 Thomas Aquinas from the University of Naples defended reason along with belief, and that the physical world was a book written by God, and in science the ancient Greeks had revealed the world of God, while in theology the Catholic Church and the Bible were the divine authorities. Thus, it has unearthed Greek science with Christianity. Although Thomas Aquinas adopted the Ptolemy's system with the Earth at the center of the universe, it was insufficient not to be considered as a heretic his ideas by the Catholic Summit, for it presented theories of the Greeks in which the Earth was outside the center of the universe and also Aristotle's ideas that were considered heresy by it.

Around 1265, by the position of ideas contrary to the professors of power, in which the prospect of these new ideas was seen as frontal to the ideas of the installed power, the decree of the Inquisition should be applied. So Roger Bacon was imprisoned and Thomas Aquinas convicted by the Inquisition, that it was the sector of the Catholic Church that judged the "heretics".

High Studies – Around 1420, far from "scientific persecution", and away from the Inquisition, in Persia Ulugh Beg founded the

Institute of High Studies of Samarkand. Studies of Astronomy and Mathematics were developed by Beg, Al Kachi and Al Rumi. The book "Elementary Encyclopedia" was written by Al Kachi on practical mathematics for astronomers, surveyors, architects, clerks and merchants.

Pure Heresy! – Around 1440 Nicholas of the City of Cusa (now Bernkastel-Kues-Germany), Cardinal of the Catholic Church, as written in "De Docta Ignorantia": *"The Earth moves as the stars do, revolves around its axis and around the Sun, and there is life in every universe that is infinite"*. But emphasizing the creation of the universe by God and carrying his cardinal title, he managed to minimize the attack of the Inquisition.

Religious Science – In 1449, Islam was a vision of science as subordinate to religion, in antagonism to the realization of science by Islamic Arabs and independent scientific thinking. And for political-religious matters was eliminated the founder of the Institute of High Studies of Samarkand in Persia. And in 1500 religious fanatics destroyed the astronomical observatory of Samarkand in Persia, culminating in the stagnation of science by the Islamic Arabs.

Astral Center and Celestial Disharmony – In 1500 Leonardo da Vinci, Italy, in his notebook, wrote that the Sun illuminated both the Earth and other stars like the Moon, this then with its image obtained by the reflection of sunlight. So the Sun could be a center of an astral system. And Domenico Novao, in Italy, criticized the central system of the immovable Earth, with epicycles and equantes by the planets of Ptolemy, for coming against the celestial harmony advocated by Pythagoras and Plato.

"Publications Heretics" – For the heartbreak of Religious Power Summits, were edited in 1501 scientific publications, and the worst were the Aristarchus written by Plutarch of Chaeronea, that indicated the Earth revolving around the Sun, by the publisher Georg Valla. A few years before works of Aristotle were edited between 20 and 40 years after the publication of the Bible.

Compatible Heliocentric System – Around 1530 Tiedemann Giese of Poland, presented a treaty of defense of the heliocentric system seeking to reconcile with the Bible, but this treaty was "lost".

Two Religions and Four Opinions – In 1540, the Catholic Church and the Protestant Reformation in addition to its doctrinal divergences also differed as to the heliocentric system. Pope Clement VII and other members of the Catholic Church showed to be favorable to the heliocentric system in initial presentation by the canon of the Catholic Church Nicholas Copernicus of Poland, but other members of the Catholic Church were contrary. Among members of the Protestant Reformation there were proponents of Copernicus' work, but others like Martin Luther widely opposed. As Luther: *"A new astronomer wants to prove that the Earth moves. Who is this idiot who wants to turn all science of astronomy upside down, for by the holy Book of the Bible it was in the Sun and not to the Earth that Joshua (the successor of Moses) had ordered him to stop (so that during a battle with the Sun stopped he could overcome it).* But another member of the Protestant Reformation, Georg Joachim Von Lauchen, of Austria, published a small treatise of Copernicus' work, in which the premise of the same wrote that: *"The Earth is on the move. And still in the midst of everything is the Sun, illuminating everything at the same time, and for that reason called by some people of lantern of the universe".*

"Index Librorum Prohibitorum" – In 1557 the Catholic Church launched the first list of books that could not be seen by catholics, through the institution of the "Index Librorum Prohibitorum".

Lenses for Long Range of Vision – In 1558 while the vision was restricted to books prohibited by the Inquisition of the Catholic Church, Giabattista Della Porta, from Italy, drew lenses to increase what was far away from sight!

VII. NEW HYPOTHESES AND NEW RESISTANCES – **500 to 1633**

Turning Earth, Rotating Sphere, Rotating Astros, Elliptical, Heliocentric System and Extraterrestrial Systems – Around 500, Ariabat I, he had presented that the sphere of the fixed stars was stationary and that the Earth when turning produced day and night. However, around 600, Isidoro of Seville and Beda, the Venerable, presented the daily rotation of the sphere of fixed stars. And around 750 Alcuin of York presented a model of the motion of the stars, contrary to the epicycles of Ptolemy. Around 1000, Al Biruni determined the angle of the orbit between the Earth's plane and the solar plane, that is, the angle of the ecliptic at 23° 34' 0''. And in 1440, Cardinal Nicholas of Cusa had presented that the Earth revolved on its axis and around the Sun, in the infinite universe, in which the stars were other suns with planets inhabited as the Earth.

Orbits and Celestial Spheres – And in 1543 Copernicus, of Poland, had his work "De Revolutionibus Orbium Caelestium" ("Of The Revolutions of the Orbits or Celestial Bodies"), in which he claimed to be the Earth in circular motion natural and inevitable, around the immobile Sun, explaining the errant motion, of the planets, and all the positions of the stars, simplifying the system of Ptolemy of 80 celestial spheres to 34 or maximum of 48. By the positions of the planets in their annual motions, the Sun was fixed in a position at a displaced distance from the center of the orbit of the planets around the Sun. In the case of Earth, the Sun was shifted from the center of the Earth's orbit around the Sun at 1/50 from the distance from that orbital center of the Earth to the Sun. The center of Copernicus' system was the center of Earth's orbit around the Sun. Copernicus presented the astronomical calculations for the heliocentric system in which the Earth was only the center of the lunar sphere, contrary to the Ptolemy's system and the religious scriptures that the Earth was the center of the universe.

By this system, Copernicus determined precisely, beyond the periods of the planets around the Sun, also the relationship between the distances of the planets to the Sun, based on the distance admitted from the Earth at the time of 8.4 million km. Through the triangle formed by the Planet, Earth and Sun, was obtained the exact distance between the planet and the Sun.

As Copernicus: *"Each variation of the star positioning would be observed by the observer motion, or by the motion of the heavenly body, or by both motions. Thus, the apparent motion of the sphere of the stars seen by the observer on Earth, could be in reality the motion of the observer himself, as if each star passed leaving behind the Earth, similar to the writing of Virgil said by Aeneas: "We left by the boat of the bay and we saw the fields and cities go backwards".*

Copernicus explained that the Earth revolved around the Sun and around itself in the direction of the East, that is, from West to East, through which the whole world seems to turn dizzy in a vortex. Thus the Earth revolved with the water surrounding it and the air near. Thus, it explained the reason for an object that fell vertically, not to be West of the point where it had gone as it would if the Earth rotated and the object did not participate in that motion. Also, explaining that air and water also turning along with the Earth there would be no occurrence of tornadoes and tidal waves. Copernicus accepted the calculation of Aristarchus for the distance from Earth to the Sun in ~ 4.3 million nautical miles, equivalent to ~ 8,000,000 km (current 149,600,000 km).

Considering the circular orbit of the Earth around the Sun, by Copernicus, it was obtained an orbital length C of $2\pi d_{ES}$ equal to ~ 27,004,000 nautical miles equal to ~ 50,240,000 km, in which d_{ES} it was the distance from the Earth to the Sun. And the time to scroll through it in 365 days x 24 hours x 60 minutes x 60 seconds equals ~ 31,536,000 seconds. Therefore, the Earth's orbital velocity around the Sun at ~ 27,004,000/31,536,000 seconds equals ~ 0.85 nautical miles/second, equal to ~ 1.6 km/second. The Earth rotating on its axis at 24 hours x 60 minutes x 60 seconds equal to 86,400 seconds, shifting its circumference C_E equal to 2 $\pi D_E/2$ equal to 2 x 3,14... x (~ 7.560/2) nautical miles equal to ~ 23,700 nautical miles or ~ 43,800 km, where D_E it was the diameter of the Earth, therefore a speed around its axis at ~ 0.27 nautical miles/second or ~ 0.5 km/second (current 0,46 km/s). Thus, a maximum speed of the Earth in orbit around the Sun of ~ 1.1 nautical miles/second or ~ 2.1 km/second (current 30 km/s), which Copernicus considered dizzying, because the velocity of a bird was only 0.016 nautical miles/second or 0.03 km/second. E, the fall of an object at the height of 10 m in ~ 10 seconds, in which if the object did not rotate with the Earth would be at a distance of 2.1 km x 10 seconds equal to 21 km of the beginning of the fall or ~ 11.3 nautical miles.

Copernicus also explained that there was the parallax effect of the stars by the Earth, that is, angle difference when the stars observed in a position of the Earth in space and in another position while rotating around its axis and around the Sun , because the distances of the stars are far apart, thus preventing a difference being observed that could be measured.

And should the Earth be too close to the sphere of the fixed stars, at night it would be observed only half the stars.

Copernicus' heliostatic system contained mathematical calculations that explained the motions of the Moon, planets and stars, but without directly proving the motion of the Earth. Thus, in 1543 when being edited this system with the motion of the Earth around the Sun, assistant coach Andreas Osiander cleric of the Protestant Reformation, to editor John Petrejus of Nuremberg, placed a preface, unsigned, focusing on the work of Copernicus as a system of mathematical possibility that adapted to the observations of the errant motions of the planets in relation to the Earth, Sun and stars. This preface as Osiander continued to maintain the reality of the Earth as the center of the universe without contradicting the dogma of the Bible, and it was mentioned that the Copernicus system was just a mathematical hypothesis. It would thus preserve Copernicus from future accusations of heresy, mainly by the opposition previously issued by cleric Luther.

Copernicus he had left enlightened in his work that the Earth because it was not the center of the universe was moving around the Sun. However, there was no evidence to prove that it was the Earth that revolved around the Sun, and not the Sun and the fixed stars that spun around the Earth. The "sacred" scriptures said that the Earth was the center of the universe, so any contrary theory would be pure heresy.

As the dogmatics, the central part of the universe where the Earth was, could turn around the Sun, and in this case, the Earth would continue to be the center of the universe. Also, it would not be the Earth that would move around its axis, but a sphere around it would turn to explain day and night. The Earth would remain motionless within that sphere, being the sphere that would turn around its axis. The Sun and fixed stars would be real estate in the universe. And the central part of the universe, where the Earth would be within a sphere, is that it would revolve around the Sun.

So the Earth could apparently move around the Sun, through the turning of the central part of the universe where the Earth would be motionless within a sphere. But by the system of Copernicus that removed the Earth from the center of the universe and still making it move around the Sun would be a great blasphemy because it is contrary to the sacred scriptures.

Motion Center of the Rotating Circle – In 1544 Luther's defender, cleric Filipe Schwarzerd Melanchthon, mentioned in his book "Inita Doctrine Physicae": *"Just as the center of a rotating circle remains motionless, the center of the universe, the Earth, also remains still"*. And that Copernicus' Heliostatic System was a simple copy of the Aristarchus System. In this book, Melanchthon did not write Copernicus' fundamental calculations that justified the heliocentric system of Aristarchus. Nor was Copernicus's basic argument that the immobility of the Earth was supported in the appearance that it was immobile, and it was this same appearance of motion when it was admitted that the surrounding bodies were motioning with the Earth. In reality, Melanchthon remained programmed in which the Bible was presenting the sacred scripture of the reality of nature, in which the Earth was the center of the universe, created by God so that on Earth man worshipped his Lord, his God, above all things. The fact that Copernicus put the center of the Earth's orbit into the Sun, as the center of the system in which the planets revolved around the Sun, was of no acceptance by the dogmatics. For, the Earth would have to be the center of any system and still be motionless in heaven as it was "prayed" by the sacred scriptures.

"Hypotheses Orbium Caelestium" – In 1551 Erasmus Reinhold, of Germany, published "Hypotheses Orbium Caelestium", with astronomical tables based on Copernicus' work, but considered the system of Copernicus as a simple hypothesis copied from Aristarchus. The evidence of this criticism was to minimize Copernicus' work, when the fundamental difference between the two works was that Copernicus contained the physical and mathematical apparatus that justified the heliocentrism or heliostatic.

The Earth that Does Not Move – In 1555 Reiner Gemma Frisius of Holland criticized the Earth that was moving through the Copernicus System. One of the main objections was precisely that the Earth could present a dizzying velocity around a nautical mile/second or ≈ 2 km/second, when a bird flew with 0.016 nautical miles/second or ≈ 0.03 km/second.

"And it could hardly be expected that men connected to the interpretation Bible, accept the contrary ideas to religious convictions, only because the celestial spheres of 80 to 48 or 34 were simplified", according to Jeans.

"Hypotheses Astronomicae" – In 1571, Caspar Peucer, son-in-law of the Melanchthon clergyman favorable to Luther, affirmed in his book "Hypotheses Astronomicae" that:*"He only presented the astronomical tables and observations of Copernicus' heliocentric work, but he would not present Copernicus' hypothesis so as not to offend or disturb his beginners".*

New Star in the Celestial Sphere Makes the Sky Fall – In the year 185 had appeared in China's firmament a superbright star, and in 1054 also seen in China an explosion in the sky originated a superstar that shone 23 days even during the days in the constellation of Taurus, and then again in 1181 arose another superbright star. In 1572, Thomas digges of England and Tycho Brahe of Denmark and other astronomers in China and Korea observed the appearance of a supershiny new star. But as Aristotle and the Bible reported with the creation of heaven by God, the sky was immutable. So it should just be some astro circling the lunar sphere.

But measurements of the position of the supershiny new star appeared in 1573, by Brahe, showed this new star held in the celestial sphere of the stars, so far beyond the lunar sphere. Thus, "falling the sky" with the appearance of this new star, according to his book "of Nova Estella". And another superbrilliant star in 1604, confirming the fall of the heavens!

Geo-Heliocentric System – In 1574, Brahe presented a model of the universe in which the Earth was still, being orbited by the Moon, Sun and fixed stars, while the planets orbited the Sun, it was a Geo-Heliocentric System.

Infinite Universe – In 1576 T. Digges presented in the updated edition of the Almanac "Eternal Prognosis" of his father Leonard Digges, the Heliocentric System of Copernicus, and adding to be the infinite universe. In this universe, the stars stretched all over space and not just in a celestial sphere. This theory was emphasized after astronomical observations carried out with juxtaposed lenses that increased the sharpness of the stars.

Comet Paralyzes the Earth – Between 1577 and 1582, Brahe performed measurements in comets that appeared in those years, looking for the opportunity to check parallax in the face of its route through the sky. Even though these two comets were at a great distance from the lunar sphere, they predicted that their instruments could detect this parallax. But, no angular deviation between two Earth positions was detected by Brahe. His conclusion was that the Earth was still in the universe, and the stars, the Sun, the Moon, whirled around the Earth, and the planets around the Sun, and the daily rotation between night and day was attributed to the sphere of fixed stars, according to their system presented in the book "De Mundi Aetheri Recentioribus Phaenomenis Liber Secundus" in 1588.

"Friend Ursus" – In 1584, Brahe had received the visit of the astronomer Ursus Dithmarsus, and in 1588 he published "Ursi Dithmarsi Fundamentum Astronomicum" (Astronomy Foundations of Ursus Ditmarsus), with similar Geo-Heliocentric System of Brahe, but attributing to the daily rotation of the Earth around its axis. In this case, the stars were considered fixed in the firmament, so without turning daily on the Earth. And the Sun revolving only annually around the Earth. Brahe didn't like any of this "Fundamentum" from his "friend Ursus", because he thought it was a copy of his work and the Earth revolved around his axis for daily day and night.

About Infinity, the Universe and the Worlds – Giodarno Bruno, friar of the Catholic Church, of Italy, expanded the theory of T. Digges from an infinite universe and also from another author, Nicholas of Cryftz, of life throughout the universe. In 1584, Bruno admitted the plurality of planets with civilization: in his book "De Infinity, Universe et Mondi" said: *"The universe must be infinite, by the capacity of infinite space. So there is the possibility of countless worlds like this. Why would divinity prefer to remain only in a tiny sphere? It should rather shine in an unlimited mirror, following its infinite and immense way".*

The book soon goes on the "Index" as prohibited, because it contradicted the dogma of the Catholic Church in which Christ was born by divine conception and not by nature.

Between 1584 and 1591 Bruno published dialogues in which one of them frontalally attacked the dogmas of the Catholic Church, as "Spaccio de la Bestia Trionfante", and "Dialoguei Idiota Triuphans", ie the "Triumph of the Beast" and "Dialogue of the Triumphant Idiot", being interpreted the beast and the idiot as being Pope Gregory XIII of the Summit of the Catholic Church, provoking the wrath of it against Bruno. Also, Bruno published the Defense of Copernicus' System that did not consider the Earth to be the center of the universe.

Calculations Yes, No theory – In 1588 Michael Mästlin of Germany, professor of the astronomer Johannes Kepler, also used the astronomical tables and data of Copernicus, but rejected the theory of the Heliocentric System, in his book "Epitome Astronomye". And Giuseppe Magini, from Italy, adopted the astronomical calculations and tables of Copernicus, but considered the Heliocentric System as an absurd theory.

Microscopy – In 1590, different from the philosophical microscopy of the inquisition of the Catholic Church in limiting the vision of religious doctrine, Hans Jessen and Zacharias Jessen, in Holland, built the microscope, allowing to increase the normal vision of parts of any substance.

The Triumph of the Beast – In 1600 Giodarno Bruno, on the charge of heresy, is condemned to death at the stake by the inquisition of the Catholic Church, without clemency, in the papacy of "Clement" VIII.

Cosmographic Mystery – In 1596 Johannes Kepler, from Germany, published "Mysterium Cosmographicum" (Cosmographic Mystery) seeking to explain the positions and motions of the planets in relation to the Sun admitting it as the still center of the universe by the Copernicus System. He searched for the six planets: Mercury, Venus, Earth, Mars, Jupiter and Saturn, an order of harmony: *"An architectural order befitting the divine perfection by which the Lord of the Universe had created him. God should have effected his creation of the universe through a determined plan and not as the work of a fluke without any correlations".*

Conjectured that there should be a reason why there were only five polyhedra that presented equal faces: tetrahedron, cube, octahedron, dodecahedron and icosahedron. He set them between inscribed and circumscribed spheres, including thicknesses to the spheres for the variations observed in the motion of the planets, and after calculations between the real and the model of polyhedra, he observed that the proportion between an subscribed circle and another circumscribed seemed almost equal to that between the orbit of Saturn and that of Jupiter. Then, with the positions adjusted between the five polyhedra, and the observation by the instruments of measurements of parameters of the planets, would come to be possible with a single measure more precise of any distance between a planet and the Sun, in glimpse the "secret of the universe" by its size from that proposed model.

Sun, the Physical Center – Kepler analyzed the motion of Mars and the other planets, considering the heliocentric theory as correct, but focusing on the Sun the center of the system and not the center of the Earth's orbit around the Sun, as Copernicus had determined. Copernicus had observed that the planets passed at a nearer distance (perihelion) and another farther (aphelion) from the Sun, and Copernicus putting the Sun at a distance from the center of the Universe (the center of the Earth's orbit and the other planets around the Sun) in 1/25 about the distance from the radius of the Earth's orbit around the Sun. To Copernicus the motion of the planets was seemingly uniform and circular around the center of the Earth's orbit, with epicycles around the orbit of the planets to justify greater speed in the perihelion and minor in the aphelion.

But between 1607 and 1621, for Kepler: *"The sun is immobile in the universe; It is the mathematical center describing the orbits, speeds and periods of the motion of the planets; The physical center where it emanates bodily entity magnetic* (of origin in stones that are attracted discovered by Magnes in 2000 BC) *that move the planets in tangent to their trajectories; and a center built into the creator of the universe, God. The Sun is formed around it and lines magnetics which they rotate by conducting the planets in their orbits. Also, the Earth presents these lines magnetics that by the rotation of the planet drags the Moon around it. The Sun also has its rotation, like the Earth, producing these magnetics lines"*.

Although Kepler regarded the Sun as the physical and mathematical center for moving the planets, his calculations of the positions of the planets placed him, as well as Copernicus, outside the center of the Earth's orbit around the Sun. But for Kepler was the Sun more next, less than twice the distance, from this center of Earth's orbit around the Sun. Kepler studied the orbit of Mars and observed that between the perihelion and the aphelion there were different speeds between the passage of Mars through these points.

He decided to effect measurements of Earth's orbit, but as if he were on Mars. observed that the motion of the planets should be governed by the Sun, through bodily entity solar that would be higher when in the perihelion and smaller in the aphelion. And not the motion of the planets governed from the center of their orbits, that is, from the center of the Earth's orbit considered the center of the universe.

It was then concluded that each planet in the perihelion ran, at a certain time, a surface formed by the radius that went from the planet to the Sun, with a greater distance in the orbit and of less extension to the Sun. And in the aphelion, for that same time, forming a surface with a smaller distance traveled in the orbit, but larger in extension to the Sun. In both cases, for the same time, they were identical to the surfaces traveled by the planets in their orbits around the Sun. According to Kepler: *"The line that goes from the Sun to the planets travels in its trajectory equal areas in equal times"*.

The uniformity of the motion of the stars, that is, their constant speed, around a center of the universe, by the perfect symmetry of the circle, was toppled, along with the epicycles of the age of Ptolemy and Copernicus. The epicycless made oscillating the planets in their orbits, to explain their wandering motions and also the different speeds in their orbits, with the purpose of preserving the uniform and circular motion.

The Surprising Non-Circular Curve of the Orbits of the Planets – For Kepler, the Sun was the mathematical and physical center that ruled the motion of the planets. But Kepler was trying to understand why the Sun was not also the geometric center of the orbit of the planets around it. By the motion of the planets the Sun was shifted from the center of the orbit of the planets around it.

Between 1604 and 1605 with data of Brahe's observations, he determined a curve of the orbit of the planet Mars in its motion. This curve is different from the circular orbit that was considered to be the most uniform and perfect curve of nature. But also Kepler considered the circle as the most perfect figure in nature and decided to geometrically confer positions of the planet Mars in its motion. Again it was different from circular orbit. In principle, he admitted to being more of an oval, meaning the genesis of life. Then the curve indicated it was an ellipse. He then decided to move as if he were on Mars and to measure distances from there to the Sun and the Earth, because by the measures from the Earth could the motion of this indicate an ellipse for Mars and not the circular motion around the Sun. The preflight was precisely the elliptical orbit of Mars around the Sun, concluding in 1609 in the book "Physica Coelestis" (New Astronomy), in which: *"The planets describe elliptical orbits around the Sun, and the Sun is shifted from the center in exactly one of the outbreaks of the ellipse"*.

The distance from the Earth to the Sun allowed 4.3 million miles calculated by Aristarchus and accepted by Copernicus, as an astronomical unit equal to 1 AU for the Earth, indicated to the other planets the ratio of 0.4 AU to Mercury, 0.7 AU to Venus, 1.5 AU to Mars, 5.3 AU for Jupiter and 9.5 AU for Saturn. However, the periods of the planets around the Sun followed different proportionality between them in relation to the distances of the Sun. Thus, the periods of the planets were not directly related to their distances from the Sun, being: Mercury in 0.24 years, Venus 0.60 Years, Earth 1 year, Mars 1.88 years, Jupiter 12 years and Saturn 30 years. Distance/Period ratio: Mercury 1.66; Venus 1.16; Earth 1; Mars 0.79; Jupiter 0.44; Saturn 0.31. That is, the speed (length of circular orbit by the period) was smaller the farther away was the planet of the Sun, but without a direct proportionality with the distance. Kepler tried to find out how these variables of speed, period and distance were interconnected in the solar system. By his calculations he obtained: *"The periods of the orbits of the planets around the Sun in their squares are proportional to the average distances of their distances to the Sun in their cubes"*. ie: [Orbital Period of the Planets]^2 /[Distance from the Planets to the Sun]^3= [Constant for all Planets].

Transporting to the Moon – In "Somnium" work on a fictional site in Lavania, on the Moon, of inhospitable underground life in which its inhabitants lived under caves and waters to shelter themselves from the heat or the cold. By transporting to the Moon through this fiction, Kepler explained the illusion of common sense about the immobility of the Earth. Previously, Copernicus emphasized the apparent motion of the Sun and the stars around the Earth and its immobility.

According to Kepler: *"Levnia, for its inhabitants, seems to be still between the stars and the Earth moving in the sky. The goal of my dream is to be worthy of the case of the Moon to build an argument in favor of the motion of the Earth".*

Force of the Sun, Mathematics and Intelligence – In 1607 Johann Georg Brengger stated that: *"Although the orbits and motions of the planets are described without the epicycles, circles, and the uniformity of the motion, but with a new Physics of the sky, in which magnetic forces enter to trigger these motions, I cannot imagine and understand what the mathematical procedure involving these magnetic and intelligent forces that trigger the planets around the Sun".*

Miniature Solar System and the Imperfect Sun – In 1610, through a spyglass produced by "Galileo Galilei" (Galileo) of Italy, he achieved an increase of up to 30 times of the distant object, and published "Siderus Nuncius" (The Messenger of the Stars), reporting that the planet Jupiter featured four heavenly bodies rotating its surroundings. He also commented that Venus presented phases similar to that of the Moon, which would indicate that Venus revolved around the Sun, as by analogy to the "smaller stars" of Jupiter rotating around it. Thus, the planets Venus, Earth, Mars, Mercury, Jupiter and Saturn would revolve around the greatest "star" among them, the Sun. And reported that the Sun presented stains, going against the perfection of the Sun admitted as God's creation. He wrote that the stars presented undefined contours by flickering, differing from the planets that present form of well-defined round globules. So the stars would be much farther away than was admitted. Simon Marius of Italy, shortly before also observed four heavenly bodies revolving around Jupiter, attributing the names of Io, Europa, Ganymede and Callisto.

Sunspots – In 1613, in "This and Dimonstrazione allathe alle Macchie Solari" (History and Demonstration on Sunspots) Galileo observed by the spyglass that sunspots were present in the Sun. They were restricted to the disc of the Sun, without escaping ever out of the solar disk, which evinced to be the Sun's own constitution and not any kind of star that came through the Sun.

If heavenly bodies revolved around the planet, then the planets would also revolve around the Sun, going against the proposed by Copernicus. And as the stars would be extremely distant, there would be a need for immense speed of them to daily revolve around the Earth. So the "friend Ursus" system from Brahe would keep the stars really fixed without this incredible daily speed around the Earth.

But by Ursus' system to admit that the Earth was moving by rotating around its axis, the dogmatics immediately resolved to accept the Brahe System in which all planets, with the exception of the Earth, would revolve around the Sun and this one around the Earth. With this keeping what was in the sacred scriptures in which the Earth was immovable in the center of the universe.

And sunspots were soon admitted as a result of standing heavenly bodies in front of the Sun, since the belief of the dogmatics was that the Sun created by God should be of total uniformity, perfect as its creator, free from any stain.

However, observations by the spyglass confirmed that these stains when existing in the Sun's equator were in motion and repeated every 25 days, indicating the Sun's own rotation. This imperfection of the Sun was then to be treated by the dogmatics as more a reason for the Earth and in it the man created by God to be proven to be in the center of the universe, still in the sky.

Lockdown by the "Holy Mother" of the Heliocentric System – And in 1616 the "Holy Mother Catholic Church" prohibited, by "Index" the heliocentric work of Copernicus.

Phases of Venus – In 1632 Galileo published "Dialogo sopra i due Massimi the Tolemaico and Copernican". "Dialogue on the two Maximum Systems of the World Ptolemaic and Copernicus", containing ample defense of the Copernicus System, by arguments like heavenly bodies rotating on Jupiter, phases of Venus, sunspots, which would prove the Earth to be moving on its axis and spinning around the Sun. Just as the moon presented phases by solar illumination eclipsed by the planet, also the planet was eclipsed by moons of Venus, proving as soon as the moon revolves around the planet, also the planet revolves around the Sun.

Prohibition by the "Holy Inquisition" of Galileo's Work – In 1633 the Holy Inquisition forbade the work of Galileo, and he also condemned him to the retreaty in which he accepted that the Earth was immovable in the center of the universe, so that he would not move around the Sun. And Galileo did so, but completing in murmur (almost in utter silence , as with the motion of the Earth in the space where in the absence of the air there is no sound propagation): *"And pur si muove"* (*"But moves"*). And with the prison for "blasphemy" that the Earth moved.

PART 2 – EVOLUTION OF BASIC CONCEPTS OF PHYSICS UNTIL THE QUANTUM

VIII. FORCE, MOTION AGENT, VACUUM, SPEED, VELOCITY, IMPETUS, SPECIES OF MATTER, MOMENTUM, INERTIA AND MASS, IMPULSE, ACCELERATION AND ACTION, ENERGY – <u>350 AC to 1896</u>

Force, Motion Agent, Vacuum, Speed, Velocity, Infinite Speed – As Aristotle around 350 BC force was a motion agent to push and pull. And to move a body that was standing on a frame would require an agent of that motion, that is, a force. The change of position of all the trajectory, the distance by the time, the speed, or the displacement, this is, the length of the shortest trajectory between these positions by time, the velocity. And all the bodies would have their natural place on earth, water, air and volatile as fire. And to go against these natural sites would be exerted forces, and only after all ceased the bodies would return to the natural position. So a medium like vacuum would be impossible. For, without any resistance, after the force was applied and ceased, the bodies would continue with speed preventing them from returning to their natural places. As Aristotle, the speed of a body would be proportional to the force applied and conversely to the resistance of the medium. So, if the resistance were zero, the body would acquire an infinite speed! So Aristotle's logic was that an infinite speed was absurd, so there would never be a vacuum! It was the "horror of the vacuum", because the vacuum would present the infinite speed of its distance.

Instantaneous Speed – As Aristotle, around 350 AC, the speed at which the tip of a spear would, when receiving a drive at the opposite end, was instantaneous, because it was admitted that the spear would behave like a rigid body, so at the same instant that the end to the tip was displaced also it did move it in the middle with air. Also, the light would be emitted with instantaneous speed, as the light being emitted at one point would shift an opposite point instantaneously in the ether medium.

"Impetus" and Impressed Force – In Greek antiquity with Hipparchus, and around 530 with John Philoponus of Alexandria it was formulated the printed force, "impetus", impetus, impulse, impressed or force motivates as an agent of instantaneous variation of speed in a body, which dissipated with time, and of qualitative explanation.

Impressed Force – Around 1020 as Ibn Sina (Avicenna) a body would acquire a motion by being printed a force to it, which was in the body after the contact was finished. This imprinted force would only be diminished by being resisted by the middle. In the case of the middle zero resistance, the body would present an eternal motion. In the case of two bodies receiving the same impressed force, the highest weight would have a lower speed.

Around 1340, also as J. Buridan the impetus applied it was permanently in the body.

Divine Force – Around 350 BC as Aristotle: *"The Earth is the center of the universe, where spheres in which the Sun, the Moon, the planets and the stars have their motions produced by a driving force, a vortex contained in an external sphere triggered by God, through the ether that fills all the space and all the bodies".*

Force of the Outer Sphere and Fragment of the Universe – The motion of some celestial bodies, in a given sphere, as similar were observed presenting different speeds. Because they are similar in dimensions and matter, the same speed should occur for these bodies in this sphere. Thus, it was questioned the driving force (vortex) of this sphere triggered by God, force that according to Aristotle, according to Theophrastus, around 330 BC in "Opines", the first book published on History of Science.

"But it prevailed that the Earth was the center of the universe, and the driving force of the outer sphere rotating all the bodies placed men in the grandeur of a great Earth, and not in a miniature fragment of the universe, besides a great Earth being impossible to move in space without any driving force acting, that is, without Gods", as quoted by Jeans.

Motions Without Forces: Exchange of Motions – When a ball of much greater mass is thrown over another of a much lower mass at rest, it was observed that the ball of larger mass decreased a little its speed, while the ball of lower mass presented a certain speed. Around 1300 according to William of Ockham, the two balls were contacting each other, without any one transmitting any force to the other, but they simply moved by changing motions, without any forces imprinted on each other.

Transient Force – Around 1350 Nicole Oresme of the University of Paris explained that the impetus given by the applied force would not always be inherent to the body. The impetus would be a transitional force, self consumable, which after a while would result in the cessation of motion. In the case of a body that was always falling vertically, in which Aristotle said that this occurred because the Earth had no motion, Oresme explained that this fact occurred through the motion of the Earth that applied a force transmitting an impetus to the body, coming then to make the body to accompany the Earth as it fell vertically.

Intangible Agent – Around 1480 according to Da Vinci the force was an intangible agent, an agent of invisible power that varied from position. And an arrow thrown from the arc received intangible motions, momentum of all the positions of the strained (stretched force about a body) rope from the position of higher tension (pressure, force about area).

Force and Acceleration – Around 1480 da Vinci, in his notes, he wrote that the force applied continuously in a body produced increased speed, this is, the acceleration, not a single motion. And, ceasing the force applied the motion was to continue, coming to stop only when by force contrary to friction, contrary to the idea of Aristotle.

Bodily Agent – Around 1600 according to Kepler, force was a bodily agent, contrary to Da Vinci.

Species of Matter and Ether – Around 1630 according to René P. Descartes, the matter was made up of different species, one forming the luminous bodies like the Sun and other stars, which were centres of immense vortexes, consisting of subtle matter, a subtle floating fluid or ethereal element, a ether, which pressed the bodies in the direction of the center of the vortexes and other bodies. Another kind of material was the interplanetary space composed of globules, which were constantly tensioned. And another species was the one that formed the planets, the dense matter.

Momentum and Force – Around 1600, as Galileo when a stone was thrown or when its speed acquired at a given time would be a force propelling, named as "virtu impellente", "impressed impetus", "momentum", "potenze", "possanza", "moment", "momentum", "movimentum", "forza", "efficia", "momentum della potenze", and ascended as to the power of gravity. And that force continuously imprinted on the stone could explain the matter at rest or on motion. Then, to Galileo force was the variation of the momentum by time.

Conservation of the Momentum – The matter through a distance in space and time, velocity, denoted the quantity of motion, the momentum, as a product of the matter m by speed v. And Descartes around 1630 introduced the conservation of the quantity of motion, mv, proved in 1668 by J. Wallis.

Mathematical Force and Inertia – In 1644, as Descartes, the force took place as exclusively mathematical conception by geometry or mathematical abstraction, and introduced the notion of inertia in which a body maintained its speed after an applied force was withdrawn.

Inertia and Reference System – And, in 1638, Galileo also affirmed that the uniform motion in a horizontal plane was eternal, that is, it was maintained an impetus, significand inertia. And how consequence would be necessary to fix an external reference for which a body would be at rest or in motion with respect to that reference, that is, a relative reference of motion or referential Galilean.

Differentials and Integrals – In 1672 Newton and at 1684 G. W. Leibniz presented rates of differential minimums of variables, and the calculation of the sum of these rates, to obtain the highest accuracy of the mathematical result. So, Newton indicated these differentials as $\dot{x}, \dot{y}$..., and Leibniz as dx, dy... and sum S or $\int$ to $dx, dy,....$

"Vis Centrifuge" – In 1673 C. Huygens introduced and applied the notion of "vis centrifuge", centrifugal force as tension on a rope for which a weight is fixed, where the other side of the rope was held by an observer fixed on the edge of a revolving wheel. In this sense it was called centrifugal force, as static force, while signifying a force that did not produce motion, but only tension in the rope. However, Huygens did not present a distinction between what was called real or fictitious force. Huygens treated the centrifugal force to be a real force, force due to the rotation of a body that actually compared centrifugal force to gravity, due to the centrifugal force of a rotating flowing vortex.

Matter, Density, Mass, Space and Time – In 1687 for Newton the matter was measured by the quantity of its density and volume, which it called body or mass, occupying parts of the space which was unrelated to any external entity and that remained always motionless, as well as the time flowing without relation to any external entity.

Inertial Force, Force of Impulse, Impressed Force – In 1687, Newton in his "Philosophie Naturalis Principia Mathematica" ("Mathematical Principles of Natural Philosophy") established the "vis insita" or "innate force" or "vis inertiae" of matter, i.e. the inertial mass, the power to resist in remaining in its present state, be the body at rest or in uniform motion in the direction of the straight line. The more material, the more mass, the more inertia, the more force. The impressive force, impetus or impulse was an action exerted on the body to vary its state of rest or uniform motion in a straight line. The impetus had been the primitive origin of the variation of the momentum, the impulse, which ceased the action of the impetus would not remain in the body. The impressed "vis printed" force or impulse was proportional to the variation of momentum. Also, the impressed force was an admitted force only as in the applied action in which it would not remain in the body when this action was terminated. That is, according to Newton the impressed force was a force like shock, pressure, tension, but that was not fixed on the body, it was only to relate the motions between bodies. And a print force in which the bodies were driven or impelled to a center was defined as centripetal force.

And that ahead, with quantity, direction and sense, that is, vectored, like mv, where m is the mass and v the velocity, or later written as m**v** or with arrow in the upper part of the letter, the momentum. And the variation of the momentum: $\Delta mv = (m_2v_2 - m_1v_1) = I =$ Impulse, and if $m_1v_1 = 0$, then $I = m_2v_2$. And the variation of force, the accelerating, was a series of successive force that resulted in the motion of successive velocity increases of the bodies. And for every action was always exercised equal opposite reaction, or mutual actions between two bodies were always equal and directly contrary.

Centripetal Force and Centrifugal Force – By the action and reaction of forces, a centripetal force had as a reaction in an accelerated system a force of inertia, the centrifugal force of Huygens. As Newton, the mass of a body received the gravitational force of the Earth and could receive the tensile strength (elastic force) of a compressed spring by expanding or of a tensioned wire, also receiving the force of friction and if this mass was magnetic could receive the magnetic force of a magnet. However, the centrifugal force, which fled from the center, acted on an observer on a rotating platform without having a force on the platform that pushed the observer out. The spinning platform it presented a tangential velocity, with an observer on the edge of a rope connected to its center and there would be an acceleration to the center, centripetal, from the rope that pulled the observer to the center. But the rope did not push the observer out of the center of the platform. According to Newton the centrifugal force was the force of inertia that pushed out the observer of the platform, and as there was no material force acting in the observer would be a fictional force of inertia.

Relative and Absolute Motion – Besides the motion between bodies that moved away, approached or remained immovable, which were relative motion in which forces were impressed on them by material bodies or among them, there was also as Newton the absolute motion of a body that received an inertial force, fictitious, without any material body to print a force on this body. In the case of a water container, for example, a bowl with a handle suspended by a wire that showed a rotation would result in raising the water level to the rim of the bowl. As the bowl and the water were with equal motions, then the cup would not exert force to raise the water on the edge. The Earth acting in the water only by gravity would also not cause water to rise. And as the fixed stars would be distributed throughout the space, the gravitational force resulting in the water would be nil. Therefore, for Newton, the force that would cause the motion of the water that rose on the edge of the cup was the force of inertia in space. And as Newton, this space would be absolute, unchanging, in which time flowed with or without any kind of time-pulsation count.

It was observed by the Ptolemy System that the Sun and the fixed stars revolved around the Earth, and the Copernicus System that the Earth revolved around the Sun and the fixed stars. So there was a relative rotation between the Earth and the Sun and the fixed stars. These two rotations were identical, with no possibility of observing what would be the real one. However, as Newton, in the case of the Earth's rotational motion if it were real, the centrifugal force of inertia would appear, which would lengthen the horizontal axis of the motion.

Between 1715 and 1716 Leibniz presented criticism of Newton's absolute space, mentioning that not even an angel could determine which body would be at rest and which in motion. The motion was just a change of position between the bodies, and the space without matter was only imaginary. While the force would be something absolutely real in the substances, space, time and motion would only be real because they involve infinity, eternity and the ability to perform force in the substances. Although the force was real, the motion would belong to the relationships and it would be necessary to look for its causes. The bodies would have a certain quantity of motion or force, and there would be something more in nature than was determined by geometry. It would be necessary to recognize something greater as force. But in relation to the circular motion, in which Newton showed that the force, acting on the bodies outside the center of the axis of rotation, would indicate the recognition of the absolute motion. But Leibniz said that the fact that the immediate cause of the motion was in the body would only indicate an absolute motion that left another body with relative motion in relation to the body with absolute motion. However, S. Clarke replied to Leibniz, who between two alternatives, for the circular motion, in a case if it ceased to exist an external body, the centrifugal force would not exist, which would be absurd because it was always active being near or far from an external body. And in the other case, if another outer body distant or absent, the body in the circular motion would be with the absolute motion itself, which would mean that there would be absolute motion regardless of whether there is another relative body for comparison of motion.

In 1721, G. Berkeley presented that a body could actually be moving by the force that acted on the body, in which the force was something absolute. In the case of the circular motion, in which a body tended outward from its axis, it was only a force that acted on the tangent, while an additional centripetal force was not applied equally to balance a centrifugal radial force out.

In 1851, J. B. L. Foucault presented a proof of the rotation of the Earth on its axis by a 30 kg pendulum suspended by a wire of five in the dome of the "Invalides" in Paris. The pendulum that oscillated always on a plane, while the Earth revolved around its axis was explained by the gravity force of the pendulum $F = mg$, m its mass, g the gravity acceleration, and the motion of mass of inertia of the pendulum in relation to the absolute space, the force of G. G. Coriolis = $2mwv$, in which m = mass of the pendulum, w = angular velocity of the Earth-pendulum, v = tangential velocity of the Earth-pendulum, observed in 1829 by Coriolis, in accelerated reference. This Coriolis force acted on the pendulum in the direction perpendicular to the axis of rotation of the Earth's motion, in relation to the inertial reference of the absolute space, as Newton, as a force of inertia of this absolute reference of space. And because this force of inertia acted in an absolute space without any mass, there would be no real force, so this force of inertia would be regarded as fictitious. But in this frame of inertia, in the case of the absolute reference of space, this force of inertia would act as if it were a real force, and for the frame of the pendulum that force would appear to be real. However, this force of inertia was regarded as only apparently exerted by a real body by Newton's mechanics, so it would be a fictional force. Then, the pendulum with ≈ zero friction, along with the bracket fixed on the Earth with the velocity around its axis with respect to the fixed stars, would follow the inertia of the absolute space, despite being driven by a nonexistent, fictitious or just seemingly real force.

And the elongation of the equatorial diameter in relation to the axis between the poles of planets such as the Earth, a ellipsoid, was explained as Newton by the motion of Earth's inertia mass in relation to the absolute space, by centrifugal force = mw^2r, fictitious, where m = mass of Earth, w = angular velocity of Earth and r = radius of Earth's orbit, in which in an absolute space with no mass acting, no real force could be exerted.

It was observed by E. Mach between 1868 to 1912 that after a real rotation of the Earth in relation to the Sun and the fixed stars, the Earth performed a rotation in relation to the inertia of the absolute space in identical times, as the Foucault pendulum in an identical time. That is, there was an impressive coincidence, meaning that the motion of inertia in relation to the absolute space was equal to the motion relative to the Sun and the fixed stars. That is, the two motions were equal by the fact that all the stars fixed as a whole, that is, the universe, remained immutable, without turning in relation to the absolute space. As Newton in the case of the actual rotation motion of the Earth would appear the force of inertia, centrifugal, fictitious, which would lengthen the horizontal axis of the motion. And in 1851, Foucault's pendulum by moving by the Coriolis inertia force in relation to Newton's absolute space, also showed the actual motion of the Earth revolving around its axis and the pendulum's oscillation plan, always following the same direction in the absolute space.

And between 1868 to 1912, E. Mach had the realization that the fact that the set of fixed stars did not rotate would coincide with the absolute space. And if there was no mass the absolute space would not have any force that could be exerted on the oscillation of the pendulum or over any other mass of the universe. Therefore, the mass of the fixed stars is that it would be the real cause acting on the oscillation of the pendulum and over all the masses of the universe, not the absolute space that would be responsible for the oscillation of the pendulum or the motion of masses of the universe. Then it would be a force of the interaction between the motion of the planets, like the Earth, and the oscillation of Foucault's pendulum, with the distant masses of the universe. And as Mach the resistance of the mass to the change of motion, the inertia, or to a uniform motion, that is, an inertial motion would have as origin the interaction with all the masses of the universe. That is, the forces of inertia would be originated entirely by the other masses, which would later be called the Principle of Mach.

Variation of Speed, Motion Quantity and Impulse – It was formalized the F force relating resistant m mass, inertial, with the variation of velocity, the a acceleration, as $F = ma$, which was due to the variation in the quantity of motion $\Delta mv = I$, the impulse, by time, that is, $F = I/time$. Newton emphasized that he used the notion of impulse as the quantity of force for mathematical proportions, and that he would not define in the "Principia" treaty the nature of that impulse that caused force.

Principles of Motion – According to Newton, the principles of the motion being of a very large extent left their causes to be discovered.

Motion, Object and Particles – Newton claimed that all objects would appear to have been composed of particles. These particles would have a force accompanied by motion laws and would also be governed by active principles of the cohesion of objects.

Impressed Force and Resistance – Contrary to Newton, as I. Kant, in the 1700, it was not the inertia of matter with its inability to move itself that would produce resistance to the motion of force. Therefore, the forces of action and reaction between bodies should be applied both in motion and relatively at rest. And the motions would be counterposed between the bodies. As Kant, "vis inertia" was a superfluous concept, only convenient for calculus, and it was a word without any meaning. For Kant inertia it was only a state of a moving body or relative rest. The resistance to motion was a force that was similar to the printed force.

Convenient Force – In the 1700, also for Berkeley force was a convenient term for calculating the motions of bodies, but not for the understanding of nature itself, for it was not the physical truth. And the real and final cause of the motion could not be of mechanical consideration or physical science, for it was a divine concept. Science could not assume, without explanation, that the bodies possessed or exercised real forces that acted between them. Therefore, the force for being only partially visible should be disregarded.

Force of Virtual Velocity or Virtual Work – In 1717, John Bernoulli presented a concept of force with velocity applied in the force itself or called the principle of virtual velocity or virtual work.

Motion Property – In 1743, as D'Alembert force was only a property that moved bodies, and mass was the fundamental concept in mechanics. But in 1760, Euler considered the concept of force more fundamental.

Action – Around 1660, P. Fermat had presented a concept of minimum time for a shorter path between two points, i.e. displacement/velocity. And in 1744, P. L. M. Maupertius presented a concept of action as a mass m with v velocity running a minimum displacement d, ie mvd. But, Euler replaced d by vt, re-introducing minimum time: mv^2t as minimal action for particle motion and initiating analytical mechanics. And around 1760, J. L. Lagrange introduced time t to virtual work: $T_{virtual}$ x time = minimum action.

In 1788 Lagrange applied this concept of virtual work to coordinates (n, p) Resulting $T = T_{virtual}(dn/dt)$ and with conjugated momentums $p = mv$ and for the coordinated n in x, with concepts of Euler and S. D. Poisson: $p = mv = \partial T/\partial(x/t)$.

And later in 1834 and 1835, W. R. Hamilton introduced systems with different states and variables, obtaining a new expression.

Metaphysical Force and Impulses – For L. Carnot, in 1783, the force as cause was a dark metaphysical notion and treated the motion by mass and velocity, the quantity of motion, in a conventional way as the product of mass and velocity, the momentum mv or increments of momentum by time, that is, impulses. So force was conceived as a series of infinitesimal increments of momentum, that is, impulses. That is, for the variation of the motion was applied constantly a force, $F = d(mv)/dt$.

Also, J. B. C. Bélanger in 1847 presented the impulse $I = \Delta mv$, as the product of force and time, $I = F \times t$, that is, the greater the time, the greater the impulse applied.

Force Only one Formula – To Lagrange in 1788 force was only a measurement of the pressure needed to print a speed on a body. For mathematical science, only general formulas were needed.

Force Only one Expression – In 1860, for Kirchhoff also force was only an expression of calculation, relating mass and acceleration in space and time, without greater physical significance.

Force Only a Pure Mathematical Expression – And to Mach, in 1870, coherently with his ideas the physical phenomena could be explained macroscopically, considered force and mass only as a purely mathematical conception of expression.

Force and Energy – Between 1690 to 1700 as Leibniz a uniform moving body of m mass with a speed v and under a continuous force equivalent to the same speed v was called "vis viva", "living force". And later in the vector symbology that is, with direction and intensity, $m\mathbf{v} \times \mathbf{v} = mv^2$, where **v** the velocity was represented as vector, or admitting the spelling without the bold or with arrow on the physical component. In 1744, Euler introduced in mechanics the concept of "vis potentials", force of potentials, for motions retained as elastic body when folded. In 1777, Lagrange had shown that the components of the force of attraction, in the motion of gravitation of the bodies, at any point could be expressed simply by the infinitesimal variations of these components and the coordinates of each point of attraction. Variations that were obtained by adding together to the masses of all the particles of an attraction system related to the distance of the point.

And in 1782 Laplace showed that these variations free of the attraction of matter would $(\partial^2 V/\partial^2 x^2) + (\partial^2 V/\partial^2 y^2) + (\partial^2 V/\partial^2 z^2) = 0$, where V is a potential function. And in 1813, Poisson showed the usefulness of this potential V function, when a point (x, y, z) is inside a body of attraction, this Laplace equation should be replaced by $(\partial^2 V/\partial^2 x^2) + (\partial^2 V/\partial^2 y^2) + (\partial^2 V/\partial^2 z^2) = 4\pi\rho$, where ρ was the density of the matter of attraction at the point.

Young in 1807 emphasized the notion of "vis viva" mv^2 like energy. In 1826, J. V. Poncelet presented the concept of work W as a force F acting on a body by distance d, this is, W = F x d. In 1829, Coriolis emphasized the notion of work W as force x distance. In 1853, Rankine applied the concept of energy as the ability to perform a work W. Then, force x distance, resulting in energy as the ability to perform a work W. For energy E_{energy} = 1g x 1 cm/s^2 x 1cm, later named in 1874 of 1 erg, of the Greek ergon as energy or work, and 1 kg x 1 m/s^2 x 1 m, the energy called later in 1948 of 1 joule. And in 1862, Thomson-Lord Kelvin and P. G. Tait called the energy of the motion as $mv^2/2$ as kinetic energy. In 1837, Mohr sought to determine a mechanical equivalent of heat, through work to expand an quantity of air and the quantity of heat Q for this expansion. Being that E_{energy}= F x d = JQ, where J was the equivalent of mechanical energy and heat energy Q. So it was found that the heat was motion, vibration. And the product of force and distance produced the more distance travelled and more velocity, more energy, more heat.

And between 1845 to 1847 Joule showed the mechanical equivalence of heat, through motion of reeds. In the 1840, Joule, Helmholtz, J. R. Mayer and L. Colding concluded that in all physical phenomena the energy was preserved. And the heat was conceived as a type of energy, the thermal energy. Also, the light when raising the temperature of the matter came to be admitted as a thermal energy. The equivalence of heat and work has been characterized as the conservation of energy or first law of thermodynamics and the conservation of thermodynamic energy would be represented in a system by the U = Energy ($Q - W$). So this conservation of energy would represent the unification of all the phenomena of Physics. But around 1880, Mach and P. Duhem only admitted mathematical relationships between the physical magnitudes obtained from the experiments without reductions to a certain concept, even that of mechanical energy. In 1842 Mayer also measured the mechanical equivalent of heat through mechanical work. And in 1843, Colding, through friction of materials. And in 1850, Clausius stated that the production of work by heat resulted from the displacement of the heat from the higher source to the lowest temperature, with the consumption of the heat. The passage of the heat was always from the higher source to the lowest temperature, featuring a second law of thermodynamics and represented by W = ΔQ. In 1851, J. Waterston stated that the "vis viva" of the chaotic motion of a gas constituted the warmth contained therein. In 1856, A. K. Krönig stated that the temperature was related to the "motion energy" of the gas particles. In 1860, Kirchhoff introduced the concept of a black body, which absorbed all radiation of light and or heat that focused on it. In 1865, Clausius presented the first law of thermodynamics as being that the energy of the universe was constant and the second law of the relationship between heat Q and temperature T, in which the energy of transformation, that is, energy of stability $S_{entropy}$, or entropy that tended to a maximum value, and the heat distributed to the stability of the physical phenomenon, resulting in the decrease or stability of the temperature that measured the heat. Thus, Clausius enunciated the second law of thermodynamics in which the entropy of the universe tended to a maximum value. And in 1868, L. E. Boltzmann came to admit entropy interconnected to molecular chaos. Between 1872 to 1877, Boltzmann presented the probabilistic interpretation of entropy, molecular or atomic disorder. The entropy would be due to microscopic conditions of the particles that would constitute a macroscopic system determined by pressure, volume and temperature. The entropy $S_{entropy}$ of a system with N particles would be proportional to a total number W of microscopic configurations. For this, he considered that the particles would be distributed in microphases of the system with each with energy E.

And obtained $S_{entropy} = k \ln W$, where $k = R/N_A$, with $R = (PV/T)/$molecule-gram, $P = 1$ atmosphere, $V = $ Volume of 22.4 liters, $T = 273\,°K$, $N_A = $ Number of Avogadro $\approx 6 \times 10^{23}$ atoms or molecules/atom-gram or molecule-gram.

In 1876, a. Bartoli sought to correlate the radiation of heat and light with thermodynamics. In 1876, J. W. Gibbs presented the concept of free energy between the internal energy U and the enthalpy $H = (U + PV)$ of the thermodynamic system. In 1878, E. Lommel formulated that thermal radiation was function of wavelength and temperature. In 1879, J. Stefan described experiments in which the radiation intensity by area and time was proportional to the fourth temperature power. In 1884, Boltzmann showed the equivalence of the thermal radiation of Stefan's spectrum by thermodynamics, through treating as a gas to thermal or spectrum radiation. In 1884, C. Christiansen described experiments on radiation cavities of bodies that absorbed and emitted radiation. In 1893, W. Wien noted that the peak of maximum radiation energy density corresponding to a wavelength was inversely proportional to the temperature. And determined the wavelength for the radiation energy above and below the visible wavelengths, that is, for ultraviolet and infrared radiation. And In 1896, Wien admitted that thermal radiation stems from the vibration of molecular or atomic oscillators.

IX. GRAVITY – 350 BC to 1920

Mass and Falling Velocity – Around 350 BC, for Aristotle the fall of the bodies, gravity, was a natural motion to the center of the Earth. The greater the mass of the matter measured, for example, in a scale, the higher the gravitational drop velocity. By the scales, with a small iron cube in one of the plates and the other dish a larger iron cube of larger size, it was seen that the plate with the largest iron cube had a sharp drop in the scale. And then concluded that the bodies that went down the scale, that is, called higher weights would have a higher velocity of gravitational drop, because it is visible the fall of the dish of the scales with the heavier bodies.

A Same Time Falling from the Same Height to All the Bodies – But, around 530, as Philoponus the bodies of very different weights presented a slight difference of time of fall, so were almost equal to the velocities in the direction of the fall. After more than 1000 years, at 1553 G. B. Benedetti of Italy, he admitted as correct the Philoponus experiments, in which heavier and lighter bodies fall from the same height at the same time. Between 1557 and 1585, variations of the momentum were observed in the fall of the bodies by A. Piccolomini, J. C. Scalinger and Benedetti.

Gravity Motion and Magnetic Force – In 1600, Gilbert conceived gravity as a magnetic emanation. And also Kepler, in the 1600, attributed to the Sun a magnetic force responsible for the attraction of the planets, while other bodies were also attracted by some magnetic effect.

Vortex of Fluid or Ethereal – In 1644 as Descartes there was a subtle vortex or ethereal floating celestial element that pressed the bodies in the direction of the center of this vortex and other bodies. Thus, in the case of the Copernicus Heliocentric System, in which the Earth revolved around its axis, it should form an extended cycloid in the vertical direction to the axis of rotation.

Different Weights, Equal Falls – In 1576, G. Triletta released from the top of a tower a twenty-pound lead ball (around 9 kilograms) along with a one-pound lead ball (around 453 grams), and after launched a wooden ball along with a same dimension of lead ball. and noted that in both cases all four balls hit the ground at one time.

Also in 1576, and T. Digges and L. Digges presented that a body that fell from the mast of a ship that was in motion, would also have the same motion of the ship, so reaching the floor next to the mast, and also the body would reach the floor by the mast if the it was at rest.

In 1586, S. Stevin and H. I. C. Groot dropped from the same height two spheres, around 10 times heavier than the other, and noted that their impact on the floor produced a single sound at the same instant, so the same terminal velocity of fall.

In 1594, Bruno presented that an observer on the bank of a river would see a body fall from the mast of a moving ship making a curve in relation to the riverbank. And an observer on the moving ship, would see a body fall from a pole on the riverbank as a curve in relation to the ship,

Between 1590 to 1623, Galileo observed that a pendulum with a larger mass on the tip and another of smaller mass, with the same length, presented a same oscillation period. The pendulum motion was the fall of a deflected body from the initial vertical direction at the point of support, and in the position of higher deviation from the vertical the two different weights came down with equal velocity. Therefore, different masses of equal or different materials fall with the same variation of velocity. And Galileo obtained the distances traveled that varied over time, that is, the acceleration with which all the bodies presented by the Earth's gravity.

With the proportionality between the v velocity, t time and gravity acceleration g was obtained geometrically, and subsequently by the laws of the accelerated rectilinear motion, the distance $d = vt/2$, $v = gt$, therefore $d = (½) gt^2$.

In 1634, J. B. Morin, observing that a body falling from a ship's mast reached the floor by the mast, decided to interpret for being Aristotelian that the body had been thrown from the top of the mast with the velocity in the direction of the ship.

In 1660, Boyle and Hooke they built a glass tube with thin air and placed a feather and a coin in its interior, a device until now known as Newton's tube. Thus, they were the first to show that when the air resistance was negligible, the bodies dropped at the same height fell with the same acceleration.

Particle Impulses – In 1669, Huygens proposed that a fluid involved the Earth and its atmosphere. The particles of this fluid described large circumferences around the center of the Earth in all directions. And the bodies fell due to the rapid succession of impulses of these particles.

Moon and Apple Fall – In 1686, Newton determined the acceleration with which the Moon would fall on the Earth in the same way as the apple fall.

From the displacement d_M of the Moon in space with orbital velocity v_M around the Earth in time t, ie $d_M = v_M t$, in the distance of R_M = radius of the orbit of the Moon, between two infinitesimal positions forming a rectangle triangle of sides R_M, d_M, being $R'_M = R_M + r_M$, it was getting $r_M = (v_M^2/R_M) \, t^2/2$, in more current simplified expression. And in the expression of the Galileo equation, $d = gt^2/2$, it was equivalent to $d_M = a_M t^2/2$, where a_M = acceleration of Moon fall to Earth. Matching d_M with r_M resulted $a_M = (v_M^2/R_M) = (v_M/R_M)^2 R_M = \omega_M^2 R_M$, where $\omega_M = v_M/R_M$ = angular velocity of the Moon in Earth's orbit = rotation period of orbit. For the period T_M from a turn of the Moon around the Earth, the circular length was $2\pi R_M$, so $\omega_M = 2\pi v_M /2\pi R_M = 2\pi/T_M$. The orbital time of the Moon that Newton had of astronomers, according to Proposition IV, book III of Principia, T_M = 27dias, 7 hours and 43min = 2.36×10^6 s, so ω_M = 2.66×10^{-6} s^{-1}. And with the Earth radius $R_E = 123249600/2\pi$ feet of Paris × 32.48 cm = 6,371 km. And the relationship between the radius R_E of the Earth and the distance R_M from the Earth to the Moon, which Newton had of astronomers, was around 60 times, obtained R_M = 382,260 km. Thus it was obtained $a_M = \omega_M^2 R_M$ = 0.27 cm/s^2. For the acceleration of a body like an apple on Earth, the drop velocity was 15 feet, 1 inch and 1 line $^{4/9}$ per second = 490.04 cm/s, i.e. $d = gt^2/2$, $g = 490.04$ cm × 2/s^2 = 980.09 cm/s^2. How Newton had the relationship R_E/R_M = 60 and the ratio g/a_M = 980.09 cm s^{-2}/0.27 cm s^{-2} = 3,629, around 60 × 60 as Newton, with the square root of 60.2, noted that the relationship between the accelerations of gravity was equivalent to the square of the distances, which was valid for all other planets and also for the apple and all other bodies. And by the power ratio, Newton's law, with the acceleration, this is, $F = ma$, in which the force was directly proportional to the mass of the body, presented the law of the attraction of the bodies, Law of Universal Gravitation, with the expression in $F = (m_1 m_2 /d^2) K_G$, where K_G it was the constant of gravitation that provided the effective value or power, for the value of the force that involved together the attraction of the product of the masses and the distance squared between them, and be obtained experimentally. What came to be determined by measuring the Earth's density in 1798 by H. Cavendish, using a torsional balance between lead masses that by the gravitational pull between them caused a twist in the hanging wire of the scales.

Greater Weight, Greater Inertia – By Newton's Law of motion, in which the force of inertia was the strenght force, every body continues in the present state, either at rest or in uniform motion in the direction of a straight line, and the larger the mass, the greater the inertia, so for gravity a body heavier should fall with lower velocity, opposite the concept of Aristotle's weight. However, as the force of gravity, by Newton's Law of Gravitation, was directly proportional to the mass and the acceleration g of gravity, and in this case, the expression was that $F = m_{iron} g$, and $f = m_{apple} g$, in which the force F was higher to pull a larger mass of iron and the force f was smaller in pulling the smallest mass of the apple, therefore the forces compensating the masses and therefore maintaining the same acceleration g gravitational drop for iron and apple.

Mathematical Gravitational Force and a Dense Ether Propelling the Fall of the Bodies – Newton had stated that he had provided the necessary instrument for mathematical prediction of gravity in the "Principia". But in 1704 Newton conjectured on the existence of a species of ether whose density would be greater with the distance from the Earth, and the flow of that ether would propel the bodies towards the less dense regions.

Also, as Newton, in 1704 in "Optica" in that gravitational attraction, the attraction and repulsion of imantad materials, and of bodies when irritated as amber producing and conducting electricity, could be by impulse or other means that would still be unknown. For, the bodies act on each other by gravity, magnetism and electricity, and nature would be very consonant, attuned to a common pattern. And complemented Newton, formulating the question that the small particles of the bodies would not have certain virtues or forces that act the distance between them and on the light in reflection, refraction and inflection?

However, Newton said that if the ether was true or not, in any case, suppose that one body could act on another at a distance through the vacuum, without the mediation of any one, it was to him of immense absurdity, which he considered that no being who had in a philosophical question a competent college to think could accept that proposition.

Pressure Severity – At 1690 N. F. Duillier, at 1746 C. Colden and after 1756 G. L. Sage proposed particle pressure as a cause of gravity. According to Le Sage the particles came from all directions, originated from the depth of space, and one body would shield the other resulting in attraction between them. Le Sage argued that the real cause of some variation in the state of a body was caused by the impulse of some kind of invisible matter. So the attraction of two bodies was caused by the impulse of some kind of invisible matter. But the speed of this impulse would be around 10^{13} or 10^{24} times the speed of light, and evidence of this incomparable speed greater than light was inconceivable.

Mathematical Science for Unknown Causes – The fact of gravitation was demonstrated by Newton's Gravitational Law, without presenting how the attraction between the masses of bodies occurred. And almost all physicists vetoed mechanical hypotheses and there was minimal interest in the cause of gravitation, and considered without prospect to be discovered the cause of gravitation.

Around 1665, Newton considered the universe with matter occupying a finite volume, but space being infinite and eternal in time, and at the end of "Principia" Newton mentioned that: *"This wonderful system of the Sun, planets and comets could only come from an intelligent and powerful Being. And to prevent other fixed-star systems from being attracted to each other by their gravity, this Being put these systems at immense distances from each other"*. However, questioned by Bentley between 1692 and 1693, Newton realized that after a finite time fixed star systems would tend to be drawn to the middle of all space and finally collapse.

Thus, Newton modified his cosmology to a universe with infinite systems of matter through infinite space, which would be in equilibrium with gravitational forces.

Between 1894 and 1896, H. J. Seeliger and C. Neumann they observed that the mass M of a sphere of radius r, density ρ, centered at one point of the universe, attracting a mass m in the distance r of the radius had the mass: $M = (4\pi\rho r^3)/3$ that attracts the mass m with the force $F = (K_G Mm)/r^2 = (4\pi K_G \rho mr)/3$. In other words, the force applied in m was greater the greater the radius of the sphere centered at different points of the universe, contrary to the force of gravitation in which the greater the distance of a M mass, smaller was applied force in a mass m. Then they performed the introduction of a parameter αr, where α = constant interval of interaction in Newton's law: $F = [(K_G Mm)/r^2](1 + \alpha r)\exp^{-\alpha r}$.

Space Curvature – In 1854, G. F. B. Riemann presented space curvature geometry, where flat geometry was a particular case of infinite radius curvature. And in 1870, W. K. Clifford conceived the motion of masses as originated by the variation of curvature of space. It was the unification of Physics with Geometry, through the physical fact of a cause create an effect. The Moon would cease to fall into the Earth by the attraction of a gravitational force by Newton's law, to fall by the curvature of space. According to D. Berlinski, the Moon has the inertia of maintaining its rectilinear trajectory and bending by the action of gravitational force according to Newton, and stating: *"For it is so"*. But in 1915, Einstein presented the principle of equivalence, in which the acceleration of a reference system would be perceived to an observer in this reference as an equivalent action of gravity. Gravitation would be a particular case of acceleration originating from mass of inertia. And in the curved space-time, forces would be replaced by the variation of the curved spatial coordinates. And so, Einstein presented the theory of general relativity in which mass would exert distortion in space-time, resulting in acceleration of bodies. Also, in the cosmology of the curvature of space, by the attraction between the bodies of the universe system there would be the collapse of the system. So, like Newton, Einstein also modified his theory by introducing a cosmological constant, that is, a repulsion to negate the collapse by the gravity.

Gravitational Fluxes of Matter – In 1920, according to Q. Majorana, the matter itself would be the source of gravitational fluxes, by a certain area that would continue to emanate.

X. MASS, MOLECULE AND CHEMICAL ELEMENTS – 1630 to 1862

Mass and Molecule – The quantity of matter or mass was called mols of Latin. And around 1630, as P. Gassendios atoms would be assembled in groups formed by small quantities of matter or mass denominating molecules. In 1661, R. Boyle called the element all accepted substance as simple, which would constitute all composite substances.

Primordial Elements – In 1789, A. L. Lavoisier presented on the basis of the experiments of the chemical reactions that the masses remained constant before and after these reactions and that the chemical elements were primordial when they reached the last decomposition. Thus presented a list with primordial elements consisting of metals: Iron, Copper, Tin, Silver, Gold, Lead, Mercury, Bismuth, Antimony, Arsenic, Cobalt, Manganese, Molibdeny, Nickel, Platinum, Tungsten and Zinc, plus the non-metals: Oxygen, Azote (subsequently called Nitrogen), Hydrogen, Sulfur, Phosphorus and Carbon. And substance as yet simple, but which could result in primordial elements such as muriatic acid, fluent, boracic, lime, magnesia, corita, alumina, silica, soda and potash. Thus, the water, the air and the earth were no longer elements of nature, but formed of primordial elements, which would be obtained by division and subdivisions until the end of their trajectory, as Lavoisier.

Chemical Reactions – In 1687 Newton had formulated the law of universal gravitation in which masses were attracted in the direct reason of their masses and inverse squared of distance, and between 900 BC and 1800 AC experiments on attractions between bodies of origin of the resin called amber, of the Greek "elektron", resulted also in an equivalent law of gravitation in which charged masses of electron that is, charged with electricity were attracted or repulsive in direct reason of their electrical charges and the inverse ratio of the square away. And similarly applied for attraction or repulsion between materials such as magnets originated from magnetite, ie the magnetism. Thus, between 1700 to 1800, J. L. Doembert, C. Berthollet, J. Priestley and others considered that these attractions of mass and or attraction and repulsion by electricity and or magnetism, along with the specifics of each element or compound, could explain the chemical reactions, although according to G. E. Stahl and G. L. L. Buffon, the specifics of each substance were the causes of chemical reactions.

In 1792, J. B. Richter had observed that the relative masses of compounds that were combined remained in relation to whole numbers between the products obtained in which other compounds occurred. And on 1797, J. L. Proust determined the law of the constant proportions of substances. These proportions and for being of whole numbers could indicate the atomicity of the chemical elements, in the chemical reactions.

Proportions of Elements in Chemical Reactions and the Proof of Atoms – In 1801, T. Young had presented a calculation of the eventual dimension of the atom. By the development of the microscope produced in 1590 by H. Jenssen it would be possible to see particles around 0.01 cm. But, Young could identify the reddish color of colored water with carmin, in a film of 1 x 0.0000001 cm, and the diameter of the atom should have a lower value and close to that value.

And in 1808, J. Dalton presented the multiple proportions of the chemical elements, in which when two chemical elements could form more than one compound, the masses of these elements were in relation to whole numbers, according to Table 1, which shows the combination of a fixed element with the same relative mass forming different products with other elements, with the interpretation of this chemical combination by atoms of equal masses for each element. In the case of water, a relative mass Hydrogen atom 1 would combine with a relative mass 7 oxygen atom as determined by Dalton, and a relative mass hydrogen atom 1 would combine with a relative mass carbon atom 12 obtaining different compounds related to whole numbers.

Tab. 1 – Combination of a fixed element forming different products with other elements.

		Hydrogen ☉	Oxygen ◯	Carbon ●
Water	☉◯ 1/7	1	7	
Swamp Gas (Methane)	☉ ☉●☉ ☉ 4/12	1		3
Carburet (Ethylene)	☉●☉ 2/12	1		6
Olefinic Gas (Acetylene)	☉● 1/12	1		12
Carbon Oxide (Carbon Monoxide)	◯● 7/12		7	12
Carbon Acid (Carbon Dioxide)	◯●◯ 14/12		14	12

Because these compounds formed by multiple parts defined and not by random quantities, the existence of atoms of the chemical elements was proved, after more than 2,300 years of the theory of atoms of Kanap, Leucippus and Democritus.

Proportion of Volumes – Between 1788 to 1808 J. A. C. Charles and J. G. Lussac presented gas laws for volume, pressure and temperature. And in 1809 Lussac discovered the proportions of whole numbers in volumes of gases, for the same pressure and temperature, in the formation of compounds. But these whole numbers showed smaller volumes than the reagents. As noted in the Figure 10, a volume of hydrogen reacted with a volume of chlorine to form hydrochloric acid. However, a volume of oxygen reacted with two volumes of hydrogen forming two volumes of water. If it were admitted that the chemical reaction occurred by attraction between masses and or electrical elements, the hydrogen could be occupying a greater space between its atoms, but that would not be the case, because while to form water needed two volumes, to form the hydrochloric acid needed only one volume.

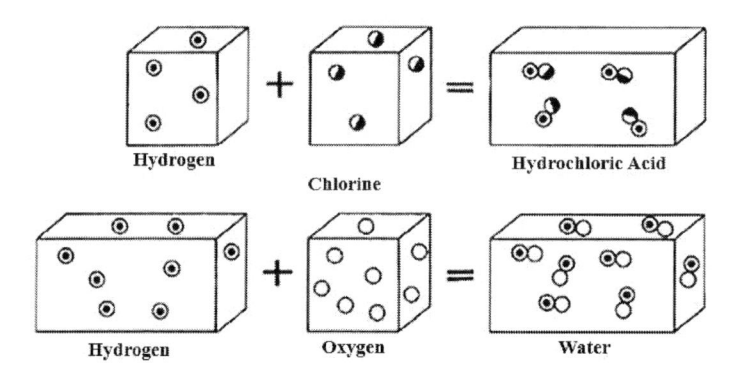

Fig. 10 – Chemical combination between volumes of chemical elements.

Molecules – To explain the constant proportions of masses and volumes, of Dalton and Lussac, came in 1811 A. Avogadro formulate the hypothesis that the atoms of an element could be grouped, that is, in elementary molecules, in which equal volumes of gases in equal conditions of pressure and temperature would have the same number of molecules. Figure 11.

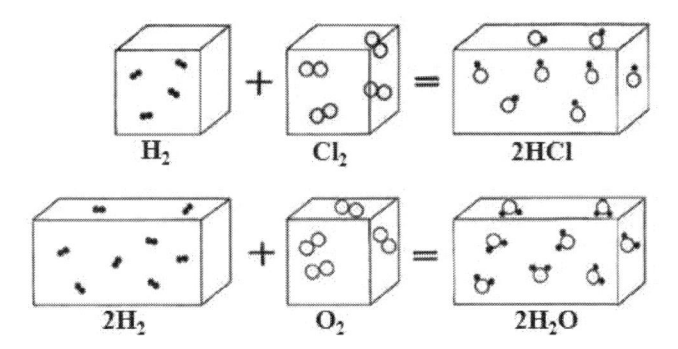

Fig. 11 – For the same conditions of volume, temperature and pressure, the chemical elements would have the same number of molecules.

But contrary to the elementary molecules of Avogadro, Dalton claimed that just as hydrogen and oxygen together formed liquid, also two atoms together should form liquid which never occurred in all gases. Although there was no response at the time to this statement by Dalton, by the deepest knowledge of the atom, the hypothesis of Avogadro explained widely the formation of compounds and volumes obtained.

Chemical Symbology – In 1813, J. J. Berzelius presented new symbology of the chemical elements replacing that of Dalton, in which the molecules of hydrogen H_2 and the oxygen O_2 formed two volumes of water molecule, i.e. $2 H_2 + O_2 = 2 H_2O$.

Fundamental Atom-Prototype – Since all the chemical elements were multiples of Hydrogen, in 1815 W. Prout concluded that it was the fundamental atom and the others would be formed by different quantities of Hydrogen in their atomic masses.

Atomic and Molecular Masses – With the proportion of Lussac volumes and the hypothesis of Avogadro, S. Cannizar in 1858 established a table of atomic and molecular masses of various gases.

Light Decomposition Lines – In 1752, T. Melvill had observed that each substance being placed in a flame produced a characteristic color, such as the salt for food that showed the orange color. In 1814 J. Fraunhofer observed dark lines in the decomposition of sunlight. In 1859, R. Bunsen and G. R. Kirchhoff built a spectroscope to accurately measure the wavelengths of light emitted by a substance, and Kirchhoff observed that the emission and absorption of light and heat radiation were function of the temperature and length of radiation wave, which would explain the dark lines of the solar spectrum, and come to result in the discovery of other chemical elements through the spectral lines.

Valence – In 1789, W. Higgins noted that different chemical substances could combine with different quantities and named the particle combinations. In 1852, E. Frankland observed combinations among the chemical elements as replaceable value. In 1865, A. W. Hofmann named this replaceable combination power value as valence quantity. And in 1867, F. A. Kekulé treated it as "valenz". In 1868, H. Wichelhaus presented this combination value of the chemical elements as valence.

So for compounds: HCl, H_2O, CH_4Nh_3, hydrogen H combines with only 1 atom of another element, so valence 1.

For NaCl, $MgCl_2Ccl_4$, $AlCl_3$, chlorine also combines with only 1 atom of another element, valence 1.

With the Na_2O, NaCl, sodium in also combines with only the 1 atom take from another element, valence 1.

The compounds in Na_2O, H_2O, the oxygen O can combine with 2 atoms of another element, so valence 2.

For the $MgCl_2$, magnesium Mg can combine with 2 atoms of another element and its valence is 2.

Valence for the NH_3NO_5, in which nitrogen can be combined with 3 atoms or up to 5 atoms of another element, so its valence can be 3 or 5.

Gravitational Atomic Model – In 1828, G. T. Fechner presented an atom model that consisted of a massive central part that will atracct gravitationally a cloud of almost imponderable particles.

Electrified Atomic Model – With the discovery of a ore by Magnes in Asia Minor in 2000 BC, subsequently called magnetite, which attracted the iron, that is, stone of attraction, the magnet, conceptualizing the magnetism. And around 900 BC was observed in Greece that the amber or electron, the resin of vegetable fossil, when rubbed with fabrics, attracted straws and leaves dry, that is, was with electricity. And W. Gilbert in 1600 noted that this amber attraction also occurred when they were rubbed with glass, crystal, seal, precious stones and other materials, which he called "electrics". And N. Cabe, in 1629, observed that the straw and dried leaves were indeed attracted by the amber and the glass, but could also occur after the attraction a repulsion. Similar fact observed by Newton in 1675 in the case of rubbed glass showed attraction in papers and then there was a repulsion. As Newton, in his "Questions of Optics": *"The bodies do not act on each other by the actions of gravity, magnetism and electricity? And the magnetic, electrical and gravity attraction between the bodies, could not be by impulse or other means that is still unknown, because nature is very attuned to a common standard?"*. So In

1862, W. E. Weber formulated that these almost imponderable particles were eletrizadas and attracted electrically to the central part of the atom.

XI. PRESSURE, VOLUME, TEMPERATURE, MOTION, HEAT, PARTICLES, SHOCK SECTION, PARTICLE FLOW – **1661 to 1908**

Pressure, Volume and Temperature – In 1661, Boyle formulated experiments obtaining that a pressure (P) – force on a surface – of the air varied with the inverse of the volume (V) When the temperature was constant. $P_1/P_2 = V_2/V_1$, and in 1676 E. Mariotte proved these results. And in 1702 G. Amontons observed that the volume of the air was proportional to the temperature variation $V_2/V_1 = T_2/T_1$. In 1727, L. Eumler comsidered that the air was composed of rotating spherical molecules and the pressure was the centrifugal force of this rotation, explaining Boyle's experiment. In 1733, D. Bernoulli considered that the air was made up of spherical particles moving in all directions. The air pressure would be due to particle shocks. and explaining Boyle's experiment to constant temperature. In 1787, Charles observed that all gases followed a ratio of volume and temperature, $V_2/V_1 = T_2/T_1$. And between 1788 to 1808 Charles and Lussac obtained for the gases the relationship between pressure, volume and temperature, $P_1V_1/T_1 = P_2V_2/T_2$.

In 1802, Lussac and Dalton, both determined the fraction of the volume increase of a gas at 1/273 at 0°C, under constant pressure for an increase of 1°C. Whatever, $V_T = V_0[1 + (T/273)]$, where V_T = volume in temperature T and V_0 = Volume in T = 0°C. For an increase of 1° C the pressure increased from 1/273 at 0°C, in constant volume. And at the temperature of – 273°C the volume should theoretically be null, called this absolute zero temperature by Dalton.

Vibration and Heat – Between 1670 to 1703 Newton had presented in its "Questions of Optics" the interaction between light, vibrations of particles and heat: *"Isn't it true that the bodies and the light interact with each other, the bodies over the light, emitting it, reflecting it, refracting it and difrating it, and the light on the bodies, heating them up and putting their particles into vibration, which is what is the heat?"*. And heat transport would be effected through an ether: *"Won't the heat be transported through the vacuum by the vibrations of a much more subtle medium than the air?"*.

In 1702, Amontons presented an indication of an absolute zero temperature for a null pressure of air, that is, when there was an absolute zero of heat. In 1750, M. V. Lomonosov admitted that there should be a zero-motion state of particles, hence the heat, and at 1856 M. Thomson-Lord Kelvin demonstrated that the absolute zero of -273°C would be reached when the motion of the molecules or atoms were totally null.

In 1740, Lomonosov conceived that the particle motion was the cause of the heat.

For a heat measure, that is, a temperature, from around 20 BC, Heron had obtained this measure by varying pressure and volume of air as heating or cooling. And around the year 170, the temperature was obtained by Galen of Pergamon through some scale between the ice and the boiling of the water. And after modifications of the scale and materials used, according to J. Hasler, Galileo, S. Santorio, J. Ray, Mahesh, Boyle, R. Hooke, H. Fabri, Mr. Hubin, Crams, O. Roemer, G. D. Fahrenheit, P. Elfwius, S. Klingenstierna, R. A. F. Réaumur, C. Linnis, A. Celsius, up to the inverse scale of Celsius by Linnis in 1743.

In 1772 according to J. Black and J. C. Wilcke, the quantity of heat Q would be measured by the product of temperature variation, mass and a constant k_h specific heat of each substance, this is $Q = mk_h\Delta T$. And in 1838, P. L. Dulong introduced the heat unit as the quantity needed to elevate a degree °C to the mass of a gram of water.

In 1775, Higgins admitted that atoms had an atmosphere of heat.

In 1798, B. Thompson-Conde Rumford observed friction, motions, work on a metal, producing continuous heat through the heating of water. And in 1799, H. Davy observed the friction of several bodies producing heat, concluding that the particle motion was the heat. Therefore, it showed a result contrary to Lavoisier's caloric fluid, which passed from one to another body.

In 1804, Thompson-Conde Rumford noted that by submerging in heated water a closed glass containing vacuum and a thermometer, there was a temperature increase indicating that the heat was propagated in the vacuum. Also, J. Leslie performed experiments indicating that the heat was also manifested by a vibration of the radiation that propagated in the vacuum.

But also in 1804, J. B. Biot observed the spread of heat in metals, through a driving in the material. And this driving, in relation to the time, being effected by J. B. Fourier, which was contrary to the identification of the nature of the heat, such as a fluid, molecular motion or radiation, being only determined by the mathematical equation relating thermal magnitudes macroscopic. However, in 1822, Fourier considered fundamental dimensional analysis among the members of the equations for consistency of the physical units involved.

In 1816, J. Herapath formulated that there was a proportionality between the temperature and the momentum of gas molecules.

In 1817, the caloric received a theory that atoms would be involved in a caloric atmosphere, according to J. B. Emmett.

In 1832, a. M. Ampère demonstrated that heat and light should be considered as waves in an ether.

In 1850, W. J. M. Rankine formulated a hypothesis that the motion of molecular vortexes of the particles generated heat. And in 1858, Rankine observed that when the heat was isolated in a container containing gas there was an increase in temperature when the volume was decreased.

Pressure, Temperature, Volume and Molecules – The experiments with gases showed their expandability and diffusion between them, which indicated that they should be composed of tiny independent particles. In considering the attraction and repulsion between particles, it could in the expansion occur that there was repulsion between them. However, in the liquid or solid state, a particle attraction was observed. This attraction would be overcome if there was a continuous motion in the gaseous state, which was the theory of the motion of gases. Thus, the particles would be the molecules or atoms, colliding in chaotic motion elastically, remaining an average velocity of the particles, for a certain temperature. The heat would be the molecular or atomic motion and at the temperature of absolute zero-273 °c all molecular or atomic motion should cease.

In 1834, E. Clapeyron equated the results of Boyle's experiments, Amontons, Charles, Lussac. That is, PV/T = $P_{1atm} V_{22.4l}/T_{273K}$ With the total quantity of n molecules in PV/T and the number of Avogadro N_A of molecules existing in a molecule-gram of $P_{1atm} V_{22,4l}/T_{273K}$. So, n = N/N_A being the number of molecules-gram in the PV/T system, obtaining PV = nRT, where R = $P_{1atm} V_{22, 4l}/T_{273K}$ /molecule-gram.

Between 1847 and 1859, with the theory of motion of molecules or atoms, J. P. Joule, R. Clausius and J. C. Maxwell deduced the equation of Clapeyron, from the total number of molecules N, pressure P, volume V, temperature T, the momentum p = mv_n of molecules, with mass m and velocities v_n, obtaining the expression P = 1/3 Nmv^2/V = 2/3 Nmv^2/V. And in 1860, Maxwell demonstrated in gases, the relationship between the motion of molecules and absolute temperature, by a probabilistic distribution of velocity of the molecules. And admitted that the molecules present in their collisions a great variation of velocity, the molecular chaos.

Brownian Motion – On 1828, R. Brown observed in the microscope that particles, which he called primitive molecules, presented a motion continuously.

In 1843, J. J. Waterston introduced the concept of free middle path between molecules as inverse to the density of the medium.

In 1851, Joule calculated the velocity in different directions of gaseous molecules, in which he observed that the gas pressure in a container was proportional to the square of the velocity of the molecules.

In 1857, Clausius admitted to each molecule an average velocity and in 1858 formulated the free middle path between molecules through the number and dimension of the molecule.

In 1858, Rankine admitted that the Brownian motion was caused by the increase in temperature through the incidence of light.

In 1863, C. Wiener stated that the Brownian particle motion was due to the motion of the medium itself in which the particles were.

In 1867, G. Cantoni admitted that the Brownian motion was derived from the motion of the medium which was produced by heat supply.

In 1870, W. S. Jevons came to formulate the electrical origin for the Brownian motion.

In 1877, W. Ramsay stated that the Brownian motion was due to molecular or atomic collisions.

In 1881, L. J. Bodoszewski observed in addition to the Brownian motion in a medium like water, also in gases.

In 1888, L. G. Gouy presented experimental measurements of the Brownian motion, measuring the velocity of the particles, but could not find an explanation for this continuous motion.

In 1905 A. Einstein presented the Brownian motion as diffusion between solute and solvent.

So, relating a fraction of volume V_α of the total volume solute V of the number of molecules N of the solute radius r: $V_\alpha = [N\,(4/3)\,\pi\,r^3\,]/V$, and being $N/N_{Avogadro} = M/m_{mol}$, where N_A is the number of Avogadro, M = total mass of the molecules and m_{mol} molecule mass; The density of the solute $\rho = M/V$, therefore $v_\alpha = [N_A\,(\rho/m_{mol})\,(4/3)\,\pi\,r^3] = (1/k)\,[(n'/n) - 1]$, where n' is the viscosity of the solvent + solute solution, n = the viscosity of the solute, k = constant proportionality. Through the motion of the solute which was a function of the friction force in the solution which was proportional to the v velocity of the molecules, in which $f_{friction} = k_{friction}\,v$, where $k_{friction}$ it was the coefficient of friction = $6\pi n r$. And was obtained the diffusion coefficient $K_D = (RT)/(6\pi n N_A r)$, resulting $r^3\,N_A = \frac{3}{4}\,m_{mol}$ and $rN_A = RT/6\pi n k_D$. And for a sugar solution with $m_{mol} = 342$ g/mol, n = 0.0135 g cm^{-1} s^{-1}, $\rho = 1.00388$ g/cm^3, $k_D = 0.384$ cm^2/day and T = 9.5 °C = 282.5 ° K, was obtained: $r = 9.9 \times 10^{-8}$ cm and $N_A = 2.1 \times 10^{23}$.

In 1908 J. B. Perrin observed that the particles of the Brownian motion were distributed in greater quantity at the bottom of a solution column than at the top and behaved as gas molecules.

So by the pressures between blade differences in the column determined the number of molecules per unit per volume $N = N_A/V$, from $PV/N_A = RT/N_A$, then $P = [(N_A/V)\,(R/N_A)]\,T$. For a particle of ρ density, mass m and volume of displaced liquid V_{Liquid}, $\rho = m/V_{Liquid}$; for displaced liquid with ρ' density, mass m_{liquid} and the same volume of displaced liquid V_{liquid}, $\rho' = m_{liquid}/V_{liquid}$; so, the mass of the liquid $m_{liquid} = \rho' V_{liquid} = \rho' (m/\rho)$, and thrust force $F_{thrust} = mg (\rho'/\rho)$, where g is the acceleration of gravity, and the difference in particle and liquid forces is mg-mg (ρ'/ρ) = mg $[(\rho - \rho')/\rho]$. And differentiating the quantity of molecules per volume unit in the infinitesimal blades between two levels l_2 and l_1 was obtained $N_2 = N_1 \exp [mgN_A(l_2$ and $l_1)/RT$ and replacing mg by the difference of particle and liquid forces: $N_2 = N_1 \exp [mg (\rho - \rho') N_A (l_2$ and $l_1)/\rho$ RT that was the equilibrium equation of the solute in the suspension resulting from the Brownian motion. And determined r = 4.9 x 10^{-8} cm and $N_A = 6.56 \times 10^{23}$. The current value of N_A it's from 6.022... x 10^{23} atoms or molecules by atom or molecule-gram.

Speed, Shock and Particle Flow Section – By experiments on motions, fluxes and shocks of molecules were certain expressions relating these variables.

Being that n = number of molecules/volume then the quantity of (collisions/time) = n (area = πr^2) v t, in which area = πd^2, v = velocity of the molecules and t = time, with (v t) = d = distance or free middle path. and was obtained the attenuated intensity by the middle of the particle beam, $I = I_0 \exp^{-n\delta d}$, where I_0 it was the initial intensity of the beam and the ratio $(I/I_0) = \delta$, shock section for presenting an area dimension.

The flow Φ in mechanics is defined by physical variable as volume, mass, of a fluid that passes through an area S_{area} In a certain time. So, Φ_{vol} = (vol/t) = S_{area} (l/time) = S_{area} v, where v is velocity. That is, the flow could be seen as the product of velocity by the area it traversed. And it could also be observed by the m mass with its ρ density at one time: Φ_{vol} = (vol/t) = $[(m/\rho)/t]$. Through a particle incident flow $\Phi_{particles}$ = $N_{particles}/S_{area}$ /t, where $N_{particles}$ = number of particles incidents with v velocity and ρ density, could occur between the diluted particles a shock in which there was deviation or return of the particles, elastic shock. Or between the particles a shock with absorption or variation of the momentum mv incident, inelastic shock. And the relationship between $\Phi_{particles\ incidents}/\Phi_{diverted\ particles}$ it was the δ shock section.

XII. PARTICLES AND WAVES OF LIGHT – <u>350 BC to 1885</u>

Vibration of Light – For Aristotle, in 350 BC, the light was similar to the sound vibrations, in which a luminous object had vibration that was propagated through a transparent medium.

Primordial Matter of the Universe – Around 1200, Grosseteste presented that light was the primordial matter of the universe, the first base form of creation, because it was emitted by all the bodies and propagated in all directions.

Instantaneous Pressure – In the 1600, for Descartes the light was a pressure transmitted by the bodies through a subspecies of matter, a type of ether in globules in contact tension, behaving as rigid body and thus the light being transmitted instantly. And the colors were caused by different rotation speeds, that is, periods or frequencies of vibrations.

Waves of Light, Sound and of Water – In the 1650, F. M. Grimaldi observed that the light showed deviations in obstacles, resulting in light and dark stripes, the diffraction, explaining how similar waves of sound and water.

Light Particle in Ether – Newton, between 1664 to 1690, indicated that the light was made up of streams of particles emitted by vibrations of brilliant bodies with violent agitation of ether motion which excited the incandescent substances to emit light which reflected, refracted and diffracted in other bodies, and the vibrations of these streams would be propagated through this ether. The colors of the light were associated with the quality of the particles or vibrations of the ether, also of periodicity as the colors of the light of the theory of Descartes. Upon entering a denser medium, the light in the air should increase its speed by gravitational force, modifying your direction, the refraction.

Longitudinal Wave of Light in the Ether – Between 1665 and 1678, Hooke and Huygens observed the light as a wave, in which a luminous body generated a vibration, the light, propagated by a means, an ether. Just as the wave of the sea or sound exists a means like water and air, for light it would be an ether. And the colours were diversions of refractions. The wave of light would vibrate lengthwise as the sound wave, as Huygens.

Particle or Wave – At 1601 T. Harriot, in 1621 W. Snell, and in 1637 Descartes, observed that in a denser medium the light was refracted according to the ratio between the distances of sine from one medium to the other. As Huygens, in the most refracted medium, the light as wave would be an extension, running a shorter distance at the same time from the previous medium, would have a lower speed and according to the ratio of speed v, frequency $\mathcal{V}$ and wavelength λ, $v = \mathcal{V}\lambda$, the light would also have a smaller wavelength. According to Newton, by the attraction between particles the speed of light and the line between two particles would be higher in the most refracted medium. Newton came to determine the interval between two points of light, later called wavelength, through light and shadow color rings, by the relationship between the diameters of the rings of the light colours and the intervals between the light points, until inferring the light intervals in the first dark ring that was the smallest range of light colors, i.e. the wavelength of each light color.

For 183 inches of the diameter of the glass sphere that transmitted the light colors, the diameter of the sixth circle was 55/100 of an inch. and the diameter of the fifth ring at 8/79 inches, which inferred at 32/567931 inch for the light range, and for the first dark ring the actual length of the light at 1/88739 inch, ie around 0.000286 mm between the yellow and orange. Subsequently with interferometry would be obtained 0.00057 mm.

For the red would be greater the wavelength than the blue color. But in the refraction that could indicate a change in light color, by wavelength, Newton had observed an imperceptible change of color to homogeneous colors or later called monochromatic.

Transmutation of Light and Matter – Between 1664 and 1690 and published in 1704 in "Optics", as Newton, *"Wouldn't the bodies and the light convertible into each other? And the bodies could not owe a large part of their activity to the particles of light that would enter into their composition, because all the bodies emit light? Transmutations follow nature, and why nature could not transform the bodies in light and light on bodies? The bodies and the light don't act each other? And the light on the bodies while heating them would not print in their parts a vibrating motion in which the heat would consist?"*. However, Lavoisier added, in his list of primordial elements, the light, and as an element of the heat, the caloric, of the chemical reactions.

Polarization of Light – In 1669 E. Bartholin, in 1678 Huygens and then in 1808 E. L. Malus, observed that the light presented double images as it crossed a crystalline mineral. And in one direction, for example, to the north the light went through the crystal and in the transverse position the plane of the crystal blocked the light. It was the polarization of the light.

Speed of Light – In 1675, Roemer obtained the time difference from the reflected light from a satellite of Jupiter, between two opposite points of Earth in orbit of the Sun, the speed of light at ≈ 215,000 km/sec.

Aberration of Light – In 1728, J. Bradley pointing his telescope at the star γ of the constellation Dragon in N51° 29', he noted that to see the light, the telescope described an annual circle with an angle of inclination of 20"25, deduced that this slope was the resultant between the telescope speed set on Earth at the speed of light. It was the discovery of the aberration of starlight. And with the tangent of this angle, tg 20"25 = V/c, between the speed of Earth V ≈ 29.3 km/s in the Sun's orbit and the speed c of light, obtained c ≈ 298,500 km/sec.

Cracks and the Drawing of the Wave of Light – In 1803 Young presented an experiment in which the light when crossing two close slits showed a typical pattern of design of wave interference, such as water, enhancing the wave theory of light.

Prism and Earth Velocity – In 1810, D. F. Arago performed an experiment with prism, in which a stellar light entered the prism in the same direction of the velocity V_{Earth} of the Earth in the solar orbit and after six months in the opposite direction to the motion of the Earth. As the medium where the light ran showed refraction depending on the speed in the medium, the expected forecast would be that the speed c of light entered the prism with $c' = (c - V_{Earth})$ when it was in the same direction as the Earth motion and $c' = (c + V_{Earth})$ when in the opposite direction to the motion of the Earth, thus showing a difference in the angle of refraction by that difference of speed. But the result was that the refraction was identical in both cases. Figure 12.

Arago Experiment about Earth Motion and Refraction Index
Prediction: The refraction index would be greater for a smaller speed of the light in the prism when the Earth "flees" of the light

Result: The refraction index went equals

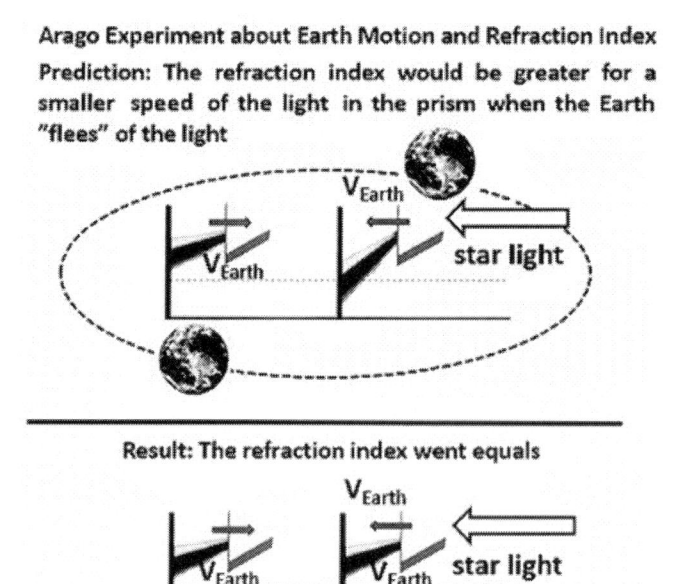

Fig. 12 – The prediction of refraction in the prism for the entry of the stellar light in the direction of the Earth's orbital motion was that it should be different when the Earth was six months later in the opposite direction to the entrance of the stellar light in the prism. However, the experiment showed that the refractions in both cases were the same. Adapted from the clipart of *Editora Europa*.

Refractive Index, Ether and Earth Velocity – Then, between 1817 to 1818, A. J. Fresnel presented a wave theory in which the refractive power of transparent bodies would depend on the concentration of ether within them.

The density of ether in space being ρ and in the prism ρ_1 resulting $\rho_1 = n^2\rho = (c/c_1)^2$ where $n = c/c_1$ Is the refractive index, c the speed of light in the vacuum and c_1 the speed of light in the prism at rest. And with motion the prism would carry a part of the ether inside it. That is, the ether carried by the prism was $(\rho_1 - \rho) = (n^2\rho - \rho)$. And the speed of ether within the prism was $V_{ether} = (n^2 - 1)/n^2 \times V_{prism}$, and $V_{prism} = V_{Earth}$. Therefore, it resulted in the speed of light within the moving prism: $c_1' = c_1 \pm (n^2 - 1)/n^2 \times V_{prism}$, in which $V_{prism} = V_{Earth}$. Then, the refractive index of the prism was independent of the motion of the Earth in the direction or direction contrary to the entry of the stellar light in the prism.

Transverse Wave, Rigid Ether and Velocity of Light – The wave of light admitted as longitudinal should cross the planes of the crystal, similar to a sound wave that went through a horizontal or vertical opening. However, in 1817 to 1824, according to Young and Fresnel, the explanation would be that the wave of light should vibrate in all transverse planes, propagating perpendicular to these planes, by an ether that impregnated in all the bodies and that it should be of a type that acts as rigid body for propagation with the high speed of light.

Wave of Light and Stellar Aberration – Between 1817 and 1824, by the wave theory of Fresnel and Young, the stellar light crossed the length of the telescope in the same direction as it traversed the atmosphere, because the ether permeated the air of the atmosphere and from inside the telescope, thus keeping the angle of the aberration stellar.

Ether Drag – In 1845, G. Stokes presented an ether as a incompressible fluid, which was dragged by the bodies on the surface of the Earth, but remained without dragging above the surface. However, this ether should be without rotation so that the light crosses it so as to follow the stellar aberration. This ether should present the tangential speed of the Earth, for example, the speed
 directed around the Sun, but being rotational and as a incompressible fluid, and according to H. A. Lorentz the ether would have only a speed. But, if it was compressible the ether would have a higher density and the speed of light would be lower, which would be contrary to the aberration of light.

Partial Ether Drag – With Foucault's experiments and A. H. L. Fizeau between 1849 to 1851, the speed of light in a denser medium than the air, in the water, showed the speed of 225,000 km/sec. In glass or crystal, in a prism, the speed of light was around 200,000 km/sec. These results would show that light followed the wave theory and still showed that the speed of light, in the denser environment like water, was only given a lower part of the speed of motion water. The ether would only drag the light partially. Or it would be the medium that would partially drag the light if there were no ether, but the light wave by the Fresnel theory and Young needed of the ether to propagate the wave. And Fresnel predicted that in a water telescope, the light would present a lower speed, but the partial drag of the ether would increase equaling the same speed of light in the telescope without water. And G. B. Airy in 1871 proved this Fresnel prediction.

Radiation Spectra – In 1849, Foucault when observing spectra of radiation, by the sunlight, concluded that the substance emitting light of a certain wavelength, also absorbed the light at that wavelength, that is, there would be a vibration of radiation and atoms, that is, an resonance. In 1853, A. J. Angstrom observed that the Hydrogen spectrum presented 4 lines of radiation.

In 1885, J. J. Balmer presented an empirical expression of the 4 wavelengths of hydrogen as a function of the radiation lines: $\lambda = 3645.6\ n^2/(n^2 - 4)$, with n = 3, 4, 5, 6). In 1888, J. R. Rydberg showed another empirical expression for the 4 lengths of hydrogen waves: $1/\lambda = R_H (1/2 - 1/n^2)$, $R_H = 1.09737 \times 10^{-5}$ cm.

XIII. MAGNETISM, ELECTRICITY AND OPTICS – 2000 BC to 1908

Magnet, "Efflux" and Connections – With the discovery of a ore by Magnes in Asia Minor in 2000 BC, after being called magnetite, which attracts the iron, that is, stone of attraction, the magnet. Plato reported in 450 BC that the magnet when coming into contact with a fragment of iron magnetized this fragment which was going to act like a magnet. That is, the induction of magnetization. Empedocles of Sicily around 450 BC tried to explain these facts by "effluvia", "efflux", which emanated from both materials and that were located in the internal spaces. Other scholars who dealt with science in antiquity attributed to a means of connection involving these bodies, such as a type of air and blow, "pneuma" of elastic property. In 1225, G. Auvergne sought to explain that the attraction and repulsion of magnets and this magnetized induction occurred by connections in the proximity between magnets and iron. But without explaining how the connection did this interaction between two bodies. These connections would later be in 1845 called the field where these magnetic actions occurred.

In 1269, P. Marincourt presented the first publication that opposite poles of the magnets were attracted and equal poles repelled.

Magnetic Compass Needle Declination and Magnetic Structure – In 1581 R. Norman observed that the needle of a compass beyond its motion on a horizontal axis pointing to the north direction of the Earth, also showed a slope on that axis. And in experiments, like that of a magnetic needle, supported on a cork base placed at the center of the surface of the water that filled a cylindrical container, it was observed that the needle kept pointing to the north pole of the Earth, but that remained motionless in the center of the surface in the water of the container. That is, if there was attraction to the north pole, the needle should move in the direction of the edge of the container in the direction of the pole, however as the needle remained still it followed the structural property of the needle magnet. That is, the structure as subsequently in 1845 called the magnetic field.

"Orbis Virtutis", Magnetic Efflux and Ether – But in 1600 Gilbert showed that the direction of the magnetic needle followed a "orbis virtutis", that is, a magnetic property involving the Earth and the magnet, that is, a sphere later in 1845 called the magnetic field. And when between two magnets, or the magnet and an iron, fragment, were placed thin pieces of wood, paper and other materials, this wrap continued to exist and only attenuated by these materials. And Newton, in 1717, in "Optics" mentioned that the efflux or ether of a magnet was rather rarefied to the point of crossing a glass blade without any resistance or diminishing of its force, and potent to rotate a magnetic needle beyond the glass.

Amber Attraction – Around 900AC, there were been observad in Greece that the resin of the vegetable fossil, the amber or electron (word derived from the Greek), when rubbed with tissues attracts straws and dried leaves, that is, it was with electricity. And in 1600 Gilbert noted that this amber attraction also occurred when they were rubbed with glass, crystal, seal, precious stones and other materials, which he named them electric.

Attraction and Repulsion – Cabe, in 1629, observed that the straw and dried leaves were indeed attracted by the amber and the glass, but could also occur after the attraction a repulsion. Similar fact also observed in 1675 by Newton in rubbed glass showed attraction in papers and then there was a repulsion. As Newton, in his "Questions of Optics": *"The bodies do not act on each other by the actions of gravity, magnetism and electricity? And the magnetic, electrical and gravity attraction between the bodies, could not be by impulse or other means that is still unknown, because nature is very attuned to a common standard?".*

Electricity Charging – In 1663, O. Guaricke managed through a sulfur globe that rubbed over a tissue, turning on an axle, got a shipment of electricity to send to conductive wires.

Electricity Driving – And in 1729, S. Gray realized that this electricity accumulated in bodies could be conducted through metals, indicating some efflux, or electrical fluid that constituted the materials.

Two Types of Electric and Magnetic Charges – Between 1733 and 1734, C. F. Fay verified that a gold leaf in contact with the amber had caused an attraction and then a repulsion, but a glass that was rubbed as it approached that gold leaf presented attraction. He also noted that amber, sulfur, silk, paper and other resins showed actions contrary to glass, crystals, gems and wool, when they were rubbed with the same material. Then, he concluded that there were two types of electricity, two effluxs, defining what quantity of electricity or electric charge of the same type is repelled and of opposite types are attracted, similar to the equal poles of magnets that repelled and opposed that were attracted.

Quantity or Electric Charge of Positive and Negative Electricity – But between 1746 and 1750, W. Watson and B. Franklin proposed for these and other similar experiments, that friction removed electricity from one body to another, getting one with excess and the other with less electricity, that is, only an electric fluid was transferred. According to the excess or lower quantity of efflux or electric charge that the bodies presented there would be two types of electricity, positive or negative, and Franklin called it positive electricity that had excess fluid. But in both cases, one or two fluids, the quantity of electricity, the electric charge of these fluids was preserved in transfers between bodies.

Electric Induction – In 1755, Franklin, J. Canton and Wilcke observed that if a conductor is approximated to a rubbed body, the furthest part of the conductor acquires an electric charge of the same signal as the body rubbed, while the nearest part acquires an electricity from the contrary type, which is the induction of electricity or electric. This is, besides friction, also electricity was effected by just approaching a electrified body to the electrification of another body. And the attraction or repulsion of the electric fluid was caused by an essential external part of the electric fluid, which would later be named the field of the electric charge.

Electric Fluid Without Envelope – But, Gray in 1729 when doing an experiment with a solid oak cube and the other just with thin strips, it showed that when rubbed the same way was produced identical effects. Concluding that it was only the surfaces that had participated in the physical phenomenon, and that the air and glass were impervious to electricity between surfaces laden with opposite charges. So, F. U. T. Aepinus and Wilcke excluded this essential external part of the electric fluid and proposed to follow the mathematical formulation of action at the distance of Newton's "Principia" of 1687. Although Newton said that supposing that a body could act on another distance through the vacuum, without the mediation of any other one, it was to stop it a great absurdity and that no being that had in philosophical matters a competent faculty to think, he could accept that explanation. And to avoid talking about pseudo-science and occultism, like angels triggering celestial stars, Newton treated the existence of ether.

Magnet Attraction and Repulsion with Inverse Square Distance – In 1750 J. Michell published a Treaty of Magnetism, which with its observations and of B. Taylor and P. Musschenbroek was found that the attraction and repulsion of magnets grew with the square of distance between the poles.

Attraction and Electric Repulsion with Inverse Square Distance – And in 1767 Priestley with experiments showing similarities to Newton's law of gravitation concluded that the attraction of electric charges was subject to the square of distance between charges. And in 1769 J. Robison determined by experiment the force between charges and counter with a value around the inverse of the square distance between the charges.

Electric Potential – In 1777, Lagrange had shown that the components of the attractive force at some point, in the motion of gravitational bodies, could be simply expressed by the infinitesimal variations of these components and the coordinates of each point of attraction, which was obtained by the addition of all the masses of the particles of an attraction system related to their point distances, being similar to the electric potential.

In 1782 P. S. Laplace showed that these variations of attractive force would be expressed by $(\partial^2 V/\partial^2 x^2) + (\partial^2 V/\partial^2 y^2) + (\partial^2 V/\partial^2 z^2) = 0$. And in 1813, Poisson showed the usefulness of the potential V function, when a point (x, y, z) is within a body of attraction the Laplace equation could be replaced by $(\partial^2 V/\partial^2 x^2) + (\partial^2 V/\partial^2 y^2) + (\partial^2 V/\partial^2 z^2) = 4\pi\rho$, where ρ is the density of the matter of attraction at the point.

In 1828, G. Green extended the properties of the function representing the sum of all electric or magnetic charges in the field, divided by their respective distances from some points, which named the potential similar to "vis potentialis" which in 1744 Euler had introduced in mechanical for motions retained as elastic body when it was curved.

Attraction and Repulsion: Electric and Magnetic, and the Gravitational Inverse of the Square of Distance – At 1785 C. A. Coulomb accurately verified the attraction and repulsion of electric charges with a spherical device, metallic rod and two gold sheets, the electroscope built by Coulomb. And the attraction and repulsion of magnetic masses, in which the force of attraction or repulsion was directly proportional to the masses and the inverse of the square of distances, similar to the force of gravitational attraction. And the electric charges were of two types, because in experiments with mixtures of gases there could be repulsion or attraction and should be by two different types of electricity.

And the law, subsequently to 1850, dealt with $F = (k_e Q_1 q_2 /d^2)$ d and $F = (k_m m_1 m_2 /d^2)$ d where q_1 and q_2 they're electrical charges, m_1 and m_2 are magnetic masses, d the distance between them, k_e electricity constant and k_m constant of magnetism, of attraction or repulsion, and d the distance in the space of the three conventional dimensions x, y and z. So for a force F applied in q at different distances d_n, $F = k\ Qq/d_n^2$, as formulated by Coulomb and others, originated from the mechanical mv/t and gravitational force of Newton. And the work or energy done defined between 1826 to 1853 by the product of the force by the distance, i.e. $E_{energy} = F \times d = kQq/d_n^2$. And the energy between d and d_n is expressed by $E_{energy} = k\ (Q/d - Q/d_n)$, in which when d_n → ∞ energy was the maximum, this is, the potential energy $E_{potential} = k\ Qq/d$, in which $kQ/d = V_1$ it is the electric potential in point 1 and $kQ/d_n = V_2$ is the electric potential in point 2 in the distance d_n of Q. Therefore, $E_{energy} = q\ (V_1 - V_2)$, where $(V_1 - V_2)$ is the potential difference and called 1 volt when $E_{energy} = 1$ joule and $q = 1$ coulomb ; $i = q/t$, i = current or electrical charge that passes in a given time t. and admitting a unit intensity of current or electric charges, subsequently in 1880 called of ampère, which is passing in two parallel conductors in the distance of 1 metre between them, produced a force of 2×10^{-27} kg x m/s² by meter of conductor, in 1 second, the passage of a unitary quantity of electrical charge q, subsequently in 1880 called 1 coulomb. That is, the coulomb being a charge quantity passing in a second, by a cross area in two parallel conductors in the distance of 1 metre between them, produces a force of 2×10^{-27} kg x m/s² per conductor's meter. To $E_{energy} = 1$ kg x 1 m/s² x 1 m = 1 joule and the charge $q = 1$ coulomb, is obtained $E_{energy} = q\ (V_1 - V_2)$, $(V_1 - V_2) = 1$ joule/1 coulomb = 1 volt, which happened to be named in lieu of the so-called automotive force $F_{electromotive} = [(V_{P1} - V_{P2})/q]$, where V_{P1} and V_{P2} is the potential energy in the P_1 and P_2 points. And according to G. S. Ohms, in 1827, the passage of electricity on a conductor wire, with electrical current intensity i with an electrical resistance R, at a time t, for a charge q_1 at one point P_1 and another q_2 in P_2 in distances d_1 and d_2, with an electric potential of $\Delta V = (kq_1/d_1 - kq_2/d_2)$ is expressed in $\Delta V = R \times i$. And the work or energy to shift a q charge between those two points is the $E_{energia\Delta V} = F \times d = \Delta V \times q$.

Battery – In 1800 A. Volta observed that two different metals placed on separate sides in a solution of water and salt or acid, retained by cardboard, forming a stacking or battery, came to result in an electric current circulating between the two metals and the solution of the products.

Electricity Uniting Compounds – In 1800, J. W. Nicholson and A. Carlisle performed electric current in acidified water obtaining hydrogen and oxygen, which J. W. Ritter came to collect in two separate tubes within the electrolytic equipment. In 1671, Boyle had obtained, from a reaction between iron and acid filings, a gas that generated water, hydrogen gas, and also observed as inflammable gas by Cavendish in 1766. And in 1774 Priestley when decomposing a product obtained mercury and a gas that presented a sharp combustion, which was called oxygen. The conclusion of the Nicholson and Carlisle experiment by Davy and W. H. Wollaston was that electricity formed the union of compounds or elements.

A Proof of the Existence of Atoms – Proust in 1797 had determined the law of the constant proportions of masses between known chemical elements at the time, and in 1803 Dalton obtained the relative atomic masses among the chemical elements, being that in 1815 it was by Prout admitted in 1 gram the mass reference of the hydrogen atom to combination with other elements, forming molecule with the mass in grams, the molecule-gram. And in 1811 Avogadro discovered that equal volumes of gases in equal conditions of pressure and temperature contained the same number of atoms in an atom-gram or molecule-gram, in the molar volume of 22.4 liters. Thus, these combinations allowed to prove that the matter was made up of atoms, as Kanad, Leucippus and Democritus had predicted more than 2000 years ago.

Electrical Disturbance – In 1812, according to Poisson the uniform distribution of two fluids was called a natural state and when that state presented a disturbance in some body, the body was said to be electrified, and several physical phenomena of electricity were formed.

Electromagnetic Unification – At 1820 H. C. Oersted observed that an electric current in a conductor showed a deviation from the magnetized needle of a compass. It was the unification of electricity with magnetism.

Unknown Force Around the Electricity – Oersted tried to explain that the electric current presented electric forces of unknown cause acting widely dispersed in the space around the conductor wire.

Minimum Electric Current in Magnet – Ampère in 1820 came to explain the deflection of magnetic needle as a minimum electric current in the magnet interacting with the electric current of the conductor, and that the magnetic needle would be orthogonal to the conductor wire if it were to separation the Earth's magnetism.

Mathematical Electrical Force and One and Occasional Ether – Also in 1820, Ampère performed a fundamental experiment on electricity, with two wires conducting electrical current, obtaining attraction between the currents when they were flowing in the same direction and repulsion when in opposite directions, which termed as electro-dynamic this motion of the electric current. Ampère considered the attraction and repulsion of the electric current as a mathematical electric force, in which the intensity i of the electric current running an elementary distance ds, ids in the conductors was calculated at all locations of the coordinates x, y, z around the conductors. However, in which two elements of electric current $i.ds$ were in the same direction the calculation of the mathematical force indicated that there would be repulsion, contrary to all other points of the space where there was always attraction, surprising Ampère. But, by experimenting in which a first wire diving on mercury, which led the electric current to a second wire in the same line as the first, noted that in fact there was repulsion between the wires. Similar to Newton who presented the "Principia" as a mathematical treatise, also in "Théorie des Phénomenes Électro-Dynamiques", from 1820 to 1825, Ampère followed Newton in deducing laws of physical facts solely from experience and the value of the forces of nature by mathematical calculus, regardless of every hypothesis about the nature of these forces that produced the physical phenomena. However, also following Newton, he came to analyze the occasionality that the electrodynamic interactions could occur through a medium as a type of ether in conjunction with mathematical laws, which would try to explain the action of force at the distance of the mathematical law through of a medium as support for these interactions.

Magnetic Needle Revolution Motion – In 1821, M. Faraday concluded from the Oersted experiment that the magnetic needle revolved around the electric conductor wire by a revolution motion caused by forces acting between the magnetic needle and the electric current, because placing the magnetic needle in a circle around the conductor wire was always tangent to that circle. These forces would be caused by magnetic curves, later named magnetic force lines and magnetic field. The poles of the magnetic needle did not point to the conductor wire with the electric current passing, so there was no force of attraction or repulsion between the poles of the magnetic needle and the electric current. It was a similar remark to that of Norman of 1581, in which there was no direct attraction of the magnetic needle with the magnet, but a structure around the magnet.

Visuality of Magnetic Curves – Magnetic curves were shown in 1821 by Faraday through iron filings scattered around conductive wires with electric current and magnets.

Revolution Motion On Electric Wire – The conductor wire with the electric current caused the motion of the magnetic needle. And by the action and reaction of Newton's law forces there should be a force of the needle for the conductor wire. Thus, if the lead wire was free for motion it should have a revolution motion when a magnet was fixed, as predicted by Wollaston and Davy. And Faraday in 1821 showed this fact of these two rotating motions, of a magnet rotating continuously around a fixed conductor wire with the electric current, and the fixed magnet with the conductor wire in which the electric current revolved continuously around the magnet.

Magnetic Virtue of Electricity – In 1821, Biot and F. Savart proposed that the electric current, that is, the electricity in motion, produced the magnetic virtue, subsequently called the magnetic field, in the conductive material, then by interaction with the magnetic needle causing its deflection.

Magnetization in Non-Magnetized Iron – In 1821, Arago when passing an electric current in a coil, around a non-magnetized iron obtained the magnetization, ie the first magnet for electricity, the electromagnet. The explanations of revolution motion by magnetic curves, magnetic virtue, mathematical forces between current forces and electric current minimum of distance action and unknown forces were presented and conflicting between them.

Magnetic Needle and Non-Magnetic Copper Disk – And in 1824 Arago observed that a fixed copper disk in one place, when being rotated made a magnetic needle also rotate on its axis, accompanying the turning of the disc, that is, the drag by the disc. And when the copper disk was fixed in a place without turning and the magnetic needle was rotated, the copper disk also rotated, that is, it was dragged by the magnetic needle, which was surprising, because the copper although it was conductor of electricity was not magnetic. The explanations of revolution motion, magnetic virtue, unknown forces and mathematical forces between current forces and electric current minimum of distance action continued to be presented as cause of these facts.

Electric Current Induction – In 1831 Faraday, installed an iron ring with an electric wire coil attached to one pile, and on the other side a secondary coil of electric wire without contact with the first coil with a current meter. He observed that by turning on and off the electric current of the first coil, in addition to magnetize the iron ring as it had been obtained by Arago, an electric current was obtained in the secondary coil, the induced current. Also, approaching this secondary coil or metallic wires and helices of a magnet, were obtained electric currents, which stopped when the motion of these electric conductors stopped. And if the magnet were to be removed, these electric currents were produced in the drivers. J. Henry and Ampère had also previously observed the formation of the induced current, but reporting after Faraday which presented an explanation for the emergence of the induced electric current. According to Faraday, with the magnetic curves without motion of magnet, secondary coil or other electric conductors, there would be a balance of forces between these magnetic curves and electric conductors. Magnetic curves would provide an energy to electric conductors. And by the motions of magnetic curves it would result in induced electric current to achieve a new energy balance. However, Faraday noted that the induction of electric current was an identity between the electric current that caused the displacement of the magnetized needle and the magnet that caused the electric current. That is, the electric current was the resulting between electricity and magnetism, and Faraday considered it a considerable proof of the Ampère theory in which magnetism was formed of electric current minimum interacting with the electric current of the conductors of electricity. Faraday observed that the induction occurred when there was variation in magnetic curves, i.e. by turning the electric circuit on and off or moving the electric conductors or the magnet. But this variation would be equivalent to the motion of the electric currents of the magnet which caused interaction with the electric charges of the conductors and induced the electric current according to the theory of Ampère. Faraday imagined magnetic force lines in which the closer they were to each other, the more intense the magnetism was, and supposing that these lines tended to shorten and repel each other. And in 1837, Faraday also introduced the idea was going to power lines.

Intersection of Magnetic Curves with the Conductors – In another Faraday experiment in 1831, in the case of a magnet, placed next to the center of a fixed copper disk in the same place and without rotating or rotating, the magnetic curves would remain fixed, because the power disc also remained in the same location. That is, it would remain the balance of energy between the magnet and the electricity. Therefore, the electric current should not occur on the disc by magnetic induction. However, the rotating disc resulted in electric current forming on the disc. Faraday found that another cause was needed to gain another balance of energy. Then it was considered as cause of the appearance of the electric current on the disc, as previously on the reels, by the motion of the electricity of the disc, the conductor wire of the coils or the other wires of the electric circuit, crossing perpendicularly with curves magnetic curves. So what would vary in magnetic curves was their motion equivalent to the motion of the conductive wires crossing these curves. Thus, Faraday explained the Arago experiment, the motions of the magnetic needle and the copper disk, by crossing the magnetic curves of the needle with the disc, which induced electric circular current on the disc. And this electric circular current produced the magnetic curves, the magnetic virtue of Bio and Savart, the magnetic field, which interacted with the magnetic needle, creating the drag of the disc and the magnetic needle. He also showed a version of this experiment in which a copper disk being rotated between the poles of a horseshoe-shaped magnet, the electric current appeared on the disc that was observed in the electric circuit with two points against the disc. P. Barlow and Henry showed another version, passing electric current on the disc and obtaining the motion of the disc. But it did not explain why this intersection of magnetic curves with the drivers caused the electric current induced.

Electricity in Water, Insulation and All Materials – In 1800, Nicholson and Carlisle performed electric current in acidified water obtaining hydrogen and oxygen, which Ritter came to collect in two separate tubes within the electrolytic equipment. And Wollaston and Davy observed by the passage of electricity in decomposed substances in the water, that chemical affinity was essentially electrical in nature. In 1807 and 1808, Davy by decomposition in solution with electric current isolated chemical elements such as potassium, sodium, barium, magnesium, calcium and strontium.

Between 1831 and 1834, Faraday found that an electric current was conducted in water with chemical compounds, the electrolysis, in which these compounds were dissociated into electrolytes, i.e. in components of positive electricity called cation and the negative of anion, which were accumulated in plates, the electrodes, with the positive cation toward the negative or cathode electrode, and the negative anion going to the positive or anode electrode.

The mass m deposited in the electrodes was directly proportional to the quantity q of electricity led by the solution or to the intensity of electric current and to the time of electrolysis: $m = K_e q$, where K_e it was the electrochemical equivalent that was the mass deposited in the electrolysis by the q charge in the electrolysis. And the atomic mass A and valence n, deposited in the electrode, were related as $A/n = f_q K_e$, where f_q = constant factor representing the charge placed on the electrode by molecule-gram or atom-gram, i.e. the relative molecular or atomic mass in grams. This quantity of electric charge was subsequently determined around the current value of 96,489 coulombs occurring in the electrolyte to deposit a mass equal to the molecule-gram or atom-gram of the atomic element. So 1 gram of hydrogen equals a charge quantity of 96,489 coulombs.

Showed that insulating materials, which when rubbed remained without crowding the electricity in the electrolysis showed the passage of electricity and concentration of negative and positive electricity in the electrodes, that is, the electricity constituted all the materials. Thus, the existence of electric current minimum would also be found in non-magnetic materials such as copper which would explain in the Arago experiment that the non-magnetic copper disk was dragged by the rotation of the magnetic needle, if the theory of Ampère so that these electric current minimum also act in the interaction between all electric currents.

Faraday's experiments with electrolysis, in which precise quantities of metals from the solutions were deposited in the metallic rods of the electrodes of the cathode and the anode, showed the chemical affinity as of electric origin and the first experimental evidence on the structure of the atom by the decomposition of the substances.

And in 1874, S. A. Arrhenius explained the decomposition of the electrolytes in the water in which one of the atoms of the compound lost electric charge, the cation, and another atom receiving this charge, the anion. In 1881, H. L. F. Helmholtz claimed that electricity was formed of elementary particles that behaved as an atom of electricity. And it called the charge particles charged with positive or negative electricity as ion, in which previously according to Faraday the anions were negative charges and the cations were particles with positive charge. And in 1887, Arrhenius presented the ionic theory in which the ions, positive or negative electric charge components deposited in the unitary quantity of the valence, in the electrodes, which constituted the electric current through the solution, in the phenomenon of electrolysis, They were nothing but atoms charged with electricity.

Real Forces of Matter, Electricity, Magnetism, Optics and Gravity – Faraday also considered power lines and they could be obtained in bodies with charges by electrified materials, in oil films. As Faraday, publications between 1834 to 1845, forces of electric lines and magnetic curves when concentrated resulted in mass of the atoms, therefore the force being originated of mass, and would be the fundamental constituent related to electricity, magnetism, light, mass and gravity. So, Faraday concluded that electricity, magnetism, optic phenomena, chemical matter and gravity would all be interconnected. Between 1939 and 1847 C. F.

Mohr, W. R. Grove and J. Liebig emphasized force as coming from a single nature to act on all physical phenomena.

Lines of Force and Unification of the Forces of Nature – Between 1834 and 1845 as Faraday, the lines of force allowed the deduction, interpretation and prediction of physical phenomena and which should be expanded and investigated for a unified theory of the forces of nature.

Electric and Magnetic Fields – And the structure of the lines real (physical) of electric and magnetic force as Faraday in 1845 would constitute the field of force lines or fields of electric and magnetic forces.

Magnetic Field Interaction and Light – And in 1846, Faraday observed that a ray of light, as it passed through a glass prism that was in the presence of a magnetic field, presented modification of the polarization plan. But, a maximum magnetic field at the time applied on a candle, the light remained without any difference as its spectrum relative to the sail without acting of that magnetic field.

Propagation of Electric, Magnetic, Gravitational and Light Fields – Just as a material line of a rope is propagated as a wave, also the electric, magnetic, gravitational and light field, in which the polarization plan was modified by the magnetic field, which would consist of lines of force, would also be propagated like a wave, according to Faraday around 1846.

Force and Charges in Motion– In 1846, Weber presented the force acting between electric charges in relative motion.

Where: $F = (q_1 q_2)/(4\pi\varepsilon_0 d^2) [1 - (1/c^2)(dd/dt)^2 + (d/c^2)(d^2d/dt^2)]$, where q_1 and q_2 are the charges in interaction, $v = (dr/dt)$ is the relative velocity between the charges and $a = (d^2d/dt^2)$ is the relative acceleration between the charges, $c = (\varepsilon_0\mu_0)^{-1/2}$ = light velocity, where ε_0 it is the permittivity, dielectric capacity, specific capacity for electrostatic induction or dielectric constant, and μ_0 the permeability of the free space, magnetic permeability or magnetic constant

Electricity and Temperature – In 1821, T. J. Seebeck had observed that in a junction of a metal wire with another wire of another metal, the temperature variation between the ends of this junction occurred the passage of electric current. In 1834, J. C. A. Peltier observed that in the passage of electric current, in a junction of a metal wire with another wire of another metal, a temperature difference occurred at the two ends of the junction. And the temperature increased at one end of the junction when the chain was in one direction and there was decrease of temperature in the other direction of the electric current. That is, different metals, different electric resistances, different potentials. And in 1856, Thomson-Lord Kelvin noted that in a metal with temperature difference between two positions there was electric current formation.

Dielectric Capacity and Magnetic Permeability – In 1856, the electricity constant $k_e = 1/\varepsilon_0$, where ε_0 it was the permittivity, dielectric capacity, specific capacity for electrostatic induction or dielectric constant, and $k_m = 1/\mu_0$, where μ_0 it was the permeability of the free space, magnetic permeability or magnetic constant, Coulomb's law was measured by R. H. A. Kabala and Weber getting that the inverse of the product $(\varepsilon_0\mu_0)^{1/2}$ it resulted in the value close to the speed of light. And in 1857, Kirchhoff mentioned that forces of attraction and repulsion of electric and magnetic charges caused an electromagnetic disturbance with a vacuum propagation at the speed calculated by Kabala and Weber which was the speed of light.

Field, Energy and Electric Flow – By the attraction and repulsion, of particles, were related electric field E with electric force F and charge q.

For $E = (F/q_1)$ and to $F = (k\, q_1\, q)/d^2$, $E = (k\, q)/d^2$, and $(q/S_{area}) = \rho_{electric}$ = electric density, where k = constant electric, S_{area} = area, was obtained $E = (k\, \rho_{electric}\, S_{area})/d^2$. In 1867 J. K. F. Gauss presented the flow Φ_E of an electric field E by the mathematical relationship $\Phi_E = (E\, S_{area}) = [(k\, q)/d^2)\, S_{area}] = k\, q\, \omega$, where $\omega = 4\pi$ sphereradians, so $\Phi_E = 4\pi\, k\, q$. And applying in a dielectric field under an electric potential was obtained an electric flow that resulted in energy from the electric field expressed by the energy per unit of volume, that is, the energy density $(E_{energy}/vol)_E = k\, E^2/8\pi$ and the energy density for magnetic field B as $E_{energy}vol)_B = k'\, B^2/8\pi$. For the electromagnetic field was obtained $(E_{energy}\, vol)_{EB} = kk'\, 1/8\, \pi\, (E^2 + B^2)$.

Disturbances in Electric and Magnetic Fields – By Coulomb's experiments and others there were forces of attraction and repulsion acting between electric charges and magnetic masses. And as Faraday, there was around an electric charge q_e or a magnetic mass M_m, power lines forming an electric field E or magnetic H.

And for an electric charge q_1 placed in the electric field E at a point P_1 of coordinates x_1, y_1, z_1 in distance d_1 of q_e acted a force F_1, and for larger charges q_2, q_3, q_n, they acted higher forces $F_2, F_3\, F_n$, where F_n/q_n it was always constant for that point P_{E1} of the electric field E. So, $E_{P1} = F/q$. Then to $F_P = k_e q_e\, Q/d^2$, $E_P = k_e q_e\, /d^2$ and similarly to the magnetic field at a point P, $H_P = k_m\, M_m/d^2$ in the theoretical case of considering the magnet poles enormously distanced. The electricity constant $k_e = 1/\varepsilon_0$, where ε_0 it was the dielectric constant, the magnetic constant $k_m = 1/\mu_0$, where μ_0 it was the permeability of the free space.

And as Maxwell, between 1855 to 1873, if an electrified body were placed in any part of an electric field, it would produce a sensitive disturbance in the electrification of another body. Then, for any additional electric charge, magnetic mass, electric field or magnetic field, it would come into an electric field and, magnetic H or magnetic induction $B = \mu_0 H$, there would be a modification at a certain time in these fields, that is, a disturbance, a vibration, and consequent propagation of this vibration in the field. This propagation of this disturbance in the field would be through a mathematical action at a distance or by a type of ether in space, according to the existing theories. For Maxwell, the mass was present or hidden in force: $F = qE = mv/t$, and as the light passed through the electric and magnetic fields would show that there were spaces in the field and then there should be a means of support for the field, the ether, in which by the action of contiguous parts of ether the energy was transmitted from one part to the next. The ether would be a vessel of kinetic energy such as vibration or distortion as potential.

Momentum and Absolute Force Unit– A considerable part of Maxwell's "Treatise on Electricity and Magnetism" was about the precise definitions of the electrical and magnetic variables for the physical and mathematical explanation of the existing electromagnetic phenomena. And on the first page of this treaty was defined three fundamental units for all the dynamic sciences: length, time and mass. In which the mass follows a standard of a platinum cylinder adopted in 1799 as a prototype of the preserved kilogram in France, and after 90 years replaced by another equivalent standard. The distance pattern, the meter, was also adopted in 1799 by standard platinum and iridium and subsequently by 1,650,763.73 wavelengths of radiation in the vacuum of the transition between the levels of the atom of the yellow-red line. The time by measurement carried out by standard adopted as solar day units system of 86,400 seconds.

And the units derived from: velocity, acceleration, density, momentum, absolute unit of force as the unit at the momentum by the unit of time, work defined by force acting through the unit of length, and the capacity of work or energy, that is, also set to force acting on the unit of length, and the other units that they are referred to. And in 1877, Maxwell added that: *"The unit of force acting on the unit of mass by the unit of time generates the unit of velocity and the variation of the momentum of a body is numerically equal to the impulse in the same direction"*. Perhaps, by mass being key to Maxwell, he considered that also the electric and magnetic forces were caused by the motion of hidden masses.

So, distance = d, time = t, mass = m, velocity v = (dd/dt); Acceleration $a = (dv/dt) = \{[d(dd/dt)]/dt\} = (d^2d/dt^2)$; density = m/d^3; momentum p = mv; $\Delta mv = \Delta p = I$, where I is impulse I; F = (I/t); F = $(\Delta p/t)$, F = (dp/dt) = m (dv/dt) = ma and W_{work} = work = F x d = energy E_{energy}.

The Unit of Electricity, the Electric Charge – Maxwell has expressed the electricity unit as an electric charge.

And the quantity of electricity, a q charge, when it acted a unit of force by a certain volume and ρ density, it was expressed by q = $\iiint \rho\ dxdydz$.

Field Force – Maxwell expressed that an electric field E, where if a charge q was placed at a point P of that field there was a force acting on the charge q: F = E q, being E = electric field or electric intensity or even resulting electro-motive intensity. And in the case of the charge unit q = 1, F = E, that is, the field would be a force acting on a unitary charge.

And the resulting force between two electrically charged bodies in a direction in space, for example, for x: $F_x = \iiiiii \{[\rho\rho'(x-x')\ dxdydz\ dx'dy'dz']/[(x-x')^2 +(y-y')^2 +(z-z')^2]^{3/2}\}$, where x, y, z are coordinates of a point in the first body in which the electric density is ρ, and x', y', z' and ρ' are the corresponding quantities for the second body, and the integration is extended first over one body and over the other.

Potential Energy of the Electric Field and Electromotive Force – Defined the potential energy acting on a charge q as the electromotive force $F_{Electromotive}$.

So $F_{Electromotive}$ = $[(V_{P1} - V_{P2})/q]$, being V_{P1} and V_{P2} potential energy in P_1 and P_2 points. And for a displaced distance d, such as: [(F/q) d] = (E d) = $[(V_{P1} - V_{P2})]/q]$, where E = electric field. So $F_{Electromotive}$ = [(F/q) x d] and $F_{Electromotive}$ = (E d). And relates the electric current i between charge q and time t, i = q/t, with the potential energy $[(V_{P1} - V_{P2})]$ and the electric charge q, in which in 1827 as Ohms, $(V_{P1} - V_{P2})/i$ = R where R = electric current resistance. And for the charge q, $[(V_{P1} - V_{P2})/q]$ = R i, is obtained: $F_{Electromotive}$ = R i = [(F/q) d] = [E d].

Forces: Automotive and Mechanical – According to Maxwell, the electric force, that is, the energy of the electric field: *"The automotive force is always to be understood to act on only electricity, not on the bodies in which electricity operates. And never be confused with the mechanical force, which acts on only the bodies, not on their electricity. If we come to know the formal relationship between electricity and ordinary matter, we should* *probably also know the relationship between electric force (electric field energy) and ordinary force".*

Energy and Electromagnetic Force – Maxwell defined the mechanical kinetic energy by the work $W_{mechanical}$.

So, $W_{mechanical} = mv^2/2$ and the force of that energy for a distance x, as $F_{mechanical} = dW/dx$ or $F_{mechanical} = m\ [(dv/dt)]dx/dx = m\ (dv/dt)$. And similarly defined the electromagnetic force as $F_i = dW/dx$ and the electrokinetics or the electrokinetic momentum as $p_i = (mv/q)\ d$.

Electric and Magnetic Field Energy – The energy E_{Energy} of the electric field E_Q of a charge Q can be defined by the force of the field acting on charges.

So the energy E_{Energy} of the electric field E_Q of a charge Q expressed by $E_{Energy} = F \times d = Q\ E_Q\ d$, being d = distance, in which $E_Q = (Q/k)d^2$, k = electric constant, being $d^2 = S_{area}$ = area and for a charge dq: $dE = (dq/k)\ r^2$, so $dq = dE\ k\ S$. The force F acting on an additional charge dq at a distance d: $dE_{Energy} = Fdd$ e $dE_{Energy} = dq\ E\ d$. So, $dE_{Energy} = dE\ k\ S\ E\ d$, e $E_{Energy} = k\ A\ d\int_0^{E(Q)} EdE = ½\ k\ S\ d\ E_Q^2$. Similarly for magnetic induction field $E_{Energy(m)} = ½\ k'\ S\ d\ B^2$.

Despite the law of force between quantities of magnetism to have exactly the same mathematical form of the law of force between quantities of electricity, in which the mathematical treatment should be similar to that of electricity there are other properties of the magnets that must be observed, because while the electric charge are unitary, the magnets are dipoles.

Fundamental Field of Electric and Magnetic Energy – In the induction of electric current by the variation of the magnetic field, as noted by Faraday, Maxwell sought to interpret this physical fact of the formation of the electric current. emphasized the lines of electric and magnetic force, presented by Faraday, as streams containing n lines of force: φ = n lines of force/A, where A = area, as field of force lines or force field as a fundamental state of energy, which Faraday emphasized before to prefer the explanation of electromagnetism by crossing magnetic power lines with electric conduits.

Electric Fields, Magnetic Fields and the Electric and Magnetic Forces – Maxwell defined the electric and magnetic fields, and the electric and magnetic forces.

Maxwell set in an electric circuit the electro kinetic moment of electric current as $p_{ek} = (mv/q)\ d$, mass = m, v = velocity, q = charge, d = distance, equivalent to a quantity of magnetism m_m of a mass m_m of a magnetic polo. The electromagnetic momentum p_{em} as (dp_{ek}/dt). And by the definition of force $F = dp/dt$ applied to $d\mathrm{p_{ek}}$ resulted in the electromotive force: $F_{Eletromotive} = [(d\mathrm{p_{ek}}/dt)\,d\,]/q$, to keep the current i against the resistance R and increase the electromagnetic momentum $\mathrm{p_{em}}$. That is, the electromotive force was to be supplied with the induction of the electric current and of independent sources. So $F_{Eletromotrice} = Ri + [(d\mathrm{p_{ek}}/dt)d]/q$.

And to $B = \mu_0 H$ and $H = \mu_0\,(1/\mu_0\,\mathrm{m_m}\,/d^2)$, for a charge q in an area d^2 and a time t: $B = \mathrm{m_m}/qt$, that came from the kinetic momentum $\mathrm{p_{ek}} = (\mathrm{m_m}\,v/q)d$ applied in an area d^2. This is, $\mathrm{p_{ek}} = (\mathrm{m_m}\,v/q)d / d^2 = (\mathrm{m_m}\,v/dq) = \mathrm{m_m}/qt = B$.

For the electric field $E = F/q$ it was the kinetic momentum applied in time: mv/qt. And for a potential $V_{potencial}$ of electric current was the kinetic moment at one point (x, y, z): $d\mathrm{p_{ek}} = (mv/q)\,d / d = V_{potencial}$.

And was obtained $\mathrm{p_{ek}} = \int n\,V_{potencial}\,\cos\alpha\,dd = \iint n\,B\,\cos\theta\,dd^2$, where α was the angle between $V_{potencial}$ e dd, and θ is the angle between B and the normal to dd^2 . That is, $V_{potencial}$ and B related, and in Maxwell's notation: $\int n\,\mathfrak{A}\cos\alpha\,dd = \iint n\,\mathfrak{B}\,\cos\theta\,dd$.

And on a moving circuit with velocity v at the coordinates *x, y, and z*, in the magnetic field to produce the induction of electric current, got the electromotive force $F_{Electromotive}$ In the shifting of the magnetic field by the variation dd at the coordinates *dx, dy and dz*, which was related to the electric intensity, that is, the electric field and, depending on the electric potential $[(V_{P1} - V_{P2})]/q = F_{Electromotive} = (F/q)\,d = E\,d$. It was this electromotive force that generated the induced electric current related to the magnetic field B, as Biot and Savart, the crossing of magnetic curves is, $F_{Electromotive} = \int n\,E\,\cos\alpha\,dd$, where $E = (E_1 + E_2 + E_3)$ was the electric field resulting from three parts of electric field acting on the magnetic induction on the element dl of the circuit. As Maxwell, the energy of the magnetic field and the additional energy supplied to the circuit caused an additional electric field to the electrostatic field in the balance of the circuit. E_1 it was the electric field that resulted from the magnetic field B and the velocity v of the element dd of the circuit, $E_1 = vB$. The electric field $E_2 = dV_{Potential}/dt$ was a potential $V_{Potential}$ electric current due to the operating time of the magnetic field B or the electric current. And the electric field E_3 it was the result of the modification of an electric potential in different parts of the field itself.

In Maxwell's original:

$\mathfrak{E} = \mathfrak{E}_1 + \mathfrak{E}_2 + \mathfrak{E}_3$

$\mathfrak{E}_1 = V \cdot \mathfrak{G}\,\mathfrak{B}$

$\mathfrak{E}_2 = -\mathfrak{A}/dt$

$\mathfrak{E}_3 = -j\Psi$

where $\mathfrak{E}_1 = E_1$, $\mathfrak{E}_2 = E_2$, $\mathfrak{E}_3 = E_3$, $\mathfrak{G} = v$, $\mathfrak{B} = B$, $V =$ part of product vector $\mathfrak{G}\,\mathfrak{B}$, $\mathfrak{A}/dt = V_{potencial}/dt$ e $(d\Psi/dx)\mathbf{i} + (d\Psi/dy)\mathbf{j} + (d\Psi/dy)\mathbf{k} = j\Psi$.

And $\mathfrak{A}$ was reported to $\mathfrak{B}$, This is, $\mathfrak{B} = j\mathfrak{A}$ in which $\mathfrak{B}$ it was the magnetic induction, the current magnetic field, was the rotational vector of the electric potential.

That is, in the current writing:

$E = E_1 + E_2 + E_3$
$E_1 = V \times G\,B$
$E_2 = -A/dt$
$E_3 = -j\Psi$

where E1 = E_1, E2 = E_2, E3 = E_3, G = v, B = B, V = part of product vector G x B, $V_{potential}$ = potential, and $(d\Psi/dx)i + (d\Psi/dy)j + (d\Psi/dy)k = j\Psi$.

And the potential $V_{potential}$ reported to magnetic field or magnetic induction B, isto é, B = $jV_{potential}$, the rotational vector of the electric potential.

As E = F/q, this expression in the current denomination is presented as, F = qE +q(v X B), where qE = $F_{electric}$, the electric force and q (v X B) = $F_{magnetic}$, the magnetic force, similarly expressed by J. J. Thomson em 1881, O. Heaviside in 1889 and Lorentz in 1892, which was called the force of Lorentz. So, for the electric force $F_{electric}$ = q E = $F_{electric}$ /q = mv/t = ma, and for the magnetic force $F_{magnetic}$ = q v B = q v . m/q t = mv /t = ma.

And the electric force between electric conductors as Maxwell originated from the internal alteration of these conductors when the electric current passed. However, in 1877 Clausius indicated that the electric force between electric conductors would be caused entirely by the electric current.

Electric Current Displacement and Electric Shift Field– By the Ampère theory, an electric current, passing through a surface and going to a second surface where there was a charge accumulator, by ceasing the current caused several interpretations. As Biot and Savart this electric current caused on the first surface an electric force $F_{electric}$ equivalent to a force $F_{magnetic}$ of the magnetic field B. According to Faraday occurred the intersection of magnetic curves, and per Ampère were caused electric forces between currents or microscopic currents on an element of the surface. And by the electric current when stopping in the accumulator in which the current ceased, no magnetization should occur on the second surface.

But based on the correlation of the magnetic field, the magnetic curves of the electric currents minimum and or crossing these curves with electric conductors, which induced the formation of electric current, by Faraday. And according to Maxwell by the connections between the charges of accumulators and electric current, where if an electrified body were placed in some part of the electric field would be produced a sensitive disturbance in the electrification of the other bodies, then considered that in the accumulator of charges, by the variation of charges that were piling up in a field and with an electromotive force $F_{Electromotive}$ = E d, an electric shift would arise with motions that would constitute an electric current $J_D = q/d^2$, forming an electric field of displacement $E_{displacement}$, which would pass on that second surface. And that electric field $E_{displacement}$ it would be correlated with a magnetic induction field B, similarly to a field B resulting from an electric current as Biot and Savart, the crossing of magnetic curves as Faraday or electric forces as Ampère. Therefore, the magnetic induction field B would be associated beyond the electric current *i* that created a B field, but also with the electric of displacement J_D that formed the electric field $E_{displacemen}$: $J_D = ¼ \varepsilon_0 E_{displacement}$.

In Maxwell's original:
𝔇 = ¼ k 𝔈   where 𝔇 it was the current of displacement, k the electric constant and 𝔈 the electric field.
Or by the current writing:
D = ¼ kE where D was the displacement current, k the electric constant and E the electric field. Among the main concepts of the Maxwell treaty for electromagnetism were these electric currents, the total electric current i composed of the driving current i_C and the displacement current J_D that were related to the fields E and $E_{displacement}$, and these with the kinetic momentum p_{ek} through $\{[dp_{ek} = (mv/q) d]/dt\} = dV_{potential}/dt$ as a potential electric current and a modification of electric potential Ψ from different parts of the electric field. The electric current could be related to the electro-kinetic momentum through the $dV_{potential}/dt$ which was related to the magnetic field B, and with the modification of a Ψ electric potential.

So, for Maxwell, it was the charge and the electric current with its densities that acted as sources of E and B. Because, the electric fields were produced by charges, and magnetic fields were produced by currents which were precisely moving charges. In electromagnetic induction, the variation of magnetic field would be produced by the current motion or electric current minimum of the magnet causing an induced electric current which would create an electric field, and that electric field by varying by the motion of electric current would result in a magnetic field. Therefore, it would be this variation of electric and magnetic field a consequence of the motion of electric charges in motion.

Propagation of Electromagnetic Disturbances and Electro-Magnetic Wave – The induction of electric current by the motion of a magnet or electric conductor was an electric and magnetic field oscillation that followed the behavior of a wave equation between the axes x, y, z.. This motion of the magnet or conductor resulted in a disturbance of the fields which propagated with a certain velocity $V_{propagation}$. By applying the relations between electric current, electric field and magnetic field in relation to time, through the differentials of these variables, Maxwell obtained the expression of the propagation of the electromagnetic disturbance through a medium-to-rest uniform, which could be involved in this disturbance, with a constant k_c of conductivity of that medium, with the dielectric constant ε_0 for vacuum and with magnetic permeability constant μ_0 for vacuum, in that for vacuum $k_c = 0$. Figure 13.

In current notation: $[(\partial/\partial x)x + (\partial/\partial y)y + (\partial/\partial z)z] \times E = \partial B/\partial t$, and
$[(\partial/\partial x)x + (\partial/\partial y)y + (\partial/\partial z)z] \times B = \mu_0\varepsilon_0\partial E/\partial t + \mu_0 J_0$
Where, $(\mu_0\varepsilon_0)^{1/2} = 1/V_{propagation}$ where $V_{propagation}$ it was the c speed of light.

So getting electromagnetic waves of velocity $V_{propagation}$ with frequencies $\mathcal{V}$ and λ wavelength. The electromagnetic disturbance was propagated, in which the electric and magnetic field presented variation according to the equation of the wave, with the physical explanation that they were originated by the motion of charges and electric currents. And with the forces involved between these electromagnetic fields forming radiant energy.

The acceleration of a charge q resulted in a component of the field in which for the electron electric field of e charge and r radius of charge action, k = electric constant, with acceleration a: $E_{p'} = ea_{p'}/4\pi krc^2$ and the magnetic field $B_{p'} = E_{p'} = ea_{p'}/4\pi krc^2$, which shows the emission of the radiant electromagnetic field.

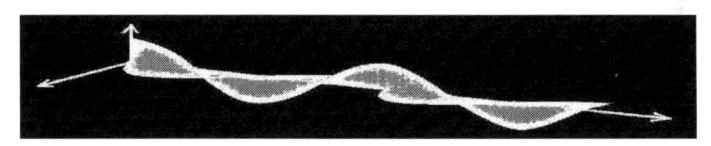

Fig. 13 – Schematic representation of the propagation of an electromagnetic field, forming a wave by the ether in the vacuum, as Maxwell.

Therefore, while Oersted, Faraday, Bio, Savart and Ampère united Electricity and Magnetism, Maxwell united the Optics with Electromagnetism.

Electromagnetic Wave and Magnetism – Maxwell sought to explain Faraday's negative result on the absence of any modification in the emission of light when the emitting source of a candle was placed in a magnetic field: *"In the case of circular polarized light, the magnitude of this vector always remains the same, but its direction rotates in the direction of the radius to complete a revolution in the periodic wave time. When a circular polarized light beam falls into a medium under the action of magnetic force* (field)*, its propagation within the medium is affected by the ratio of the direction of rotation of the light to the direction of the magnetic force* (field)*. In the middle, when under the magnetic force* (field) *action, some rotation motion is performed, with the rotation axis in the direction of the magnetic force* (field)*. The only similarity that we can locate, between a means by which the circular polarized light is propagated, and a means through which they pass lines of magnetic force* (field)*, is* *that in both there is a motion of rotation on an axis. But here the likeness is finalized, because the rotation in the optical phenomenon is that of the vector that represents the disturbance. This vector is always perpendicular to the direction of the radius, and rotates over a known number of times per second. In the magnetic phenomenon, the rotating vector presents no property by which its sides can be distinguished, so that we cannot determine how many times it rotates per second. So there is nothing in the magnetic phenomenon that corresponds to the wavelength and the wave propagation in the optical phenomenon. A medium in which a constant magnetic force (field) is acting is not, because of this force, filled with waves traveling in one direction, as when the light is propagated through that medium. The only similarity between the optical and magnetic phenomena is, that at each point in the middle, something exists of the nature of an angular velocity on an axis in the direction of the magnetic force (field).*

So Maxwell stated that there was no force in nature that could alter the mass and frequency of small bodies that make up the matter.

But later on Maxwell, the current physics eliminated the sources of currents from the electromagnetic field equations and related only to the variation of the electric field and magnetic field.

So:

$[(\partial/\partial x)x + (\partial/\partial y)y + (\partial/\partial z)z] \times E = \partial B /\partial t$

$[(\partial/\partial x)x + (\partial/\partial y)y + (\partial/\partial z)z] \times B = \mu_0 \varepsilon_0 \partial E /\partial t$, getting

$\partial^2 E /\partial x^2 + \partial^2 E /\partial y^2 + \partial^2 E /\partial z^2 = \mu_0 \varepsilon_0 \partial^2 E /\partial t^2$

$\partial^2 B /\partial x^2 + \partial^2 B /\partial y^2 + \partial^2 B /\partial z^2 = \mu_0 \varepsilon_0 \partial^2 B /\partial t^2$

In this case, electric fields could be generated by the variation of magnetic fields, and magnetic fields by the variation of electric fields, according to these modified Maxwell equations. Thus, electric fields in the vacuum could be created as by magic when the variation of magnetic fields! And although Maxwell considered the electromagnetic disturbance to be propagated through a medium like ether, these modified expressions that relate these fields are independent of ether, resulting only as support of the fields, without being included in the equations. As in these equations there is not also the ether, then out of nowhere would be created electric and magnetic fields! With these modified equations, in the oscillation of electric charge in which along with it there is always an electric field and, would be generated a magnetic field B which varies would also generate an electric field and, and so on. Therefore, the electromagnetic wave would be characterized by the independent oscillation quantities, the electric field and the magnetic field in the vacuum.

But with the motion charges producing magnetic field, and the motion of the magnetic field producing current induction and electric field, it would not be possible to explain how these moving charges and fields produce electric and magnetic fields, in addition to the active presence of electromagnetic force. And also without explaining, by modified Maxwell equations, as of nothing would be created electric and magnetic fields.

Light, Magnetism, Strength and Frequency – Maxwell commented that Faraday's experiment on polarization of light in the magnetic field showed that the only similarity between light and magnetic field was the rotational motion. But, Faraday's experiment in not finding any other modification of the light, was that there was no magnetic phenomenon that corresponded to wavelength and wave propagation in the optic phenomenon. As Maxwell would appear that there was no force in nature that could alter the mass and the frequency of small bodies that make up the matter. But the constitution of light and magnetism could indicate a link factor between light and magnetism, and as Maxwell: *"The motion that constitutes light could enter as a factor in terms involving the motion that constitutes magnetis"*.

Electric Current and Magnetic Field – As Maxwell the electric current would not be dependent on magnetic field and would not produce a sensitive effect. But in 1879 E. H. Hall presented an experiment in which the magnetic field exerted an action on the electric current, and showed that magnetism was actually a phenomenon of rotation.

Minimum Charge, the Electron – In 1874 G. J. Stoney indicated the existence of electricity as a particle containing electric charge and estimating its value.

By electrolysis, $m = K_{EQ} Q$, where m is the mass deposited in the electrode, K_{EQ} it is the electrochemical equivalent for a charge Q. And the atomic mass A and valence n_v, deposited on the electrode, being related as $A/n_v = f_Q K_{EQ}$, where f_Q = constant factor representing the charge, of the chemical equivalent, deposited in the electrode by atom-gram = 9.65 x 10^4 coulombs or 1 faraday in current unit. So $f_Q = (A/n_v) (Q/m) = (A/m) (Q/n_v)$ and $Q = f_Q n_v m/A$; $(A/m) = (N_A/N)$, where N_A is the number of Avogadro and N is the numbers of ions in the atom-gram or molecule-gram, so $N = N_A m/A$; $q_{Ion} = Q/N$, where q_{Ion} it is the charge of an ion; So $q_{Ion} = (f_Q/N_A) n_v$, where $e = (f_Q/N_A)$ is the minimum charge with the value determined by Stoney from 10^{-20} of the quantity of electricity which was at that time by the current of one ampère, this is, the quantity of electricity passing through each second in a current of one ampère or 3 x 10^{-11} CGS system electrostatic unit.

And in 1894, Stoney named the minimum charge of electricity as electron, and reported article in Philosophical Magazine. Document 1.

Philosophical Magazine
Series 5, Volume 38, p. 418-420
October 1894

Of the "Electron", or Atom of Electricit
To the Editors of the Philosophical Magazine
Gentlemen:
Professor Ebert, in his paper on the Heat of Dissociation in last month's (September) Phil. Mag., says on p. 332 :--"Von Helmholtz, on the basis of Faraday's Law of Electrolysis, *was the first* to show in the case of electrolytes that each valency must be considered charged with a minimum quantity of electricity, the "valency-charge", which like an electrical atom is no longer divisible".
Now I had already twice pointed out this remarkable fact : first, at the Belfast meeting of the British Association in August 1874, in a paper "On the Physical Units of Nature", in which I called attention to this minimum quantity of electricity as one of three* physical units, the absolute amounts of which are furnished to us by Nature, and which may be made the basis of a complete body of systematic units in which there shall be nothing arbitrary. This same paper was again read before the Royal Dublin Society on the 16th of February, 1881, and is printed both in the Proceedings of that meeting and in the Phil. Mag. of the following May.
* The two other units being (1) the "Maxwell" of velocity, that velocity which connects electrostatic with electromagnetic units in a medium of which the inductive capacity is unity, and which, under the electromagnetic theory of light is also the maximum velocity of light ; and (2) the "Newton" of gravitation, that coefficient of universal gravitation of the amount of which Professor Boys has lately made so accurate a determination.
Professor Helmholtz"s announcement was made in his Faraday Lecture delivered on the 5th of April, 1881, subsequent to both the announcements I had made. See the Journal of the Chemical Society for 1881. The announcements as made by him and by myself are as follows :

G. Johnstone Stoney in Aug. 1874, and again in Feb. 1881.

Professor von Helmholtz in April 1881.

"And, finally, Nature presents us, in the phenomenon of electrolysis, with a single definite quantity of electricity which is independent of the particular bodies acted on. To make this clear I shall express "Faraday's Law" in the following terms, which, as I shall show, will give it precision, viz.:-- *"For each*

"Now the most startling result of Faraday's Law is perhaps this. If we accept the hypothesis that the elementary substances are composed of atoms, we cannot avoid concluding that electricity also, positive as well as negative, is divided into definite elementary portions which behave like atoms of electricity. As long as it moves about on the

chemical bond which is ruptured within an electrolyte a certain quantity of electricity traverses the electrolyte which is the same in all cases". This definite quantity of electricity I shall call E_r. If we make this our unit quantity of electricity, we shall probably have made a very important step in our study of molecular phenomena. "

electrolytic liquid each ion remains united with its electric equivalent or equivalent or equivalents. At the surface of the electrodes decomposition can take place if there is sufficient electromotive force, and then the ions give off their electric charges and become electrically neutral".

(Readers of my paper are requested to bear in mind that it was written so long ago that the Ampère was still understood to mean the electromagnetic unit *quantity* of electricity of the Ohm series. The term is now always applied to the unit *current*. Moreover the paper, having been written before C.G.S units came into use, employs those which had been at an earlier date adopted by the Committee of the British Association on Electrical Standards, viz. the G.M.S., or gram-metre-second system.)

In this paper an estimate was made of the actual amount of this most remarkable fundamental unit of electricity, for which I have since ventured to suggest the name *electron*. According to this determination the electron = a twentiethot (that is 10^{-20}) of the quantity of electricity which was at that time called the ampere, viz: the quantity of electricity which passes each second in a current of one ampere, using this term here in its modern acceptation. This quantity of electricity is the same as three eleventhets (3×10^{-11}) of the C.G.S. electrostatic unit of quantity.

That determination was made twenty years ago and is founded on the estimate I had previously announced in 1867 of the number of molecules present in a gas, viz.: 10^{18} in each cubic millimetre of a gas at standard temperature and pressure (see Phil. Mag. for August 1868, vol. xxxvi. p. 141).

An estimate of this same remarkable unit of electricity was made in 1891 by Professor Richarz, as quoted by Professor Ebert (see p. 335 of the September number of the Phil. Mag.). His determination makes the electron = 12.9 eleventhets (12.9×10^{-11}) of the C.G.S. electrostatic unit. This appears to be in sufficiently satisfactory agreement with my previous determination, having regard to the amounts of the probable errors of some of the data.

Doc. 1 – Stoney article of 1874 on the minimum charge of electricity which it called electron, from "Philosophical Magazine", Series 5, Volume 38, page 418-420, October 1894.

Mechanical Pressure and Electricity – The metals were observed as the best conductors of electricity, but also the coal came to show a good electric conductivity or little resistivity in relation to insulators such as glass, rubber and wood. However, in the case of coal powder, it could receive air pressure, i.e. air-transmitted sound, and present driving power, resulting in 1876 in the technology of phone for A. G. Bell and A. S. G. Meucci. And

J. P. Hansen

in 1880, P. Curie observed that a mechanical pressure on a quartz crystal caused an electric potential that at high pressure resulted in the ultrassound. That is, the mechanical force $F = ma = mv/t = \Delta mv/t$ = Impulse/t resulting in increased electron velocity, therefore electric current.

Electromagnetic wave Flow – In 1884, J. H. Poynting determined for an electromagnetic wave the flow of energy that passed through an area.

So the energy flow Φ_P that passed through an area S_{area} as being expressed by $\Phi_P = (c/4\pi)(E \times B)$, where c is the speed of light, and the electric field and B the magnetic field. And for a certain direction $\Phi_P = (c/4\pi) E_0^2$, and being (*Energy* /volume)$_E$ = $E_0^2 / 4\pi$ came to be obtained the flux vector $\Phi_P = $ (*Energy* / volume)$_E$ c. And the intensity $I_{\Phi P}$ of the vector flow to the maximum value of the field E_0, expressed by $I_{\Phi P} = (E_0^2 / 8\pi)$ c.

Unit Charge Evidence – By electrolysis, a determined quantity of electricity passed in a solution.

Thus for the hydrochloric acid HCl, it released 1.0081 grams of hydrogen and 35.457 grams of chlorine and in the solution of silver nitrate $AgNO_3$ it liberated 107.88 grams of silver, i.e. it was released by the same quantity of electricity the atomic mass in grams of the chemical element. However, with that same quantity of electricity in the copper chloride solution, the copper mass deposited corresponded only to half of the atomic mass expressed in grams. Between 1884 and 1887, Arrherius observed that electrolyte molecules were dissociated, in solution, in atoms or in groups of atoms charged with opposite electricity called ions, with positive ions going to the cathode that was the negative pole and negative ions going to the anode it was the positive pole. It was observed that hydrogen had the valence 1, the sodium chloride NaCl with valence 1 for Na^+ and Cl^-, potassium nitrate KNO_3 with valence 1 to K^+ and NO_3^-, $AgNO_3$ silver nitrate$_3$ with valence 1 for Ag^+ and NO_3^-, magnesium sulfate $MgSO_4$ with valence 2 for Mg^{++} and SO_4^{--}, calcium carbonate $CaCO_3$ with valence 2 for Ca^{++} and CO_3^{--}, copper chlorite $CuCl_2$ with valence 2 for Cu^{++} and valence 1 for Cl^-. So, in the case of a quantity of electricity releasing only half of the atomic mass in grams of copper would be releasing the electrochemical equivalent, as well as releasing the electrochemical equivalent of all other elements. As silver combined with other atoms with valence 1 and copper with greater combination power with valence 2, the released copper was half of the atomic mass in grams for the quantity of electricity contained in a Cl atom$^-$ of 2 Cl^-, that is, the same quantity of electricity NO_3^- to the atomic mass in grams of the silver atom. So the ion charge was proportional to its valence. And these ions that would constitute the electric current through the solution, in the phenomenon of electrolysis, were nothing more than atoms charged with electricity.

And Arrhenius observed that in the electrolysis the components of negative electricity of the atoms, the anions, carried different quanties of electricity in whole numbers, which would indicate the evidence of a unitary electric charge.

Emission and Reception of Electromagnetic Waves – In 1887, H. R. Hertz sought to observe the behavior of electric current induction in insulating bodies. It was used in a primary circuit an electric coil that could produce on an oscillator, with a device consisting of two metal spheres of 3 cm diameter and distance less than 1 cm, with a metallic strip between the spheres where the charges were passed electric and which always emitted a spark, light or radiation from this electric origin. And with a frequency around 3×10^8 cycles/sec it would induce an alternating electric current of that same frequency on a secondary circuit and or on a conductor wire. In the secondary circuit was also installed a micro device similar to that of the oscillator, and that by receiving the induction of electric origin from the conductor wire or directly from the oscillator, could through resonating vibration also produce sparks, light or radiation, that is, a resonator that was shaped like a rim or rectangular. The distances between the primary circuit, secondary and conductor wire were observed so that the sparks of the resonator were perceptible, as they were between 1 to 0.01 mm, more observable by a small click than by the light. One of the results showed that even though there was an insulating block between the primary circuit and half of the ring of the resonator in the perpendicular, there was the passage of the induction of electric origin. He then observed the effect between the oscillator and the resonator which was displaced in several positions near the conductor wire. And it obtained greater sparks in a peak position and no spark in the knot position, with spacing between the peaks or nodes of around 3 meters, evincing an inductive wave, aerial, and also progressive waves that were the furthest from the conductor wire that they transmitted the induction of electrical origin, that is, radiations of electric origin or the electromagnetic waves of Maxwell. And then in 1888 without using the lead wire, in the air, it obtained the propagation of the electromagnetic wave that called the electro-dynamic wave.

In 1890 E. Brandy presented an electromagnetic wave receiver consisting of spiral metal wires. In 1893, R. L. Moura presented the first sound transmission of voice and music by electromagnetic wave of radio, wireless, at a distance of 8,000 meters, in Sao Paulo-Brazil, i.e. wireless telephony. And in 1894, O. J. Lodge presented another type for reception and emission of electromagnetic waves. In 1895 another receiver was presented with two parallel wires by A. S. Popov and G. Marconi, one with connection in the electric power source for amplification of the electromagnetic wave and another connected to the receiver of the electromagnetic wave, i.e.

the antenna. In 1895, Marconi presented the first transmission of telegraphic signal by electromagnetic wave of radio, wireless, at a distance around 100 meters, in proximity to Bologna in Italy, ie the telegraphy wireless. And in 1898 N.Tesla presented the first remote control by radio waves.

Electron Theory – In 1892, Lorentz presented, based on the electromagnetic experiments performed, a theory of electric and magnetic charges, "The Theory of Electrons", and its applications, between 1893 to 1904, for the motion of electric charges that generated the electromagnetics waves and forces between electric and magnetic charges and fields. Thus, he admitted that particles were charged with positive or negative charge and formed the material bodies. These charges called electron when oscillating emitted light at a certain frequency. Then it explained the refractive index of the light dispersion through the frequency. And when it focused on an electromagnetic wave over the electrons, they would vibrate at the wave frequency and reemit radiation. And the electromagnetic wave produced by the electrons by oscillating in a medium, the ether, equating in the medium as the air for the sound and the earth and water for the seismic or maritime wave. In the case of the emission of light being effected by the electrons under a strong magnetic field, the interaction between that field and the electron field should cause variation of the emission frequency of the spectral line.

Electron Theory and Ether – As Lorentz the ether penetrated the matter completely and was the receptacle for electromagnetic energy, the means for all forces acting on the weighted matter and the absolute referential always at rest. So, the energy of the electromagnetic field in the ether results that the ether could exert a force acting on other charges, that is, the ether coming to transmit a resulting force, the ponderomotive force, subsequently named Lorentz force or total electric force. If in an ether site the energy increased by the introduction of an electric charge, the ether could transmit that energy to another site of the ether and act a ponderomotive force on another charge. So it was the state of ether with greater or lesser electromagnetic energy that would exert this ponderomotive force. This ether by being always at rest would not receive any pressure, therefore without speed, acceleration and also no need to define its mass and nor forces applied to it. Therefore, the charge of the electrons could be displaced by the total electric force exerted by the ether, and this displacement would immediately cause a new force to which the particle would be pushed back in the direction of its original position, an elastic force. Thus, from an ether of small density and rigid to a fluid medium, elastic, partially dragged or fully dragged, the ether of Maxwell that permeated, entangled and or penetrated the matter and received the transmission of motion by the matter, and it was the medium that it propagated the electromagnetic wave, until the Lorentz ether that did not receive any pressure of electric charges, but that could exert a force acting on other charges, that is, the ponderomotive force, and a motionless ether without the need to define its mass. But in 1900, P. Drude conceived a motionless ether, but without mass, only the space endowed with certain physical properties. And in 1901, E. Cohn admitted that the ether could be of some mass or even the vacuum.

Electron Theory and Elastic Force – As Lorentz: *"Again, under the influence of elastic forces, electrons can vibrate over their equilibrium positions. Doing so, and perhaps also because of other more irregular motions, they become the centers of waves that travel in the ether and can be observed as light if the frequency is quite high. That way we can respond to the emission of light and heat. On the opposite phenomenon of absorption, this is explained by considering the vibrations that are communicated to the electrons by the periodic forces that exist in a light incident ray. If so the motion of the vibrating electrons does not go into disturbance, but is converted one way or another into an irregular agitation that we call heat, it is clear that part of the incident energy will be accumulated in the body, in other conditions that there is certain absorption. Nor is the only absorption that can be considered for a communication of motion to electrons. That optical resonance, as it can in many cases be admitted, can make also be felt if there is no resistance, so that the body is perfectly transparent. Also in this case will be placed the electrons contained within the molecules in motion, however no vibratory energy is lost, the oscillating particles exert an influence on the velocity with which the vibrations are propagated by the body".*

Electron Theory and its Dimensions – According to Lorentz: *"In the first place, we will assign to each electron certain finite dimensions, but smaller ones that may be, and we'll set our attention in the outer field and also in the inner space in which there is room for many elements of volume and in which the state of the particle can vary from one point to another".*

Electron Theory and Lorentz Force in the Coordinates of the Ether and the Earth – Lorentz presented the expression of the ponderomotiva force, in the coordinates of the ether and the Earth.

Through the displacement of the electric current or dielectric displacement D and the total electric current, the density ρ of the electric charge q with the mass m, the velocity v in relation to ether and v_1 in relation to the motion Earth in the ether, ω the velocity of the Earth in relation to ether, the potential energy of an electromagnetic system in which there is transmission with velocity c in relation to spatial coordinates (x, y, z) in the time t considered real and universal, it resulted in the expression of the ponderomotive force, the total electric force or subsequently the Lorentz force: $f = qD + 1/c\, q\, (v \times B)$, where $qD = f_{electric}$ and $q\,(V \times B) = f_{magnetic}$, and $f = qD_1' + 1/c\, q\, (v_1 \times B_1')$, where q = charge, B = magnetic field, $v_1 = (v - \omega)$ and c the velocity of light for the coordinates (x, y, z) and also (x', y', z') on Earth in a time t regarded as if it were dilated as a mathematical auxiliary quantity, in relation to the actual t time of the ether's reference.

Electron Theory and Coordinate Transformations – In the coordinate transformations of a system (x, y, z) with a v velocity relative to another (x', y', z'), such as velocity and distance, where both change as referentials are at rest or in motion, also the time that is related to velocity and space appears with different values in the coordinate transformations, seeming times apparent in relation to a system considered standard as sidereal time with 24 hours, 60 minutes and 60 seconds.

In 1887 W. Voigt presented for the wave of light transformations of the coordinates of a system (x, y, z) with a v velocity relative to another (x', y', z'), and between 1886 to 1904, J. J. Larmor and Lorentz obtained the definitive transformations: $x' = [1 - (V^2/c^2)]^{\frac{1}{2}} (x - vt)$; $y' = y$; $z' = z$; so $t' = [1 - (v^2/c^2)]^{\frac{1}{2}} [t - (v_x/c^2)]$.

Electron Theory and Particle Dimension Variation – As Lorentz: *"We can understand the possibility of the allowed change of dimensions, if we remember that the shape of a solid body depends on the forces between its molecules, and that, in all probability, these forces are propagated in some way by the intervening ether more or less resembling those in which electromagnetic actions are transmitted through that medium. From this point of view it is natural to assume that, equal to electromagnetic forces, the attractions and molecular repulsions are modified by a translation provided to the body, and this may well result in a change of its dimensions. We could assume that the electron changes the shape and volume of them".*

To a point in the system S moving the coordinates being $(x_P y_P z_P)$ and that point corresponding to the coordinates (x'', y', z') in the system S_0 without translation, they present a correspondence of the existing F electric forces through k_sd dimensions on the axle x and d in y and z: $F(S) = (d^2, d^2/k_s, d^2/k_s)\, F\,(S_0)$.

Compared to two systems of molecules S and S_0 in which the particles have the same effective coordinates, we admit to the molecular forces the ratio expressed on $F(S) = (d^2, d^2/k_s, d^2/k_s) F(S_0)$. Actually, this equation implies that if $F\,(S_0) = 0$, $F\,(S)$ is likewise, so that when, in the system S_0, each molecule is in equilibrium under the actions exerted around it, the same will be true for the system S. So in that there is one more equilibrium position of the particles, we can assert that in the motion system S, the molecules will take from them the corresponding arrangement, in the manner specified by $x' = kd_xP$, $y' = d_yP$, $z' = d_zP$, for the configuration that exists in S_0. Since that x', y', z' are the true coordinates in this system and xP, yP, zP, are the relative coordinates in S, changing dimensions in different directions is determined by the coefficients $x' = \mathrm{k}dxP$, $y' = \mathrm{d}yP$, $z' = \mathrm{d}zP$. And between two reference systems L_1 and L_2 with w velocity on L_2 is: $L_2 = L_1 \mathrm{k}P = L_1[c^2/(c^2 - w^2)^{1/2}]$.

So, in considering the speeds, ω of light in relation to ether and c in relation to the observer, therefore without inertia of the wave of light, Lorentz explained the experiment of A. A. Michelson-E. W. Morley for the contraction of matter, similar to the proposition of G. F. Fitzgerald in 1887.

Electron Theory and the Harmonic Motion of the Electron – The emission of electromagnetic waves such as light was attributed to the oscillations of electric charge, according to Maxwell's magnetic theory and H. R. Hertz. In atoms, these vibrations of electrons, applying expressions of the displacement of the balance position of the charge of mass m, in relation to the elastic force in the atom and on the influence of a magnetic field, resulted in directions for the emission of light with different frequencies. The electron radiation was emitted whenever the p momentum was varying and that it emitted regular vibrations where p was a periodic function, in which the p components for the e charge of the electron were: $p_x = \Sigma ex$, $p_y = \Sigma ey$, $P_z = \Sigma ez$.

Electron Theory and the Harmonic Motion of the Electron with the Magnetic Field – As Lorentz: *"I will present first elementary explanation of the decomposition of the spectral lines found in electron theory, and for which it was even possible to predict certain peculiarities of the phenomenon. We already know that the emission of light is caused by vibrational motions of electric charges contained in the atoms of the ponderable bodies, of a sodium flame, for example, or the luminescent gas in a vacuum tube. The distribution of these charges and their vibrations can be very complicated. But in the production of a single spectral line, we may have a simple hypothesis. Let each atom or molecule contain a single electron, while having a position of equilibrium for which it comes back when removed by an elastic force, as soon as it is moved through one cause or another".*

And for components of the harmonic motion of the electron under the influence of a magnetic field B, the force that acts is $F = e/c$ (v X B), where v is the speed of the electron, e the charge of the electron and c the speed of light. And by applying the electron motion expressions to their mass m to these components, they are obtained their displacements as oscillators in the atom, and resulting frequencies by three directions of the magnetic field: $m (d^2d/dt^2) = f(x) - e[(v/c) \times B]$, d is the position of the electron in the ether at time t by the differential dd and dt; without the presence of the magnetic field: $m\omega_0^2 r = f(r)$, r = orbital radius of electron in atom, $\omega_0 =$

J. P. Hansen

v/r is the angular frequency; with the magnetic field the angular frequency increases to $\omega = \omega_0 [1 + (\Delta\omega/\omega_0)]$, where $\Delta\omega = (\omega - \omega_0)$ and is obtained $\Delta\omega = (e/2mc) B$, in which $\Delta\omega$ is the cause of the new spectrum lines, as resulting from electron radiation interaction.

The elastic force of the ether holds the electrons to the atoms. And in the presence of a magnetic field B, the electrons oscillate in the direction of B with frequency ω_0, and rotating with ω frequency in circular orbits in planes perpendicular to the direction of B. In the direction of B, perpendicular to the spin of the electrons the motion of electrons is uniform in relation to the motion of the magnetic momentum in the direction of B, therefore without issuing radiation. For the observation carried out in the direction of B, the two ω frequency emissions are detected. And with electron rotations, the result would be a decomposition of the spectral lines in more than three components.

As Lorentz: *"A body whose particles can perform free vibrations from certain periods may be able to absorb light from the same periods, that is, the law of resonance. So if in a sodium flame under the influence of a magnetic field there are three periods of free vibrations instead of one, we can expect that the flame can produce in a continuous spectrum three absorption lines corresponding to the periods, and in general, if we want to know what types of light are emitted by a body under certain conditions, we only have to examine the absorption in a beam of light sent".*

Electron Theory and the Dispersion of Light – About the refraction of the light in the Lorentz prism sought its cause.

Through d_x, d_y and d_z displacements, with a constant f for the particle of m mass electron with v speed of vibration, it was represented the elastic force by fd_x, fd_y and fd_z, which pushed back the electron towards its equilibrium position; and in a medium as the prism occurred another resistance constant, g, to any other motion of the electron, in which this resistance could be proportional to that motion of the electron at a time t, $g_x = g(dd_x/dt)$, which would result in obtaining for the coordinated x by the equation of motion: $m (d^2 d_x/dt^2) = [e(E_x + k_p p_x) - fd_x - g(dd_x/dt) + H]$, where e electron charge, E_x electric force in x, H was the force of the external magnetic field, k_p and p_x momentum in x. Since N was the number of electrons per unit volume was obtained $(m/Ne^2) = m_1$, and $(f/Ne^2) = f'$ for the coordinated x, and with $\alpha = f' - k_p - m_1 \mathcal{V}^2$ where $\mathcal{V}$ it was the frequency of vibration of the electron, being obtained E_x = function of (α).

So the bigger $\mathcal{V}$, the smaller the field would be, and therefore the smaller the power on the electron. However, this electrical force on the electron was much smaller to correspond to the absorption of light in the prism.

And as Lorentz: *"Then we shall look for some other explanation. It occurred to me that it can be found in the assumption that the vibrations inside a ponderable particle that are aroused by light-wave incidents, may not always remain without some disturbance. It is conceivable that the particles of a gaseous body are vibrated so deeply by their mutual impacts that any regular vibration that has been effected is transformed into the disordered motion that we call heat. The temperature elevation produced in this way must be due to a part of the energy of the incident rays, so that there is a real absorption of light. It is also clear that the accumulation of vibrating energy in a particle, in the period of the vibrating electrons and the incident light, would never be over, and would be kept within limits of disturbance by the collisions of the electrons, in the same way it could be a resistance in the ordinary sense of the word".*

As Lorentz: *"Working out this idea, we can still use the formulas preceded that we establish, provided if we understand by a coefficient g of the equation of motion, the quantity $2m/\tau$ in which τ it is the average time during which vibrations in a particle can occur without disturbance. For, the vibrations of the electrons are again in disturbance by the occurrence of impacts, and with N the total number of particles, n_c being the number of collisions per unit of time, then $N/n_c = \tau$ and by equation of electron motion and by $f = mn_0^2$ is obtained $g = 2m/\tau$. According to the previous idea, the interval t, in gaseous bodies, should be equal to the interval of time that runs between two successive encounters of a molecule. But, it is found that the value of t deducted from experimental data is less than the interval between two encounters".*

As Lorentz: *"We must conclude that there are causes within a molecule, whereby the regularity of vibrations goes into disturbance in less time than it would be by the molecular impacts. We may not have elucidated the phenomenon of absorption satisfactorily, so its true cause still remains to be discovered".*

Electron Theory and Refractive Index in Ether – As Fresnel the light was partially dragged into a medium. And as Lorentz the electric forces of the referential charges in the motion were those that dragged the light in the medium in an all-penetrating ether and at rest, and was the dielectric constant, permittivity, dielectric capacity, specific capacity electrostatic or at present the electric constant, That increased when a body was moving in the ether. Also, as the ether was all penetrating into the bodies, it could not be dragged.

Light Pressure – In 1892, according to Maxwell, the light exerts a pressure, through a plane at right angles for both electric and magnetic fields, as a ponderomotive force, (Lorentz force), which would be named light radiation pressure. In 1899. P. Lebedev observed that luminous rays, focusing on a thin metallic disc suspended in the vacuum, caused its rotation, attributing to the light pressure. In 1903, E. F. Nichols and G. F. Hull attributed these rotations as fluxes of change of electromagnetic momentum. And in 1909, Poyting and Barlow observed that there was a radiation pressure as an indentation against the source, produced in the emission of light.

Electric Field and Spectral Lines – In 1845, Faraday had obtained the polarization of the light under magnetic field and in 1896, P. Zeeman in a higher intensity magnetic field detected a magnetic influence over the emission of the sodium spectrum, modifying the emission frequency of the radiation electromagnetic, while in 1902, C. Runge and F. Paschen obtained more than three spectral lines. And between 1913 and 1914, J. Stark, through a strong electric field applied to the radiation emission material, obtained modifications of spectral lines of chemical elements of the emission atoms.

Energy, The Central Concept of Physics – In 1840 Joule came to establish the equivalence of work or mechanical energy with the energy E_Q of the heat Q, through electricity.

Being observed that the electric current i resulted in dissipation of the energy $E_{\Delta V}$ in heat Q at a time t:
$E_{\Delta V}/t = i\,\Delta V$, and being $\Delta V = Ri$, where R was the electrical resistance, was obtained $E_{\Delta V} = R\,i^2\,t$, and turning this mechanical energy into heat energy Q, was obtained $E_{\Delta V} = R\,i^2\,t = JQ$. Later with greater accuracy was obtained $J = 4.18 \times 10^7$ erg/calorie in the CGS system in which erg is the mechanical and electric power unit.

In 1847, Helmhotz stated that the mechanical energy, the ability to work, force x distance, was the central concept of physics, and that all kinds of energy: electrical, chemical, thermal and gravitational, could be reduced to an equivalent mechanical energy, to unify mechanics with heat, thermodynamics, chemistry, electricity and magnetism. Between 1845 to 1850 Joule showed beyond thermal equivalence, also that of electrical, magnetic and chemical energy. In 1846, Groves showed the conservation of all energy.

Essence of the Forces- In 1834, Faraday sought the essence of the forces and had the realization that the forces that produced mechanical motion, electricity, magnetism, heat, light, were of a single cause, emphasized between 1839 to 1847 by Mohr, Grove and Liebig.

Origin of All Forces – In 1892, as Lorentz: *"All forces exerted by one particle or another, all molecular motion and gravity itself,* *are transmitted by some means by the ether, so that the tension of a stretched rope and the elasticity of an iron rod must find its explanation in what occurs in the ether between the molecules. Therefore, as long as it is possible to admit that one and the same means are capable of transmitting two or more motions by broad different mechanisms, all forces can be observed as connected more or less intimately with those studied in electromagnetism. For the present, however the nature of this connection is entirely unknown and must be continued to say of many kinds of forces without it being possible to point to the origin of them. The electrons are subject to unknown forces. Such are, for example, the forces by which the electrons in a dielectric are driven back to equilibrium positions, and the forces that are at stake when an electron moving in a metal has its path modified by an impact against an atom. "*

Separate Mass Energy- As Poynting, in 1884, an electrostatic charge forming between plates a potential energy difference V, acted forces and fields electric E, magnetic H or magnetic induction B.

With the charge density $Q_c = Q$/area, the energy density $d_{energy} = E_{energy}$ / volume. and the acceleration charge propagating an electromagnetic wave at a given time and to the area unit resulted in the electromagnetic energy $E_{em} = d_{energy}$ c = (E X B/μ_0) =E X H, which was the Poynting vector, where μ_0 = magnetic permeability of vacuum = 4 π 10^{-7} newton/ampere^2.

Therefore, as Poynting, the electromagnetic energy was a propagation with the c speed of light, and as it was separated from the mass of the electric conductors, then it was independent of action at the distance by the masses, contrary to what occurred with the universal gravitation in which a gravitational energy acting on a mass of a body instantly appeared in the mass of another body that was at any distance as Newton.

Force Electromagnetic-Mechanical and Motion – In 1861, Maxwell had presented that the electromagnetic force was due to the motion of cloaked masses that would be by the definitions of Maxwell's force: F = q E = mv/t, where F was the force, q the electric charge, m and v the mass and speed of the q charge particle. On 1894 , for H. R. Hertz this motion of cloaked masses should also, in addition to electromagnetic forces, explain gravitational force, distance and mechanical actions. But, if he was able to eliminate those mysterious forces coming from the mechanics, it would be possible to entirely avoid its entry into the mechanics. For H. R. Hertz if a force applied through a rope caused the circular motion of a stone, it was inconceivable that that motion would cause the opposing force. In the laws of Newton's motion, force was the cause of the motion. The force was present before the motion. How force could be consequence of motion? So he proposed principles of mechanics, only with mass, space and time. And the force would be the consequence of these variables, in which the force would produce motion in a body and that body would also produce force that would cause motion. The conceptual obscurities involved in such considerations would have to be clarified with the mass and speed according to mechanics principles. These variables would appear to be like the mv motion quantity.

Electric Force and Mass Increase – In 1881, J. J. Thomson observed that a sphere, with electric charge of m mass, under the action of an electric force caused the motion of this charge and also a magnetic field around the charge, forming an electromagnetic field and resulting in an electromagnetic energy. And admitted a dissipation by friction of energy in the medium as an insulator, then it should be equivalent to a mass increase of the motion charge. As J. J. Thomson: *"As if the mass increased"*.

Between 1886 and 1904, as Lorentz, by the coordinate transformations of a system (x, y, z, t) with mass m_0 and v speed in relation to the system (x', y', z', t') with m mass, was obtained m = (v function) m_0, being in longitudinal equivalent to $m_{longitudinal} = m_0 / [1 - (v^2/c^2)]^{1/2}$ and $m_{transversal} = m_0 / [1 - (v^2/c^2)]^{3/2}$. In 1906, as Lorentz, an electromagnetic force was a function of the E electric field and of the B magnetic field, F = function (E, B), resulting $F = [(1/c^2)] (f (E, B)]/t$, where $G = [(1/c^2)$ (function (E, B)]. And that at 1903 M. Abraham named it an electromagnetic momentum. So, $F = dG/dt$ and consequent $G = (m_{em}w)$, where w was the speed of the electron.

And as Lorentz: *"If it is confined to a rectilinear motion of an electron, the notion of electromagnetic mass may be derived from this electromagnetic energy. In fact, the electromagnetic energy is higher for an electron motion than for an immovable electron. Therefore, if the particle is put in motion by an external F force, it is produced the kinetic energy ½ m_0w^2, but in addition to the part of the electromagnetic energy that is due to speed. The effect of the field will then be greater than a quantity of energy that is required than if it had with a material particle still m_0; It's the same as if the mass was greater than m_0".*

And Lorentz sought to explain the cause of the mass increase: *"The mass of a free electron is a function of speed, so if the corpuscle has already the translation of the body in which it belongs, the force required for the modification of the speed will be altered. Now that it is extended to the molecular interactions of the electric forces of the bodies, it may be possible to imagine a change of masses of the molecules caused by the motion of translation, and even if it would prove necessary to conceive two* *different masses, a m$_{longitudinal}$ longitudinal mass with which it must find when it is considered the accelerations parallel to a coordinate axis OX, and another m$_{transversal}$ transversal mass that comes to play when it is with respect to an axis OX or OZ".* That is, although the coordinate transformation presents the mass increase in function of speed, this increase would be as Lorentz due to a real electromagnetic force applied, that is, the electromagnetic momentum (m$_{em}$v)/tempo.

Electromagnetic Mass – According to G. Mie 1898, about the inertial electromagnetic mass in motion, no one had manifested itself. And as Mie a fluid was carried by transport and energy conduction, and the deformation of the bodies ascended from a flow of energy. All energy exchanges would be the consequences of actual energy currents. But, Heaviside came to manifest in 1899, in which the motion of a sphere of r radium and charge q caused an additional energy of electromagnetic field, which resulted in a resistance equivalent to an inertial effect of an electromagnetic field, that is, a force of inertia of an electromagnetic mass. Although Mie mentioned that no one had spoken about inertial electromagnetic mass, H. Poincaré in 1900 stated that all the inertia of the mass was of electromagnetic origin and that the inertial mass would be an appearance. And that the m mass was related to the energy E by m = (E/c^2) (c/v), where c was the speed of light and v the speed of the charge. For Poincaré, in 1900, the electromagnetic energy was a fluid endowed with electromagnetic inertial mass m$_{em}$, in which m$_{em}$ = (E/c^2) (c/v). And that this energy fluid was destructive. In 1907, N. R. Campbell concluded that the atomic mass was of purely electron origin, while in 1947 G. Racah again concluded that the mass of the electron was of electromagnetic origin.

Increased Mass and Energy of the Electromagnetic Field – In 1900, also Wien, who researched the black body irradiation, concluded that it was more promising for the future development of theoretical physics to consider electromagnetic equations as the basis for the derivation of the laws of mechanics. The increase in electromagnetic mass m$_{em}$ was obtained by the energy E_{em} of the electromagnetic field, at first approximation, as E_{em} = m$_{em}$ v^2/2, where v was the speeds of the charge. The emission of radiation from a black body indicated that the higher the frequency or shorter wavelength of the radiations, the greater the quantity of energy obtained. However, in the ultraviolet range, a minimal energy was obtained in this radiation range, a sharp energy drop, the ultraviolet catastrophe. And M. K. E. Planck in 1900 presented the quanta of energy $\mathcal{E}$ = h$\mathcal{V}$, where h was a constant and $\mathcal{V}$ the quanta frequency, in which a minimal energy density, but with quantum high frequency explained the ultraviolet catastrophe.

Mass Increase and Electromagnetic Field – In 1902 W. Kaufmann through experiments on the deflection of electrons by electric and magnetic fields observed an increase of the electron mass by the ratio between charge and electron mass. The higher the electron speed was the increased electromagnetic mass and even exceeded the m_e electron initial mass, concluding that all electron mass was electromagnetic. This experiment received the approval of Abraham in 1902, Poincaré in 1908, A. W. Conway in 1907, D. F. Comstock in 1908, and A. H. Bucherer that corroborated the Kaufmann experiments stating that the electromagnetic mass should be replaced by the mechanical mass of atoms. Abraham, on the basis of Kaufmann's experiments, presented that the mass of the electron originated in the electromagnetic field.

Electromagnetic Force and Momentum – By Kaufmann's experiments in 1902, Abraham presented the concept of electromagnetic momentum of the field $\mathfrak{M}_{In}$, in which $\mathfrak{M}_{In} = m_{em}v$, where m_{em} was the electromagnetic mass and v the speed of the charge.

And resulting in the electromagnetic force $F_{em} = d\mathfrak{M}_{em}/dt = d\mathfrak{M}_{em}/dv$. $dv/dt = m_{em} w$, where w was the acceleration.

Energy and Mass – In 1904, F. Hasenöhrl, applying the laws of motion to a box with reflective walls containing electromagnetic radiation from the light, presented that the electromagnetic energy behaved as if there was a mass m proportional to the energy E, that is, $m = E/c^2$, where c was the speed of light. And as a result, $E = mc^2$ as Hasenöhrl, before Einstein presented in 1905 the theory of relativity.

Ether and the Michelson and Morley Experiment – Since Magnes when observing the attraction or repulsion of stones, magnetite, was connected to a medium that acted in this attraction or repulsion, similar to what occurred with the attraction of electrified bodies, and in 1873, Maxwell and between 1881 and 1889 J. J. Thomson and Heaviside they considered for electrical and magnetic phenomena a medium, an ether that permeated, entangled and or penetrated the matter and received the transmission of motion through the matter, and it was the medium that propagated the electromagnetic wave, as the air propagated the sound wave, the water propagating the wave of the sea and the earth the seismic wave. And in 1892, Lorentz through its electron theory, considered that the ether penetrated the whole matter and was the receptacle for electromagnetic energy, the means for all forces acting on the ponderous matter and the absolute referential always at rest. So, the energy of the electromagnetic field in the ether resulted in the ether being able to exert a force acting on other charges, that is, the ether coming to transmit a resulting force, the Lorentz force. If in an ether site the energy increased by the introduction of an electric charge, the ether could transmit that energy to another site of the ether and act the Lorentz force on another charge. So it was the state of the ether with greater or lesser electromagnetic energy that would exert this Lorentz force. This ether by being always at rest would not receive any pressure, therefore without speed, acceleration and also no need to define its mass and nor forces applied to it.

A lake, with the water completely leveled, would remain leveled if the Earth was fixed or in motion, and when a stone was thrown into that water the formed wave would be propagated at the same speed in relation to the lake, it was the Earth fixed or in motion. That is, the speed of the wave it is independent of the speed of the source, because it was only the water that propagated the wave. Also, the seismic wave by the earth and the sound wave through the air were independent of the velocity of the emitting source. A violin rope when triggered moves the air moving with the sound speed of $\approx$ 330 m/s, regardless of whether the violinist is stopped or at speed. For it is the air that transmits the vibration, the sound wave. The light showed a constant speed of $\approx$ 300,000 km/s in relation to vacuum-space-ether, and would come out in all directions, with measurements on Earth. And in that medium of the ether, which was immovable and all penetrating in the bodies, the Earth or any source that emitted the wave of light passed free through the ether that was who would actually propagate this wave of light. Thus, little mattered the speed of the emitting source of the wave of light.

With the waves of the sea, earth and air, the Earth in its motion in vacuum-space-ether, "carried" the sea, the earth and the air. Then, the propagations of the waves of the sea, the earth or the air, propagated in these means without depending on the velocity of the Earth in any direction, was the Earth standing or in motion in vacuum-space-ether. In the case of the light wave in the ether, the Earth in its motion in the space-vacuum in the ether, did not carry the ether, but the speed of the wave of light remained the same in relation to the ether. However as the Earth was "fleeing" from the wave of light in the ether, the light wave would reach a target, which is on the motion Earth, in a larger time, so its velocity would be lower compared to that target.

Then, in the case of the wave theory of light, the ether by being always motionless did not carry the light, but only propagating the wave, or as Fresnel was only partially. In the seismic wave, the epicenter of the earthquake or tsunami transmits the seismic vibration, but it is the earth that propagates. In the sea wave, the wind propels the water, but it is the water that propagates the wave.

In the sound wave, the vibrating emitter, that is, the rope, does not carry the air, but just shifting the air which transmits the vibration to the continuity of propagation through the air while the rope returns to its initial position. Figures 14 and 15.

Fig. 14 – If the Earth was stopped in vacuum-space-ether, the sea wave by the ocean, seismic by the earth, sound by the air, and light or any other electromagnetic by the ether, all these waves would be received in a mirror with the respective speeds of 5 to 100 m/s of the sea waves; 200 to 5,000 m/s of seismic waves; ≈ 333 m/s of sound waves and ≈ 300,000 km/s of electromagnetic waves. Adapted from the clipart of *Editora Europa*.

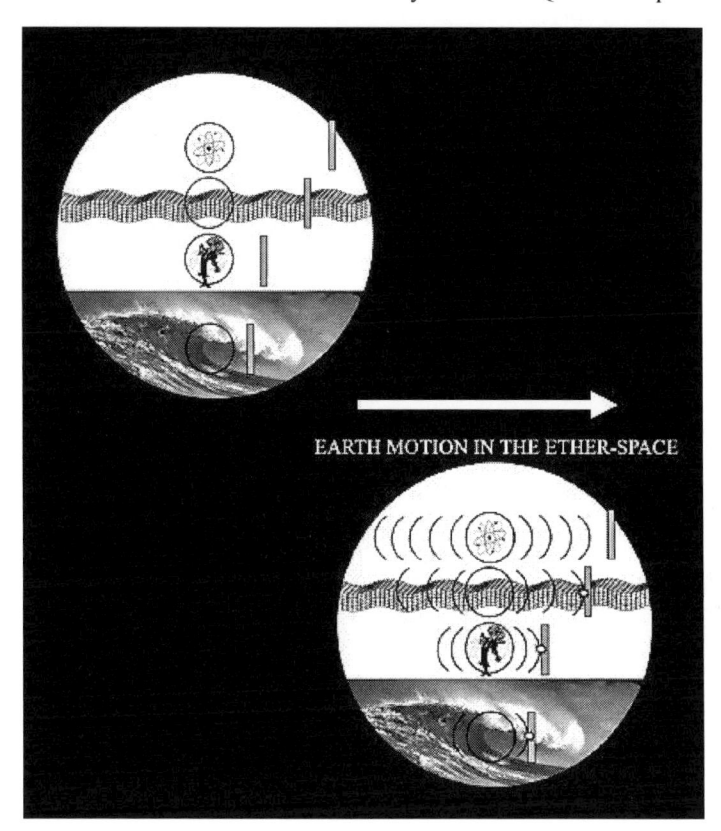

Fig. 15 – The Earth moves in vacuum-space-ether carrying the sea, the earth and the air, while the ether as immobile in space and permeable in the bodies resulted in the Earth motion in space without carry or dragging the ether. In this way, the mirrors that by the motion of the Earth in vacuum-space-ether "fled" from the position of their emission origins, would detect these waves with their characteristic speeds of 5 to 100 m/s of sea waves; 200 to 5,000 m/s of seismic waves; ≈ 333 m/s of the sound waves. However, since the Earth did not carry the ether, the electromagnetic wave would reach a larger time the mirror that "fled", that is, its final speed to reach the mirror would be slightly smaller than its speed of ≈ 300,000 km/s. Adapted from the clipart of *Editora Europa*

Also by wave theory, the speed of a boat does not change the speed of the wave that is transmitted by the water and not by the helix of the boat that only increases the speed of the boat. A stumble of horses in speed propagates the wave across the earth at the same speed if the horses were stopped beating their paws, for only the earth transmits the wave. A locomotive stopped and another in motion, the whistle emitted is propagated at the same speed, because it is the air that transmits the vibration. And a locomotive stopped and the other in motion, when they lighthouse on at the same instant, the speed of light by propagating through the ether and not by the locomotive, is the same for both cases, therefore independent of the source velocity in any direction. Figure 16.

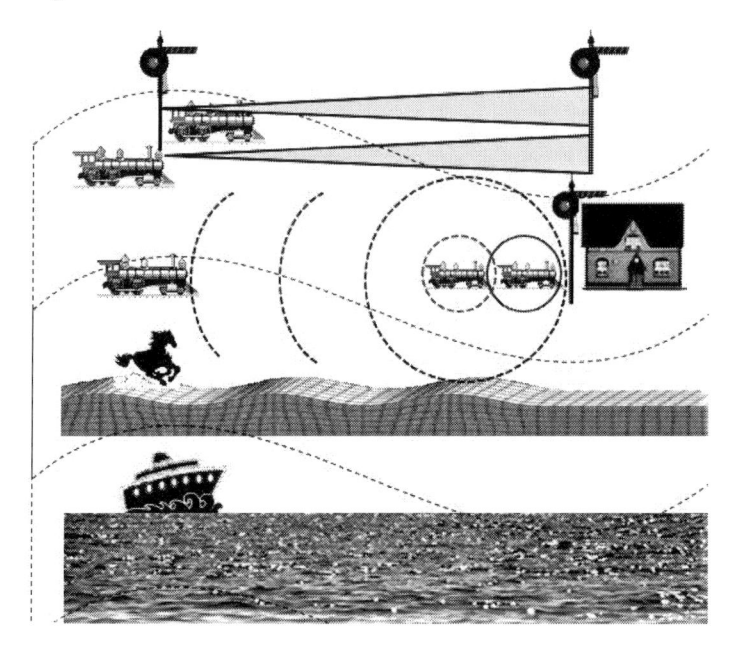

Fig. 16 – By wave theory, the speed of the waves is independent of the speed of the emitting source. The speed of the boat does not increase the speed of the wave, because what transmits the wave in the sea is the medium of the water. A stumble of the motion or stopped horses transmits the wave of vibration at the same speed, as it is the earth that propagates the wave, only changing the intensity of vibration. A locomotive stopping or motion transmits the sound of a whistle at the same speed as the sound wave, in which the waves approach resulting in larger intensity when the locomotive is on the motion. And one locomotive stopped and the other in motion as they turned on the lighthouse at the same instant in a signal marker, the speed of light of the two locomotives are equal, because what transmits the light is the ether and not the locomotive, then received at the same time in a signal marker. Adapted from the clipart of *Editora Europa*.

In the experiment of Michelson-Morley, in 1887, as the Earth did not drag the ether, and the speed of the light wave was $c \approx 300{,}000$ km/s in the ether, then if placed a mirror to receive the wave of light in the direction that the Earth in motion in the vacuum-space-ether "fled" the wave of light, it would be the end velocity of the light wave in the reference system Earth-Sun-Interferometer. In that case, how would there be a larger time for the wave of light to reach the "running mirror", the end velocity of the light wave would be less than the velocity of $\approx 300{,}000$ km/s. Then, as Michelson-Morley, to the light with c velocity go through the distance arm D of the interferometer in the direction of $v = V_{ESI}$ velocity of the Earth-Sun-Interferometer system in different directions in the ether, traversing the distance S, at the same time as it travels D, as seen from the space the path would be $D' = (D + S)$ for the T_{going} going time, so $D = (D' - S) = [(T_{going}\,c) - (T_{going}\,V_{ESI})]$, then $T_{going} = D/(c - V_{ESI})$ in which the Earth "flees" from the light. On the back, $D' = (D - S)$, $D = (D' + S) = [(T_{back}\,c) - (T_{back}\,V_{ESI})]$, then $T_{back} = D/(c + V_{ESI})$ in which the Earth goes against the light. Diagonally, as Michelson-Morley the light with velocity c diagonally at a $T_{diagonal}$ time would follow the path $(D^2 + S^2)^{1/2}$, that is, $c = (D^2 + S^2)^{1/2}\,T_{diagonal}$. Therefore, the three paths and the three times of the wave of light in space would be different, and would cause fringe interference.

As Michelson-Morley, let c = velocity of light; v = velocity of the Earth in its orbit; D = arm distance; $T_{going} = D/(c - v)$, $T_{back} = D/(c + v)$. Adding these two, $T_{going} + T_{back} = 2Dc/(c^2 - v^2)$. The distance travelled in this time of going and back = $2Dc^2/(c^2 - v^2) = 2D(1 + v^2/c^2)$, up to 2nd order. The length of the diagonal path, according to the light admitted with its velocity c, that enters in the telescope inclined in the Earth (aberration stellar) due to the velocity of the Earth in space-ether, that is, $= 2(D^2 + v^2(D/c)^2)^{1/2} = 2D(1 + v^2/2c^2)$, up to 2nd order. Therefore, the difference between the two paths = Dv^2/c^2. If, the whole apparatus is turned through 90°, the difference would be in the opposite direction. Then, the total displacement of the interference fringes should be $2Dv^2/c^2$. Using $v \sim 30$ km/s, $c \sim 3 \times 10^8$ m/s, the expected displacement = $2D \times 10^{-8}$. The displacement was estimated to be $4Dv^2/c^2$, a factor of two higher, with D = 11 m $\sim 2 \times 10^7$ wavelengths of yellow light and hence the displacement to be expected was 0.4 of the distance between interference fringes. Michelson and Morley estimated that their interferometer was capable of measuring a shift of about 0.01 fringe. However, the result was null or very distant from the calculated interference. Figure 17.

J. P. Hansen

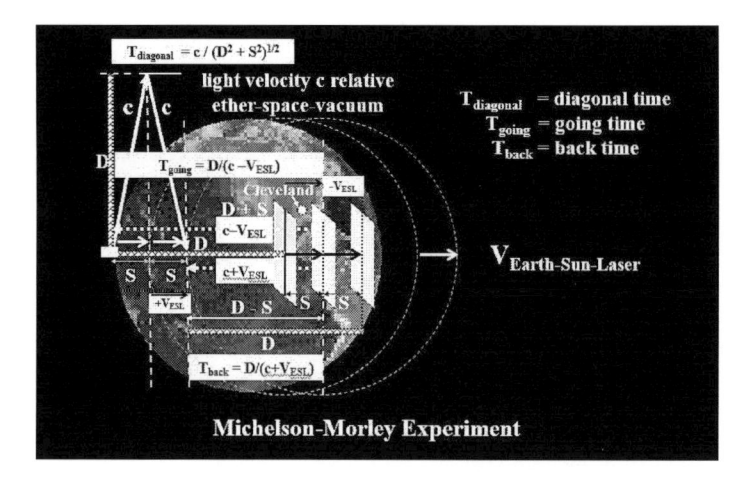

Fig. 17 – In the Michelson-Morley experiment the three paths and the three times of the light wave in space would be different, and would cause fringe interference. Earth clipart from *Editora Europa*.

XIV. THE CATHODE RAYS AND THE DISCOVERY OF THE ELECTRON
1745 to 1911

Stored Charge – In 1745, Musschenbroet discharged electric charge from an electric conductor, in water inside a sphere or bottle at Leyden University, obtaining storage of these charge in this bottle.

Atmospheric Click – In 1751, Franklin noted that when an electric charge was removed from a sphere it emitted a miniature ray of light and a click, and it was equivalent to the rays emitted in atmospheric discharges.

Electrical Discharge in Tubes – In 1835, Faraday observed for the first time the electric discharge in low-thinning gases in tubes with cathode and anode. And that presented a greenish luminosity inside the tube.

Fluorescence and Phosphorescence – In 1852, Stokes called fluorescence this green light emission in tubes with cathode and anode, which appeared during electric discharge. The phosphorescence occurred only after a light emission was ceased, for example, the sunlight.

Electrical Discharge Diverted by Magnet – In 1858, J. Plücker noted that the greenish luminosity within a tube, with a certain vacuum, was deflected by motion of a magnet, so the electric discharge was diverted in its trajectory. And in 1869, J. Hittorf showed that this displacement of electric discharges into tubes was in a straight line, looking like shaped by particles.

Cathode Rays – Between 1871 and 1876, E. Goldstein observed that the electric discharge came from the cathode, the negative pole, to the direction of the anode and reaching the inner surface of the glass tube, showing the greenish luminosity, being named cathode rays. Therefore, electricity flowed from the negative pole, contrary to what Franklin had indicated in 1752.

Rays of Radiant Matter – In 1880, W. Crookes, using a glass tube with a larger vacuum, observed the sharp deviation of the cathode rays under the presence of a magnet, concluding that they were rays formed of particles of a type of negatively charged gas, and named rays of radiant matter. According to the experiments of electromagnetism, an electromagnetic field was connected to the electric charge, so there was an interaction between a magnet and the electric charge. However, the light being considered, as Maxwell, as an electromagnetic field, crossed an electric and or magnetic field directly applied to the light without any deviation, so the light was exempt of charge. So, the conclusion of Crookes was that the cathode rays were formed of electric charge particles by the deviation of the magnet and not an electromagnetic wave that was free of charge.

Wave Rays – In 1882 to 1883, Hertz, Goldstein, Hittorf. G. H. Wiedmann and P. E. Von Lenard using glass tubes with a minimum vacuum observed that a magnet did not divert the cathodic radius. And as the cathodic ray crossed gold leaves and other materials, they concluded that it was a kind of wave that circumvented these materials.

Unit Charge of Electricity –Between 1884 and 1887, Arrhenius observed that electrolyte molecules were dissociated, in solution, in atoms or in groups of atoms charged with opposite electricity called ions, with positive ions going to the cathode which was the negative pole and negative ions going to the anode it was the positive pole. And the ions that would constitute the electric current through the solution, in the phenomenon of electrolysis, were nothing more than atoms charged with electricity. And Arrhenius also observed that in electrolysis the components of negative electricity of the atoms, the anions, charged different quantity of electricity in whole numbers, which would indicate the evidence of a unitary electric charge and, therefore, particles.

Positive Rays – In 1886, Goldstein noted that in perforated channels, they flowed the cations of atoms of positive charges of rarefied gases in the cathode ray tubes.

Charge and Particle Mass – In 1890, A. Schuster established a relationship between charge and mass for the permitted particles of the cathode rays.

External Rays of Cathodic Tube – In 1894, experiments by P. E. Von Lenard, Goldstein and Hittorf with cathode rays, showed that in addition to these rays sensitize photographic plates it was also observed that substances that were near external to the cathode ray tube were bright. Also, Lenard noted that, using an aluminium plate window with 0.0026 mm thick on the wall of the cathodic tube that was kept the vacuum, around the window was detected a discharge of the electroscope, and that Lenard interpreted as cathode rays coming out of the aluminium window and electrifying the atmospheric air.

Particle Rays – In 1895, Perrin, when sending cathode rays to a cylinder, observed that there was storage of a large quantity of negative charges in that cylinder, even if it ceased to send these rays, so the evidence that they were negative particles. If cathode rays were only a wave, a continuous decrease of vibration energy would occur, until a perceptible motion ceased.

Power of Radiation, Mass and Electric Charge Radius – According to the electromagnetism of Maxwell and Lorentz, an electromagnetic wave propagating in one direction, its energy interacted with particle of electric charge that moved with velocity v through mass m with charge Q in the field E and B.

And resulting in force $F = [(q\ E) + q(v \times B)]$, at that momentum p in time: $(dp/dt) = [d(mv)]/dt = q[E + (v \times B)]$. And for the c velocity of the electromagnetic wave in one direction, the energy of the electromagnetic wave absorbed by the particle at a given time was expressed by $(dE_{energy\text{-}wave} / dt) = [(dp/dt)\ c] = (dp/dt)\ c$, and power P_{power} of the wave: $P_{power} = E_{energy\text{-}wave} / c$.

In 1897 Larmor came to determine the intensity and potency of radiation of an electric charge q with velocity v traversing a distance d of a component of transversal electric field $E_{transversal}$ in a time t_v, em que $d = vt_v$. And in a time $t_i = v/a$ occurred an a acceleration of the electric charge q emitting electromagnetic radiation of c velocity in the radial component $E_{radial} = k'\ (q/d_r^2)$, where $k' = 1$, where in that time t_i the radiation traveled a radial distance $d_x = ct_i$. And during the time t_v the radiation with c velocity traveled a distance d_r in the radial component of the field, this is $t_v = d_r/c$. So an accelerated charge emitted an electromagnetic wave equal to an oscillator. So, $(E_{radial}/E_{transversal}) = (v\ t_v\ sen\alpha) / (c\ t_i)$, resulting: $E_{transversal} = (a\ sen\alpha\ /c^2) = (q\ a\ sen\alpha\ /c^2 d_r)$ e como $I = (E_0^2/8\pi)c$, the radiation intensity was obtained: $I_\alpha = c/4\pi\ E^2_{(\text{transversal average})} = (q^2 a^2_{(\text{average})}) \text{sen}^2\alpha / (4\pi c^3 d_r^2)$, and $I_\alpha = (q^2 a^2_{(\text{average})}) \text{sen}^2\alpha / 4\pi c^3 d_r^2)$. And the power P_{power} of the electromagnetic wave irradiated in an area dS, $P = \int I\ dS = (2 q^2 a^2_{(\text{average})} /3c^3)$. Being that $F = ma = E_0\ q$, and as by the c transversal component $E_{\text{transversal}} = (q\ a\ \text{sen}\alpha/c^2 d_r)$, was obtained: $E_{\text{transversal}} = (q^2\ E_0 \text{sen}\theta/ d_r m\ c^2)$. And by the intensity $I_\alpha = (q^2 a^2_{(\text{average})}) \text{sen}^2\alpha /(4\pi c^3 d_r^2)$ and the incident intensity $I = (E_0^2/8\pi)c$, was obtained: $(I_\theta / I) = q^2\ E_0\ \text{sen}^2\theta /m^2 c^4\ d_r^2\ E_0$ and in sequence the radiated power $P_{\text{power}} = (8\pi/3)\ (q^2/mc^2)^2\ I$, in which $P\ I / I = \delta$ it was the shock section $= (8\pi/3)\ (q^2/mc^2)^2$. And being a section sphere πr^2, $\pi r^2 = (8\pi/3)\ (q^2/mc^2)^2$ and $d_r = 8/3\ (q^2/mc^2)$, and it would be necessary to get the value of the charge q of the mass of the electric charge to calculate d_r

Charge and Mass of the Electron – In 1897, E. Wiechert through the deviation of electric charges in magnetic field obtained a ratio q/m between charge q and mass m, which was much greater than the ratio q_H/m_H between the q_H charge of the ion hydrogen and its mass m_H, that is, the q mass should be much smaller than that of the hydrogen and as the rays of the charges were diverted by this magnetic field different from the electromagnetic waves that did not show deviation when they passed through this field, therefore without charge, then being seen as particle, and independent of the atoms of matter, or only as atoms of electricity as Helmhotz.

In 1897, a month after Wiechert, came J. J. Thomson to carry out experiment of electric charges in magnetic field and also in electric field.

Being $F_{\text{electric}} = qE = ma$, in which q is the charge, m the mass of a particle that receives the electric force of the field E, a = charge acceleration; $F_{\text{magnetic}} = q(v \times B)$, v = charge velocity, B= magnetic field, and a force $F_{\text{circular}} = mv^2/r = qvB$, where r = action radius of the charge, so $mv = qBr$ and $r = mv / qB$. And for a beam of charges between cathode and the anode, crossing an electric field to the end of the cathode ray tube, the distance travelled is $x = v_x t$, and for the vertical acceleration was obtained from: $F_{\text{electric}} = ma_y = qE$, so $a_y = qE/m$. Being that the hourly equation of the accelerated motion was $d = (d_0 + VT + \frac{1}{2}\ at^2)$, where d and d_0 = distances, so: $y = \frac{1}{2}\ (qEt^2/m) = (qEx^2)/(2mv_x^2)$, and with L = length of plate and D = length of the plate to the end of the tube, the total deflection was $y_E = (qE)/v_x^2\ [(L^2/2) + LD]$. And with $r = mv_x / qB$ and the equation of the circular trajectory $r^2 = [x^2 + (r + y)^2]$, $r = (-x-y^2)/\ 2y \approx (-x^2) / 2y$, was obtained $y = - qBx^2 / 2mv_x$ and the total magnetic deflection $y_B = - [(qB / mv_x)(L^2/2 + LD)]$. For the balance between electric and magnetic force $F = (qE - qv_x B) = 0$, so $v_x = E/B \approx 1/10$ of the c velocity of the electromagnetic wave. J. J. Thomson obtained for the electron e charge, $(e/m) \approx 1{,}75 \times 10^7$ emu/g or $\approx 1{,}75 \times 10^8$ coulombs/g.

And it was obtained q/m for any material of the cathode and residual gas, meaning that the cathode rays were made of a single type for all chemical elements, and that were removed from the atoms which then would be divisible.

Measurement of the Electron Charge by Drops of Water – In 1874 Stone had estimated the charge of the electron through the total quantity of charge that in the electrolysis was deposited in the electrode.

In the case, f_Q = constant factor representing the charge placed on the electrode by molecule-gram or atom-gram = 9.65 x 10^4 coulombs or 1 faraday in current unit. So $e = (f_Q/N_A)$ was the minimum charge with the value determined by G. J. Stoney of 10^{-20} of the quantity of ampere electricity by time, that is, the quantity of electricity that passed each second in a current of an ampere, ie 3 x 10^{-11} of the CGS electrostatic unit and since 1985 as coulomb.

In 1898, C. T. R. Wilson had observed that the ions formed condensation of oversaturated water vapor. In the electrolysis the value of the electron charge could be estimated by the relationship between the quantity of electricity and the mass deposited in the electrode through the molecule-gram or atom-gram with the number of Avogadro, resulting in the elemental charge. Then, in a droplet of water, with the measure of its mass was determined the charge of that droplet. And by the relationship with the gram molecule and the number of Avogadro would be obtained the value of the elementary charge, or the value of multiple charges between several droplets of very close masses and consequent value of the elementary charge.

The mass of the droplet could be measured by the velocity of the drop in gravitational drop and by the difference between an acid that absorbed the droplet that in 1897 J. S. Townsend obtained 0.9 to 1.0 x 10^{-19} coulombs; By the velocity in gravitational drop and involving measurements of electric conductivity of the air and the temperature at which J. J. Thomson obtained 1.1 to 2 x 10^{-19} coulombs; By the velocity in gravitational drop and electric force in which H. A. Wilson obtained 1.03 x 10^{-19} coulombs.

Negative and Positive Charges – In 1900, J. J. Thomson formulated the hypothesis that in the electrical conductors the negative charges were the particles of the cathode rays and the positive charges were fixed on the atoms of the conductors. And in 1906, it presented a juncture that the ether receiving external vibration transmitted this vibration to the electron, contrary to the Lorentz electron theory.

Electron Radius – The electron radius could be obtained according to its charge $q = e$, mass m_e and r ray, obtained from particle shock to charge q, mass m, radial distance d_r, c = speed of electromagnetic radiation.

Being that $F = ma = E_0 q$, and as by the transversal component of an electric field $E_{transversal} = (q\ a\ sen\alpha/c^2 d_r)$, was obtained: $E_{transversal} = (q^2\ E_0\ sen\theta/\ d_r\ m\ c^2)$, and by intensity $I_\alpha = (q^2 a^2_{(average)}\ sen^2\ \alpha)\ /(4\pi c^3 d_r^2)$ and the incident intensity $I = (E_0^2/8\pi)c$, was obtained: $(I_\theta / I) = q^2\ E_0\ sen^2\theta\ /m^2\ c^4\ d_r^2\ E_0$ and in sequence the radiated power $P_{power} = (8\pi/3)\ (q^2/mc^2)^2\ I$, in which $P\ I\ /I = \delta$ it was the shock section $= (8\pi/3)\ (q^2/mc^2)^2$. And being an r ray sphere and section πr^2: $\pi r^2 = (8\pi/3)\ (q^2/mc^2)^2$ and $r = 8/3\ (q^2/mc^2)^2$ that in 1906 J. J. Thomson obtained for the beta-rays in the X-ray scattering. And with e/m_e and e obtained, and therefore m_e, the beta particle radius was calculated in $\approx 2.8 \times 10^{-13}$ cm.

Measurement of Electron Charge by Oil Drops – The mass of a droplet could also be measured by the speed in gravitational drop and electric force, but with oil droplets, in which there would be reduced evaporation of droplets, therefore its constant mass that would allow to observe a droplet for a better determination of the speed and consequent measurement of its mass, which was obtained between 1906 to 1911 by H. Fletcher and R. A. Millikan.

The droplet fell by the gravitational force $F_{droplet}$ in an electric field $E = F_{Electric}/q = mv/t$, so $q = F_{electric}/(mv/t)$. The resistance of the air in the drop of the droplet was $F_{air} = k_d v$, and $k_d = 6\pi\eta r$, where k_d = constant proportionality, η = viscosity coefficient and r = droplet radius. In the droplet acted the forces: gravitational $F_g = m_d g$, m_g = droplet mass and g = gravity acceleration; F_{thrust} = air thrust force and $F_{air} = k_d v$; the total force resulting $F = (F_g - F_{thrust} - F_{air})$, and $(F_g - F_{thrust}) = k_d V_{end}$; with ρ_{oil} = oil density and ρ_{air} = air density, $F_g = 4/3\ \pi r^3\ \rho_{oil}\ g$; $F_{thrust} = 4/3\ \pi r^3\ \rho_{air}\ g$, therefore: $4/3\ \pi r^3\ (\rho_{oil} - \rho_{air})\ G = 6\pi\eta r V_{end}$, then: $r^2 = 3/2\ \eta V_{end}\ /(\rho_{oil} - \rho_{air})g$. And with the experimental determination of speed, was obtained the mass $m_d = 4/3\ \pi r^3 \rho_{air}$ and the charge of the q droplet.

For each droplet with a tenth of a thousandth in each inch in diameter, a speed was obtained, thus a set of integer values of an elementary charge. To obtain the smallest quantity of charge in the droplet was used the radio activity for the removal of electrons, which altered the speed of the droplet, and what was seen was a bright point of light. There was a certain minimum speed that was double or triple that speed, and the lowest speed was the result of the loss of an electron. This fact proved conclusively that the smallest invisible charge that was able to be removed from the droplet was actually a single electron, and that any electron consisted of exactly the same quantity of negative electricity. Measurements and results of droplet charges showed values as Table 2.

Tab. 2 – Results of the quantity of charges in oil droplets, as extracted from *Great Experiments in Physics*, Shamus, M. H., Dover Publicationa, Mineola, NY, 1987, of Holt-Rinehartand, New York, 1959.

Charge in the drop	Difference between sequential measurements	Number of elementary charge in the drop	Charge in the drop	Difference between sequential measurements	Number of elementary charge in the drop
34,47		7	24,60	5,02	5
39,45	4,98	8	24,60	0	5
44,42	4,97	9	24,60	0	5
49,41	4,99	10	24,60	0	5
39,45	9,96	8	24,60	0	5
59,12	19,67	12	19,66	4,94	4
44,42	14,7	9	24,60	4,94	5
49,41	4,99	9	29,62	5,02	6
53,92	4,51	11	34,47	4,85	7
49,41	4,51	10	39,45	4,98	8
68,65	19,24	11	44,42	4,97	9
83,22	14,57	17	49,41	4,99	10
78,34	4,88	16	53,92	4,51	11
63,68	14,66	13	59,12	5,20	12
24,60	39,08	5	63,68	4,56	13
29,62	5,02	6	68,65	4,97	14
29,62	0	6	78,34	9,69	16
29,62	0	6	83,22	4,88	17

The closest difference between the charges was an average of 4.89, resulting in multiples for all measurements of droplet charges. Was obtained the elementary charge of the electron $e = 1.592 \times 10^{-19}$ coulombs, its mass $m_e = 9,1084.10^{-28}$ g, and the number of Avogadro by the total quantity of charge that in the electrolysis that was deposited in the electrode, where f_Q = constant factor representing the charge placed on the electrode by molecule-gram or atom-gram = 9.65×10^4 coulombs or 1 faraday in current unit. So $e = (f_Q/N_A)$ and $N_A = 6.062 \times 10^{23}$ molecules per molecule-gram or atom-gram.

XV. X-RAYS – 1895 to 1913

Impact Rays – In 1895, W. C. Röntgen sought to verify the origin of the luminosity occurring in substances external to the cathode ray tube, according to experiments of Lenard, Goldstein and Hittorf. Isolating the cathodic tube with black cardboard and having fluorescent paper of platinum-barium cyanide for eventual detection of external luminosities, when connecting the electric circuit he observed a black line on the paper. This line was similar to a light that projected on a wire held the passage of light forming the black shadow in the paper of revelation. However, there was no light coming from the tube by the black cardboard shield. Also, the ultraviolet and infrared radiations and the electric arc light, the circuit induction coil in the emission of electric circuit discharge, were isolated by the black cardboard. The brightness was observed around two meters away from the cathodic tube. And it could not be escaping from cathode rays, because when an exit was opened in the glass of the cathodic tube, there was external detection of the cathode rays up to only three centimeters. He concluded that these rays of luminosity, which he named X-rays, came from the point on the wall of the tube which it showed the greatest fluorescence to be considered as the center of which radiated the X-rays in all directions. That is, the X-rays came from where the cathode rays passed through the glass wall, or also from the equipment of the cathode rays on the aluminium plate or other substances. and Röntgen interpreted that through the aluminium window of the Lenard experiment were the X-rays emitted by the cathode rays while colliding with the aluminium window. Also, P. Heen, R. Lawrence, G. Meslin, J. A. Imbert and H. Bertin-Sans found that the X-rays were originated in the impact between the cathode rays and the inner part of the glass tube, and Perrin that they were coming from the collision with any solid obstacle. And as Röntgen, a later difference, the most important between the behavior of the cathode rays and the X-rays was that there was no deflexing X-rays by a magnet, even with intense field applied.

Penetrating Rays – Also, Röntgen noted that X-rays penetrated larger material thickness than cathode rays. So he reported that X-rays were passing a thousand-page books, blocks of two to three centimeters of wood, plates of 0.2 mm copper, silver, gold and platinum and 15 mm aluminium, and the shadow of bones that was between the tube and the screen or photographic plate. And it did not obtain reflection, refraction or diffraction of the X-rays, unlike the electromagnetic waves of light, ultraviolet, infrared and those obtained by Hertz. Ultraviolet rays also went through thin silver films, as A. Cornu in 1880 to realize that silver-coated mirrors presented minimal reflection, and so X-rays should have a higher frequency than ultraviolet rays and present a strong refraction, but Röntgen had not detected any refraction. And in 1893, Helmholtz had formulated a theory of dispersion whereby the refractive index increased with frequency to a certain value and then decreased to a minimal refraction. So, as Röntgen the X-rays could be transverse and longitudinal vibrations in the ether. Or else admitting high-frequency transverse electromagnetic waves superior to ultraviolet waves.

Particles, Light and Rays – In 1896, B. Galitzine and A. Karnojitsky detected that the central point of emission of the X-rays was in the inner part of the tube, but a few millimeters before the point where the cathode rays reached the glass wall. This experiment seemed to show what Newton had formulated in his "Optical Questions" between 1664 to 1690: *"The bodies did not act on each other by gravity, magnetism and electricity, and nature would be very consonant, attuned to a common pattern? And the small particles of the bodies would not have certain virtues or forces that act the distance between them and on the light in reflection, refraction and inflection?"*.

Waves of X-Rays – In 1905, C. G. Barkla through crystal polarization observed that the X-rays behaved as transverse wave. In 1912, M. T. F. Von Laue observed that as electromagnetic waves interacted with electric charges of atoms, according to Lorentz's electromagnetic theory, then the X-rays if they were electromagnetic waves should interact with atoms in the form of crystals, Which constituted most of the chemical elements in the solid state, as was observed in 1849 by A. Brava, and that causing the oscillation of these atoms and radiating electromagnetic waves would result in interference of the radiations of the various atoms, That is, it would be observed the diffraction of the X-rays and the wavelength of X-rays. And in 1912, it was verified experiment in crystals by W. Friedrich and P. Knipping that the photographic plates showed the characteristic wave interference to the X-rays a type of X-ray wave, with a length of 10^{-8} to 10^{-11} m. In 1913, W. H. Bragg and W. L. Bragg obtained the equation of the number of waves of length λ in the deviation of the X-rays, by diffraction at a $2d\,sen\theta$ distance from the crystal, ie $n\lambda = 2d\,sen\theta$.

XVI. RADIOACTIVITY – ALPHA, BETA AND GAMMA RAYS
1896 to 1925

Fluorescent Sources and Radiation Phosphorescent – In 1896, as Poincaré X-rays are emitted by the highest fluorescence point in the glass, it could also be that materials with intense fluorescence could emit X-rays, whatever the cause of this fluorescence as achieved by cathodic ray or light absorption. And despite the emission of X-rays, it could also come from the aluminium plate instead of the glass wall without presenting fluorescence, and there could be some connection between aluminium and light absorption. According to E. Becquerel, in 1859, even in non-visible phosphorescence bodies, the incident light could cause molecular vibrations that would result in thermal effects and other unknown molecular actions.

Phosphorescent Spectrum Uranium – In 1896, A. H. Becquerel initiated experiments with uranium compounds that presented greater fluorescence or phosphorescence in relation to other substances. But some uranium compounds were not fluorescent or phosphorescent. However, E. Becquerel had observed between 1867 to 1868 that all uranium compounds showed phosphorescence spectra with thin, bright stripes as a harmonic series and the other compounds were wide and irregular bands. But, in 1852, Stokes had formulated that the emission of light from a substance should be of a frequency less than that of the incident light. However, between 1871 to 1880, E. Lommel formulated that an incident light could present harmonic frequencies with higher and lower emission frequencies, that is, to cause resonances, for example, a solution of red of mothball alcohol under the incidence of red or yellow light emitted greenish light.

So, A. H. Becquerel admitted that higher frequency radiations could be emitted by harmonics of the phosphorescent radiations of all uranium compounds and even other non-luminous substances, through internal or intermolecular vibrations of substances, permanent or under incident light. Thus, after receiving sunlight, if there was the influence of fluorescence or phosphorescence these compounds could emit X-rays, crossing a thin plate of glass, to isolate eventual vapor that came out of the compounds, and intercepting a coin or a plate pierced metallic, which would present the image in the photo paper. And obtained radiations by presenting the images of objects in the photo paper. Also in 1896, S. Thompson observed that a phosphorescent compound of uranium when receiving sunlight emitted a radiation that crossed a thin aluminium plate and reached the photographic paper, that is, a similar effect of the X-rays that was obtained by cathode rays it would also be obtained by the phosphorescence produced, which it named the compound of hyper phosphorescence.

Invisible Phosphorescence Uranium Radiation – And also in 1896, A. H. Becquerel by storing a uranium compound next to the photographic paper, in a closed drawer for several days, so without receiving sunlight, he observed that there was the high-sharp outline of a copper cross that had stood between the compound and the photo paper. They were photographic prints equal to that obtained with the Sun exposure of the same uranium compound. But although penetration in materials is different from that obtained by X-rays, it observed similarity with that of X-rays and that should be radiation emitted by a phosphorescence not visible for long duration in relation to the phosphorescence of other substances, as E. Becquerel in 1859. Still in 1896, F. P. Le Roux noted that the metals were more permeable to the rays emitted by the uranium compounds than the X-rays. It also detected that the metals that were traversed by the radiation as coming from the phosphorescence, emitted another type of radiation, a secondary radiation, concluding that the metals had some kind of fluorescence or phosphorescence.

Uranium Rays – And A. H. Becquerel to discover that the metallic uranium, obtained for the first time by H. Moissan in 1896, and which was not phosphorescent also emitted the penetrating radiation, either combined with fluorescent or phosphorescent or non-phosphorescent or fluorescent compound, He reported that it was the first example of a metal that would present a class phenomenon of an invisible phosphorescence. And that would be related to the harmonic frequencies of emission by specific spectral bands of the uranium previously reported by A. H. Becquerel, and named this radiation from uranium-rays.

Uranium Activity – As uranium emitted radiation, then if it was used as a target of cathode rays it could present a greater X-ray intensity. And S. Thompson used uranium as a cathode ray target and claimed to be more active than aluminum or platinum.

Phosphorescence and Fluorescence: Only Light Radiation – In 1896, it was observed, through L. Benoiset, D. Hurmuzescu, E. M. L. Bouty, R. Swyngedauw, H. Dufour, J. J. Thomson, C. T. R. Wilson, J. A. McClelland, E. Rutherford, I. I. Borgman, A. L. Gerchun, A. Righi. C. Raveau, C. D. Child, Perrin, G. Sagnac, that X-rays discharged charges of the electroscope even when not directly focused on the electroscope, evincing that they conduct electricity in the air by separation of charges as a type of electrolysis. In 1897 J. P. L. J. Elster and H. Geitel observed that also the radiation emitted by the uranium led to the electricity in the air. And that with or without incidence of sunlight on the metallic uranium was obtained an identical radiation, which showed the lack of any relation with phosphorescence, hyper phosphorecence or phosphorescence not visible, which came to be proved when in 1898, G. C. N. Schmidt observed that common phosphorescent materials did not ionize the air and neither emitted X-rays or other similar radiation, but only the wave visible light electromagnetic. Also M. Curie, in 1898, noted that the heating, incidence of light and X-ray irradiation in metallic uranium resulted in equal radiation, thus coming to prove the absence of any relation with phosphorescence or fluorescence.

Uranium and Thorium Coupled and Active Radiation, Radioactivity – In 1898, M. Curie observed that the radiation emitted by uranium compounds was directly proportional to the uranium contained in different compounds. However, in a compound called "pitchblend", the emitted radiation was greater than pure uranium. So, M. Curie and P. Curie searched for another chemical element that would also emit radiation. By detecting electric current in the air for different chemical elements found equal radiation in the thorium. M. Curie and P. Curie called substances such as uranium and thorium, radiation-active, radioactive or radioactivity.

Cosmic Hypothesis of Radioactivity – M. Curie formulated the hypothesis that uranium and thorium would have the property of transforming some possible kind of cosmic radiation more penetrating in this radioactivity.

Uranium and Polonium in Coupling – Also, in 1898, M. Curie and P. Curie observed that in another compound of uranium, the copper-uranyil phosphate, there was also much greater radioactivity than metal uranium, but no thorium was found in that compound. From qualitative chemical analysis by separation, M. Curie and P. Curie obtained a bismuth with higher radioactivity where the radioactive chemical element should be. However, the analysis of spectral lines performed by E. A. Demarçay indicated only non-radioactive bismuth. Considering that the quantity of the radioactive element near bismuth was minimal to be detected by the spectral line, M. Curie and P. Curie admitted that there was a new chemical element, which they named polonium.

Uranium and Radium in Coupling – In 1898, M. Curie, P. Curie and G. Bémont obtained in the compound of "pitchblend", besides the thorium with bismuth, also detected in the separation with the barium another radioactive element. But, they managed to concentrate this element together with barium through successive fractions, and Demarçay observed the existence of a new spectral line. M. Curie, P. Curie and Bémont named the radium this new chemical element.

Alpha and Beta Rays – In 1899 Rutherford with radiation experiments by uranium observed that a type of radiation was absorbed with 50% of its intensity by an aluminium foil of only 0.0005 cm, and which named alpha, and another more penetrating type in which 50% of radiation was absorbed to a thickness of 0.05 cm that named beta.

Identical Rays: Beta and Cathode – Elster, Geltel, F. O. Giesel, S. Meyer and R. Schweidler, in radium-beam experiments, concluded that beta rays were deflected by a magnetic field and identical to cathode rays. After, in 1900, P. Curie obtained these two radiations emitted by the radium and also a deviation of the beta radiation in a magnetic field.

Atomic Origin of Radioactivity – In 1899 Elster and Geitel claimed that the origin of the radiations of uranium, thorium, polonium and radium were from the interior of these atoms and not by absorption of light from phosphorescence or cosmic radiation. They had detected that at 850 m deep, in which the eventual type of cosmic radiation would be partially absorbed by the Earth, the radiations of these elements were identical to those emitted on the surface. and also M. Curie and P. Curie observed that at noon or midnight there was no difference in the emission of uranium in which a possible type of cosmic solar radiation could act during the day on uranium and other radioactive elements.

Gamma-Rays – In 1900, P. U. Villard came to study one of the uranium properties from the radium. The radiation emitted was focused on a photographic plate protected by a thin layer of lead, sufficient to stop the advance of the alpha particles. Villard showed that the radiation that went through the lead layer was of two different types. One was deflected when subjected to a magnetic field, identified as the beta radiation discovered a year earlier by Rutherford. However, the other radiation was not deflected by electric and magnetic fields. In addition, the radiation unknown had a penetrating power far superior to the alpha and beta radiation. It was actually Rutherford following the discovery of Villard which identified as the third type of radiation arising from the radioactive decay and designated it, in 1903, by gamma radiation.

Alpha, Beta and Gamma Rays – And in 1902, Rutherford and S. G. Grier observed the alpha and beta rays in the thorium radiation, and Rutherford, F. Soddy and E. Andrade began the studies on the rays, α, β and γ: the α (alpha)-rays and β (beta) were formed by electrically charged particles; β rays were a hundred times more penetrating than α rays; and the γ-rays were similar to light and a hundred times more penetrating than the β-rays.

Alpha-Rays – On 1901 J. W. STrutt-Lord Rayleigh and Crookes in 1902, through electric discharges in the air, admitted that the alpha rays of lesser penetration of the radioactive elements should be charged particles with positive charge and at great speed. In 1902, Rutherford detected that the radio alpha rays were diverted on the opposite side of the beta rays, by magnetic field, indicating that they were positively charged particles, and by the ratio obtained charge and mass, $q/m = 6.4 \times 10^3$ in CGS unit, the deviation from the magnetic field and the length traveled by the alpha rays, determined the speed of 2.5×10^9 cm/s. relating q/m of hydrogen released in electrolysis = 10^3 with e/m alpha-rays = 6.4 x 10^3, and estimating that hydrogen would carry a same charge of the same type as the alpha particle, the alpha mass would be twice the hydrogen atom.

Inactive and Reactivated Uranium – In 1898, M. Curie, P. Curie and Bémont had found new chemical elements, such as thorium, polonium and radium, along with uranium compounds. But in 1900, Crookes when conducting chemical separation analysis on uranium also obtained a minimal residue with intense radioactive activity, but the uranium was detected as inactive by nothing revealing in a photographic plate, and named that residue of UrX. Similarly, A. H. Becquerel, in 1901, found that a barium residue was very active, while the uranium was almost inactive. However, after 50 days the uranium emitted 80% of the maximum radioactivity and the barium residue fell to 20%.

Atomic Disintegration – And in 1902, Rutherford and Soddy found in thorium compounds an analogous residue of uranium, which they named ThX. So, Rutherford hypothesized that the radioactive element residue would produce a constant average of other radioactive elements. The UrX residue was separated with a chemical reagent, while ThX was with another reagent, evincing different chemical elements. They were miniature quantities, which were minimal to be detected by scale or spectroscopy. And in experimenting with the thorium compound, after several ThX removals he would come back to form. Soddy and Rutherford concluded that radium-element atoms were susceptible to spontaneous disintegration. The alpha particle output should be the result of the disintegration of radium-elements atoms and the formation of other radium-elements, in which a definite small quantity of the atoms of each radium-element became unstable in a given time. So there were a number of chemicals produced. Since the thorium had a mass of 237, and the mass of the alpha particle was around 2, it was evident that if only one alpha particle was expelled in each exchange, the process of disintegration could pass through a number of successive stages to the last in which it was hit when the beta and gamma rays would appear, and this process was similar to uranium, radium and other radium-elements. And in 1903, M. Curie, P. Curie and A. Laborde observed that near the radium compound there was a higher temperature at varying degrees of ambient temperature. According to Rutherford, the energy emitted from the radium was demonstrated by its direct heating effect. And there would be no doubt that the effect of the radium would be a result of the succession of radium-active modifications occurring in it. Table 3 shows the disintegration of elements as Rutherford.

Tab. 3 – Disintegration of elements, as Rutherford's original, in *Radio-activity*, E. Rutherford, Cambridge University Press-1904 e Dover Phoenix Publications-2004.

Radioactive Products.	T.	Rays.	Some Chemical and Physical Properties.
URANIUM. ↓	5×10^9 years.	α	Soluble in excess of ammonium carbonate.
Uranium X. ↓	22 days.	β, γ	Insoluble in excess of ammonium carbonate.
Final product.			
THORIUM. ↓	10^9 years.	α	Insoluble in ammonia.
Thorium X. ↓	4 days.	α	Soluble in ammonia.
Emanation. ↓	1 minute.	α	Inert gas condenses about $-120°$ C.
Thorium A. ↓	11 hours.	no rays.	Attaches itself to negative electrode, soluble in strong acids.
Thorium B. ↓	55 minutes.	α, β, γ	Separable from A by electrolysis.
Final product.			
ACTINIUM. ↓			
Actinium X?. ↓			
Emanation. ↓	3.9 seconds.	α	Gaseous product.
Actinium A. ↓	41 minutes.	no rays.	Attaches itself to negative electrode, soluble in strong acids.
Actinium B. ↓	1.5 minutes.	α	Separable from A by electrolysis.
Final product.			
RADIUM. ↓	1,000 years.	α	
Emanation. ↓	4 days.	α	Inert gas, condenses $-150°$ C.
Radium A. ↓	3 minutes.	α	Attaches itself to negative electrode, soluble in strong acids.
Radium B. ↓	21 minutes.	no rays.	Volatile at $500°$ C.
Radium C. ↓	28 minutes.	α, β, γ	Volatile at about $1,100°$ C.
Radium D. ↓	About 40 years.	β, γ	Soluble in sulphuric acid.
Radium E. ↓	about 1 year.	α	Attaches itself to bismuth plate in solution, volatilizes at $1,000°$ C.

Earth Age – J. Hutton, in 1785, observed the impossibility of obtaining traces of a beginning and prospect of an end to Earth based on the science of the time. C. Lyell in 1830 to follow that the present was the key of the past concluded that it would only be possible to explain the evolution of the Earth's crust if it had large intervals of time for erosion and sedimentation that would require untold millions of years. C. Darwin in 1851 in presenting the theory of evolution of species, which would require a long time for biological development and transformation, estimated the age of the Earth in 300 million years from chemical elements in fusion. In 1854, Thompson-Lord Kelvin, with thermodynamics refuted Hutton and Lyell's theory, because it was possible to calculate the entropy of the system within a time far below the indeterminate or infinite that they had completed. Thus, it calculated for the age of the Sun, less than 100 to 500 million years, so the Earth should have its crust solidified around 100 million years. And subsequently, Thompson-Lord Kelvin reestimated around 25 million years. But in 1903, as mentioned by J. Joly and R. J. Strutt, by radioactivity there was an intense heat production located in the radioactive atoms that would negate the thermodynamic calculation of heat flux, thus involving the age of the Earth. In 1905, Rutherford presented a method of how the Earth's age could be evaluated by radioactivity, in which uranium was transformed at the end of lead disintegration, as it had been found around 1904 by B. Boltwood, the percentage of lead in radioactive minerals would indicate the age of the mineral when it was at the maximum initial percentage. And in 1907, Boltwood obtained for uranium minerals the age of 0.410 to 2.2 billion years.

Alpha Rays: the Helium Atom – At 1908 Rutherford and H. W. Geiger through a two-pipe equipment, one inside the other, and filled the inner tube with extremely fine thickness with radium before sealing to the outer tube. After two days, in the space between the two tubes in which vacuum was made, the spectroscopy detected the presence of helium, which had been discovered in 1868 by J. N. Lockyer and P. J. C. Janssen by the spectrum of the solar atmosphere. The following year in 1909, T. Royds and Rutherford identified the alpha rays as atoms of helium gas.

Atomic Instability – On the cause of the instability of atoms for the disintegration of radio atoms, Rutherford stated in 1904 that it might be supposed to be provoked by the action of external forces or inherent forces between the atoms, and that the application of some small external force could cause instability and consequent disintegration accompanied by the release of a large amount of energy. And also about the instability of the radium atoms, as Rutherford, despite the gravitation that was not under our control, that instability could be influenced by gravitation, if there was a way to change this instability of the radium-atoms. And if the vision was that gravitation was the result of forces that had the origin of them in the atom, it would be possible that, if the atom were disintegrated, the weight of the parts could not be equal to that of the original atom. It would seem likely that the primary cause of the disintegration of the atom should be sought in the energy loss of the atomic system due to electromagnetic radiation.

Atomic Stability – In 1900, Larmor had shown that the condition to be largely in order that a system move fast electrons could persisted loss of energy, because the sum vector of the accelerations in the direction of the center would be permanently null. While a single electron moving in a circular orbit was a powerful energy radiator, it was remarkable how quickly the energy radiation decreased if several electrons were revolving around a ring. It had been shown by J. J. Thomson in 1903, which mathematically examined the case of a negatively electrified corpuscles system, set at equal intervals around the circumference of a circle, and rotating in a plane with uniform speed around of your center. He found that the radiation of a group of six particles moving at a speed of 1/10 of the speed of light was less than 1/1,000,000 of part of the radiation of a single particle that described the same orbit at the same speed. When the speed was 1/100 of that of light, the quantity of radiation was only 10^{-16} of a single particle that moved at the same speed in the same orbit. These results would indicate that an atom consisting of a large number of revolving electrons could radiate extremely slow energy, and yet, finally, that minimal output, but continuous energy of the atom should result or in a rearrangement of its components in a new system, or an expulsion of electrons or groups of electrons from the atom.

More than a Thousand Electrons for the Mass of Hydrogen – Also, in 1878, A. M. Mayer had observed that small magnets like needles were organized in equilibrium settings around a central magnetic field and repeated at regular intervals, which could explain the properties of the chemical elements in the periodic table. So J. J. Thomson between 1889 to 1905 based on A. M. Mayer conceived that the atom, consisting of concentrated corpuscles rings, electrons, rotating safely in equilibrium positions by a positive fluid, a sphere of positive electricity, then the atom would be stable. But this positive fluid was an electromagnetic design without mass, and it would be necessary that there were thousands of corpuscles, electrons, to obtain the total mass of the atom. According to J. J. Thomson, in the case of the hydrogen atom it would be formed by more than a thousand electrons to obtain the mass of this atom, according to its book "Electricity and Matter" in 1904.

Negative and Positive Electron – In 1901, Jeans conceived the atom with an almost infinite number of massless electrons concentrated on the outer layer, with positive electrons cloaked inside the atom. In 1903, Lodge presented a proposition that the center of the atom would be a set of a large quantity of positive and negative electrons.

Electrons as Saturn Rings – Between 1901 and 1904, Perrin and H. Nagaoka presented the image of the atom as a kind of solar system, with the concentrated center of positive electricity and the revolving electrons, without attention to the problem of the stability of the atom, because the atom of the Nagaoka model was inspired by the stability of Saturn's rings. And as Nagaoka, the emission of radioactivity would be by rupturing the electron rings in the atom.

Negative and Positive Electricity, Inseparable – And between 1903 to 1906, Lenard presented the idea that the atom consisted of a kind of positive and negative energy association indivisible and inseparable in units, that is, a dual electric one.

Radioactivity and Gravity – Between 1905 and 1906, J. J. Thomson and Sagnac observed experiments in which the radioactive emission occurred without the influence of gravity. Through a lead pendulum and the other with uranium, with equal masses, J. J. Thomson observed that the motions of the pendulums were identical. and Sagnac sought deviations on pendulums behind a shield in which was placed on the other side of the shell masses of uranium and non-radioactive material, in which no difference was found from deflection of the pendulum to the two materials.

Negative and Positive Charge, Equal in Atom– In 1909, Royds and Rutherford identified the alpha rays as atoms of the helium gas that lost their two electrons. Therefore, dual and positive electric charge ions. And in 1908/1909, Rutherford after experiments with Geiger confirmed the identity of the alpha particles as a positive charge consisting of a positive double charge of helium, and showed by measuring the electrical deviation that this positive double charge was equal to two electron charges, that is, the positive charge was equal to the negative charge of the atom.

Nucleus and Electrons Around – Also, Rutherford calculated, by experiments, the high speed of alpha at 30,000 km/s and its mass around 8,000 times that of the electron. So Rutherford could use the alpha particle to observe, by the model of atoms as a sphere of mass containing positive and negative charges proposed by J. J. Thomson, its impact on these target atoms. Also, by virtue of radioactivity by the alpha particle forming new atomic elements discovered by Rutherford, the direct attack of alpha particles on atoms might be able to form a new atomic element. In 1903, Crookes had reported sparks produced by radium alpha particles reaching zinc sulfide and using this process, Rutherford with Geiger and E. Marsden could observe the effect produced by the impact of the alpha particles. The thickness of the gold leaves used in the experiment was 0.00005 cm. As the atomic diameter was of the order of 5×10^{-8} cm, measured according to the mass, density, volume and estimated number of Avogadro, by kinetic theory, would fit a thousand layers of metal atoms. And since the momentum of the alpha particle was much larger than the momentum of the diluted mass in the target atom, the alpha particle would cross the gold leaf with repulsion by the positive charge and the attraction for the negative charge of the electrons in the atoms, resulting in minimal deviations multiples, until they reach the chemical target of zinc sulfide. And it was observed that the alpha particles penetrated thicker metal sheets until they were absorbed, so the alpha particle passed through the atoms, because it had high speed, that is, high momentum. In 1911, Rutherford and Geiger demonstrated that if it were correct the atomic model of J. J. Thomson, of the atom consisting of an atomic mass with positive charge and with electrons in rings rotating within the atomic mass and positive charge, to have balance of the charges, So the number of alpha particles diverted at angles greater than about 10 degrees would be null. But, the deviations were observed in directions away from the incident particle beam and even the particles being thrown back. The conclusion was that the deviations were due to the repulsions of a massive central charge Ze surrounded by a cloud of opposite electricity. And perhaps because of the difficulty of seeing how the component parts of the positive charge in the center were insured, i.e. atomic stability, Rutterford considered that the atom could be a concentration of electrons in a positive fluid, practically with all the positive charge $+Ne$ and negative $-Ne$ where for convenience the central charge would be assumed positive, and the concentrated mass in the center of the atom, the nucleus as a sphere of radius no greater than 10^{-12} cm, which in 1913 Rutterford calculated from its hyperbolic spreading formula obtained from the experiments with the alpha particles, in which this nucleus was involved by a sphere of electricity containing negative charge $-Ne$ and positive $+Ne$, and for convenience that outer nucleus sphere would be negative, supposedly evenly distributed through a sphere of the atom of radius 10^{-8} cm. In 1911 Nicholson presented the atom as a positive charge formed from an electric fluid in the mass nucleus, with the electrons rotating around the nucleus.

Nuclear Transmutation – Between 1914 and 1915, Marsden had observed that the hydrogen subjected to alpha particle radiation acquired a speed of 1.6 times that of the alpha particle and was easily detected by the flickering method of a zinc target, far from the ratio of the particles alpha. And also noted that alpha rays over nickel amounted to a number of flickering like those coming from hydrogen, but the number of the hydrogen flicker was much higher to be pointed at as the possible presence of hydrogen in the material or if it was a strong evidence that the hydrogen ascended from the radioactive source of the radio that it emitted the alpha particles. Then, in 1919 Rutherford noted that perhaps it might be that the large number of scintillations observed by Marsden were due, not to the hydrogen atoms, but the high speed of the nitrogen and oxygen atoms between the source and the target, which by passing alpha particles in nitrogen and oxygen, in the air, got a number of glittering twinkle. But, an astonishing effect was reported when the dry air replaced the air. With steam water, that is, with hydrogen, there should be a greater number of flickering. But, with dry air there was an increase, about twice that of what was observed with steamed water, with a long percentage of flickering that was equal in brightness to the sparkles of hydrogen. The greatest number of flickering was observed in the dry air, but not in oxygen or carbon dioxide. Therefore, this increase in flicker should be due to nitrogen by the collision of the alpha particles with the nitrogen atoms. And another surprising observation was that the percentage of the nitrogen atoms was the same as the oxygen atoms, although it would be expected a difference around 19% of a larger mass for nitrogen between the masses of oxygen, alpha particle, hydrogen and nitrogen. The conclusion was that the nitrogen atom was disintegrated under the intense force developed in the vicinity of the collision with a rapid alpha particle releasing the hydrogen atom that formed the nucleus of the nitrogen and resulting in the formation of the oxygen atom. The nitrogen was disintegrated and transmuted into oxygen. That is, eternity is nonexistent for atomic elements.

Proton – Between 1911 and 1919, Rutherford and C. T. R. Wilson experimented on alpha particle collisions in sodium, gold, aluminum and phosphorus, and was observed by spectroscope that positively charged hydrogen atoms had been obtained. So the hydrogen atom should be present in the nucleus and form the atomic nuclei. And the intensity of the positive charge of hydrogen was equal to that of the electron with the opposite signal, with the big difference in its mass for almost twice as heavy as that of the electron. And in 1920, Rutherford then named this positive charge of hydrogen, which would constitute all the nucleus of atoms of the chemical elements such as proton, of "prots" of the Greek, the first component of all others, and also of the name Prout which in 1815 had proposed that all atoms were hydrogen aggregates, the lightest of the chemical elements.

Neutral Mass – In 1920, as Rutherford protons would constitute the atomic nucleus, but for nuclear stability there should also be electrons in the nucleus as seen by the radioactive emission. But, he observed that the oxygen that he had obtained in 1919 of 8 positive charges of protons and atomic mass 17, could be oxygen of 8 positive charges of protons with atomic mass 16 and another neutral mass equal to that of the proton.

Nuclear Origin of Gamma Rays – By radioactivity it was observed that the unstable nucleus occurred the alpha decay, beta and there could also be emission gamma. But it would be necessary to know the real origin of the gamma emission, whether it would actually occur after the alpha or beta, or whether it was independent of those emissions.

In 1925 Rutherford and W. A. Wooster observed that in the decay of the B-radium with an average life of 26.8 minutes there was emission of electrons and gamma rays. And when the nucleus changed the atomic number, the electron in conversion to another level emitted X-rays that were emitted after the variation in atomic number. Also, in 1925, C. D. Ellis and Wooster, they observed that the B-radium in a platinum tube occurred the emission of the gamma rays that caused the emission of electrons from the platinum. And on the outside of the tube with B-radium layer was the emission of gamma rays accompanied by conversion of energy level of the electrons. The energy difference showed that the gamma rays were emitted after the change of the atomic number, that is, after the disintegration of the alpha particle and or beta emitted. That is, the gamma emission was linked to the electric and magnetic charges of the atomic nucleus.

XVII. COSMIC RAYS – **1900 to 1920**

Radioactive Rays – For Back from 1900, C. T. R. Wilson, Elster and Geitel, noted that the conductivity of air contained in an electric charge meter, the electroscope of gold leaves remained constant, though they were withdrawn ions through an electric field. The two gold sheets of the electroscope were charged with the same negative charges, resulting in their removal. But with the ionization of the air, the positive ions formed caused the attraction with the negative charge, resulting in the approximation of the leaves. They concluded that some unknown agent would constantly produce new ions, and should be caused by radioactivity from of the Earth's crust. Therefore, ionization should decrease with the increase in altitude, as radioactive absorption should be obtained by up to 1,000 meters.

Rays of the Sun – In 1906, G. W. Richardson suggested that the effects of electroscopic discharge should be related to the Sun. But it was verified that the effects were equally intense both in the day and in the night.

Atmospheric Rays – In 1909, K. Kürz summarized all investigations citing three possible origins of the electroscopic discharge effect, in which the rays would come from the Earth's crust, the atmosphere or places beyond the atmosphere, but discarding those of atmospheric origin .

However, on 1909 T. B. Wulf, with a meter of the ion charges produced inside a hermetically sealed tube, showed a higher level of radiation at the top of the Eiffel tower than at its base.

In 1910, A. Gockel launched an electric charge meter using the default Wulf setting, in a balloon that reached 4,500 m altitude, in which there was a slight increase in the residual charge with a smaller height. But at higher altitudes he noticed that the meter was discharged faster than on terrestrial soil, concluding that the rays would originate in the regions beyond the atmosphere. But still maintaining a hypothesis of radiation in the atmosphere or a solar type.

On 1911 D. Pacini observed simultaneous variations of the ionization rate on a lake, over the sea, in depth of 3 meters from the surface. Pacini concluded the decrease in radioactivity under the water that a certain part of the ionization should be due to other sources than the radioactivity of the Earth.

Between 1911 a 1912, V. F. Hess, to verify the observations of Gockel with the altitude, the eventual radiation would vary with soil sites and solar influence, carried three precise charge meters and a Geiger counter in a balloon flight up to an altitude of 5,300 meters. He noted that when the Geiger counter was moving away from the surface of the Earth and the radioactive chemical elements occurring in it, the radiation count decreased. In top of the Eiffel tower, for example, the count was lower than the surface, contrary to measurements in 1909 of Wulf. To investigate this phenomenon, he was taking the counters to increasing altitudes, by means of balloons charged with detectors of radiation, and then from a certain height, he verified that the radiation count increased strongly and could not be leaving of the Earth, it should come from some point in the space. Hess concluded, after his investigations, that the ionisation observed was due to the action of an unknown radiation, highly penetrating, coming from space. You gave her the name "cosmic rays". He found an increased ionization rate at roughly the quadruple rate at the ground level. Hess also observed that during a total eclipse of the Sun, in which the Moon blocked the visible radiation from the Sun, that the source of radiation increased as the balloon increased altitude. And Hess concluded that: *"The results of my observation are best explained by assuming that a radiation of great penetration power enters our atmosphere from above".*

Cosmic Rays Disorder – In 1920, H. T. Wolff considered that radioactivity would be associated with the fluctuations or internal disorder of the atom caused by high-velocity external particles like cosmic rays.

XVIII. QUANTUM – 1872 to 1928

Hertz Effect: Electric Spark – In 1872, A. G. Stoletov had observed in an experiment that ultraviolet light caused an electric current detected by the galvanometer in a voltage of 0.01 volts. At 1887 Hertz had experimentally observed that a electrified body with negative charges when focusing ultraviolet radiation had increased emission of sparks of light along with electromagnetic waves, the Hertz effect. In 1888, Righi showed that two electrodes in the presence of ultraviolet radiation formed a light-electric effect. In 1888, W. Hallwachs observed that a zinc plate discharged from electricity, when receiving ultraviolet radiation, was charged with positive charge, and was also observed with pure metals such as rubidium, potassium, sodium, lithium, magnesium and thorium. In 1889, Stoletov detected that negative charges leaved of a metal, in the presence of ultraviolet radiation, by the electric current in a circuit connected to that metal. And in 1899, Lenard detected electrons emitted from metallic surface when in the presence of ultraviolet radiation.

Quantum – In 1900, Planck observed that the radiation of a cavity or an oven with a minimal opening, that is, a black body, only absorbed radiation, which would be emitted by an quantity of energy, a quantum, proportional to the frequency emitted. That is, $\mathcal{E} = h\nu$ where $\mathcal{E}$ it was the energy, ν the frequency of the emission of radiation and h a constant where when the frequency ν increased, the energy $\mathcal{E}$ quantum also increased, that is, quantum was more "massive", because energy is mass x acceleration x distance.

In 1900, according to Planck the total energy of radiation E_t of a black body was:
$E_t = 0{,}0731$ [watt/cm^2] $= 0{,}0731$ [erg/cm^2 s]$= 7{,}31 \times 10^5$ [erg/cm^2 s]
And the specific intensity K of the black body radiation:
$K = (c/4\pi)\, u$ [erg/cm^2 s] $= (k_a c /4\pi)\, T^4$ [erg/cm^2 s], where:
c = speed of light = 3×10^{10} [cm/s]
u = energy density [erg/cm^3]
k_a = constant = 7.28×10^{-15} [erg/cm^3 °C^4]
T = temperature [°C]

It was observed, in the two formulas, of E_t and of K, the unit of the [second] in the denominator.

But, Planck after obtaining his radiation formula:
$u = (8\pi h \nu^3/c^3)\,(1/(\exp^{h\nu/kT} - 1))$, where exp is the exponential function,
$k = 1{,}34 \times 10^{-16}$ [erg/°C],

with dimensions: u [erg/cm^3] $= (8\pi \nu^3/c^3)[s^{-3}/(cm^3/s^3)]$ h [erg x s]

And the term $(1/(\exp^{h\nu/kT\,[\text{erg s s-1}/\text{erg °C-1 °C}]} - 1))$ was dimensionless

So: $u_{energia}$ [erg] $= (8\pi \nu^3/c^3)[s^{-3}/(cm^3/s^3)][cm^3]$ h [erg x s]) $(1/(e^{h\nu/kT} - 1)$

This is, u_{energy} [erg] $= (8\pi \nu^3/c^3)\,(1/(e^{h\nu/kT} - 1))$ h [erg x s]

But when this formula was differentiated by dV and after integrated was obtained the unit [erg] on both sides of the formula.

And Planck obtained the value of the constant h through $h = kC_2 / 48\pi\alpha c$, in that k = constant = 7.28×10^{-5} [erg/cm^3 °C^4]; C_2 = 1,436 cm °C; α = constant = 1,0823; c = speed of light = 3×10^{10} [cm/s], getting: h = 6.415×10^{-27} erg x s.

Mathematical Artifice – In 1905, Jeans It considered a mathematical artifice and without any physical concept the Planck quantum.

Quantum Effect on Hertz Effect – By the classical theory wave, a higher intensity of light of less wavelength should provide a higher energy density of the electromagnetic field in propagation at the speed of light, sufficient for the electron to be emitted from the metal atom, which didn't occur. And for any length of light with higher energy densities and or greater times of light absorption should emit electrons, which also did not occur. The fact was explained by the quantum light theory, in 1905, by Einstein, similar on 1900 to the Planck quantum of energy $\mathcal{E} = hV$, but that it was according to Einstein a radiation quantum located in a unit of volume in space and moving away from the emitting source with speed c, contrary to Planck which recommended that the quantum hV it was propagated as a classic electromagnetic wave. A quantum hV_1 focusing on an electron resulted in the energies involved: $hV_1 = E_{kinetic} + E_{connection}$; $E_{connection}$ = energy hV_0 of the quantum to remove the electron; as in the general case $E_{kinetic}$ = 0, then $hV_1 = hV_0$. That is, the quantum hV_1 would have to have a minimum energy hV_2 to move the electron. A quantity of NhV_1 it would increase the energy density in the surrounding area, but without the minimal localized energy of quantum to move the electron. where N = average number of quanta by time in an area perpendicular to the propagation direction of the quanta. In the theory of electromagnetic wave the intensity I was proportional to the average value of the field $E^2_{average}$ of the electromagnetic wave and $E^2_{average}$ it is proportional to radiant energy in a volume unit. And for the quantum the intensity was $I = NhV$. These waves with $E^2_{average}$ propagated the quanta and was proportional to the radiant energy of the quanta. With $I = k\ E^2_{average} = NhV$, with k = constant and N for being an average value, it measured also the probability of flow of the quanta per unit area at a given time. So Einstein interpreted $E^2_{average}$ as a probability of quanta density.

According to R. Eisberg and R. Resnick: *"These electromagnetic waves, whose intensity can be measured by $E^2_{average}$ they can be seen as conductive waves of the quanta; the waves themselves have no energy, there is only so much, but the waves are a greatness whose intensity measures the average number of quant per unit volume".*

Einstein considered that a variation of electromagnetic field resulted in radiation emission, that is, energy, applied this energy as a particle, the quantum of light and the quantum of energy.

Quantum in Emission and Absorption, Wave in Propagation – In 1907, Planck stated that the significance of Einstein's quantum of action or quantum light was only in the emitter and radiation absorber, because in the space-vacuum was the electromagnetic wave described by Maxwell's equations. And for Einstein in 1909 the quanta were punctual that motioned independently with energy $h\nu$ with the speed of light c. In 1911 stated that the quanta should have a provisional concept until it is reconcilable with the wave theory.

Constant h as Momentum – In 1912, Nicholson when looking for the explanation for the emission of spectral lines by the atoms noted that as Planck's constant h that has the unit of energy x time, but also the unit of angular momentum, that is, mass x velocity x distance (radius) from the orbit , so for the atom this angular moment could rise or fall only in distinct quantities when the electrons departed or returned from their original orbits. Thus, Nicholson realized that Planck's constant h determined the angular momentum, which was from the orbit of the electrons in the atom and could change only in discrete quantities.

Similar Particles of Different Energies – In 1925, by the statistics of S. N. Bose-Einstein, treating the quanta $h\nu$ as all similar as particles, it was obtained also the Planck equation. But with the same physical inequality of units.

Piloting Waves – In 1926 as Einstein the waves were to indicate the path to the quanta punctual $h\nu$ which were propagated with the c speed of light or electromagnetic radiation.

Wave-Particle – And in 1928, N. H. D. Bohr presented the principle of complementarity of light as a propagation now as a wave, now as a particle.

PART 3 – THEORY OF RELATIVITY, RELATIONAL MECHANICS AND QUANTUM MECHANICS

IXX. THEORY OF RELATIVITY – 1887 to 1918

Origin of the Theory of Relativity – Although Einstein mentioned that it was the electromotive force acting as an electric field that led him to the theory of relativity, however he himself stated in 1922 that the first model for him to create the theory of relativity was the null result of Michelson-Morley in 1887, in his lecture "As I Created the Theory of Relativity" at the University of Kyoto, in which admitted: *"The invariance of the speed of light to any velocity of the emitting source"*, as cause of the null result and consequent lack of the ether, which would permeate the bodies, the medium for the propagation of the electromagnetic field. So postulated the principle of relativity: *"In all coordinate systems, the laws of optics, electrodynamics and mechanics were valid, and the speed of light in the vacuum was always constant when measured in any system of coordinates"*. In 1898, Poincaré through the coordinates transformations of Voigt-Larmor-Lorentz had observed that the time intervals t and t' were different between two referentials with different velocities, and that the velocity was dependent on the intervals of space and of time. A resting frame with a v velocity body in relation to this reference, and another referential with very higt V velocity with a body in v velocity in relation to this reference, there would appear to be simultaneity, because it was the same velocity v of the body for the two referentials. However, the interval of space traversed was immense by the higher velocity frame, for simultaneity. Then, the time interval should also present some relation to the simultaneity, without the time interval remaining equal for the two referentials at completely different velocities. The Poincaré conclusion: *"We cannot measure such velocity without measuring a time interval"*. The definition of a single time, as Poincarè: *"Definitions are nothing more than the result of an unconscious opportunism, and the equality of two time intervals is to intuit the illusion"*. Between 1900 and 1904, Einstein stated in the "Akademie of Bern" that *"Poincaré's articles on coordinate transformations and the illusory time impressed a lot on the Akademie and left those who read without breath for weeks and weeks"*.

Coordinate Independence – According to Faraday, Maxwell and Weber, the physical phenomenon of electric current induction in a circuit occurred independently of the motion of the coordinates where the observer was, but only of the motion between a magnet or a circuit representing the magnet in relation to another circuit. A magnet in motion and a resting circuit resulted in a circuit-induced current, and the magnet at rest and the other circuit in motion also resulted in the induced current in the circuit.

Electric Field observed as Magnetic and this as Electric – By the electrodynamic theory of Maxwell and Lorentz, the electric and magnetic charge could produce electric and magnetic field according to its motion in relation to the permeable ether that passed through the mass, through the electromagnetic force $F = q[(E + (v \times B)]$, where q was the charge, and the electric field, v the velocity of the charge and B the magnetic field. Similarly to Maxwell, Einstein also observed that the electromagnetic force was the potential energy of an electric field and and therefore what would act on an electric charge in a magnetic field was the electromotive force as an electric field. So for an electromagnetic force $F = q[(E + (v \times B)]$ in a system S, and $F' = q[(E' + (v \times B')]$ in S', by transforming a system to another of coordinates $F = F'[(1 - (v^2/c^2)]$. And in 1905, Einstein came to explain by the theory of relativity, through the coordinates of each reference system, how the electric and magnetic fields are perceived, but excluding the ether. An observer, in a referential system S at rest with an electric charge q fixed, measured the electric field E and a magnetic field $B = 0$. To another observer in a system S' with v velocity, he measured in that system S' a magnetic field $B' = (v \times E)/c^2$ perpendicular to the motion and an electric field E'. After, in S was placed a fixed magnet without any electric charge and the observer measured a magnetic field B and an electric field $E = 0$. In S' with v velocity, the observer measured in S' an electric field $E' = v \times B_{S'}$ by motion and a magnetic field $B' = B_S$. That is, the electric field E in a system S resulted in an electric field E' and another magnetic B' in S'. E a magnetic field B in S it resulted in a magnetic field B' and another electric E' in S'. Thus, the electromotive force as the potential energy of an electric field and was also a magnetic field as the transformation of the coordinate system. Therefore, by Einstein's relativity, the motion of charges and coordinate systems was the cause of the appeararance of an electric magnetic field and a magnetic field in electric. In the reference of the electron in motion was seen an electric or magnetic field and in the other reference of the observer a magnetic or electric field, that is, a same phenomenon seen from different causes by the coordinates sites.

Increase in Mass by Relativity – From the coordinates of a reference system (S_1) x_1, y_1, z_1 and time t_1 and its transformation to a coordinate system (S_2) x_2, y_2, z_2 and time t_2 with v velocity in relation to the system S_1 on the axle x_1x_2, it was admitted at the origin of the two systems the emission of a radiation with c speed in all directions, forming a sphere radius, with $r_1 = ct_1$ and $r_2 = ct_2$, that is, $x_1^2 + y_1^2 + z_1^2 = ct_1^2$ and $x_2^2 + y_2^2 + z_2^2 = ct_2^2$. and transforming the coordinates of one system to the other, so c always remain constant resulted in the case of the axis x_1x_2 where $x_{1\ (0)} = 0$ and $x_{2\ (0)} = 0$: $x_2 = x_1\sqrt{1 - (v^2/c^2)}$, which was the contraction of space in the frame S_2 that was at v velocity; and with $t_{1\ (0)} = 0$ and $t_{2\ (0)} = 0$; and for $t_2 = t_1\sqrt{1 - (v^2/c^2)}$, slower for the reference S_2 with v velocity compared to S_1. And that Einstein-Larmor-Lorentz reference transformation applied to it was effected to c constant independent of the velocity of the source. And to preserve the conservation of the momentum, in particle shock, $p_1 = m_1v_1$ in the system S_1 that would be in the system S_2 as $p_2 = m_1v_2$, it would be necessary that m_1 was also modified by passing from one reference to another, so that there was conservation of the moments $p_1 = p_2$, that is, $m_1v_1 = m_2v_2$. As space, time and velocities appeared in the transformation of coordinates with the factor $\sqrt{1 - (v^2/c^2)}$, then the mass could also depend on the velocity in a reference system so that the conservation of the momentum was preserved. So when multiplying by m in the transformation of velocities, ie the momentum mv, was obtained $mv = m_0V/\sqrt{1 - (v^2/c^2)}$, so $m = m_0/\sqrt{1 - (v^2/c^2)}$ being named m relativistic mass that increased when the velocity v_0, of the mass at rest, was increased to v. And so Einstein's relativity determined that the particle mass was a function of velocity, m = (v function) when changing the frame or the actual reference in which there was increasing velocity.

Impulse by the Theory of Relativity – As L. Infeld and Einstein in 1938, the application of an impulse, i.e., Δmv in a body of mass m_1 and velocity v_1 it then resulted in modification of its velocity to v_2, that is, your acceleration $(v_2 - v_1)/t$, with the mass $m_2 = m_1 / \sqrt{1 - (v^2/c^2)}$.

Force and Motion of the Electron by the Theory of Relativity – According to Einstein the ponderomotive force or Lorentz force acting on the electron caused its motion and for the coordinates x, y, z, on the axle x: $m\ [(d^2x)/(dt^2)] = ma = eE = F_{em}$, where m = mass of the electron in motion; a = electron acceleration; e = charge of the electron; E = electric field; F_{em} = Lorentz electromagnetic force which was the force extracted from electrostatic or electromagnetic field, obtaining the masses: $M_{longitudinal}$ = longitudinal mass of the electron = $m_{0e}/[\sqrt{1 - (v^2/c^2)}]^3$ and $M_{transversal}$ = transversal mass of electron = $m_{0e} / \sqrt{1 - (v^2/c^2)}$, where m_0 = immovable mass or electron rest.

In 1906, according to Planck the ponderomotive or Lorentz force applied to a point of mass m, following the law of Newton's mechanics: $F_{em} = ma = eE$, it was obtained for the mass m_{0e} of the electron at rest and m mass with the v velocity, on the axle x, the expression of force:

$$F_x = d[(m_{0e}/\sqrt{1-(v^2/c^2)})v_x]/dt.$$

And in 1913, according to Einstein the Force F_{em} it was just simplifiedly defined as the product of mass and acceleration, different from a velocity and mass ratio for the momentum or velocity and acceleration for force. That force F_{em} which caused the motion of the electron and particles, by the transformation of coordinates presented in a reference system the force acting jointly, as the momentum mv, $F = mv/t$ and also as the variation of energy by the distance d: $F = E_{energy}/d$, or also as power of force: $P_{power} = (E_{energy}/t) = F(x/t) = F v$. By the theory of relativity with the transformation of coordinates, when the force was applied in a particle, its mass was increased from v^2/c^2, in which if $v = c$ the mass would be infinite and it would be impossible to obtain velocity greater than that of the light. So the force according to Planck was considered by Einstein as: *"There's no physical significance, but it's enough to be understood as the defined equation of force"*. Because it appeared in relativity the increase in mass as a function of velocity: $m = m_0 / \sqrt{1-(v^2/c^2)}$. According to Einstein: *"The definition of force as a product of mass and acceleration is not advantageous. It is more to the point to set force so point in a direction in which the laws of momentum and energy assume a simple form"*.

Energy by the Theory of Relativity – According to Einstein, the increase in energy was similar to the mass increase obtained from the transformation of coordinates.

According to Einstein the electron of mass at rest m_{0e} and mass m with velocity v, presented the kinetic energy $E_{kinetic} = \int F_{em}dx = m_{0e}\int_0^v \beta^3 v dv$, where $\beta = \sqrt{1-(v^2/c^2)}$ and $m = m_{0e}/\sqrt{1-(v^2/c^2)}$, where F_{em} it was the force extracted from the electrostatic or electromagnetic field, resulting $E_{kinetic} = mc^2 [1/\sqrt{1-(v^2/c^2)} - 1]$, where this energy depended on the velocity of the electron. Therefore, according to Einstein, the increase in energy was similar to the mass increase obtained from the transformation of coordinates by the factor: $[1/\sqrt{1-(v^2/c^2)}]$.

Total Energy by the Theory of Relativity – For Einstein, mass was energy and energy was mass. Although mass has unit in gram and energy in gram x cm/s² x cm, in CGS.

To move a particle of mass m_0 at rest and m mass with velocity v, momentum p between two points, and F_{em} from electrostatic, electromagnetic or mechanics: $F_{em} = ma = eE$, where m = mass, a = acceleration, E = electric field, was applied to classical motion energy:

$$E_{kinetic} = \int_0^{x1} F_{em}\, dx = \int_0^{t1} F_{em}\left(\frac{dx}{dt}\right) dt = \int_{t0}^{t1} F_{em}\, v\, dt$$

$$F_{em} = dp/dt = d(mv)/dt,$$

$$E_{kinetic} = \int_{t0}^{t1} v\left(\frac{dp}{dt}\right) dt = \int_{p0}^{p1} v\, dp = \{[(vp)]_{(0,v1)} - [\int_0^v p\, dv]\} = \{[(mv^2)]_{(0,v)} - [\int_0^v mv\, dv]\}$$

Being: $v(dv) = d(v^2)/2$, the result of integration was

$$E_{kinetic\ relativistic} = m_0c^2/\sqrt{1-(v^2/c^2)} - m_0c^2\ ;\ E_{rest} = m_0c^2$$

In the case of velocities $<< c$, the kinetic energy was the classically obtained by the equations of the accelerated uniform motion, $E_{kinetic} = mv^2/2$.

By the transformation of coordinates by varying the velocity the momentum to be preserved should vary also the mass and was obtained $m = m_0/\sqrt{1-(v^2/c^2)}$ and squared: $m^2c^2 - m^2v^2 = m_0^2c^2$, and multiplying by c^2:

$$mc^2 = m_0c^2/\sqrt{1-\left(\frac{v^2}{c^2}\right)} = E_{total}.$$

So, $E_{total} = E_{kinetic\ relativistic} + E_{rest} = E_{kinetic\ relativistic} + m_0c^2 = mc^2$
$E_{kinetic\ relativistic} = (E_{total} - E_{rest}) = (mc^2 - m_0c^2) = (m - m_0)c^2$

That is, the increase in kinetic energy through the relativistic mass caused by velocity when it varied from one to another referential system or in the frame itself when the velocity was increased.

To get a relationship between the particle motion with its $p = mv$ momentum, it was related to:

$$E_{relativistic\ kinetics} = (E_{Total} - E_{rest}) = (mc^2 - m_0c^2),$$

with $E_{relativistic\ kinetics} = (m_0c^2)/\sqrt{1-(v^2/c^2)} - m_0c^2)$ and with the momentum conserved:

$p = mv = m_0\, V/\sqrt{1-(v^2/c^2)}$, was obtained:
$(E_{relativistic\ kinetics} + m_0c^2)^2 = (cp)^2 + (m_0c^2)^2$

in the equation: $E_{Total} = mc^2 = E_{relativistic\ kinetics} + E_{rest} = E_{relativistic\ kinetics} + m_0c^2$,

replacing: $E_{relativistic\ kinetics} + m_0c^2$ by E_{Total}:
$E_{Total}^2 = mc^2 = (c^2p^2 + m_0^2c^4)$

Einstein concluded that if the inertial mass of a body was in the form of energy radiation E, the mass m of the radiation decreased by $m = E/c^2$. And that the mass of a body was the measure of her energy content, that is, $m = E/c^2$.

When he presented the equation of the momentum the physical units were accurate:

$$I_x = mq_x/\sqrt{1-q^2},$$

where q = v/c was without dimension, because v = velocity of the particle and c = speed of light, therefore, [gcm/s] on both sides of the equation.

And in sequence the same set of equations, Einstein wrote for energy:

$$E = m / \sqrt{1 - q^2},$$

E aplicando à particula imóvel (q = 0), he wrote: *"The energy, E_0 of a resting body is equal to its mass"*. Document 2.

> **PRINCETON SCIENCE LIBRARY**
>
> # The Meaning of Relativity
>
> **Albert Einstein**
>
> By equating the components, we obtain, in three-dimensional notation,
>
> (43) $\begin{cases} l_x = \dfrac{mq_x}{\sqrt{1 - q^2}} \\ \cdot \quad \cdot \quad \cdot \\ \cdot \quad \cdot \quad \cdot \\ E = \dfrac{m}{\sqrt{1 - q^2}} \end{cases}$
>
> We recognize, in fact, that these components of momentum agree with those of classical mechanics for velocities which are small compared to that of light. For large velocities the momentum increases more rapidly than linearly with the velocity, so as to become infinite on approaching the velocity of light.
>
> If we apply the last of equations (43) to a material particle at rest (q = 0), we see that the energy, E_0, of a body at rest is equal to its mass.

Doc. 2 – Einstein book page "The Meaning of Relativity", Princeton University Press, Princeton, 1956, where it is presented the final term in which the energy was mass, but in which the physical units were different on both sides of the equation.

Einstein observed that a variation of magnetic field resulted in radiation emission, that is, energy, then assigning to the field an energy deposit, and stating that energy had mass, that mass represented energy, and that there was no essential distinction some between mass and energy, for the radiation at rest. Also, stating what matter was where the energy concentration was large and the field where the energy concentration was small. And culminating emphatically that: *"Mass is energy and energy is mass"*.

That is, for $q = v^2/c^2$ by the theory of relativity would result $E = m$ with different units on both sides of the equation.

Energy of Light by the Theory of Relativity – When the energy of the light was emitted by a body results in the variation of energy.

So, $\Delta E_{energy} = E_{energy} \{[(1/\sqrt{1-(v^2/c^2)}] -1\} = mc^2 \{[(1/\sqrt{1-(v^2/c^2)}) - 1]\}$

and $\Delta E_{energy} \approx \frac{1}{2}(E_{energy}/c^2)v^2$, where $(E_{energy}/c^2) = m$, concluding that the variation of the energy results in the variation of the mass, and that the radiation transmitted inertial mass between the sending and absorbing bodies, that is, transmitted mass of the resistance force of inertia of Newton.

However, as the speed of light or electromagnetic radiation was at speed c in the vacuum, it resulted in the mass $m_{0\,(emr)}$ electromagnetic radiation at rest and mass m_{emr} of electromagnetic radiation at speed $v = c$: $m_{emr} = m_{0(emr)}/\sqrt{1-(v^2/c^2)}$ and to $v = c$, $m_{emr} = m_{0(emr)}/0$, this is, m_{emr} it would be infinite, so the mass of the light at rest needed to be zero. So at the speed c of light, it resulted $m_{emr} = 0/0$, or undetermined, because at this speed the energy of the quantum or undetermined, because at this speed the energy of the quantum $E = h\mathcal{V} = m_{emr} c^2$ it was variable as the frequency. But as $E_{total} = E_{kinetic\ relativistic} + E_{rest} = E_{knetic\ relativisticc} + m_0c^2 = mc^2$, so to $m_0 = 0$, $E_{total} = E_{kinetic\ relativistic} = m_{emr} c^2 = E = hv$, so, $m_{emr} = hv/c^2$.

That is, by the theory of relativity, the mass of the quantum was increased by the kinetics, that is, by the motion, that is, by the speed. And the higher the frequency of the electromagnetic radiation, the greater the mass of the quantum at the speed of light. And while relativity indicated that the radiation presented mass of inertia, it also showed that the radiation had zero inertia mass when at rest, for by theory of relativity $m_{quantum} = m_0 / [\sqrt{1-(v^2/c^2)}$, and to $V_{quantum} = c =$ speed of light and radiation, $m_0 =$ quantum mass at rest, $m_0 = m_{quantum} \times 0 = 0$.

Time, Gravity and Wavelength – In 1907, Einstein presented coordinates transformations in the gravitational field of space-time, obtaining to the surface of a body a time dilation that when farther away from the gravitational center, there would be a contraction of distance on the surface, with a lower frequency of radiation emitted, therefore a greater wavelength of radiation.

General Relativity – In 1830, Gauss had presented the infinitesimal variations, geometric differentials of curves, and in 1854 Riemann had presented the curvature geometry of space, in which Euclid's flat geometry was a particular case of infinite radius curvature. It related the structure of the electric and magnetic field as originated by charges, with the structure of the space originated by the mass. Around 1870 Clifford conceived the motion of masses as related by the variation of curvature of space. It was the unification of Physics with Geometry, through the physical fact of a cause always to create an effect. The Moon would cease to fall into the Earth by the attraction of a gravitational force applied to the mass of inertia of the motion of the Moon, by Newton's law, to fall by the curvature of space. As Newton the Moon had this inertia to maintain its rectilinear trajectory and bend by the action of gravitational force, because as Berlinski: *"And so it is"*. And around 1880, G. Ricci-Curbastro, T. Levi-Civita, E. B. Christoffel presented the mathematics of infinitesimal relations, the differentials and partial derivatives between the variables of geometric coordinates, with transformation of the referential system, resulting in determinants or matrices, which for a vector these partial derivatives of these coordinates constituted a tensor (the word tensor originated from a vector applied for tension – force over bodies).

In 1915, D. Hilbert presented equations of the gravitational field in the curved space of Riemann geodesi. And in 1915, Einstein presented the principle of equivalence, in which the acceleration of a reference system was perceived to an observer in this reference as an equivalent action of gravity. Gravitation would be a particular case of acceleration originating from mass of inertia, i.e. the mass of inertia m_i was equivalent to mass m_g of gravity. And this acceleration would be caused by the curvature of space-time in the presence of mass energy and its surrounding field, using also the curved space of the Riemann mathematics and the application of the calculation of tensors developed by Ricci-Curbastro, Levi-Civita And Christoffel, with the work of M. Grosmann for the development of tensors to Einstein's theory. In this field of mass, relativity was included the electromagnetic energy, because the mass formed of electrically charged particles was considered as the main part of the electromagnetic field.

So a distance dx of a point at all coordinates $dx_\mu dx_\nu$ would be obtained by the tensor $g_{\mu\nu} dx_\mu dx_\nu$, where μ e ν were the dimensions of spatial coordinates x_1, x_2, x_3, x_n. And Einstein obtained the tensor $T_{\mu\nu}$ that involved mass, velocity and density, called the mass-energy tensor, and the tensor $R_{\mu\nu}$ that related the tensor $T_{\mu\nu}$ with its displacement, resulting in the equation of the gravitational field of the general theory of relativity: $R_{\mu\nu} - \frac{1}{2} g_{\mu\nu} R = - k T_{\mu\nu}$, where $R = g^{\mu\nu} R_{\mu\nu}$ and k originated from the gravitational flow of Poisson $\phi = 4\pi G\rho$, G = Newton's gravitational constant, ρ = density of matter, $k = 8\pi G/c^2$.

The field would transmit energy and momentum to the mass, exerting force and supplying energy. Thus, this force would be the distortion of space-time and the energy would be provided by the charged particles of the electromagnetic field. In this energy field of curved space-time, forces were replaced by the variation of the curved spatial coordinates. And so, in 1915, Einstein presented the theory of general relativity in which the mass with its energy exerted distortion in space-time, resulting in acceleration of bodies in this distortion. It indicated that the luminous rays should present in the vicinity of gravitational masses a larger curving than the theory of gravitation of Newton, that is, of ≈ 83" to 1.74". Also, that light should present in gravitational field as the Earth a shift to the red. And presented the calculation of the deviation of the perihelion of the planets by the curvature of space. And in 1917 added a cosmological term in the equation of the gravitational field applied to the universe, for the purpose of keeping it static.

Gravitational Waves – According to Newton in 1686 by gravitational law, gravity was instantaneous, but Poincaré in 1905 presented that forces should be transformed through the formulas of the Voigt-Larmor-Lorentz transformations, so that gravitational waves could occur with the speed of light.

Gravitational Radiation and Quantum Mechanics – Between 1916 and 1918, Einstein presented that, to explain the electromagnetic and gravitational radiation, quantum mechanics should be modified: *"For the intra-atomic motion of electrons, atoms to radiate should have not only electromagnetic energy, but also gravitational energy, if only in miniature quantities. Since this is extremely true in nature, it seems that quantum theory should modify not only Maxwell's electrodynamics, but also the new theory of gravitation".*

Gravitational Waves of Space-Time – Between 1916 and 1918 Einstein noted that by the general relativity equation there could be specific cases in which there was acceleration of gravitational mass to generate modification of the structure of space-time, that is, gravitational waves. It was a linear equation in which the mass-energy tensor that presented a disturbance of the gravitational field was similar to the electromagnetic field equation, that is, a wave equation. And the gravitational wave would originate from two power-mass polarization tensioners with double rotation, that is, a quadrupolar disturbance. In the case of a massive body, but of symmetrical motion, such as the collapse of a symmetrically spherical star, the equation was acceleration, but without the waveform, so in that case the space-time would remain totally flat without any ripple. The gravitational waves by the general relativity equation, originated by massive bodies and or initial speed, that would allow for some relativistic effect would result in frequencies of the oscillation of these massive bodies as of 10^{12} to 10^{-18} Hz, that is, in the part of the spectrum between short and long waves. And these gravitational waves would be transmitted through the medium of the space-time field itself. And quantum gravitational radiation would just be quantum field-space-time radiation.

XX. ABSOLUTE SPACE AND FORCE AT A DISTANCE – 1687 to 1851

Between 1715 and 1716 Leibniz presented criticism of Newton's absolute space in 1687, mentioning that not even an angel could determine which body would be at rest and which in motion. The motion was just a change of position between the bodies and the space without matter was only imaginary. While the force would be something absolutely real in the substances, space, time and motion would only be real because they involve infinity, eternity and the ability to perform work or force in the substances. Although the force was real, the motion would belong to the relationships and it would be necessary to look for its causes. The bodies had a certain quantity of motion or force, and there was something more to the nature than was determined by geometry. It would be necessary to recognize something greater as force. But in relation to the circular motion in which Newton showed that the force that acted on the bodies outside the center of the axis of rotation would indicate the recognition of the absolute motion, Leibniz says that the fact that the immediate cause of the motion is in the body would only indicate an absolute motion that left another body with relative motion in relation to that of the body with absolute motion itself. But, Clarke responded to Leibniz that between two alternatives, for the circular motion in a case if there was no external body, the centrifugal force would not exist, which would be absurd, for it was always active being near or far from a body external. And in the other case, if another outer body distant or absent, the body in the circular motion would be with the absolute motion itself, which would mean that there would be absolute motion regardless of whether there is another relative body for comparison of motion.

In 1721, Berkeley presented that a body could be actually moving by the force that acted on the body, in which the force was something absolute. In the case of the circular motion, in which a body tended outward from its axis, it was only a force that acted on the tangent, while an additional centripetal force was not applied equally to balance a radial centrifugal force outward.

In 1851, Foucault presented a proof of the rotation of the Earth on its axis by a pendulum of 30 kg suspended by a wire of 67 m in the dome of the "Invalides" in Paris. The pendulum that oscillated always on a plane, while the Earth revolved around its axis was explained by the gravity force of the pendulum $F = mg$, where m was the mass and g the acceleration of gravity, and the motion of mass of inertia of the pendulum in relation to the absolute space, the Coriolis force = $2mwv$, in which m = mass of the pendulum, w = angular velocity of earth-pendulum, v = tangential velocity of Earth-pendulum observed in 1829 by Coriolis, in accelerated reference, this Coriolis force acted on the pendulum in the direction perpendicular to the axis of rotation of the Earth's motion, in relation to the inertial reference of the absolute space, as Newton, as a force of inertia of this absolute reference of space. And because this force of inertia acted in an absolute space without any mass, there would be no real force, so this force of inertia would be regarded as fictitious. But in this frame of inertia, in the case of the absolute reference of space, this force of inertia would act as if it were a real force, and for the frame of the pendulum that force would appear to be real. However, this force of inertia was regarded as only apparently exerted by a real body by Newton's mechanics, so it would be a fictional force. Then, the pendulum with ≈ zero friction, along with the bracket fixed on the Earth with the velocity around its axis with respect to the fixed stars, would follow the inertia of the absolute space, despite being driven by a nonexistent, fictitious or just seemingly real force.

And the elongation of the equatorial diameter in relation to the axis between the poles of planets such as the Earth, a ellipsoid, was explained as Newton by the motion of Earth's inertia mass in relation to the absolute space, by centrifugal force = mw^2r fictitious, where m = mass of Earth, w = angular velocity of Earth and r = radius of Earth's orbital, in which in an absolute space with no mass acting, no real force could be exerted.

However, as Newton in the case of the Earth's rotation motion was real, the force of inertia, centrifugal, fictitious, would lengthen the horizontal axis of the motion. And Foucault's pendulum by moving by the Coriolis inertia force in relation to Newton's absolute space, also showed the actual motion of the Earth revolving around its axis and the swinging plane of the pendulum always following the same direction in the absolute space.

And it was observed that after a real rotation of the Earth in relation to the Sun and the fixed stars, the Earth performed a rotation in relation to the inertia of the absolute space as the pendulum of Foucault in an identical time. That is, there was an impressive coincidence, meaning that the motion of inertia in relation to the absolute space was equal to the motion relative to the Sun and the fixed stars. That is, the two motions were equal by the fact that all the stars fixed as a whole, that is, the universe, remained immutable, without turning in relation to the absolute space.

XXI. RELATIONAL MECHANICS – **1846 to 1989**

Between 1868 to 1912, Mach had the realization that the fact that the set of fixed stars did not rotate would coincide with the absolute space. And without mass the absolute space would have no force that could be exerted on the oscillation of the pendulum or any other mass of the universe. Therefore, the mass of the fixed stars is that it would be the real cause acting on the oscillation of the pendulum and all the masses of the universe. Then it would be a force of the interaction between the motion of the planets, like the Earth, and the oscillation of Foucault's pendulum, with the distant masses of the universe. And as Mach the resistance of the mass to the change of motion, the inertia, or to a uniform motion, that is, an inertial motion would have as origin the interaction with all the masses of the universe. That is, the forces of inertia would be originated entirely by the other masses, which would later be named the principle of Mach.

Weber in 1846 had presented a force acting between electrics charges in relative motion, from the attraction and repulsion of coulomb and Ampère forces.

In which: $F = (q_1 q_2)/(4\pi\varepsilon_0 d^2) [1 - (1/c^2)(dd/dt)^2 + (r/c^2)(d^2d/dt^2)]$, where q_1 and q_2 were the charges in interaction, $v = (dd/dt)$ it was the relative velocity between the charges and $a = (d^2d/dt^2)$ was the relative acceleration between the charges, $c = (\varepsilon_0\mu_0)^{-1/2}$ = speed of light. So, the force between charges appeared to be related for the first time with velocity and acceleration. And in 1848, it presented the energy of interaction between two charges q_1 e q_2: $E_{q1q2} = q_1 q_2 / 4\pi\varepsilon_0 d \{1 - \{[(dd/dt)]^2 /2c^2]\}$, where $c = 1/(\mu_0\varepsilon_0)^{1/2} = 3 \times 10^{10}$ cm/s = speed of light, d = distance between charges, dd/dt = radial velocity between the charges.

And other expressions between force, energy, charge and velocity were presented or commented, by H. Seeliger in 1864, Helmholtz in 1872, Gauss in 1877, Riemann in 1867, Clausius in 1876, Lorentz in 1895 from A-M Liénard, Wiechert, K. Schwarzschild and C. G. Darwin; W. Ritz in 1908. In 1868, Neumann obtained through Lagrange's expressions for energy, an expression involving electric charges with kinetic energy.

The laws of attraction and repulsion of electric and magnetic charges followed an equivalent expression of Newton's gravitational attraction force, and in 1870 G. Holzmuller presented expressions of gravitation as equivalent to the expressions of electromagnetism for force, energy, speed and acceleration of gravitational masses. And after, in 1872 by F. Tisserand; and around 1875 to 1885 Weber, F. Zollner, from Young and F. O. Mossoti; in 1898 and 1917 P. Gerber; Poincaré between 1906 to 1907; in 1925 E. Schrödinger; in 1977 and 1979 P. B. Eby and in 1989 A. K. T. Assis; A. L. Xavier Jr. and Assis in 1994, subsequently others between 1950 and 1999.

And applying these expressions of electromagnetism to gravitation including the principle of Mach, in 1896 B. Friedlaender and I. Friedlaender presented a first work; in 1900 E. Höfler; on 1904 W. Hofmann; on 1914 H. Reissner; in 1922 Einstein presented that through the theory of general relativity the principle of Mach should be incorporated into relativity, in which the inertia of the motion could not exist in relation to space, but only the inertia of the motion of the masses between them. That is, all the particles in their spatial-time coordinate references that understood the gravitational field $g_{\mu\nu}$ it would be determined by the mass of the universe. But, D. W. Sciama on 1953; M. Reinhardt in 1973; D. J. Raine in 1981; A. Pais in 1982; H. Pfister on 1995; C.H. Brans in 1962; H. Thirring on 1921; B. Mashhoon, F. H. Hehl and D. S. Theiss in 1984; F. M. Peixoto and M. A. F. Rosa, P. S. Letelier and W. A. Rodrigues Jr. on 1994; and subsequently L. Bass, F. A. E. Pirani in 1955; D. R. Brill, J. M. Cohen in 1968; M. Ghins on 1991 and 1992; M. Jammer in 1969; Schrödinger on 1925; Xavier Jr. and Assis in 1994; and others, they presented that general relativity only incorporated in a certain reference the interaction between masses of the universe and of the bodies, and in the interactions appeared additional terms of non-existent forces of circular motions, by the fact of the general relativity appear the space-time independent of the distant mass and always with the inertia of the motion, also by general relativity the coordinate systems appeared unlinked with the physical reality of distant masses and when it occurred the physical reality was by transformation to another coordinate system; G. B. Brown in 1955 and 1982; W. F. Edwards on 1974; J. B. Barbour on 1974; Barbour and B. Bertoti in 1977 and 1982; Eby on 1977; Assis in 1989, 1992, 1993, 1994, 1995, Assis and P. Graneau in 1995; P. Wesley in 1990.

In 1989, Assis applied the interaction of Weber's charges with Newton's gravitation in the interaction between two gravitational masses $m_{(1)}$ and $m_{(2)}$.

As Assis: $E_{m(1)\,m(2)} = [-K_G\, m_{(1)}\, m_{(2)}\, /d]\, \{1 - [kw(dd/dt)^2/2c^2]\}$, where $E_{m(1)m(2)}$ = energy of the masses $m_{(1)}$ and $m_{(2)}$, K_G = constant of Newton's universal gravitation, kw = constant, d = distance among masses, c = speed of light.

And the interaction between a mass M and local bodies such as Earth, magnets, electric charges, friction forces and the Milky Way was expressed by the energy among its masses.

With the interaction of M with the distant galaxies, in which E. P. Hubble in 1929 discovered the first galaxies beyond the Milky Way with the fixed stars, expressed by:

$$E_{M/galaxies} = (4\pi k W K_G \rho / 3H^2_{Hubble})[(3M c^2/kw) - (M v^2_{M/galaxies}/2)],$$

equation of the universe,

where: ρ = mass density in the universe

$v_{M/galaxies}$ = M speed relative to distant galaxies

H_{Hubble} = Hubble constant = V_{galaxy}/d_{galaxy}, at which speed v_{galaxy} of remoteness of the galaxies in the universe was proportional to their distance d_{galaxy}.

It was observed that the gravitational mass, in which bodies of different masses presented the same g acceleration of gravity for different gravitational forces, was equal to the different masses of inertia, this is of resistance to motion, for the same acceleration a resulting in different forces, similar to gravity:

Force of gravity $F_g = M_g g$ and $F_g = m_g g$

Inertial Force $F_{inertial} = M_{inertial} a$ and $f_{inertial} = m_{inertial} a$

To $g = a$, is obtained $M_g g = M_{inertial} a$, e $m_g g = m_{inertial} a$

So, $M_g = M_{inertial}$ and $m_g = m_{inertial}$

This is, the gravitational masses and inertial were all the same.

But in this equation of the universe as Assis, $(m_g v^2_{M/galaxies}/2)$ was identified for a mass m_g as the kinetic energy $m_{inertial} v^2_{m(inertial)}/2$ of inertial mass $m_{inertial}$, which explains the proportionality or equality of the mass of inertia with the gravitational mass. That is, inertial mass, motion resistance, uniform motion, or inertial motion, was a gravitational interaction with the universe. The masses of inertia and kinetic energy emerged as a direct consequence of the gravitational interaction with the universe.

Thus, in the circular motion as the rotation of the Earth there was a centripetal force for the center of rotation and the force $F = m_{Earth} a$, real of gravitational reaction exerted by fixed stars and external galaxies. This acceleration a of the Earth occurred in relation to the fixed stars and external galaxies, and not in relation to the absolute space. And Assis presented brilliantly the deduction of the actual centrifugal force that would result if the fixed stars and outer galaxies revolved around the Earth.

So in the Earth's frame the centrifugal force F with angular acceleration ω_{Earth} of the Earth and radius r of orbit, got $F = m_g \omega_{Earth} (\omega_{Earth} r) = m_g \omega_{Earth}^2 r$. In Foucault's pendulum occurred the actual Coriolis force equivalent to a gravitational force $F = m_{Earth} a$, exerted by the fixed stars and external galaxies that was equivalent to force $F = m_{Earth}\{a + (2v_{Earth} \times \omega_{Earth}) + [\omega_{Earth} (\omega_{Earth} r_{Earth})]\}$. On a rope tensioned with a body of mass m, the reaction exerted by the actual gravitational force $F = ma$, of fixed stars and distant galaxies, equivalent to the gravitational centrifugal force $F = m \omega^2 r_{rope}$. Similarly in the case of Newton's cup or bucket, the water rises in the cup when rotated, in which there was a real gravitational reaction force $F = m_{cup} a$, exerted by the fixed stars and external galaxies, equivalent to the gravitational centrifugal force of $F = m_{cup} \omega_{cup}^2 r$.

XXII. CHARGES, MASS NUMBER, ATOMIC NUMBER AND ISOTOPES
1815 to 1953

Electron Model – Between 1902 and 1903 Abraham presented the first model of the electron with the electric charge e distributed in a rigid sphere of radius r_e.

The energy density for a q charge, distance d, and constant k, in the electric field E: $(E_{energy}/vol)_E = k\, E^2/8\pi$ and how $E = (k\, q)/d^2$ was obtained: $(E_{energy}/vol)_E = (k/8\pi)\, q^2/d^4$. And the total energy for the charge e electron with radius r_e it was expressed by the integration of the distance of the r_e radius to infinity, resulting $E_{total\ energy} = \frac{1}{2}\,(e^2/r_e)$. But as $r_e \rightarrow 0$ would result in energy and infinite mass, and also the repulsion force between charges e it would be manifested to the infinity in which total energy would tend to zero, which would restrict the validity of that model. However, entre 1912 and 1913 Mie using for the first time a particle field model within a mathematical formalism with the use of matrices managed to transform the electromagnetic equations into generalized equations for larger and smaller distances of charges like the electron to circumvent the masses and repulsions or endless attractions.

Quantity of Electrons in Atoms – The particles could through the momentum mv/t = force acting on a free middle path d, with the energy E_{energy} = (force x distance) have a flow Φ_{energy} per an area S_{area} in a time t_f, ie, $\Phi_{energy} = E_{energy}/(S_{area}\, t_f)$, or the particles present a flow incident $\Phi_{particles}$ with the number of $N_{particles}$ particles in an area S_{area} in a temp t_f, with $\Phi_{particles} = N_{particles}/(S_{area} \times t_f)$ with velocity v and mass m. This flow focusing on a target could present a deviation when there was an elastic shock or a return when there was an inelastic shock by absorption or other energy emission by the variation of the incident momentum. And the relationship between the particle flow after the shock and the incident flow, by having an area dimension, was called the shock section. According to the free middle path, shock section and number of particles incident in molecules, could be calculated the quantity of particles that received the shock, that is, the number of particles of the target. In this way, it could be measured the number of negative charges, the electrons, the atoms.

By kinetic theory, where n = number of molecules/volume, r = radius of the molecule, $S_{area} = \pi r^2$ = area or cross section of shock, v = velocity of the molecule, the frequency of shocks by time (f_{shock}/time) = $nS_{area}\, v$, and $f_{shock} = n\, S_{area}\, vt = n\, S_{area}\, d$ where d was the free middle way. According to the free middle path, shock section and number of particles incident in molecules, could be calculated the quantity of particles that received the shock, that is, the number of particles of the target. Thus, it could be measured the number of negative charges, the electrons, in the atoms. And by transformations, the intensity I of the attenuated particle beam, as $I = I_0 \exp^{-n'S'k}$, where n' = number of particles in the shock, k = shock thickness, S' = shock section, I_0 = initial beam intensity. Then, in 1904 Barkla came to measure the attenuation of X-rays by the electrons of these target atoms. So $I/I_0 = \exp^{-n'S'k}$. Being, N_A = number of Avogadro, m_{atom} = target atom mass, ρ' = target atomic density, A_m = atomic mass, n' = target atoms by volume = $[(N_A m_{atom})/A_m] = [(N_A \rho')/A_m]$. And if each target atom has Z electrons in the shock, then $n = Z n' = [Z N_A \rho')/A_m]$ And with the shock section S' obtained experimentally = $[(8\pi/3)(e^2/m_{atom}c^2)^2]$, got $Z = A_m [(3 m_{atom} c^4)/(40 \pi e^4 N_A)]$ And in 1911 gets $Z \approx \frac{1}{2} A_m$. That is, the number of electrons in atoms of smaller atomic masses would be around half of that atomic mass.

Electrons in chemical reactions – In 1904, R. W. H. Abegg presented that the reactions chemicals would be due to the electrons transferred from one to another atom.

Nucleus Charge, Atomic Mass, X-Ray Emission, Wavelength and Atomic Number – Between 1907 and 1909, Barkla, C. A. Sadler and J. Nicol observed that there was a relationship between atomic mass and X-ray emission by each chemical element. In 1907, A. J. Broek had observed that the atomic mass was around double the atomic number and in 1913 found that in several chemical elements the atomic mass was twice the positive charge of the atomic nucleus. For example, helium with atomic mass 4 and 2 positive charges, then concluding that the nucleus charge of all chemical elements would be equal to the atomic number of the periodic table. And also in 1913, H. G. J. Moseley sought to verify the Broek hypothesis, using cathode rays on each chemical element that could present different spacings of the crystals and properties according to their different nuclear charges, and in consequence emit a certain wavelength of X-ray for each chemical element. And found that the higher the atomic number, the smaller the wavelength of the X-rays. Therefore, there was a fundamental quantity that increased from one to the next chemical element, that is, the quantity of positive charge of the nucleus as responsible for the type of the chemical element. In 1914, W. L. J. Kossel observed that the wavelength emitted by the X-rays was due to the transitions of electrons between the more internal atomic orbits of the chemical elements.

Isotopes – In 1815, Prout had realized that all chemical elements could be expressed as multiples of the atomic mass of hydrogen. But, fractional atomic masses were observed in chemical elements. In 1913, Soddy noted that in the radioactivity there was modification of the atomic number and also of the atomic mass, and for the same atomic number there were different atomic masses that called isotopes. and J. J. Thomson by the deviation in electric field found two neon masses with 20 and 22 gram-molecules, differing from the determined value of 20.2 molecules-gram. So as J. J. Thomson the neon 22 would be the neon 20 with 2 hydrogen atoms or another type of gas coupled. However, F. A. Aston determined by spectrographic that the neon 20 and the neon 22 were the same chemical element without mixing another element, therefore non-radioactive isotopes. Other non-radioactive chemical elements also presented isotopes and T. W. Richards with new measurements as in lead confirmed the existence of the isotopes.

Isotopes and Age of the Earth – And in 1929 Rutherford estimated the age of the Earth as less than 3.4 billion years, and in 1931 A. Holmes estimated to be less than 3 billion years old. In 1942, E. K. Gerling, from lead isotopes determined the age of the Earth in 3.2 to 3.9 billion years. And in 1953, with studies by Gerling, H. Brown, F. Houtermans, C. Patterson, G. Tilton, M. Inghram, from lead isotopes in meteorites, was obtained the age of the Earth in 4.5 billion years.

XXIII. QUANTUM ATOM – 1910 to 1999

Quantization in the Atom – In 1910, A. E. Hass presented an atomic structure, in which for the hydrogen atom, of spherical charge ^+e in the r ray nucleus, with an electron in the charge orbit ^-e, mass m_e and frequency $\mathcal{V}$ around the nucleus with charge ^+e, with the potential energy $E_{potential} = [(^-e)(^+e)]/4\pi\varepsilon r = h\mathcal{V}$, where ε = dielectric constant, in which Planck's constant h was first used in the atom.

With the centripetal force of the electron $f_c = m_e v^2/r$, where v = orbital velocity, r = radius of the orbit, and the angular momentum of the electron $\mathfrak{M}_{angular} = m_e v 2\pi r$, with angular velocity $\omega = v/r$, and with Coulomb's attraction force, $F_{Coulomb} = (1/4\pi\varepsilon r)(e^2/r^2)$ or to $4\pi\varepsilon = 1$, $F_{Coulomb} = e^2/r^2$; then equaling the centripetal force: $F_c = mv^2/r = m\omega^2 r^2/r = mr(2\pi f)^2$, f = orbital frequency of electron, with Coulomb force, $F_{Coulomb} = e^2/r^2$ for potential energy with $4\pi\varepsilon = 1$: $E_{potential} = e^2/r = h\mathcal{V}$, $r = e^2/h\mathcal{V}$, so: $mr(2\pi f)^2 = e^2/r^2 = e^2/(e^4/h^2f^2)$, therefore: $r = h^2/(m4\pi^2 e^2)$.

Quantum Orbits – With the discovery that the quantity of charge of the nucleus defined the chemical element of the atoms and there was the balance between positive and negative charge, in 1913 Bohr presented the atoms with positive charges in the nucleus and equal number of negative charges in orbits circulars such as K, L,

M, N, O, P and Q and sub-orbits s, p, d and f, obtained through the X-ray spectra, and which were at stable energy levels with the multiple angular momentum of Planck's constant h, that is, $\mathfrak{M}_{Angular\,(e)} = n(h/2\pi)$. In 1916, Kossel and G. N. Lewis showed that valence, the ability to combine chemical elements, was due to a pair of electrons shared by the atoms of these elements. And in 1915, A. J. W. Sommerfeld presented the elliptical orbits. And in 1916 Lewis, in 1919 I. Langmuir and at 1921 C. R. Bury presented similar electron arrangements for the number of electrons in the orbits in 2, 8, 18, 32 in the last layers which explained the combination valence between chemical elements, applied in 1868. In 1921, I. Langmuir presented that the chemical bond occurred by electro-valence, i.e., electron sharing, or by covalence by the electrostatic attraction between positive and negative ions.

Quantum Hydrogen Atom – Adopting Rutherford's model, in 1913 Bohr proposed to the hydrogen atom, a nucleus that was formed by a positive particle, and around it spinning, an electron. This was the planetary model, where the nucleus was the Sun and the electrons the planets. He admitted that Newton and Coulomb's laws were valid, and matched with centripetal and electrostatic force.

The Coulomb force of electrostatic attraction between two charges q and q' in a medium of dielectric constant or permissiveness in the vacuum ε_0, at radius distance r:
$F_{Coulomb} = (1/4\pi\varepsilon_0\,d)\,(e^2/r^2)$ or to $4\pi\varepsilon_0 = 1$, $F_{Coulomb} = e^2/r^2$
$\varepsilon_0 = 8,854 \times 10^{-12}$ coulomb^2 newton^{-1} metro^{-2} = 7.96×10^{-2} ues^{2}dina^{-1}cm^{-2}

also, μ_0 = magnetic permeability of vacuum = $4\pi\,10^{-7}$ newton/ampere^2
electron rest mass: m_{oe} = 9.11×10^{-28} g = 9.11×10^{-31} kg
electron charge: $e = 1.6 \times 10^{-19}$ coulomb = 4.8×10^{-10} ues
1 coulomb = 3×10^9 ues
electron-volt: eV = 1.6×10^{-19} joule = 1.6×10^{-12} erg
Planck constant: h = 6.625×10^{-34} joule second = 6.625×10^{-27} erg sec
speed of light in vacuum: c = 2.998×10^8 meters/second = 2.998×10^{10} cm/sec

---

But, this model of atom was contrary to electromagnetism, in which the electron with angular acceleration rotating around the nucleus should emit continuous radiation and spiral trajectory to reach the nucleus, and as were also identified specific wavelengths of spectral lines that were emitted with discontinuity, then Bohr presented a postulate for the atom through the unification of classical physics with quantum theory. This postulate understood that the emission of radiation could only be effected by discrete orbits for electrons, with the multiple angular momentum of Planck's constant h: $\mathfrak{M}_{angular(e)} = n(h/2\pi) = n\hbar$, where n = quantum number of energy levels, only radiating by jumping from one energy level to another according to $(\mathcal{E}_2 - \mathcal{E}_1) = h\mathcal{V}$, and maintaining its orbit to the nucleus under the electrical attraction of Coulomb and the laws of classical mechanics. In the hydrogen atom, the attraction between the charge electron ^-e and the charge proton ^+e in the nucleus was compensated by the orbital velocity of the electron.

For a circular orbit round (C) orbital radius (r): linear velocity tangential orbital $v = C/t = 2\pi r/t$, where t = time; $(C/r) = \theta = 2\pi r/r$ radians = 360°; angular velocity $\omega = \theta/t = (C/r)/t = [(2\pi r/r) \text{ radians}]/t = (vt/r)/t = v/r$. The linear velocity being variable of the orbital r radius, and the variation of angular velocity resulting in a centripetal acceleration a_c to the nucleus being variable of linear velocity, by simplified comparison to minimum values: $v/r = a/v$, so $a_c = v^2 r$ electron centripetal force $f_c = m_e v^2/r$ m_e electron mass, electron Coulomb electrostatic force $f_{electrostatic} = (1/4\pi\varepsilon)e^2/r^2$, where the factor 4π was coming from the solid angle of a charge to a spherical surface $S_{area\text{-}sphere} = 4\pi r^2$, radius r around a charge, that is, $\Omega = S_{area\text{-}sphere}/r^2 = 4\pi$ sphere-radians and ε = dielectric constant in vacuum; for the CGS system (electrostatic) $\varepsilon_{CGS} = 4\pi\varepsilon = 1$ unit ε_{CGS}; angular momentum of the electron: $\mathfrak{M}_{angular} = m_e v 2\pi r = nh$, n = 1, 2.3,..., or $\mathfrak{M}_{angular}/2\pi = m_e vr = n\,(h/2\pi) = n\hbar$; equalizing forces and replacing: $v = nh/m_e r = e^2/4\pi\varepsilon\,(n\hbar)$; $r = 4\pi\varepsilon\,(n^2\hbar^2)/m_e e^2$. The total energy of the electron in the case of hydrogen can be obtained by the addition of the kinetic energy $E_{kinetic} = m_e v^2/2$ + potential energy $E_{potential} = [(^-e)(^+e)]/4\pi\varepsilon r$; $E_{total} = m_e e^4/(4\pi\varepsilon)^2\, 2n^2\hbar^2$.

Helium Atom – In 1914, Rutherford noted that the helium atom would consist of four positive electrons and two negative charges, in which hydrogen was made up of a positive charge and a negative charge. And it's named positive electron charge.

Orbital Quantum Energy – Between 1913 to 1916, J. Franck and G. L. Hertz conducted gases ionization experiments through electrons emitted by the cathode under an electric potential. And they detected that when the potential reached 4.9 volt or multiple, the electrons of the cathode were absorbed by the electrons of the atoms by inelastic shock, and emitting light. With the lower value of this voltage, the electrons of the cathode reached the electrons of the atoms with elastic shock, keeping the electrons of the atoms at the same initial level of energy. It was proven that electrons only changed their orbits by a specific quantity of electron energy inside the atom.

Fine Structure Constant – In 1915, Sommerfeld determined a constant $\alpha = k\,h\,c/2\pi e^2 = 137$, the fine structure constant that was experimentally related between the constants c of light, Planck's h and the charge e of the electron, where k is the electric constant, with the narrow spectral lines of the atoms.

In 1916, Kossel sought to explain the stability of the noble gases, that is, no formation of compound with other chemical elements, as due to the stable layers of the electrons around the atomic nucleus. And the chemical reactions were due to the external layers of the electrons, that is, the valence, in which there would be formation of the cations (+) and the anions (–) that would join through the attraction of Coulomb.

Spin – In the atom, the motion of the electron was a minimum electric current which generated the magnetic field that formed spectral lines of the atom.

But other spectral lines resulted in considering beyond the orbital motion of the electron, of m_e mass, v_L orbital velocity, r orbital radius: $\mathfrak{M}_{angular(e)}$ = $m_e v_L 2\pi r$ = nh, n = 1, 2, 3,..., or $\mathfrak{M}_{angular(e)}$ = $m_e v_L r$ = n(h/2π) = nℏ, which was the quantum number of $\mathfrak{M}_{angular(e)}$ = $[l(l+1)]^{1/2}ℏ$, l = 0, 1, 2,..., (n-1), also the atomic position of the orbital magnetic moment of the atom, observed in deviations of atoms in magnetic field by O. Stern and W. Gerlach, in 1922, and subsequently obtained $\mu_l = eh/2m_e [l(l+1)]^{1/2}$.

With the observation of spectral lines of narrow distances, S. A. Goudsmith and G. E. Uhlenbeck, in 1925, proposed the spin of the electron, The spin of the electron, that according to A. H. Compton, in 1921, the electron rotating on itself would be the smallest magnetic particle, the origin of the magnet. Also, in 1921 E. H. Kennard presented that the electrons would have the angular moment intrinsic, the spin. Before 1925, on a spin calculus elaborated by R. L. Krönig, W. Pauli Jr. issued an opinion that in nature there was no electron spin. But in 1925 J.C. Slater presented the eventual spin and in 1926, R. L. Krönig issued the calculation of the rotating ratio of the electron with h, with the value of the spin s = ½ h/2π. Also in 1926, F. R. Bichowski and H. C. Urey presented the spin hypothesis. And then the atom would have the quantum number of the angular momentum of the electron itself, the spin $S_e = [s(s+1)]^{1/2} h/2\pi$, and the position of the angular momentum of the electron itself of $\mu_s = eh/m_e [s(s+1)]^{1/2}$. In 1927, T. E. Phipps Jr. and J. B. Taylor they observed in a hydrogen beam that would only have the state of the orbital momentum of an electron, another value for the momentum. For the nucleus with the proton, the magnetic momentum would be around $e^+h/4\pi m_p$, where m_p it was the mass of the proton, by the calculation of Bohr's quantum atom. But the value obtained from the experiment was $e^- h/4\pi m_e$, that is, a value greater than $1/m_e$ in relation to $1/m_p$ to the momentum of the proton, proving the existence of the electron spin.

The spin being an intrinsic electric current generated a magnetic field, the origin of the magnet, that is, the magnetic charge and the magnetic field. And this rotating motion of its own in motion also generated an electric field. Einstein showed that an electric field could be seen as magnetic or otherwise, and the spin of the electron is the origin of the magnet with its charge and magnetic field. The exchanges of relativity reference systems only exchanged one field for the other. Faraday had tried to explain through the formation of circular currents that would generate a magnetic field and this forming current and electric field.

But the origin of the spin remained unknown by quantum mechanics.

Electron Wavelength – By classical theory, an electric charge oscillator, such as the electron, presented an electric field of charge, and by oscillating it generated a magnetic field, resulting in a wave of the electromagnetic field. This wave would be emitted with wavelength λ_1 and frequency $\mathcal{V}_1$, which would resonate with the electron doing it oscillate on the same frequency $\mathcal{V}_1$ of the wave, which was contrary to the experimental fact.

But in 1923, Compton admitting the quantized radiation $\mathcal{E} = h\mathcal{V}$, for example, the X-rays, in which it acted as particle of momentum p over a stationary electron of mass m_e, obtained the frequency $\mathcal{V}_2$ of the photon and the momentum of the electron [$p_{electron} = m_e v_e$], v = velocity of electron, after incidence of the photon.

So, $(c/\mathcal{V}_2 - c/\mathcal{V}_1) = \lambda = h/[m_0c (1 - \cos\theta)]$, where θ it was the spreading angle of the photon and $\lambda_e = h/(m_e c)$ the wavelength of the electron = 2.4... x 10^{-10} cm and r = 3.86 x 10^{-11} cm.

Bohr's Postulate Explained by the Postulate of L. V. P. R. De Broglie – Between 1925 and 1927, as L. De Broglie, the wavelength of the electron λ in the hydrogen atom was equal to the length of the orbit λ_{orbit} of the electron. Thus, integers of electron wavelengths in the electronic layers allowed stability in the atom. In the case of fractional waves there would be irradiation of the electron by shifting the orbit. So Bohr's postulate was explained by postulate of L. De Broglie.

A quantum of energy $\mathcal{E}$ of radiation admitted as wave and particle by the expression: $\mathcal{E} = h\mathcal{V} = mc^2 = mc \times c$ and being $\mathcal{V} = c/\lambda$, $\lambda = h/mc$, where h = Planck's constant, $\mathcal{V}$ = quantum frequency, m = quantum mass, c = light speed, λ = quantum wavelength. Applying for any particle of momentum material $m_{particle} v_{particle}$: $\lambda_p = h/m_{particle} v_{particle}$, where λ_p = particle wavelength, $m_{particle}$ = particle mass, $v_{particle}$ = particle velocity.

And for the electron: $\lambda_e = h/m_e v_L$; $v_L = e/(4\pi\varepsilon m_e r)^{1/2}$; being λ_e = wavelength of electron, m_e = mass of the electron, v_L = linear velocity or electron translation; e = electron charge, ε = dielectric constant, r = electron orbital radius, then: $\lambda_e = (h/e(4\pi\varepsilon r/m_e)^{1/2} = 3.3...$ x 10^{-8} cm. And with orbital radius r = 5.3 x 10^{-9} cm, $\lambda_{orbital} = 2\pi r = 3.3...$ x 10^{-8} cm. That is, the wavelength of the electron λ_e in the hydrogen atom, it was equal to the length of the orbit λ_{orbit} of the electron.

Matrices – Between 1924 to 1925, M. Born, W. K. Heisenberg and P. Jordan presented particle positions and motions in the space that resulted in matrices, mathematical calculus elaborated by a. Cayley in 1858.

For each coordinate point was the complex exponential where $i = (-1)^{1/2}$: $\exp^{(i2\pi/h)}$ And the function of Hamilton H (p,d) = W, elaborated around 1840, where p = momentum, d = position and W = energy. $H = \sum p \, (dd dt) - E (q, (dd dt))$, with E = kinetic energy = $\sum \frac{1}{2} mv^2$, m_p = particle mass and v_p = particle velocity; was introduced the J_k conjugates and ω_k as action-angle variables that were reported to H, and ω_k reported to the frequency f of motion, with $(dJ_k/dt) = \partial H/(\partial \omega/c)$ and $(d\omega_k/dt) = v_k = \partial H/\partial J_k$, where $k = 1, 2,... f_p$), f_p = particle parameters; and as Bohr: $f(n\,m) = 1/h\,(W_n - W_m)$, where n and m they were particle states; it is obtained also pd($n\,m$) $\exp^{i2f\,(nm)\,t}$, resulting in $\exp^{(i2\pi/h)\,(H_n H_m)\,t}$ where H_n and H_m were diagonal elements of matrices Hermitian, of C. Hermitian presented in 1855, p_k and d_k formed of imaginary conjugates: $d_k\,(m\,n) = d_k^{\#}(n\,m)$ and $p_k\,(m\,n) = p_k^{\#}(n\,m)$ with kind of symmetry, where in quantum theory all matrices were assumed to be of the same species. And getting a matrix S_S: $\sum_k S_s\,(n\,k)\,S_s^{\#}(m\,k) = \sum_k S_s\,(k\,n)\,S_s^{\#}(k\,m)$.

Quantum Antisymmetry – In 1915, A. E. Noether had shown that symmetry involved laws of nature. And in 1925, Pauli observed that the electrons in the atoms to be balanced should present a antisymmetry between two quantum states of energy level, besides the angular momentum that defined orbital motion, approximation and spacing, or vibration of the nucleus, and magnetic momentum space position. And concluded that two electrons would always occupy different quantum states. In 1926, P. A. M. Dirac named the Pauli exclusion principle, whereby two electrons were excluded from occupying a same quantum state.

Quantization of Electromagnetic Field – In 1926, Born, Heisenberg and Jordan presented the quantization of the electromagnetic field, with the introduction of Planck's constant h.

For the particle and quantized field with the constant h, like the electron, by the spatial coordinates C_n and the v_n velocity = C_n/dt, within an L energy field, according to Lagrange around 1760, with a field potential energy $E_{potential}$: $L = [E_{kinetic} - E_{potential}] = L\,(C_n, V_n, t)$.

Integration resulted in $S_L = \int L(C_n, V_n, t)dt$, called integral action, according to Hamilton's work around 1840, in which the physical system evolved from t_1 to t_2.

Of the variation of this integral action resulted in the field of the electron an equation of the motion for a minimum expenditure of energy:
$(d/dt)[\partial L/(\partial C_n/dt)] = (\partial L \partial C_n)$
With the introduction of the momentum $p_n = m_n v_m = \partial L/(\partial C_n/dt)$ was obtained the Hamiltonian:
$$H = \sum_n p_n (C_n/dt) - L (C, C/dt)$$
And as there were other relations with the momentum was introduced H'ψ, where ψ was a function of the field of a particle, and obtained the quantized equation with H, Planck constant, for the particle in your field, in which
$i = (-1)^{1/2}$:
$$H'\psi = ih/2\pi (d\psi/dt)$$

Quantum Wave Equation – In 1926, Schrödinger determined the wave equation of a particle in the atom, and that as this equation would represent a real wave.

As Schrödinger, where the angular frequency of a wave of a particle was $\omega = 2\pi f$; f = particle wave frequency; wave velocity $v = \lambda f$; λ = wavelength; $\lambda = h/mv$; h = Planck's constant; m = particle mass; particle energy $E = hf$; and being $E_{potential}$ = potential particle energy, was obtained ψ = wave function of the particle, by the undulatory motion equivalent to a y coordinate of the wave oscillation component, with the being the maximum oscillation of the wave or amplitude, where in the complex exponential $i = (-1)^{1/2}$.

So: $y = A \exp^{-i2\pi[t-(x/v)]}$, and as $[\cos 2\pi f - i \sen 2\pi f] = \exp^{-i2\pi f}$
Results, $y = A \cos 2\pi f (t - x/v) - iA \sen 2\pi f [t - (x/v)]$ ψ = $A \exp^{-i2\pi[t-(x/v)]} = A \exp^{-(2\pi i/h)[ft-(x/\lambda)]}$

So, ψ = $A \exp^{-(2\pi i/h)[(Et-(px)]}$ and differentiating $\partial^2\psi/\partial x^2$ e $\partial\psi/\partial t$:
$(h/i2\pi)(\partial\psi/\partial t) = (h^2/8\pi^2 m)[(\partial^2\psi/\partial x^2) + (\partial^2\psi/\partial y^2) + (\partial^2\psi/\partial z^2)] - V\psi$
And by dividing by the exponential factor: $(\partial^2\psi/\partial x^2) + (8\pi^2 m/h^2)(E - E_V)\psi = 0$

This was the Schrödinger equation for the electron and other particles without considering the intrinsic angular momentum of particles like the electron itself, in which the spin was inserted as a complement. Also, by the Schrödinger particle wave equation, it was obtained for energy emission levels $\mathcal{E}_n = (n + ½) hf$, where for n = 0, was obtained h/2.

Relativistic Quantum Wave Equation – Next in 1926, O. B. Klein, V. A. Fock and W. Gordon introduced the Schrödinger equation to the relativistic equation.

So: $E^2 = p^2 c^2 + m_0^2 c^4$
And getting the Schrödinger equation relativistic:
$$[(\partial^2\psi/\partial x^2)+(\partial^2\psi/\partial y^2)+(\partial^2\psi/\partial z^2)] = \nabla^2\psi = [(m_0 c)/(h/2\pi)]^2 \psi + (1/c^2)(\partial^2\psi/\partial t^2),$$
But also without the spin of the particles, or particles with spin = 0.

Of Reality to the Probability – While Schrödinger regarded as real the electron wave around the atom, in 1926 Born applied the mathematical and mechanical probability of matrix in the Schrödinger equation, obtaining a wave of probability mathematical probability.

So it would be associated with a wave function $\psi(x,t) = A$ sen $2\pi(x/\lambda - ft)$ for a particle, equivalent to the electric field of the electromagnetic radiation wave $E(x,t) = A$ sen $2\pi(x/\lambda - \mathcal{V}t)$. E ψ^2 being equivalent to the electromagnetic field E_{em}^2 for the waves of electromagnetic radiation, that is, also a measure of probability. And ψ^2 representing the intensity of two superimposed particle waves ($\psi_1 + \psi_2$). And these superimposed particles could be tangled, interconnected, associated, non-separated.

Spin Up and Down – In 1926, Heisenberg with Lagrange and Hamilton energy expressions, wave function and quantized field, showed electron energy levels in their orbits, in which electrons of a parallel spin were presenting a wave function with less probability of attraction and with lower energy level in an atom, in which the functions of the particles were matrices. There were no repulsion or attraction forces when the spins were antiparallel, because in this quantum mathematics were quantum field effects of a complex number of exponential function and probability positions that did not correspond to any correlation with the classical concepts. So at the energy level, if an electron was right above (right hand up) or + 1/2, the other would be right below (right hand down) or – 1/2. Also Dirac followed the same quantum mechanics of relationships between mathematical variables to explain the repulsion or attraction of particles.

Zero Point Energy – In the particle radiation from the emission frequency determined by the energy difference of the orbitals according to Bohr In 1913, the transitions of these states would be related to the positions and momentums in which the atom was found, and that there would always be minimal quantum energy h/2.

In the atom, by positions of a state n_1 to n_2, through frequencies $\mathcal{V}_{n,z} = (\mathcal{E}_n - \mathcal{E}_z)/h$, where $\mathcal{E}$ quantum energy, h = Planck's constant, which in calculus adjusts as an matrice, from Einstein's probability of emission: $P_{n,z} = 8\pi e^2/3hc^3 (2\pi\mathcal{V}_{n,z})^3(x_{n,z})^2$, where e = charge of the electron, c = speed of light, and come get $\mathcal{E}_n = (n + \frac{1}{2})h\mathcal{V}_0$, where $\mathcal{V}_0$ = quantum frequency to the zero energy level in which n = 0, was obtained h/2, as Heisenberg and Schrödinger.

This h/2 energy emerged from the potential energy that involved the zero energy point, which was seen on the electromagnetic emission oscillator. The energy of an oscillator was determined by $F = -kx = ma = m(d^2x/dt^2)$, $(d^2x/dt^2) + (kx/m) = 0$, and $x = A \cos(2\pi vt + \varphi)$ e $\mathcal{V}_f = 1/2\pi$

$(k/m)^{1/2}$, where, F = Hooke's restoration force, k = constant, m = mass of the oscillating particle, a = acceleration, $\mathcal{V}_f$ = oscillation frequency, Λ = wavelength, φ = wave phase constant. This force F was related to the potential $E_{potencial}$, through $E_{potencial}(x) = \int_0^x F(x)dx = k\int_0^x xdx = ½ kx^2$, between $x = -A$ and $x = +A$, in that energy E was expressed by $E = ½ kA^2$, and so $E_{potencial} = ½ kx^2$

And the Schrödinger equation for $E_{potencial} = ½ kx^2$: $(\partial^2\psi/\partial x^2) + (8\pi^2 m/h^2)(E - V)\psi = 0$, or: $(\partial^2\psi/\partial y^2) + (\alpha - y^2)\psi = 0$, and introducing the non-dimensional quantities was obtained: $y = \{(2\pi/h)[(km)^{1/2}]\} x = \{2\pi [(mv)/h]^{1/2}\} x$; getting $\alpha = (4\pi E/h)(m/k)^{1/2}$, and with $\mathcal{V}_f = 1/2\pi (k/m)^{1/2}$ resulted $\alpha = 2E/h\mathcal{V}_f$

Solving the equation $(\partial^2\psi/\partial y^2) + (\alpha - y^2)\psi = 0$ to y = ∞ and ψ as a asymptotic $\psi = f(y)\exp^{-y^2/2}$, resulted in $d^2f/dy^2 = \sum_{n=0}^{\infty} n(n-1)A_n y^{n-2}$, where A_n = coefficient, and resulting $\alpha_n = 2E/h\mathcal{V}_f = 2n + 1$ or $E_n = (n + ½)h\mathcal{V}_f$ with n = 0, 1, 2, ..., and to n = 0: $E_n = ½ h\mathcal{V}_f$.

Group Theory for Quantum – Between 1926 and 1927 E. P. Wigner, in 1927 Heisenberg, at 1931 H. Weyl and in 1932 B. L. Waerden presented, for the electron system in atoms, the theory of groups that contained in a closed system of a set, for any mathematical operation, inverse and associative operations, for numbers, points and all kinds of transformations, elaborated by Lagrange in 1770 and developed by F. Klein in 1870.

Photon – In 1926, Lewis named the quantum light of Einstein, which was also the Planck quantum, of photon.

Relativistic Coordinates – In 1926, O. B. Klein and also Fock, Gordon, T. Whence and H. Dungen presented the electron Schrödinger equation with the relativistic mass and transformations between coordinate systems.

Incomplete Coordinates – In 1926, L. H. Thomas and also Y. I. Frenkel observed that in the transformation of coordinates by the theory of relativity, the electron in motion and the resting nucleus for the electron at rest and the motion nucleus, the electron acceleration was nullified and the atom lost one precession in his motion. It was necessary to include this acceleration in order to obtain the experimentally observed atomic precession.

Electron Diffraction – By classical theory, the waves presented light and dark bands of interference, horizontal vibration of the polarized crystals planes and the outline of the atoms, edges and gaps, the diffraction with the interference pattern, while the particles They presented scattering without diffraction, polarization or interference bands. In crystals with certain atomic networks occurred the outline of the atoms, edges and gaps, the diffraction, by electromagnetic waves like the X-rays, while the particles as the electron presented only a scattering according to the behavior of the classic theory. But in crystal of larger atomic networks was observed, in 1927, by C.T. Davison and G. P. Thomson, and L. Germer, the diffraction of the electron, revealing the associated wave behavior of L. V. P. R. De Broglie, in which this wave was reflected in several planes of the crystal resulting in the classic wave interference.

Quantum Potential- Between 1926 and 1927 L. De Boglie introduced the Schrödinger equation to a hidden real variable function, which resulted in a position energy term, a quantum potential energy interconnecting a particle to the entire space. This quantum potential energy $E_{quantum\ potential}$ it was function of h, m and ρ, where h is Planck's constant, m the particle mass and ρ the probability density of where the particle is located. That is, the particle could be determined to be superimposed by the probability squared. However, it did not explain what this quantum potential energy was.

Waves of L. De Broglie – In 1927 as L. De Broglie waves could vary depending on the position and motion of the particles in the atoms or in the free state, constituting a function of ψ wave.

In the case of the wave velocity of L. De Broglie in which, for example, an electron with mass m and velocity v would present the velocity of wave $v_{wave} = f_e\ \lambda_e$, where f_e = electron frequency and λ_e = h/mv the wavelength of the electron.
By quantum theory: $\mathcal{E}$ = energy = $h\mathcal{V}$, So $\mathcal{V}$ = frequency = energy/h
By the postulate of L. V. P. R. of Broglie to the electron:
E_e = energy = mc^2 ; f_e = energy/h = mc^2/h
Therefore: $v_{wave} = f_e\ \lambda_e = (mc^2/h)\ (h/mv) = mc^2/mv = c^2/v\mathcal{E}$

That is, the velocity of the wave of L. De Broglie would be much greater than c, and that velocity would then be regarded as without physical significance or even without any meaning!

Undetermined Coordinates – And in 1927, according to Heisenberg, the particle positions and velocities resulted in indetermination as $\Delta x\ \Delta p \approx h$, where p = mv, for a free particle and $\Delta x\ \Delta p \approx nh$ for the atom where n was the quantum number.

Electromagnetic Field in the Atom and Interaction of Spins – In 1927, Pauli presented the action of an electromagnetic field in an atom, and in 1929 also J. A. Gaunt and in 1930 G. Barsoum, with the interaction of spins and the electromagnetic field that formed emission of the electromagnetic spectrum.

Thermionic Effect and Potential Barrier – In 1883, T. A. Edison observed that the metallic lamp wires presented the emission of light. In 1899, J. J. Thomson identified as the emission of electrons emitting the bluish light, ionizing the atoms by the high temperature of the metallic filament in the passage of the current by the metallic resistance. In 1903, H. A. Wilson explained this effect as similar to the evaporation of gases by temperature and pressure. Between 1911 and 1913, Child and Langmuir related the emission tube with the electrical potential, the dielectric constant in the vacuum and the charge and mass of the electron. In 1923, W. Schottky noted that in metal atoms, an intense electrical field removed electrons, that is, the dependence of electric current density with the electric field. And S. Dushman observed the dependence of the density of electric current with the mass m, the charge e, the constant k_B of Boltzmann, the absolute temperature T, Planck's constant h and a function $e\psi$ of the charge, in which at 1928 Millikan noted that $e\psi$ was the energy $h\mathcal{V}$. On 1927, F. H. Hund noted that the same electron with a lower energy should be able to come out of an electric potential of higher energy. In 1928, G. Gamow and also E. U. Condon and R. W. Gurney, showed with the Schrödinger equation, and with empirical observations, relationships between disintegration constants, nucleus rays and permeability, that a smaller particle of energy could overcome a potential barrier with higher energy, by probability, that is, the tunneling effect. And between 1928 and 1929 L. W. Nordheim noted that $e\psi$ expressed the potential energy difference of the surface and the higher orbital level energy of the electron. And according to Nordheim and Sommerfeld, some electrons with less energy than the potential barrier could traverse that potential barrier of the metallic surface, by increasing temperature and by the action of an intense electric field which would nullify this potential field of energy of the atoms of the surface that prevented the electrons from coming out of these atoms from the surface. And in 1928, J. R. Oppenheimer and also R. H. Fowler and Nordheim observed electron tunneling by the electromagnetic field effect. In 1929 Born showed presentation on quantum tunneling and O. K. Rice showed that tunneling was related as an initial cause of the dissociation of molecules.

Quantum Electrodynamics – In 1927, Dirac presented a quantization of the electromagnetic field of the type ψ = function of $[ih/2\pi \, (d\psi/dt)]$ in the time when $i = (-1)^{1/2}$ with interaction between electron and photon, in which there was creation and destruction of particles through mathematics by wave functions created and destroyed, in which it originated quantum electrodynamics or QED, and then also in 1927 generalized by Jordan and O. B. Klein, in 1928 by Wigner, Gordon and C. G. Darwin, in 1929 by Heisenberg and Pauli, in 1930 by E. Fermi, in 1931 by L. Rosenfeld and in 1934 by Pauli and V. F. Weisskopf.

Thus, in this sea of particles of Dirac theory, two nearby electrons presented an interaction with photons of this vacuum field, causing the electromagnetic repulsion between the electrons. In this interaction entered the fine structure constant $\alpha = 1/137$ by a coupling constant with a better value for this balance of electromagnetic interaction. This interaction occurred in distance $> 10^{-10}$ cm and the life time of these particles between 10^{-15} to 10^{-21} seconds, that by this tiny time of creation and annihilation were considered virtual.

Mechanical Wave and Particle Force in Quantum Mechanics – In 1927 P. Ehrenfest presented that the physical magnitudes calculated by quantum mechanics followed the law of Newton $F = ma$ in which the motion of a wave was equivalent to the motion of a particle by Newton's law. Also, in 1928 A. E. Ruark presented the agreement of the wave motion with the particle.

Wave and Complementary Particle – In 1928, Bohr presented the principle of complementarity of wave and particle. When there was a wave measurement, then the particle was undetectable. And if particle measurement was carried out, it was the wave that would be undetectable. But, it was simultaneous the existence of wave and particle.

Dirac Electron Equation – In 1928, Dirac applied relativistic equation to obtain an electron equation.

So, with $E^2 = p^2c^2 + m_0^2c^4$, where $E = E^2/c^2 = p^2 + m_0c^2$, where E = energy, p = momentum, m_0 = mass of the particle and c = speed of light.

With $(E/c) = \pm (p^2 + m_0c^2)^{1/2}$ and with signal resolution + : $E/c = \alpha_0 p + m_0c\beta_0$, $\alpha_0 = v/c$ and $\beta_0 = [1 - (v^2/c^2)]^{1/2}$, and as Heisenberg the particle's positions and velocities following indetermination and defined by matrices, the constants α_0 and β_0 were replaced by α_n and β_n to the quantum field, so: $E \psi = c [\alpha_n p + m_0c\beta_n] \psi$

And by the complex part of the equation in which $(-1)^{1/2} = i$, getting: $i (h/2\pi) [\partial\psi/\partial t] = c [\alpha_n [- i (h/2\pi) (\partial/\partial x + \partial/\partial y + \partial/\partial z) + m_{0e}c\beta_n] \psi$, which is the Dirac equation for the electron.

Electron Radius by Dirac Equation – As Dirac, in 1928, the electron should have a radius equal to zero, that is, the electron should be a spot particle and the spin would not be attributable to constituting revolving parts on its axis, but it was simply an intrinsic property of the particle itself.

Electron Radius – The model of the rotating sphere electron relativistic was obtained of the spin and the magnetic momentum of the electron, resulting in a sphere of radius = 4.82 x 10^{-11}, contrary to Dirac in 1928 which the electron should have a radius equal to zero.

By model of the rotating sphere electron relativistic was obtained from the spin $S_e = ½$ and magnetic momentum of electron $\mu_e = eh/2m_0ec$, resulting in a radius sphere = 4.82×10^{-11} cm while 3.86×10^{-11} cm to the Compton radius, contrary to Dirac in 1928 in which the electron should have a radius equal to zero, that is, the electron should be a spot particle. And a component of free electron velocity was driven to velocity $\pm c$ of light. And the linear momentum of the electron should be reported with a oscillatory part whose frequency was at least $2mc^2/h$ in the distance of magnitude $h/2\pi m_0ec$.

Experiments at CERN in 1979 indicated that the electron charge was confined within a radius region $< 10^{-14}$ cm and subsequent measurements presented 10^{-16} cm. But, the calculation by the Dirac equation for the magnetic momentum of the nucleus of the atom was contrary to the experimental measurement in which the nucleus presented structure, that is, contrary to a point particle of Dirac theory.

Sea of Negative Charges and Creation of Positive Charges – Dirac in 1928 had found that of the relativistic expression $E^2 = (p^2c^2 + m_0^2c^4)$, $E = \pm (p^2c^2 + m_0^2c^4)^{1/2}$, end for the negative root concluded that there was a negative energy in the vacuum, in which therefore the space could not be completely empty that nothing could. But in the space where there were no particles, that's where the lowest energy level, the negative energy, a sea of electrons completely filling the vacuum. And by activating these electrons they could go to the positive state in which there was no electron saturation. Thus, an electron would leave the negative energy resulting in an electron of charge and positive energy and in the sea of electrons would be a localized space from which an electron had emerged. But in the atom it was the proton that had the positive charge with a mass 1,840 times greater than the electron.

Mathematical Equality of Positive and Negative Electron Masses – Despite the positive charge with greater mass of the proton than that of a positive electron with the mass of the negative electron, by mathematics, in 1928 Weyl equaled the masses of the negative and positive electron, to equal the charges of the proton and the positive electron.

Spin in the Equation – The Dirac equation of the electron in 1928 resulted in four vectors of magnetic quantum origin, which would be the spins of electrons and their antiparticle.

Of the Dirac equation solutions of 1928 resulted in obtaining matrices of the type $\begin{pmatrix} 0 & 1 \\ 1 & 0 \end{pmatrix}, \begin{pmatrix} 0 & -i \\ i & 0 \end{pmatrix}, \begin{pmatrix} 1 & 0 \\ 0 & -1 \end{pmatrix}$ and resulting $\begin{pmatrix} 1 \\ 0 \end{pmatrix}, \begin{pmatrix} 0 \\ 1 \end{pmatrix}, \begin{pmatrix} 1 \\ 0 \end{pmatrix}$ e $\begin{pmatrix} 0 \\ 1 \end{pmatrix}$ of four columns consisting of vector components or "spinors" that indicated in the directions of a particle axis of spins $S_{e-} = +½\ \hbar$ e $S_{e-} = -½\ \hbar$ to the electron, being according to experimental results of the electromagnetic spectrum. And $S_{-e} = +½\ \hbar\ S_{-e} = -½\ \hbar$ to the electron antiparticle with that negative energy of the sea of electrons in the vacuum.

Loss of Matter by Abstract Spaces – From 1926, what resulted from applying relativity in quantum mechanics was to obtain a symmetry of abstract spaces where antiparticles came in, that is, combination of charges, with junction of space-time and space-matter. Positive and negative energy exchanges should result in particle creation and annihilation. According to S. Weinberg, by the introduction of special relativity in the quantum mechanics by Dirac, the matter lost the position of a larger goal of Physics by being replaced by the principles of symmetry of space-time and abstract spaces.

Quantum Fluctuation – Then, with the quantum theory of Heisenberg, Born, Jordan, Pauli and Dirac, in the 1920, the empty space or total vacuum designed by antiquity according to Plato, Democritus, Anaximander and Epicurus which was the place where the matter was situated at rest or in motion, was replaced by a quantum vacuum-space that should contain the result of probability, uncertainty and statistical quantum mechanics. In the case of Aristotle the space should contain some physical entity, because a force applied to a body in a finite density medium would result in a finite speed, and if the space were an empty or total vacuum or zero density this shifting speed in the space would be infinite, which would be contrary to nature. In the case of quantum theory, quantum vacuum-space should contain an energy, quantum fluctuation energy, nonzero. Although in 1635 E. Torricelli, when it poured mercury into a glass tube with a closed end and turned the tube dipping the open end into a vat of mercury, in which the mercury of the tube descended by atmospheric pressure leaving a space between the liquid and the closed end of the tube, and concluding that a total vacuum-space had been obtained, and also in 1657 when Guericke when removing with a pump the air of two empty hemispheres of copper resulted in atmospheric pressure keeping the hemispheres united, concluding that there was the total vacuum-space within the two hemispheres, the total vacuum-space or total emptiness should present the energy of the zero point resulting from the quantum fluctuation by the new mechanics.

And by Heisenberg's uncertainty principle the vacuum was filled with subatomic particles that were created and annihilated by the energy of the vacuum zero point for a trillionth of a second.

This energy was absolute without any physical process being able to remove it from a place in space. And remained in space even if any matter or radiation was undetectable or unobserved. How quantum energy was $E_n = (n+1/2) h\mathcal{V}$, onde n = 0, 1,2,3..., the lowest energy level was $\mathcal{E}_0 = \frac{1}{2} h\mathcal{V}_0 \neq 0$.

So, quantum fluctuation should contain the fluctuation of the electromagnetic field, which on average was zero. In this quantum mechanics a wave presented a certain frequency $\mathcal{V}$, and for level 1 of energy, $\mathcal{E}_1 = h\mathcal{V}_1 + \frac{1}{2} h\mathcal{V}_1$ with a wave unit that was the quantum or photon $E_{wave} = h\mathcal{V}_{wave}$. And removing that photon from that energy level in that quantum vacuum space would remain the energy $\frac{1}{2} h\mathcal{V}_{wave}$ of this photon, with no electromagnetic field, but having this energy of each electromagnetic field fluctuation frequency, in which the sum of the energies that corresponded to all the frequencies $\mathcal{V}$ from 0 through ∞ was the energy of the quantum vacuum-space, $E_{\text{quantum vacuum-space}} = \sum_v \frac{1}{2}(h\mathcal{V})$, so the energy of quantum vacuum-space would be infinite! For this, quantum mechanics created boundary conditions, confinement, contour, restricting frequencies to hide the infinite energy!

Minimal Energy and Energy Action in Time – In 1929, E. C. Kemble presented a generalization of the Schrödinger equation applying minimal action. And in 1932, M. Stone presented a theorem in which time was related to energy, being introduced into the denominator of the energy equation in quantum mechanics. However, in 1949, T. Kato presented a determination of quantum energy levels by minimal action. And in 1950 Dirac presented widespread dynamic theory also by minimal action by the Hamilton equations.

Energy and Infinite Field– In 1929, Pauli and Heisenberg proposed the theory of quantum electrodynamics with relativistic invariant, quantification of radiation, waves of matter, with paradoxes and divergent quantities, and in that theory the energy of the electron was the very energy of the field electromagnetic going up to infinity, which would be an unacceptable result.

Quantization for the Gravitational Field – In 1929, Heisenberg and Pauli presented a possibility of quantification of the gravitational field through the quantities of gravitational field $g_{\mu\nu}$ in space-time coordinates with exponential $\exp^{(i/h/2\pi)\ c\psi\ (g)}$. And in 1930 Rosenfeld introduced the quantization of the gravitational field with the gravitational quantum.

In 1931, Ehrenfest, R. C. Tolman and B. Podolsky presented an experiment in which two parallel light rays did not present gravitational attraction. In 1999, V. Faraoni and R. M. Dumse studied this case and concluded that the reason for the lack of gravitational attraction was due to an exact cancellation of magnetic gravitic forces and electric gravitic between the rays of light. Also, objects would tend to unite, but by the theory of relativity, the time required for attraction would be $t' = t/[1 - (v^2/c^2)]^{1/2} > t$, where v = velocity, and c = speed of light. Since the factor $1/[1 - (v^2/c^2)]^{1/2} \to \infty$ when $v \to c$, the time for the two particles to attract each other when $v \to c$ was approaching infinity. And the time for the rays to deviate would also be infinite. The conclusion was that for two particles at the speed of light, in which time was nonexistent, then the gravitational pull was also nonexistent.

In quantum mechanics two equal spins should be repelled, based on the Pauli principle in which two electrons cannot occupy the same quantum number, so if they have equal spins they would tend to be equal in a position and thus repel each other in the approach. So the lack of attraction should be attributed to the linear approximation in which photons would generate gravitational wave spins with the same orientation that would repel each other.

Negative Infinity Sea – In 1929, O. B. Klein presented that there was a paradox in Dirac's relativistic quantum theory, because an electron in the negative energy state, in a sea of electrons, could emit photons continuously, because the negative energy was infinite, that is, $-\infty$ in that sea of electrons. In 1930 Dirac replied that since the electrons were all occupied at the same level of negative energy in this sea of electrons, all electrons should behave as if they were presenting different states by the Pauli exclusion principle.

Magnet of a Polo – In 1931 and in 1948, Dirac, to obtain symmetry in Maxwell's equations, presented the hypothesis that there was a magnetic charge of distinct poles, north or south separately, as if it were a long-length rope-shaped magnet.

Schrödinger Equation for the Photon – In 1931 Oppenheimer presented a wave equation for the photon similar to that of Schrödinger for the location of the electron.

Inconsistencies and Consistency in Quantum Electrodynamics – And between 1930 and 1931, Oppenheimer, L. Landau and R. E. Peirls pointed out divergences and inconsistencies in QED. But Bohr and Rosenfeld in 1933 showed that it was consistent with the best possible measurements of electromagnetic field quantities.

And in 1934, Infeld and Born and also the proposition of Mie, presented a theory of non-linear field, for a consistency of QED.

Quantification of Gravity and Gravitons – In 1934, D. I. Blokhintsev and F. M. Gal' Perin called the quantification of the gravitational waves of gravitons. And in 1935, M. Bronstein observed an essential difference between quantum electrodynamics and quantum gravitational field theory, showing that the general theory of relativity and quantum theory are fundamentally unlikely to be unified.

XXIV. NEUTRINO, POSITRON AND NEUTRON – 1930 to 1956

Neutrino – In the beta decay of the proton nucleus and gamma rays, the final energy measured was less than the initial energy. In 1930, Pauli had presented the existence of a minimal particle without charge, that is, neutral to preserve the conservation of charge and energy in the beta decay. This particle would have a minimum mass or even zero mass by the theory of relativity if it came out of the nucleus at the speed of light. Subsequently, in 1934 it was called a neutrino by Fermi.

And in 1956, C. Cowan and F. Reines detected the particle that was supposed to be the neutrino, next to the Savannah River reactor.

Positron – In 1932, I. Curie and F. Joliot when observing in the chamber of clouds, radiations of polonium with beryllium, treated the radioactivity of these elements without considering the action of the radiation of the cosmic rays that acted at the altitude and the surface, therefore crossing the chamber of clouds that detected radioactivity. In addition to detecting this radioactivity of the polonium and beryllium, there was the detection of what came toward the two chemical elements, and interpreted that they were electrons. But, C. D. Anderson observing the behavior of cosmic rays in the magnetic field of droplets, observed a particle of momentum equal to that of the electron, but with a curvature of trajectory opposite the electron. So he concluded that it was a positive electron, denominating positron.

Traveler of the Future – In 1934, E. C. G. Stücklberg presented the hypothesis that the positron could be an electron with negative energy coming from the future to the past.

Neutral Mass – Rutherford had observed in 1920 that the oxygen when he obtained it in 1919 with 8 positive charges of protons and atomic mass 17, could be oxygen of 8 positive charges of protons with atomic mass 16 and another neutral mass equal to that of the proton.

Neutron – Between 1928 and 1930, W. Bothe and H. Becker had directed alpha particles from the radioactive polonium to the beryllium to observe gamma-ray formation. But, they had detected a higher energy radiation emitted by the beryllium, than the alpha particles, and the same occurring by the emission of lithium and boron, and concluded that they were gamma rays of superior energy, which was the same conclusion in 1932 by F. R. D. Rasetti. But, in 1930 Pauli presented hypothesis of neutral magnetic particles. And in 1931, J. Chadwick, H. C. Webster and also Oppenheimer and J. F. Carlson considered that this higher energy radiation could be due to an particle without charge, a neutral particle. In 1932, I. Curie and F. Joliot decided to check this superior energy emitted by the beryllium. And they detected that this radiation, seen through an ionizing chamber and water bubbles, pulled protons out of an absorbent material like paraffin. And they concluded it was radiation from superior energy to gamma rays. In 1932, Chadwick observed the radiation emitted by the polonium to the beryllium and the presence of helium, through a radiation detector produced in 1929 by F. A. B. Ward, C. E. Wynn-Williams and H. M. Cave, obtaining: $_5Be^{11} + {_2}He^4 \rightarrow {_7}N^{14} + N^1$, that is, a neutral particle of mass close to the proton. and targeting helium in boron obtained neutrons with mass greater than that of the proton. And observed the neutron acting on the oxygen to form new chemical elements: $N^1 + {_8}O^{16} \rightarrow {_6}C^{13} + {_2}He^4$.

As Fermi in 1933, the neutron decay through a virtual particle action that had a spin ± 1 similar to the photon of 1 spin 1 that acted in the electron and photon interaction.

Free Neutron – The neutrons when free from the atomic nucleus, present a half-life, that is, half of them transforms into proton, electron and neutrino, in the time of 10 minutes, and half of the remaining neutron decays in another 10 minutes, and so on. So after 1 hour there would be around 1.5% of neutrons.

XXV. PAIR CREATION AND ANNIHILATION – 1931 to 2014

Photon Collision – In 1931, Dirac presented that the two gamma photons collision could by the quantity of existing energy produce a couple of charges, negative and positive, according to their negative and positive particle theory in which there was the electron sea, and depending on the equality of charge and mass by Weil in 1928, i.e. production of par: $2\gamma \rightarrow e^- + e^+$. In 1933, also Oppenheimer and M. S. Plesset observed that the incidence of high energy gamma rays in an atomic nucleus could produce an electron and positron pair.

Gamma Photons of Cosmic Rays Forming Electron and Positron – In 1933, P. M. S. Blackett and G. P. S. Occhialini observed cosmic rays in cloud chamber, detecting gamma photons that materialize in electrons and positrons in the presence of heavy nuclei: $2\gamma \rightarrow e^- + e^+$.

Gamma Photons of Radioactivity Forming Electron and Positron – Also in 1933, I. Curie and F. Joliot, through a Wilson chamber with many high current electric coils, with intense magnetic field and using radio with gamma emission obtained the electron materialization and positron: $2\gamma \rightarrow e^- + e^+$.

Then also C. D. Anderson and J. Thibaud obtained this creation of electron pairs and positron.

Vacuum Polarization – In 1934, Dirac and Heisenberg observed that in the presence of a positive electrical charge nucleus in the vacuum, the appearance or creation of particles would occur in a tiny time, called virtual, as an electron and positron pair. The electron would be attracted to this positive charge while the positron would drift away to infinity, that is, the polarization of the vacuum. Thus, the charge of the positive nucleus would result diminished by the action of the negative charge that would neutralize the positive charge.

Couple of Nothing – With the discovery of electron pair formation and positron that could also be for other pairs of particles, then the empty space of the quantum vacuum would be occupied by quantum fluctuations with the formation of virtual particle pairs and virtual antiparticles that its would continually form out of nowhere and extinguish from nothing an instant later!

Measured Charge and Theoretical Charge – In 1935, R. Serber and E. A. Uehling showed that the observed electric charge was less than the theoretical charge.

Electron Variable Radius – In 1934, Weisskopf and W. H. Furry presented computation by electron energy with its field, in which the electron radius presented dimension, contrary to the electron as a point. And in 1936, Weisskopf presented that the radius varied according to the mutual action between the production of pairs in the vacuum and the Pauli exclusion principle through a variation of the electron charge density. Also, in 1936 Heisenberg and H. Euler presented the polarization of the vacuum involving the electron radius.

Shock Section in the Nucleus with Gamma Rays, Momentum of Nucleus, Electron and Positron – Between 1934 to 1952, H. A. Bethe, W. Heitler, Racah, Y. Nishima, S. Tomonaga, S. Sakata, Breit, J. A. Wheeler, L. Simons, K. Zuber, M. Kobayasi, H. R. Hulme, J. C. Jaeger, Nordheim, Landau, E. Lifshitz, H. J. Bhabha, E. J. Williams, W. E. Lamb, V. Votruba, B. Hahn, E. Baldinger, R. L. Walker, M. Stearns, R. Just, J. M. Luttinger, M. Slotnick, J. S. Schwinger, G. E. Modesitt, H. W. Koch, L. H. Lanzl, A.O. Hanson, J. E. Hooper, D. T. King, A. H. Morish, L. C . Maximon and H. Davies, they presented mathematical theory relating variables in: shock section to the atomic nucleus of the impact of gamma rays with their energy; momentums of the nucleus, electron and positron; and angles of incidence in the nucleus.

Particle Accelerators – From 1962 to 2014, continued propositions and experiments with accelerators for the production of electrons and or positrons, through A. I. Nikishov, R. J. Gould, G. P. Schréder, S. Bonometto, M. J. Rees, Y. S. Tsai, S. J. Brodsky, P. M. Zerwas, D. L. Burke, C. Decker, A. C. Melissinos, T. Koffa, C. Bamber, S. Klein, C. Gahn, J. D. Lindl, W. P. Leemans, M. Marklund, P. K. Shukla, F. V. Hartmann, C. W. Siders, C. P. J. Barty, H. Chen, S. Kneip, R. Ruffini, G. Vereshchagin, S. Xue, H. Hu, C. Möller, C. H. Keitel, C. E. Clayton, R. P. J. Town, S. H. Glenzer, A. Di Piazza, J. Beringer, G. Sarry, D. D'Enterria, G. G. Silveira, O. Pike, F. Mackenroth, E. G. Hill, S. J. Rose and other. And they were always obtained that in the annihilation of electrons and positrons resulted in gamma rays and these creating electrons and positrons.

Annihilation of Matter and Antimatter– In 1930, Dirac presented that a positive charge particle such as proton and a negative charge particle like the electron could be annihilated resulting in two gamma photons: proton + e^- → 2 γ.

And in 1933, Fermi and Uhlenbeck calculated the shock section for annihilation $e^+ + e^-$ → 2 γ.

Annihilation of Positron and Electron – In 1933, I. Curie and F. Joliot obtained the annihilation of the positron and the electron in gamma rays.

Electron and Positron in Platinum Target – In 1933, Thibaud observed that while the electron entering the target of platinum atoms, it interacted with the electrons of the atom resulting in the emission of X-ray photons, the positron when reaching the platinum target electron produced gamma photon emission with a total energy around 1 million volt, significantly higher than the electron radiation intensity, showing the annihilation of the electron mass and the positron.

Electron and Light: Two Aspects of the Same Energy – According to Thibaud, the annihilation of particles constituted the most compelling experimental evidence presented in the reality of a conversion effect of matter into radiation. *"In fundamental matters such as the nature of matter and light, modern Physics that locates difficulty on the scale of the infinitely small atomics, recognizes only in the matter the manifestation of tiny, electric corpuscles, electrons or others; and in the light, the propagation of a wave of a rather unintelligible, enigmatic, uncertain, doubtful type, since in a large number of cases we must attribute the characteristics of a true rain of luminous corpuscles, the photons. For example, the photoelectric effect, and in which we see the photon as a true projectile, launched on the outer defences of the atoms, and able to dislodge electrons. The photon, originating from the substance of the electron, understands the need for a synthesis between the conceptions of the matter and the light. And the fundamental point of Physics is to reconcile the electron and the photon, or rather to ask why they appear with two different aspects of the same energy. And it will be necessary to revise and complete notions like the location of energy in a small space domain".*

Shock Section for Annihilation – And from 1934 until 1951, modern Physics related the shock section with momentum, energy and charge of the electron, positron and other particles, by Fermi, Uhlenbeck, Hulme, Bhabha, Nishima, Tomonaga, H. Tamaki, Bethe, Jaeger, Wheeler, F. L. Yost, M. H. L. Pryce, J. C. Ward, J. Hornbostel, H. S. Snyder, S. Pasternack, J. W. M. Dumond, D. A. Lind, B. B. Watson, C. S. Wu, I. Shaknov, S. Benedetti, W. R. Konneker, H. Primakoff and J. A. Rich.

Annihilation and Creation of Negative Energies – In 1934, Oppenheimer and Furry presented publication in which transforming by mathematics the functions of wave creation and annihilation of particles in negative energy, could be eliminated the vacuum where was the sea of electrons with the infinite energy that would form particles continuously, which would be contrary to the atomic equilibrium, of Dirac's theory of 1928.

XXVI. ATOMIC NUCLEUS – **1914 to 1961**

Particle Accelerators – In 1931, E. O. Lawrence and M. S. Livingston presented a circular particle accelerator, the cyclotron, in which with the magnetic and electric fields, through the Lorentz force, hydrogen ions were accelerated, and obtaining up to 1 Mev, that is 10^6 electron-volts. And with the accelerator designed by R. Wideröe, D. H. Sloan and Lawrence used mercury ions. In 1932,

Lawrence, Livingstone and M. G. White and also J. D. Cockcroft and E. T. S. Walton obtained the first nuclear reaction with particle accelerator, through lithium-targeted protons, resulting in two helium particles: $_{+1}H + _{+3}Li \rightarrow 2\, _{+2}He$.

Independent Charges – In 1932, Heisenberg observed the independence of charges in the nuclei that it maintained a symmetry resulting in spins of similar properties, that is, spins isotropic. In 1936, B. Cassen and Condon observed the independence of charges in the interaction proton-proton, neutron-neutron and neutron-proton. And In 1936, J. H. Bartlett presented a scattering experience proton-proton, neutron-neutron and neutron-proton, in which it had detected change of position of the spins between them, maintaining a symmetry that resulted in isotropic spins.

Braking Radiation or "Bremsstrahlung" – In 1933, Heitler and F. Sauter observed the emission of photons by electrons scattered in a field of attraction and nuclear repulsion, i.e. the braking radiation or "bremsstrahlung".

And Heitler and Bethe determined the theoretical aspect of this radiation in which the electron with momentum p_0 and energy $E_{energy\,(0)}$ passing through a field of a nucleus or by the atom when presenting an acceleration a with a momentum p_a and an energy $E_{energy(a)}$ resulting $E_{Energy(a)} - E_{energy(0)}] = h\mathcal{V}$, where h the Planck constant and $\mathcal{V}$ the frequency of the photon, in which there was an electron interaction with the energy field that produced the electron acceleration, that is the radiation field, and the electron interaction with the atomic field, then causing radiation emission $h\mathcal{V}$. And in the calculations between momentums and energies was obtained the shock section $\delta = Z^2 r_0^2 / 137$, where Z it was the atomic number, r_0 the radius of the atom and the fine structure constant $\alpha = 1/137$.

Also in 1934, P. A. Cerenkov and also Heitler, Bethe and Racah and after in 1937 I. Y. Tamm and I. M. Frank observed that electrons that had a speed in a medium, greater than the speed of light in that medium, resulted in variation of the speed and consequent emission of light, according to Maxwell's electromagnetic theory, in which the variation of electric current resulted in emission of electromagnetic field with the speed of light.

Artificial Radioactivity – In 1934, F. Joliot and I. Curie observed that on an aluminium plate irradiated by a compound of polonium, that is, with emission of alpha particles, helium $_2He^4$, the aluminium was transmuted to phosphorus with neutron emission: $_{13}Al^{27} + _2He^4 \rightarrow _{15}P^{30} + n^1$. But, phosphorus emitted a positive beta radioactivity, that is, an artificial radioactivity: $_{15}P^{30} \rightarrow _{14}Si^{30} + e^+$

+ 1 neutrino ou antineutrino. And also, it was obtained: $_5B^{10} + {}_2H^4$ → $_7N^{13} + n^1$ and $_7N^{13}$ → $_6C^{13} + e^+ +$ 1 neutrino or antineutrino. and similar effect was observed with magnesium, hydrogen, lithium, beryllium, carbon, oxygen, fluoride, sodium, silicon, which they termed as nuclear.

Neutron Projectiles – In 1934, Fermi, E. Aldi, O. Agostino, Rasetti, E. Segrè and B. Pontecorvo directed neutrons obtained from artificial radioactivity in chemical elements, which should be of high yield absorption because it is neutral charge, to observe the behavior of radioactive emission. However, there remained a stability in the atoms of hydrogen, lithium, beryllium, boron, carbon, nitrogen, lead, without radioactive emission, but obtaining radiation in fluoride and uranium. In the case of uranium it would appear that the neutron reaction to proton + electron + neutrino or antineutrino is occurring: $_{92}U^{238}$ → $_{92}Ux^{239} + e^- +$ neutrino or antineutrino.

Slow Neutrons – In 1934, Fermi, Aldi, Agostino, Rasetti, Segrè and Pontecorvo discovered that chemical elements were absorbed with slower neutrons as they passed through a paraffin plate.

Nuclear Fission – But in 1935, I. Noddack indicated that the uranium being unstable should rupture in the reaction with neutrons and result in other chemical elements of smaller numbers of protons. In 1938, O. Hahn, F. Strassmann and L. Meitner directed slow neutrons to the uranium $_{92}U^{238}$ and obtained radioactive elements that would appear to be the radium $_{88}Ra$, actin $_{89}Ac$ and thorium $_{90}Th$. But, then they obtained the radioactive barium of only 56 protons, which evinced nuclear fission. And in 1939, Meitner and O. R. Frisch then interpreted it as fission of the uranium $_{92}U^{235}$ in radioactive elements such as the 54 protons xenon and the strontium of only 38 protons. And also in 1939, O. Hahn and Strassmann proved that the radium $_{89}Ra$, actin $_{89}Ac$ and thorium $_{90}Th$ were in reality the radioactive elements barium $_{56}Ba$, lanthanum $_{57}La$ and cerium $_{58}Ec$.

Chain Reaction – In 1932, L. Szilard had proposed that the neutron, by causing the fission of the nucleus, would emit more neutrons that would continue to fission, by the largest number of neutrons in formation of each fission, i.e. the chain reaction. And with the slow neutrons, it was observed by O. Hahn, Strassmann and Meitner, O. R. Frisch, Fermi, Aldin, Agostino, Rasetti, Segrè and Pontecorvo that there was an excess of neutrons in relation to the stable nuclei of mass number. That is, they were neutrons that could continue to produce new fissions in the uranium and that the reaction could be controlled. In 1940, I. B. Zeldovich and I. B. Khariton showed that the chain reaction could be effected on natural uranium, and G. N. Flerov and K. Fiziki showed that in uranium there was also a spontaneous fission. However, slow neutrals only produced fission in the uranium isotope $_{92}U^{235}$ In less than 1% contained in the natural uranium, what would be necessary its separation, or the production of plutonium $_{94}Pu^{239}$ from natural uranium to fission by slow neutrons. In 1942, Fermi and other members of the University of Chicago team, through atomic stack control of fissile material plates, to obtain critical mass with the emission of slow neutrons, obtained the first controlled chain nuclear reaction. And in 1945, based on the slow neutrons and the chain reaction control by critical mass, the first nuclear fission explosion, the nuclear atomic bomb, was detonated.

Nuclear Fusion – In 1936, R. D. Atkinson presented on the basis of quantum mechanics and radioactivity the hypothesis of nuclear reaction in stars, in which hydrogen would form deuterium, hydrogen with one more neutron, and emission of positron and release of energy: $_1H^1 + {}_1H^1 \rightarrow {}_1D^2 + e^+$

With the discoveries of radioactivity and nuclear reactions, in 1938 Bethe and C. L. Critchfield presented the nuclear reaction to the energy generated from the stars of smaller masses in which hydrogen would form deuterium, which would result in the formation of the stable helium and with energy emission: $_1H^1 + {}_1H^1 \rightarrow {}_1D^2 + e^+ +$ neutrino; $_1H^1 + {}_1D^2 \rightarrow {}_2He^3 + \gamma$; $_2He^3 + {}_2He^3 \rightarrow {}_2He^4 + {}_1H^1 + {}_1H^1 + 24.7$ Mev. This reaction would allow the base to get a nuclear fusion explosion.

Also, hydrogen could connect to the carbon that would form nucleus of larger masses until instability and fission.

A primeira detonação de uma bomba de fusão nuclear de hidrogênio-deutério ocorreu em 1952 pelos Estados Unidos by E. Teller, e a segunda em 1961 pela União Soviética.

Emission of Electrons by the Nucleus – In the radioactivity it was observed that the nucleus of the atom emitted alpha, beta and gamma radiations. In 1914, Chadwick had found that beta radioactivity had a continuous spectrum, but along with a spectral line, so there was a difference between the energies of the electron inside and outside the nucleus. In 1922, Meitner presented that the beta rays, that is, the electrons, should always come out of the nucleus always with the same energy, but some energy would be converted into gamma rays that could produce secondary beta rays outside the nucleus. In 1924, Bohr, H. A. Kramers and Slater, concluded that energy could be variable in quantum theory, which would destroy energy conservation: *"As a small base we emphasize in the present for a theoretical treatment of the problem of disintegration of beta rays. Actually, the behavior of electron leaping within an atomic nucleus would seem to fall completely off the field of consistent application of ordinary mechanical* *concepts, even in its theoretical quantum modification. Remembering that the principles of conservation of energy and momentum are of a purely classical origin, the suggestion of their failure to emit beta rays can in the present state of quantum theory be rejected".*

In 1926, Rasetti and Fermi observed that the nucleus and the electron had similar diameters in scale, and if the electron was at the nucleus, the nuclear diameter would be around double the value of the nuclear diameter, so it would be unfeasible the model in which the electron would be forming the nucleus along with the proton. And in 1927, Wigner presented conservation laws in quantum mechanics. However, in 1929, Gamow and Landau came to agree with Bohr for the breakdown of energy conservation in quantum theory. In 1929, Rutherford presented that by radioactivity the nucleus would consist of protons and electrons. And around 1929 was verified by the radioactivity experiments that the beta rays had their origin only in the nucleus, contrary to the hypothesis of Meitner that it could also be emitted out of the nucleus in radioactivity. In 1928, L. S. Ornsteni presented a rotational spectrum experience of nitrogen in which the spin of the nitrogen nucleus was equal to 1. And in 1928 and 1930, R. L. Krönig showed that the nitrogen nucleus, which would consist of 14 protons and 7 electrons, would have 21 spins S_e (½) that is, a total of 10 ½, which would go against an even number of spins according to the Ornsteni experiment. Therefore, the proton-electron model of the nucleus contradicted an experiment of spins, that is, electrons in the nucleus. In 1929, Heitler and G. Herzberg and also Rasetti showed that the nitrogen spectrum was at odds with the proton-electron nucleus model. In 1932, Heisenberg and also D. Iwanenko and E. Majorana presented the hypothesis of the nuclear structure that introduced the exchange of forces between proton and neutron for the exchange of electrons between proton and neutron. Heisenberg predicted a single proton-neutron system of interactions with his classic example of a busy tennis ball between one and another player, applied in quantum mechanics. This interaction should be effected by a particle with charge that through its field acted on the proton-neutron by exchanging the charge continuously of the proton and neutron. Iwanenko reaffirmed the hypothesis that the neutron was a single elementary particle, so the proton and electron were independent of the neutron structure, in which Dirac considered the nucleus consisting of proton, neutron, and electron. In 1933, Perrin presented that also the neutrino would be created only at the instant of its emission by the nucleus, similar to the emission of the photon by the electron. In 1934, Chadwick and M. Goldhaber showed that the neutron mass was greater than that of the proton, and then it was presented that the neutron mass was greater than that of the proton + electron. In 1934, with the discoveries of the neutrino, positron, neutron and artificial radioactivity, Fermi presented the theory of the electron emission of the beta radioactivity, through a neutron-proton interaction by a weak force of a particle, the neutrino, in which the electrons emitted by the nucleus would be created only at the instant of the emission, in which the nucleus with the angular momentums of the proton, neutron and the particle that was formed, were correlated with constant decay, kinetic energy of particles, resistance force, function of nucleus Coulomb field effect over beta particle emitted, mechanical wave matrix element which is a measure of relative probability for nucleus to emit beta particle, i.e. electron and neutrino, statistical factor and mass in electron rest. So: neutron → proton + electron + neutrino. Also in 1934, G. C. Wick and also Bethe and Peierls presented the hypothesis of the inverse decay for the positron: proton → neutron + positron + neutrino and electron + proton → neutron + neutrino. That is, the decay would be similar to a weak interaction, such as the electromagnetic interaction where the light or other electromagnetic wave is formed only at the instant of the emission. But in 1938, O. B. Klein presented that the beta decay would be mediated by larger particle mass in the atomic nucleus that named W^-, that is, the proton-neutron nucleus → proton + proton + W^-, W^- → electron + neutrino. Or the nucleus neutron → neutron → proton + W^-, W^- → electron + neutrino.

Particle Shock Section – In 1932, E. Feenberg presented the section spread of elastic and inelastic shock between particles; in 1933, Wigner calculated the section of proton-neutron shock; in 1935 White, and in 1937 Schwinger and E. Teller presented the proton-proton, neutron-neutron and proton-protons scattering; in 1936, M. A. Tuve and N. Heydenberg and L. R. Hafstad presented the spreading proton-proton and Barsoum and Feenberg showed the neutron-proton spreading; in 1947 J. Tiomno and J. L. Lopes spreading proton-proton. In 1947, Schwinger and in 1949, J. Eisenstein and others, and also J. M. Blatt and J. D. Jackson presented studies on collisions in minimal action. In 1936, Heisenberg presented a calculation for section of inelastic shock in which intensely energetic protons in the presence of a field of attraction and or repulsion of a nucleus produced electrons, positrons and neutrinos. And M. E. Fierz presented the section of proton-protons shock that resulted in neutron and positron. In 1936, Iwanenko and A. Sokolov and also B. Kahn observed the proton-neutron interaction that produced the beta decay through a weak force of interaction by the neutrino particle, while neutron-neutron and proton-proton remained stabilized. And in 1937, Tomonaga and Tamaki presented calculations of the neutrino shock section with neutron, resulting in proton and positron. In 1947, W. W. Havens, I. I. Rabbi, and L. J. Rainwater, and also Fermi, L. Marshall presented evidence of electron-neutron interaction. In 1949, V. Bargmann and also Bethe presented quantum scattering relations. In 1949, L. J. Cook, E.M. McMillan, J. M. Peterson and D. C. Sewell, and also S. Fernbach, Serber and T. B. Taylor presented neutron-scattering experiments by nuclei resulting in obtaining the nuclear radius by the expression $r = 10^{-13}$ cm x $A^{1/3}$, where A = number and mass of the atom. In 1950, G. J. Lippmann and Schwinger presented studies on inelastic spreading. And M. N. Rosenbluth presented a section of shock to the electron by a proton with its magnetic momentum.

Momentums: Angular and Magnetic Nuclear – In 1933, Stern, I. Estermann and O. R. Frisch determined the magnetic momentum of the proton and R. F. Bacher by experiment noted that the magnetic momentum of the neutron was similar to that of the proton but with the opposite signal. In 1937 Rabi presented procedures for the magnetic momentum measurement of atomic nucleus by nuclear magnetic resonance, and in 1938, H. Yukawa, and M. Taketani, and also H. Fröhlich, Heitler and N. Kemmer, in addition to M. Kobatasi presented calculation of the momentum proton and neutron magnetic. At 1939 F. Bloch and L. W. Alvarez experimentally determined the neutron magnetic momentum. The interaction of the magnetic momentum of the atomic nucleus with the application of a magnetic field generated the formation of internal image in organic tissues by the difference of the normal magnetic momentum with the damaged one. Between 1941 and 1943 Racah presented studies on nuclear angular momentums. Subsequently, from 1942 to 1946, measurements of the nuclear magnetic momentum were carried out by magnetic resonance imaging for liquid and solid substances, by C. J. Gorter, L. F. J. Broer, E. Zavoisky, E. M. Purcell, H. C. Torrey, R. V. Pound, Bloch, M. Packard, Zavoisky and W. W. Hansen. In 1943, J. M. Jauch presented studies on magnetic momentums of the proton and neutron involving the decay of these particles. In 1948, Bloch, D. Nicodemus and H. H. Staub presented a measurement of the neutron magnetic momentum with the value of –1.9 of the proton; J. M. Luttinger and also K. M. Case, and in 1949 Slotnick and Heitler; S. D. Drell and Case; and C. H. Towns, H. M. Foley and W. Law presented calculations for the magnetic momentum of nuclei. In 1949 A. Barragan and Pryce dealt with concepts of nuclear magnetism and in 1950 by E. L. Hahn on an magnetic resonance experiment. In 1973, E. Hoffamn and M. E. Phelps presented the tomography by radionuclides, as $N^{13}F^{18}C^{11}The^{15}$ and Rb^{82} emitting positrons $e^+ + e^- \rightarrow$ that generated the formation of internal images.

Nuclear Interaction – How the natural radioactivity occurred as uncontrollable W. Lenz in 1920 presented that the nuclear force between the constituent particles of the atomic nucleus should not be of electrical and magnetic type, or of forces of attraction and repulsion of Coulomb, because there is a unstable equilibrium. And the beta decay of the atomic nucleus, in proton, electron and the neutrino that completed the charge balance, momentum and energy, indicated that the neutron would be made up of the electron and neutrino. But, Heisenberg in 1932 foresaw a single proton-neutron system of interactions, with his classic example of jogged tennis ball between one and another player, applied in quantum mechanics. This interaction was to be effected by a particle with charge, which through its field acted in the proton-neutron exchanging the charge continuously of the proton and neutron. And in 1934, Tamm and Iwanenko and A. Nordsieck presented that nuclear interaction for an energy of 1 Mev would reach around 10^{-15} cm.

Abstract Nucleus Forces – In 1936, Heisenberg, Wigner, Condon, Barsoum, Feenberg and Cassen devised theory for forces between proton-proton, proton-neutron and neutron-neutron, with mathematics of abstract spaces. And in 1946, Wigner and in 1947 Wigner and L. Eisenbud presented an extensive mathematical formalism about the nuclei. That is, space apart, separate, intangible!

XXVII. FIELD AND INSTANT COMMUNICATION, THEORETICAL MASS, PHOTON MASS, IMAGINARY MASS, NEGATIVE MASS, SPIN WAVES, PUNCTUAL SPIN, CASIMIR EFFECT, STATISTIC AND FIELD
1873 to 1948

Instant Communication or "Ghostly" – Because ψ^2 represent the intensity of two superimposed particle waves ($\psi_1 + \psi_2$), as Einstein, Schödinger, Born, Heisenberg and Bohr, these particles could then be entangled, interconnected, associated, non-separated, interwoven. But in 1935, Einstein, Podolsky and N. Rosen, (EPR), presented that if two quantum particles conjugated in interaction, P_1 and P_2, by being determined one another by being conjugated would also be determined without the need to be measured, which would be contrary to quantum indetermination.

As shown in Figure 18, as EPR, if the particle P_1 in A presents a property, the particle P_2 in B should be with the inverse conjugate property, so no need to observe this property in B. But if the particle P_1 in A reversed the property, and the particle P_2 in B instantly also reverse its property, this information from the reversal of property should be considered as "ghostly" by Einstein, by exceeding the speed of light, so would not exist entanglement.

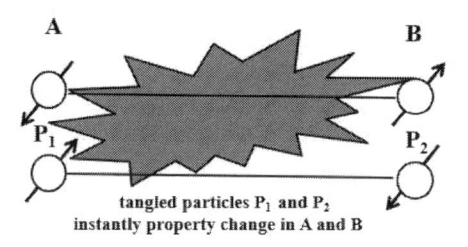

Fig. 18 – Entanglement particles.

However, Bohr considered that the indetermination consisted of large variables for local measurement, that was the interpretation also of most of the physicists at the time, the interpretation of Copenhagen, in which in a conjugate, if a of the components is modified in place A, instantly the other is modified in place B, that is, the entanglement caused by superimposed states. But without explaining how the transmission instant of these variables occurred.

Theoretical Mass and Increased Mass – In 1938, Kramers presented that with the theoretical mass of the added electron with a mass of electron interaction with its own field would be obtained the mass actually measured effectively from the electron.

Mass Photon at Rest – Between 1936 to 1941, A. Proca presented photons with mass in which the velocity of the photons could be variable, occurring superluminary velocity of electromagnetic radiation, an imaginary or negative mass!

Spin Waves – In 1946, A. I. Akhieser presented extensive study on spin waves.

Punctual Spin – In 1947, Tiomno and W. Schützer presented a field of radiation electron point with spin.

Effect Casimir – Between the atoms of a substance can occur polarity between the positive and negative charges resulting in an attraction according to Van der Walls theory in 1873, or in a localized position occur a repulsion by the electric charges. In 1948, H. B. G. Casimir and D. Polder presented an energy calculation of interaction between atoms, for two electrically neutral parallel metal plates of 1 cm^2 each and in the distance of 1 micron, obtaining a force of 0.013 dynes. And in 1948 Casimir presented another calculation with the quantum fluctuation of the electromagnetic vacuum, by the energy of point zero h/2, with h the Planck constant, in which there would be a force between two metallic plates of electromagnetic origin, by the expression: F =

[S_{area} π^2 (h/2π) c]/240d^4, where S = area of each plate, h = Planck's constant, c = speed of light in vacuum, d = distance in micron between the plates, where for the plates of 1 cm^2 and distance of 1 micron obtained the same result as the polarity repulsion by the force of Van der Walls. By calculating the energy of the zero point, the vacuum between the plates is restricted to a few fluctuation waves, by the boundary condition, therefore less energy density of the vacuum than on the external side of the plates, resulting in external pressure greater than the internal between the plates that cause the attraction between the plates. The smaller space between the plates there would be smaller possible wavelengths, therefore less energy and greater attraction between the plates.

Statistics and Statistics – According to Infeld and Einstein, in 1938, quantum Physics was not intended to describe individual objects in space and their changes in time, but the probability that the individual object had an indicated property, and there were no laws showing changes in objects individually in time. *"If a probability wave was needed in thirty dimensions to describe quantically ten particles, it would require a probability wave with an infinite number of dimensions to describe quantically the field. The quantum Physics moved away from the mechanical concept and an indentation seems unlikely and it is based on the concept of matter and field, in which it also moves away from reducing everything to the concept of field. "*

Matter: High Concentration of Energy; Field: Reduced Concentration of Energy – In 1938, Infeld and Einstein observed that a variation of magnetic field resulted in radiation emission, that is, energy, then assigning to the field an energy deposit, and stating that energy had mass, that mass represented energy, and that there was essential distinction some between mass and energy, for the radiation at rest. Also, stating what matter was where the energy concentration was large and the field where the energy concentration was small. And culminating emphatically that: *"Mass is energy and energy is mass. And the concept between matter and field, before the theory of relativity, could be defined as matter having mass and the field without matter. And by relativity, by the equation between mass and energy, energy represents matter and it would be defined which matter is a high concentration of energy and the field is where the energy concentration is small. But as well as charge and its field, also matter and field would be just a quantitative difference. If you could consider matter as a region of the space in which the field is concentrated, a Physics could be built in which the field was the only reality, a pure field Physics. Physics began with the mechanical concept of mass, force and inertia acting between distances travelled. But the behavior that involves the bodies, that* *is, the field could be the essential to the ordination and the understanding of the events. Would the light be a wave or a shower of photons? To respond we would have to abandon the description of atomic events in time and space, and further retreat from the ancient mechanical concept".*

And the field, gravitational, electric or magnetic that was a region of the space subject to the forces, was also defined as state or distortion of space, as relationship between variables, abstract mathematical functions, massless photons, and as just apparently real, according to the quantization of the field by quantum mechanics and the relativity applied to particles and the quantum of electromagnetic radiation, although Einstein and Infeld consider the field as energy and real. Although, by Einstein's own theory of relativity, The photon at rest would be without mass, in which $m_{photon} = m_{0photon}/[(1 - (v^2_{photon}/c^2)]^{1/2}$ and to $V_{photon} = c$ = speed of light and radiation, $m_{0photon}$ = photon mass at rest, so $m_{0photon} = m_{photon} \times 0 = 0$.

XXVIII. MESONS – 1935 to 2006

Mesons – Yukawa in 1935 had presented that the nuclear interaction producing a nuclear force or named strong force that stabilized the protons and neutrons was through a particle of mediation, in which the mass would be 200 m_e, i.e. 200 times the mass of the electron, calculated through the electrical potential of two nuclei and with reach around 10^{-13} cm, being this particle named U and after yukon, mesotron and meson, in which if it were free of the nucleus should decay in electron + neutrino. In 1935, W. H. Wells presented a proton-proton scattering experience in which it detected a short-range, but attractive interaction between the two protons. Also, in 1935 White and 1936 Tuve, N. P. Heydenburg and L. R. Hafstad and also Barsoum, Condon and R. D. Present observed in experiment this short-range and attractive interaction. In 1936, Gamow and Teller presented a study of coupling between particles of smaller and larger masses, and that should form atoms. In 1929 V. Skobeltzyn, Bothe and W. Kolhörster and B. B. Rossi, and in 1933 Blackett and Occhialini had observed components of larger and smaller masses in the cosmic rays, and in 1936 Anderson and S. H. Neddermeyer reported experiment in which they observed in cosmic rays the presence of intensely ionizing particles, which appeared to be protons. In 1937, Bhabha, Heiler, Oppenheimer and Carlson presented high-energy particle considerations by quantum mechanics to the components of the cosmic rays. And in 1937, C. D. Anderson and Neddermeyer also J. C. Street and E. C. Stevenson, equally Nishima, M. Takeuchi and T. Ichimiya, Oppenheimer, Serber and Stückelberg presented that in the cosmic rays were found intensely ionized particles with near mass To the predicted particle of the meson. In 1938, Yukawa, and Taketani, and also Fröhlich, Heitler and Kemmer, in addition to Kobayasi presented a field theory involving the mesons. In 1938, Bhabha, and also H. Euler and Heisenberger presented the components of the cosmic rays as mesons of larger and smaller energies. In 1939, G. Wataghin, P. A. Pompeia and M. Damy Souza Santos indicated that there was a multiple production of mesons. In 1939, Nedermeyer and Stevenson presented a new estimate for the mass of the meson that would be from 100 to 400 of the mass m_e of the electron. And L. E. Hoisington, S. S. Share and Barsoum calculated as greater than 200 m_e. And in 1940, L. Leprince-Ringt, S. Gorodetzky, E. Nageotte and R. Richard-Foy determined the mass of the meson in 240 $\pm$ M_e. In 1939 Yukawa presented that the interaction of the meson in the nucleus was around 10^{-8} s. And in 1940, the Kata and Y. Tanikawa presented that a neutral meson showed an average life of 10^{-8} s. In 1941, Rasetti found an average life of 10^{-6} s. In 1942, Rossi presented new results of the average life of meson, and in 1943, N. Nereson and Rossi found 2.15 x 10^{-6} s. In 1947, H. Bridge, Rossi and R. Williams obtained for the neutral meson $\approx 10^{-16}$ s, the same in 1948 obtained by H. W. Lewis, Oppenheimer and S. A. Wouthuysen. In 1940, G. Wentzel presented the nucleus as a field of attraction or repulsion of mesons. In 1940, R. E. Marshak presented that in the cosmic rays should occur two types of mesons. In 1941, Nordheim presented that in the decay of the meson resulted: meson $\rightarrow$ electron + 2 neutrino or meson $\rightarrow$ electron + gamma rays. In 1941, Oppenheimer and J. Seymour presented mesons-nucleus interactions. In 1942, Schwinger, and Pauli and S. M. Dancoff, and in 1943 Serber and Dancoff, and Pauli and S. Kusaka presented studies of nuclei-mesons couplings. Also, in 1946, Pauli, and Heitler. And in 1942, the Kata and T. Inoue presented a theory of two mesons. In 1945 and 1947, M. Conconverti, E. Paci and O. Piccioni and also in 1947 T. Sigurgeirson and A. Yamakawa presented experiments on mesons with negative charge that when they were blocked in carbon fall into other particles, thus disappearing the particle to the exchange of interaction in the atomic nucleus, that is, the meson of Yukawa would need to be another type of meson, in which Fermi, Teller, Weisskopf and Bethe, and also Marshak presented theory about these experiments. To have the interaction or strong nuclear force these mesons should be absorbed into which would show the force of exchange between nuclei permanently. In 1947, Oppenheimer presented a hypothesis that a neutral meson could disintegrate in two gamma rays: meson° $\rightarrow$ 2γ. Also from 1935 to 1949, D. Okayama, Z. Hai, Tanikawa, S. Hayakawa, G. Araki, T. Miyazima, Tomonaga, Möller, Rosenfeld, J. M. Lauch, J. G. Wilson, G. Nordheim and L. Nordheim, P. V. Auger, B. Gross, R. Maze, R. Chaminade, Blackett, Occhialini, M. Schenberg, Lopes,

J. C. Ribeiro presented theories and experiments with mesons of cosmic rays.

Mesons π and μ – In 1947, C. M. G. Lattes H. Muirhead, Occhialini and C. F. Powell presented that cosmic rays focusing on nuclear emulsions in mountain peaks produced two types of mesons, subsequently termed meson π or pion and meson μ or muon. and were found masses of 273 m_e for the meson π and 210 m_e to the meson μ. The meson π presented more interaction with the atomic nucleus than other mesons, so it was the most probable meson of the theory of Yukawa. The π mesons were formed at high altitudes and the mesons μ at sea level and formed by the decay of the meson π. In 1947, W. G. McMillan and Teller showed that a kinetic energy of 380 Mev would produce mesons π to the collide with the atomic nucleus. And in 1948, E. Gardner and Lattes presented through circular particle accelerator, with helium atoms in various targets, the artificial obtaining of mesons π and μ, and also in 1949, E. Gardner, Lattes, J. Burfening and W. H. Bars presented the production of mesons $\pi+$, with $\approx$ 300 m_e mass. In 1949, Lattes and W. G. MacMillan observed the experimental evidence of another meson, later referred to as $\pi°$. In 1949, J. Steinberger and also M. F. Kaplon, B. Peters and H. L. Bradt presented the average life of the meson $\pi°$ in $< 10^{-13}$ s. In 1949, H. M. Moseley and also Fermi and Yang presented the hypothesis that the mesons π could be dynamic states linked to nuclei and antinuclei. In 1950, R. F. Bjorklund, W. E. Crandall, B. J. Moyer and H. F. York and also Steinberg, W. K. H. Panofsky and J. S. Steller presented the mass of the meson $\pi°$ in $\approx$ 262 m_e. In 1947, Pontecorvo had presented hypothesis that the meson μ negative could result in: μ^- → proton + neutrino, and the meson μ in: μ → electron + gamma rays. But in 1948, E. P. Hincks and Pontecorvo; R. D. Sard and E. J. Althaus; and Piccioni; and C. D. Anderson, R. B. Leighton, A. J. Seariff presented that the meson μ remained inert in producing electron + gamma rays. And in 1950, R. F. Borklund, Crandall, Moyer and York; Panofsky, Steinberger; R. L. Aamodt and York and Steller detected again the meson $\pi°$. And also in 1950, Bjorklund, Crandall, Moyer and York presented experiment that in the collision of protons with 345 Mev in a target resulted in high intensity gamma rays, and in braking radiation was little intense the quantity of gamma rays. So it would be the decay of the neutral meson $\pi°$ that would produce the gamma rays. Steinberg, Panofsky and Steller presented experiment in which in the collision of mesons π^- in hydrogen, obtaining: π^- + protons → neutron + $\pi°$ → neutron + γ + γ. In 1948 Landau and at 1950 C. N. Yang presented that the meson $\pi°$ to decay in: $\pi°$ → γ + γ, implied to have spin, as there should be in the meson electric charge to emit gamma. In 1949 and 1950, P. T. Matthews presented that renormalization would be necessary for the mesons.

Mesons and Einstein's Relativity – In 1941, Rossi and D. B. Hall measured the decay of the meson, coming from the cosmic rays at velocities near the light, neutrino, antineutrino, electrons, positrons and gamma rays, at the top of a mountain in Colorado at 1600m up to ground level and used the time dilation of special relativity to the path to the soil. In 1963, D. H. Frisch and J. H. Smith effected the actual measurement of the half-life time. Obtained at the top of the mountain in New Hampshire at 1907 m, an average of 563 mesons and the half-life time was 0.7 milliseconds. And at the level of the soil in Cambridge would be predicted by the speed of decay of the particles at speed when next to the rest in a decay of ≈ 95% ie, obtained only ≈ 5%, that is, ≈ 28 mesons. However, 408 mesons were detected, concluding that there was a time dilation in the mesons reference, around 9 times smaller than Earth time. And by the speed of light being maximal, it would be the time that would be dilated and the space that would be contracted from the meson frame. In the experiment, in the reference of the measurement on Earth the time for the mesons was 6.4 milliseconds, equivalent to the 0.7 milliseconds of the meson reference.

But, L. Jánossi in 1971 and Phipps Jr. in 2006, considered that the mesons with higher energy and speed presented a greater time for decay.

XXIX. DIVERGENCES, MATHEMATICAL ABSTRACTIONS, NEGATIVE ENERGIES, MORE MATRICES, STATISTICS AND PROBABILITIES, FIELDS, "PHANTOM PARTICLES" WITHOUT MASS AND STRANGE PARTICLES – <u>1829 to 1977</u>

Abandon QED or Eliminate its Differences – In 1936, Dirac had concluded that QED had to be excluded because it was a theory without beauty and complicated that explained almost nothing, and had the agreement of Einstein who judged her exceptionally bad. Between 1936 to 1949, it prevailed as Heisenberg in keeping the infinite terms or eliminating the divergences as Bethe, Heitler, Fermi, Oppenheimer, Kramers, Wentzel, Möller, Dirac, Tomonaga, H. Fukuda, Y. Miyamoto, Schwinger, Pauli, F. M. Villars and H. J. Groenewold.

Negative Energy Photons – In 1939, Dirac presented the hypothesis of photons with negative energy to eliminate divergences in QED.

Calibration (Gauge) in Quantum Mechanics – In 1941, Pauli presented quantization of the electromagnetic field through the transformation of gauging or caliber ("gauge") idealized in 1918 by Weyl, with calibration fields, i.e. for the system to be invariant, for example, introducing a non-mass vector field, an electromagnetic potential with field massless, interconnected with the mathematical apparatus of quantum mechanics.

In a local transformation, a system of interaction of forces to be symmetrical, that is, that was invariant, would require a term of the Lagrange energy equation: $L_{energy} = E_{kinetic} - E_{potential}$ or the energy density: $L_{density\ of\ energy} = i(h/2\pi)c\psi_i'\gamma^\mu \partial_\mu \psi_i - mc^2\psi_i'\psi_i$, where i complex number, h Planck's constant, c speed of light, ψ_i' e ψ_i variables of the electromagnetic field related to the position and time, γ the coefficient relativistic $1/[1-(v^2/c^2)]^{1/2}$, v the velocity and m the particle mass in the system, μ index of the vectors, applied to the system, had zero mass to undo terms of the equation that presented the product of the mass with variables for each position of the system.

More Matrices, Probabilities and Statistics – In 1943 and 1944, Heisenberg and also Stückelberg; in 1946 Möller and also S. Ma; in 1947, Tomonaga and also Wigner and Eisenbud presented transition states and interactions between particles through matrices with the whole mathematical apparatus of quantum mechanics. In 1947, V. B. Berestetskii presented function and wave of electrons to calculate the probability of a nucleus transferring energy to the electrons in the atoms. V. A. Fock and S. N. Krylov presented an energy distribution function for an atomic state that was proportional to the exponential decay of that state. In 1948, R. Karplus and Schwinger presented a statistical description of a large particle system; T. Y. Wu presented potentials for expansion of the quantum matrix; and Heisenberg presented statistics in quantum atomic system turbulence for the energy transfer mechanism and minimal action with the Hamilton equations quantized in matrices for high-energy particle collisions. In 1949, M. H. Johnson and B. A. Lippmann presented electron motion studies in the magnetic field. In 1949, J. Mayal presented more statistics for calculating quantum properties. In 1950, G. Källen presented studies on quantum matrix expansion.

Standardization, Diagrams and More Mass-Free Fields and Massless Particles – In 1948, Kramers presented aspects of particle interactions adjustments, and Feynman presented calculations of particle interactions using the mathematics of Green drawn functions between 1820 and 1830, and to adjust the energy and mass calculations R. P. Feynman presented corrections, subsequently called renormalization, in which the particle decay was related to a statistical factor, to the constant h of Planck, to the masses and momentums of the particles, density of the initial and final state, particle charge that decays, momentum

𝔐 of certain momentums of particles and other scattering constants. Also, Feynman presented diagrams resulting from these calculations, in which particles could go back in time as had been formulated by Stückelberg. In 1950, J. C. Ward presented a relationship between the renormalization of electric charge and velocity of photons. And in 1954, Yang and R. Mills presented a local caliber invariance for particles in interaction, that is, introducing mass potentials into Lagrange's equation, obtaining mass free fields and massless particles. The calibration potentials, to maintain the fundamental symmetry to the laws of nature, were added in the equations of interconnected fields to the particles, within intrinsic abstract spaces outside the space-time of the real field, idealized by Yang and Mills, resulting in field without mass.

In normalization, Dirac commented that the observed mass differed from the initial mass of the equations, in which there would be interaction of the electron with the electromagnetic field that altered the mass of the electron, when it referred to the normalization of the mass in the field equations.

Synchrotron Radiation – In 1949, W. G. McMillan and Peterson observed in the synchrocyclotron the radiation of synchrotron when accelerating electrons.

A-M. Liénard in 1905 there was widespread the Larmor formula for irradiation of charged particles, for the relativistic case, that is, when the particle moved with speed near the light.

And G. A. Schott in 1906 developed a classical theory of the irradiation of accelerated particles, in the context of the classical atomic model, in which the electrons moved in circles, but compensated for the tendency to irradiation by means of various mechanisms. And in the following years he published articles on the subject, culminating in the publication of "Electromagnetic Radiation" in 1912.

Bohr in 1913 had postulated by the state of quantum energy that the electrons revolved around the atomic nucleus, and were accelerated by a centripetal force, but did not radiate as predicted the classical theory and the model of Schott. This was completely contrary to studies on energy irradiation by accelerated particles, and Schott remained a staunch opponent of quantum theory.

In 1935, scientists and engineers observed the loss of energy of the particles in the accelerators that were beginning to be manufactured, especially in those of the type betatron and cyclotron. In both models, the particles were accelerated along circles, and this circular motion was produced by an external magnetic field, perpendicular to the velocity of the particles. Initially these machines were designed to accelerate electrons, protons and other particles in nuclear physics experiments, so that the energy loss of the particles because of electromagnetic irradiation was a deleterious phenomenon for the purposes of the research. In the case of betatron, calculations suggested that the energy loss by irradiation limited in 0.5 GeV (gigaeletronvolt, one billion eletronvolt) the maximum energy of the particles. In cíclotrons, the limitation of maximum energy also had to do with the fact that the particles were accelerated, every turn, by a system of radio frequency (RF). In this case, the frequency of the accelerator system had to be the same as the particle in its circular motion. That is, the particle had to get to the point where the RF was at the time it was being activated. The problem is that in high energies, in relativistic cases, in which the mass of the particle increased, the frequency of the motion was directly proportional to the magnetic field and inversely proportional to the value of the energy. Therefore, in the case of a constant magnetic field, each time it increased the energy, it decreased the frequency, and the particle was delayed at the point of acceleration. That means the RF was going off before the particle came. There were two alternatives to solving this problem. The first was to increase the magnetic field over time so as to leave the frequency of the motion equal to the frequency of the accelerator system. This alternative originated the synchrotron. The other, which gave rise to the synchrocyclotron, was to vary the frequency of the accelerator system. In 1946, researchers from General Electric began to build a small synchrotron, with an approximate diameter of 60 centimeters, but left a glass window in the tube where the electrons circulated. In 1947, they were doing some tests when the equipment showed intermittent failures. They started to turn on and off, checking the operation of the machine, and suddenly saw through the window a bright white beam of light, which was named synchrotron light.

The electrons of the synchrotron traveling close to the speed of light in the vacuum lines make a curve, forced by a potent magnetic field generated by an electromagnet, that is, resulting in braking their speed they lose some of their energy, emitting it as light. This light emitted by the accelerator is then directed, focused and used for various purposes, in particular to study how nature operates on an atomic and molecular scale. This light has some very special properties that make it ideal for this type of study: its potency is very high, typically millions to billions of times, brighter than any lamp or other simpler source; its energy spectrum is very wide, going from infrared (thermal radiation), visible, ultraviolet to X-rays.

Asymmetries – In 1928, R. T. Cox, C. G. Mcilwraith and B. Kerrelmeyer and at 1929 C. T. Chase had observed that in beta decay experiments there would be parity asymmetries, charge conjugation and particle formation time. But in 1931, Wigner presented that opposite images of reflection or reversal should be symmetrical, that is, parity. And in 1941, Schenberg came to observe that in strong interaction could occur asymmetry in the formation of particles and antiparticles. On 1953 Schwinger; in 1954 G. Lüders; and in 1956 Pauli presented the theory that every particle and antiparticle would present equal mass, life-medium and magnetic momentum, with parity, combination of charges of particle exchanges and antiparticles and the equal times of formation and decay of particles and antiparticles, observed in the parity-charge-time set. But, there was another particle θ^+ which resulted in: $\theta^+ \rightarrow \pi^+ + \pi^-$, which presented equal mass, spin and charge that the particle $\tau^+ \rightarrow \pi^+ + \pi^+ + \pi^-$ or $\tau^+ \rightarrow \pi^+ + \pi^\circ + \pi^\circ$. Therefore, for the equality of mass, spin and charge θ^+ and τ^+ they should be the same particle, but with different images between the two particles. So, T. D. Lee and Yang presented a procedure for conducting experiments that could check parity. And in 1957, C. S. Wu, E. Ambler, R. W. Hayward, D. D. Hoppes and R. P. Hudson observed that the spin was asymmetric in the reflection of the particles. Also, L. M. Lederman, R. L. Garwin and M. Weinrich, and also J. I. Friedman and V. L. Teledgi observed parity breaking. Later θ^+ and τ^+ were called K^+. Also, the charge conjugation presented to be dependent on the particle type and antiparticle as to the helicity. But, the time remained equal between the formation and decay of particles or antiparticles in conjunction with parity and charge, i.e. parity-charge-time (PCT). In 1964 J. W. Cronin, V. L. Fitch, J. H. Christenson and R. Turlay presented parity-charge break in the mesons K° with its antiparticle $K^{\circ'}$, and Lee, J. Bernstein and G. Feinberg in electromagnetic interactions.

Free Neutron – In 1948, A. H. Snell and L. C. Miller presented an experiment in which the free neutron decay was observed. And in 1950, Snell, F. Pleasonton and R. V. McCord and also J. M. Robson presented a calculation of the free neutron half life around 12 minutes.

Plasma – In 1929, Langmuir and L. Tonks had obtained with electrical discharge highly ionized gases that called plasma. And in 1949, Fermi presented a mechanism in which the ion formed plasma and its electromagnetic field presented a displacement that approached or drifted away from the cosmic rays causing acceleration or deceleration of these rays.

Nucleus in Layers – On 1948 and 1949, M. G. Mayer and also J. H. D. Jensen, O. Haxel and H. E. Suess while observing that certain quantities of proton + neutrons of the nucleus of chemical elements presented greater stability of these atoms, presented a nuclear model similar to electronic layers, in which protons and neutrons were distributed in layers, initially being observed that nuclei with 2, 8, 20, 28, 50 and 82 protons or neutrons were the most stable chemical elements in nature. They observed that there was a spin interaction and angular momentum, and the maximum number of layers would follow the Pauli exclusion principle. And In 1950, Rainwater presented a different form of spherical to the atomic nucleus, since the nuclear layer model occurred only for one part of the atomic nuclei.

Strange Particles – In 1944, Leprince-Ringt and M. Lhéritier observed in cosmic rays a particle with mass different from all known particles at the time and in 1946 detected a particle with mass of 990 times the mass m_e of the electron. In 1949, R. B. Brode observed particles in cosmic rays with masses between 500 and 800 m_e. In 1947, G. D. Rochester and C. C. Butler presented that in the cosmic rays a neutral particle disintegrated resulting in a V-trajectory particle, called V° and subsequently called θ° and K°: θ° → $\pi^+ + \pi^-$. In 1949, R. Brown, U. Camerini, Fowler, Muirhead, Powell and D. M. Ritson observed the decay of a particle called τ^+ and subsequently K$^+$ with mass between 870 and 985 m_e in 3 mesons: $\tau^+ \rightarrow \pi^+ + \pi^+ + \pi^-$ or $\tau^+ \rightarrow \pi^+ + \pi° + \pi°$ or $\tau^+ \rightarrow \pi + \pi + \mu$ or still $\tau^+ \rightarrow \pi + \mu + \mu$. And Leprince-Ringt observed mesons of larger masses which he named τ. In 1950, J. B. Harding also detected the decay of τ in 3 mesons π, and V. D. Hopper and S. Biswas presented the detection of a particle subsequently named λ, which falls under: λ → proton + π or μ. At 1951 C. O'Ceallaigh presented particles K$^+$ and K$^-$, and K. H. Barker, Butler, A. Chacon and A. H. Chapman presented the particle V$_1$°.

Wave-Particle – In 1947, Millikan in "Electrons, Protons, Neutrons, Mesotrons and Cosmic Rays" stated that: *"The elementary and fundamental phenomena that constitute light cannot be waves and corpuscles at the same time. What are they really?"*.

Planck's Constant: Escaping Human Understanding – In 1948 A. S. Eddington in "The Nature of the Physical World" commented that: *"Something is radically wrong in the concepts of current Physics and we don't know how to fix it. But it must be Planck's constant that escapes human comprehension"*.

Ether Return – In 1951 E. Whittaker in "The History of the Theories of Ether and Electricity" in the preface presented: *"It seems absurd to continue to term a vacuum to a medium so rich in physical characteristics, so the historical denomination ether should be rehabilitated because it is more appropriate"*. And also in 1951, Dirac commented in a letter to the Magazine "Nature" that: *"So by the effect of a new theory of electrodynamics we are forced to admit the existence of an ether"*.

Par and Odd, Particles and Antiparticles, Spins on the Right and Left – The strange particles were produced in the interaction of strong force and the disintegration occurred by weak force, and in 1952, the country sought explanation by pairs of particles in the interaction and odd formation particles, as par plus par → odd + odd and observed by W. B. Fowler, R. P. Shutt, A. M. Thorndike and W. L. Whitemore, and at 1953 by R. L. Garwin. But then another particle was indeterminate and being inappropriate this model was replaced by a number $S_{strange}$ of strangeness in 1953 by M. Gell-Mann; and K. Nishijima and T. Nakano, which can vary between $0, + 1, - 1$, with the preservation of spin isotopic. And from 1953 to 1962, Le Prince-Ringt, Anderson, Blackett, Powell, Rochster, Rossi, W. B. Fretter, Aldin, B. W. Thomson, Gell-Mann, Pais, Nishijima, L. B. Okun, Tokata, Segrè, O. Chamberlain, C. E. Wiegand, T. Ypsilantis, B. Cork, G. R. Lambertson, O. Piccioni, W. A. Wentzel, T. Regge, G. F. Chew, S. C. Frautschi, Schwinger, Tiomno, M. Ikeda, S. Ogawa, Y. Ohnuki, Y. Yamaguchi, H. Fleming presented strange particle classification made by masses and charges of particulates and antiparticles, with denominations of super or hyperons that is, neutrons; strangeness, hadrons, that is, heavy or massive like the mesons π and K, and barium from the Greek, meaning heavy, like neutron and antineutron, proton and antiproton, and hyperons; leptons of Greek with meaning of small, such as electron, electron neutrino, muon, muon neutrino, and after tau and tau neutrino in 1977.

Emission of Gamma Rays, Magnetic Pole and Nucleus Mass – In 1952, Blatt and Weisskopf presented that gamma rays were related to the probability of magnetic multipole radiation from the atomic nucleus.

So the magnetic multipole of the atomic nucleus M_L by the expression: $M_L = \pi \mathcal{V}(e^2/h)10S \; [(h/2)/McR]^2 \; (R/\lambda)^{2L}$, Where $\mathcal{V}$ = gamma-ray frequency, e = electron charge, h = Planck constant, S = (L function) = 2.5×10^{-1} to 3.1×10^{-9}, L = angular momentum, M = nucleus mass, c = electromagnetic radiation speed, R = nuclear radius, λ = gamma ray wavelength.

Expanded Quantum Potential – On 1926-1927 L. De Boglie had presented the quantum potential for a particle. But in 1952, David Bohm extended this quantum potential energy to all particles. And this extension was inappropriate relativistic conditions. However, it continued without explaining what this quantum potential energy was.

Abstract Quantum Principles – In addition to the particle motion by the wave function, the particles had its motions in energy fields for each particle in which fields and particles interact. The representation of the fields was made up of energy and momentums, with the mathematical vectors of position, direction and the scalar quantity for the intensity of physical magnitudes, from infinitesimal variations between coordinates of positions, angles, electric potential, magnetic angular momentum components like spin, time and other variables that could be related, such as rotations, exchanges or permanent states, i.e. anticommuting, and or associativity as group theory, space inversions and time, additions and products of mathematical operations by differentials, integrals, matrices, potential gauging of calibration theory ("gauge"), of the coordinate systems interconnected between all references such as Lorentz transform and tensor, in the motion, creation and annihilation of particles in each particle field and in the interaction with other particles and their fields. And the equations of the motion of these particles in their fields following symmetry, with the potentials of the calibration theory for the conservation of the laws of nature, for energy, momentum and charge, according to the theorem of A. E. Noether by which the laws of nature they are only existing with symmetry, and with statistical and probabilistic application. But in 1957, as L. De Broglie: *"From around 1930, in quantum mechanics, statistical and probabilistic forecasts are used, because there is the concept of indetermination, and almost unanimously admitted as a real indetermination of the knowledge of the physical states. But it would be possible to investigate with greater depth the physical reality. Trying to stop all attempts to pass the current position of quantum Physics would be detrimental to the development of science, and contrary to the lessons of the history of science. Science teaches that the current state of knowledge is always provisional and that new directions should be observed".*

PART 4 - QUANTUM IMPULSE

XXX. THE THEORY OF QUANTUM IMPULSE – 600 BC to 1969

In 1952, I. L. Erdelyi presented the "Theory of the Momentum" or "Quantum Elementary Impulse" and other works up to 1969 on that theory. Erdelyi was born in Hungary, Graduated in Engineering from the University of Budapest, and it was the same time of Hungarian Physicists and Engineers who left abroad after the Second World War, as Teller that developed the nuclear fusion reaction and subsequently the Space Defense Project ("Star Wars") of the time Ronald Reagan, President of the United States,

Szilard which presented the initial study of the chain reaction of nuclei that generated the atomic bomb of nuclear fission and Wigner one of the greatest theorists of quantum mechanics. Erdelyi was a Teacher at the School of Engineering at the Mackenzie University in Sao Paulo, Brazil, and President of the Scientific Council of the Brazilian Interplanetary Society-SIB who was a member of the International Astronautical Federation-IAF.

In the theory of momentum or elementary impulse, the quantum impulse, from Erdelyi, the concepts and units of momentum, impulse, force and energy were effectively observed in their actual physical meanings. So, momentum = mv, m = mass and v = velocity treated with intensity, direction, that is, vectorly, as other vector physical magnitudes; the variation of the momentum with initial velocity at zero was the impulse = [(mv) – (m x 0)] = mv; force F = impulse/time = mv/t, where t is the time, and energy E = force x distance = (mv/t) x d, where d = distance.

Planck's Constant h and the Energy of the Elementary Quantum – It was the perception that Planck's constant h [erg x s], by quantum theory in 1900, in the system of units CGS (centimeter, gram, second), showed absence of a greater physical meaning, i.e. action = energy x time, which led Erdelyi to divide by time, in seconds in the CGS, on both sides of the energy quantum expression $\mathcal{E}$ [erg] = h$\mathcal{V}$ [erg] where $\mathcal{V}$ is the frequency of the quantum per second:

$\mathcal{E}$ [erg]/[s] = h[erg x s]$\mathcal{V}$[1/s]/[s] = h'[erg] $|\mathcal{V}|$/[s] ,

$\mathcal{E}$ [erg] = $|\mathcal{V}|$ h' [erg], obtaining the constant h' in [erg] of precise actual meaning of energy, contrary to what Eddington he regarded it as h [erg x s] that escaped comprehension. Because the dimension [erg x s], that is, energy x time, "action", was just an abstraction to term an event, but without a precise physical significance.

Like this to $\varepsilon_{(\mathcal{V}=1)}$: ε [erg]/[s] = h [erg x s]/[s] = h'[erg], that is, |h| [erg] = h' [erg], to $\mathcal{V}$ = 1 cycle/s, it followed physical reality because it was well defined the fundamental concept of energy.

In his works of 1952 and 1954, Erdelyi (Figure 19) analyzed Planck's equations of the black body radiation and introduced the constant herg or h' = | h | [erg], according to publications listed in documents 3, 4, 5, 6 and 7.

Fig. 19 – Engineer Istvan Laszlo Erdelyi, creator of the immortal Theory of Quantum Impulse in 1952, Teacher of the Engineering School of the Mackenzie University in Sao Paulo. Photo of "Fatos and Fotos" Magazine, Brazil, June 1967.

THE EXAMINATION AND CORRECT INTERPRETATION OF PLANCK'S CONSTANT

h

by

I. L. ERDELYI

Copyright, 1952, Istvan Laszlo Erdelyi
Published by Linotipodora Grafica Ltda., S. Paulo (Brazil)
First printing

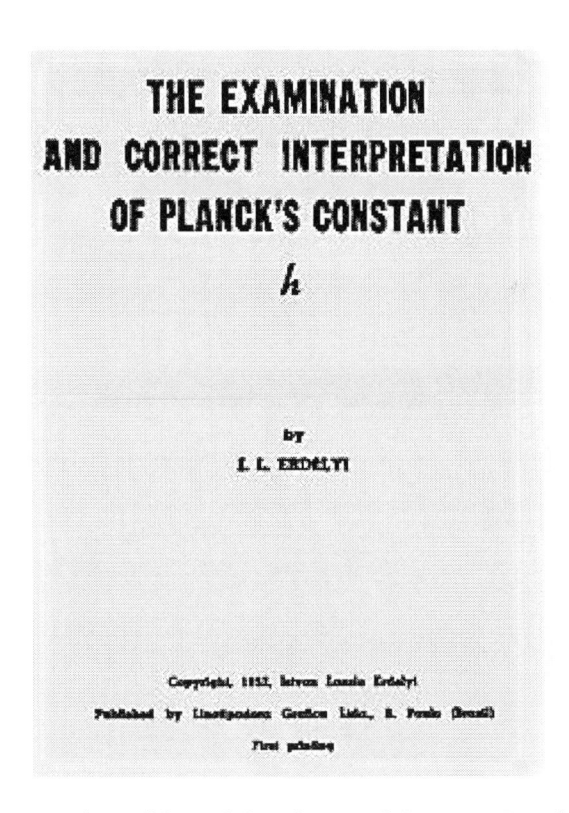

Doc. 3 – First edition of the Theory of Quantum Impulse from Erdelyi in 1952, by the interpretation of Planck's constant h. In References.

THE RESUME OF THE RESULTS OF THE EXAMINATION AND THE CORRECT INTERPRETATION OF PLANK'S CONSTANT, THE h erg sec.

1. Introduction of the idea of *quantum of momentum-transfer*, travelling in space with the velocity of light, reveals a concept which apparently antiquates Plank's differently sized quanta belonging to the various frequencies per second, and Einstein's manifoldly sized *photons*. One result of the examination is the realization that evidently there exists only one kind of *quantum of energy*, or *energy-photon* and it is expressed by the value:

27. $$herg = 6{,}622 \times 10^{-27} \text{ erg}$$

Electromagnetic radiations with different v frequencies per second can deliver only the integer multiples of the *herg* (*the primary energy-quantum*) on the surface of *material* bodies kept in the path of the radiations during any interval of observation.

We obtained the *primary energy-quantum* (the *herg*) in the third chapter entitled: "Examination of the "RIGHT TO EXIST" of Plank's h". There we also discussed the possibility of a transformation in Plank's law and equation. This transformation permitted the use of *intelligible* factors only in the well-known equation. Chapter three explained the derivation of the *herg* by logic and mathematics.

The indivisibility, and the fact of existence of this new concept was treated in the fourth chapter. There we considered the possibility of *limits* in the frequency per second of electromagnetic radiations. There also we demonstrated mathematically that one complete wave cannot represent more or less of an energy-dose than the amount represented by the *herg*. This also applies to electromagnetic radiations, which have wavelengths longer than the distance traversed by light in one second.

Doc. 4 – Excerpt from the initial article of the Quantum Impulse Theory of Erdelyi, in References.

Planck arrived at the expression furnishing the quantity of energy at constant temperature, "T", corresponding to any chosen monochromatic cavity-radiation (Planck, "Theory of Heat", pp. 275, Equation 453.) as follows:

1.) $\quad E_\nu = \dfrac{8\Pi h \nu^3}{c^3} \cdot \nu^3 \cdot (e^{h\nu/kT} - 1)^{-1} \cdot \Delta\nu$

Dividing now the amount of energy, "E_ν" by the volume of the cavity, "1^3", he obtained the expression of the energy density for the monochromatic radiation in question: (Equation 454 in Planck's above mentioned book.)

2.) $\quad \dfrac{E_\nu}{1^3} = \dfrac{8\Pi h \nu^3}{c^3} \cdot (e^{h\nu/kT} - 1)^{-1} \cdot \Delta\nu = u_\nu \cdot \Delta\nu$

He then obtained the spatial total density, "u", of cavity-radiation at a constant temperature, "T", simply by integrating the above expression for all frequencies from 0 to ∞. (Equation 462 in Planck's above mentioned book.

3.) $\quad u = \displaystyle\int_0^\infty u_\nu \cdot d\nu = \dfrac{48\Pi h}{c^3} \cdot (\dfrac{k.T}{h})^4 \cdot \dfrac{\Pi^4}{90} = a \cdot T^4$

This equation is nothing else, but the mathematical formulation of the Stefan – Boltzmann law.

It is very important to observe here, to which physical dimensions the factors of these equations (Equation 2, and 3.) correspond:

4.) $\quad \dfrac{\text{erg}}{\text{cm}^3} = \text{Dim}(\dfrac{E_\nu}{1^3}) = \text{Dim}(u_\nu) \cdot \text{Dim}(\Delta\nu) = \dfrac{\text{erg.sec}}{\text{cm}^3} \cdot \dfrac{1}{\text{sec}} = \dfrac{\text{erg}}{\text{cm}^3}$

Doc. 5 – Erdelyi article page on analysis of the physical units of Planck's constant h, in References.

16.) $\quad u = \dfrac{8\Pi}{c^3} \cdot (\dfrac{1}{\text{sec}^3}) \cdot \text{herg} \cdot \displaystyle\int_{n=0}^{n=\infty} n^3 \cdot (e^{\text{herg}.n/kT} - 1)^{-1} \cdot dn$

The symbol, "ν", – which thus became a dimensionless number has been changed to "n", merely in order to avoid any mixing up of the dimensionless form of "ν" with the one having dimension. At the same time also the symbol, "h", has been substituted by "herg", however, without presupposing anything about the dimension of this symbol.

The integrand is now composed exclusively from dimensionless numbers. This again appears quite logical, since a summation of the monochromatic energy-densities must furnish the total number of elementary energy doses present in 1 cm3.

Performing the integration the following result will be obtained:

17.) $\quad u = \dfrac{\dfrac{48\Pi^5}{90} \cdot (\dfrac{k.T}{\text{herg}})^4 \cdot \text{herg}}{c^3 \cdot (\text{sec}^3)}$

It may be asked now – "What is the physical dimension of the constant symbolised by "herg"?

18.) $\quad \text{Dim}(u) = \text{Dim}(\dfrac{48\Pi^5}{90}) \cdot \text{Dim}(\dfrac{1}{c^3} \cdot \dfrac{k^4 \cdot T^4}{\text{herg}^4} \cdot \text{herg}) \cdot \dfrac{1}{\text{sec}^3} = \dfrac{\text{erg}}{\text{cm}^3}$

The first factor on the right hand side of Equation 18.) is a dimensionless number, therefore has to be omitted from this investigation.

Doc. 6 – Erdelyi shows the introduction of the herg [erg] constant = h' [erg] = h/second [erg], replacing h [erg second] in Planck's equations. In References.

J. P. Hansen

Let us make a first attempt in explaining the mathematical meaning of Planck's radiation equation in the light of the Momentum Theory:

It has been stated in connection with Equation 27.)

27.) $u_\nu \frac{erg}{cm^3} = N_{\nu T} \frac{1}{cm^3} \cdot herg \; arg$

that the factor, "$N_{\nu T}$", is the NUMBER of the "herg"-s organized in monochromatic elementary rays of frequency, "ν", per cm3.

It has also been stated in connection with Planck's expression which he presented for the total energy-density of cavity radiation (Equation 3.):

3.) $u = \int_0^\infty u_\nu \cdot d\nu = \left[\frac{8\Pi}{c^3} \cdot (\frac{1}{sec^3}) \cdot \int_0^\infty \nu^3 \cdot (e^{\frac{herg \cdot n}{kT}} -1)^{-1} \cdot dn \right] \cdot herg$

that after moving the dimension "$\frac{1}{sec^3}$" of "ν^3" in front of the symbol of integration and then performing the operation:

Doc. 7 – Development of Planck's equations by Erdelyi in Theory of Momentum or Quantum Impulse. In References.

By the quantum theory of Planck and the quantum light of Einstein, the quantum $\varepsilon = h\nu$ it was emitted as a compact of indivisible energy ε, in which the ν frequency defined the type of radiation, and h = Planck's constant. The gamma radiation would have a quantum of higher energy and greater mass according to Einstein's relativity, by the equivalence of m mass and energy by c speed of light: m = energy/c^2, than the quanta of violet light, red light and microwave. So, according to Planck and Einstein there was a quantum for each wave frequency in the electromagnetic spectrum, that is, six trillions of different quanta between frequencies since 10^1 to 10^{21} cycles per second. Figure 20.

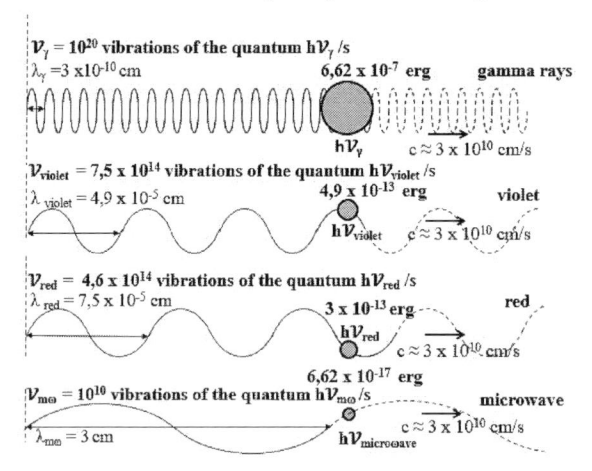

Fig. 20 – According to Planck and Einstein there is a different quantum for each wave frequency in the spectrum and electromagnetic, that is, six trillions of different quantities between frequencies since 10^1 to 10^{21} cycles per second.

And by the theory of elementary impulse only there was a quantum h' for all frequencies. That is, $\mathcal{E}_{(V=1)} = \varepsilon = h\mathcal{V}_{(V=1)} = h$. Figure 21 presents an example for the quantum of violet radiation by quantum theory and the theory of the impulse of the elemental quantum.

VIOLET QUANTUM $h\mathcal{V}_{violet}$

Violet Quantum Energy = $\mathcal{E}_{violet} = h\mathcal{V}_{violet}$
h = Planck's constant = $6{,}62 \times 10^{-27}$ (erg s)
    "action" without more physical significance
$\mathcal{V}_{violet}$ = frequency = $7{,}5 \times 10^{14}$ vibrations/s
$\mathcal{E}_{violet} = h\mathcal{V}_{violet} = 4{,}9 \times 10^{-13}$ erg

$\lambda = 4{,}1 \times 10^{-5}$ cm $c \approx 3 \times 10^{10}$ cm/s

VIOLET QUANTUM $h' |\mathcal{V}_{violet}|$

$h' = m'c \times c$
$\mathcal{E}_{violet} = h' |\mathcal{V}_{violet}|$ $\mathcal{E}_{(V=1)} = h'$

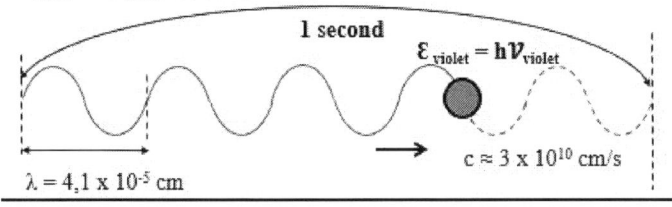

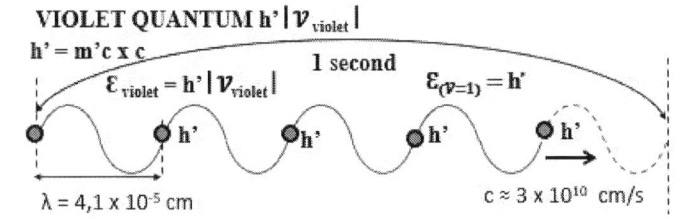

$\lambda = 4{,}1 \times 10^{-5}$ cm $c \approx 3 \times 10^{10}$ cm/s

$\mathcal{E}_{violet}$ (erg)/s = [h (erg.s) x $\mathcal{V}_{violet}$] / s = h'(erg) | $\mathcal{V}_{violet}$ |
h' = elementar quantum, this is, for $\mathcal{V} = 1 : \mathcal{E}_{(V=1)} = \varepsilon = h'$
h'(erg) | $\mathcal{V}_{violet}$ | = | $\mathcal{V}_{violet}$ | m' c x c = m' c²
h' = $6{,}62 \times 10^{-27}$ erg ; $c \approx 3 \times 10^{10}$ cm/s ; m' = $0{,}73 \times 10^{-47}$ g
$\mathcal{V}_{violet}$ = frequency = $7{,}5 \times 10^{14}$ h'/s
$\mathcal{E}_{violet} = h' \times |\mathcal{V}_{violet}| = 6{,}62 \times 10^{-27}$ erg x $7{,}5 \times 10^{14} = 4{,}9 \times 10^{-13}$ erg

Fig. 21 – By Quantum Elementary Impulse Theory of Erdelyi there is only one quantum with energy $\mathcal{E}_{(V=1)} = \varepsilon = h'$ for all frequencies.

And it was precisely the time applied in the denominator, in the expression of energy $\mathcal{E}/t = h\mathcal{V}/t = |\mathcal{V}|$ h', in which h' = h/1s is that it formed all quanta by the quantity of frequency $|\mathcal{V}|$. Therefore, the time appeared to relate admirably with the energy. And it was similar to the time introduced in the denominator of the equation of energy in quantum mechanics according to M. Stone's theorem of 1932, but that, however, there was no perception of its importance by current Physics.

J. P. Hansen

Deduction of the Constant h by Planck – The perception of Erdelyi for the meaning of the constant h was in fact brilliant when it was observed the very deduction of the quantum of radiation as Planck, in which it presented the total energy radiated from a black body in which the terms of the two sides of the equation resulted in different units, contrary to the physical phenomenon. In the case, the power unit appeared on the first side and in the second the unit of energy x time, the "action" without greater or any physical significance.

And another fact that surprised was that Planck having gotten the value of the constant h in the unit of energy x time, on the next page of his book "The Theory of Radiation" when commenting on unit system, Planck got h in [gram x cm^2/second]. That is, seeming to look for a direct relationship of h by time, but [gram x cm^2 by the time], also nothing physically representing!

So, E_{total} it was the total energy radiated by a orifice of a body that absorbed all the radiation, that is, a black body:
$E_{total} = 0,0731$ [watt/cm^2] = 0,0731 [erg/cm^2s] = 7,31 x 10^5 [erg/cm^2 s]
And the specific intensity K of the black body radiation as:
K = (c/4π) u [erg/cm^2 s] = (a c /4π) T^4 [erg/cm^2 s], where:
c = speed of light in the vacuum measured in the Earth ≈ 3 x 10^{10} [cm/s],
that is the velocity of the m' elementary mass
u = energy density [erg/cm^3]
a = constant = 7,28 x 10^{-15} [erg /cm^3 °C^4]; T = temperature [°C]
It is observed that in the two formulas, E_{total} and K there is the time unit in seconds in the denominator. But, Planck gets the radiation term:
u = (8πhν^3 /c^3) (1/ (exp $^{h\nu/kT}$ – 1), where k = 1,34 x 10^{-16} [erg/°C], with dimensions:
u [erg/cm^3] = (8πν^3/c^3)[s^{-3} / (cm^3/s^3)] h [erg s])
and the term (1/ (exp $^{h\nu/kT\ [erg\ s\ s(to\ -1)\ /\ erg°C(to\ -1)\ °C\ ]}$ – 1) right adimensional
Being u_{energy} [erg] = u x cm^3 [erg]
So, u_{energy} [erg] = (8πν^3 /c^3)[s^{-3}/(cm^3/s^3)][cm^3] h [erg s]) (1/(exp$^{h\nu/kT}$ – 1), This is,
u_{energy} [erg] = (8πν^3/c^3) (1/(exp $^{h\nu/kT}$ – 1) h [erg s]
And the physical inequality of both sides of the equation is visible.
With: $u_{energy/s}$ [erg/s] = (8πν^3/c^3) (1/ (exp $^{h\nu/kT}$ – 1) h [erg], the constant h would have the physical dimension [erg], but still resulting in the inequality of physical units of equation.
When this equation is differentiated by $d\nu$ and once integrated is where the unit [erg] is obtained on both sides of the equation, but not for the h. But, if the differential were by the wavelength

$d\lambda$, which is the inverse of frequency, would remain the inequality of the physical units.

And by the Bose-Einstein statistic, treating the quanta $h\nu$ all similar as particles, the Planck equation is obtained. But with the same physical inequality of units.

However, when quantum was admitted as a single particle it would tend to meet physical reality, which is precisely the theory of Erdelyi.

And when Planck obtained the value of the constant h through:
$h = ac_2/48\pi\alpha c$
a = constant = $7{,}28 \times 10^{-5}$ [erg/cm³ °C⁴]; $c_2 = 1{,}436$ cm °C; α = constant = $1{,}0823$;
c = speed of light = 3×10^{10} [cm/seg], obtaining:
$h = 6{,}415 \times 10^{-27}$ [erg s]

But on the next page when commenting on unit system, Planck wrote:
$h = 6{,}415 \times 10^{-27}$ [g cm²/s], ie,
looking for a direct relationship of h by the time, but [g cm²] nothing physically representing by time!

Elementary Impulse in a Space-Absolute and in Vacuum-Space-Ether – By the theory of quantum impulse there was an elementary mass m' that formed with the speed of light in relation to an absolute vacuum-space the elementary quantum impulse m'c, in which $c \approx 300{,}000$ km/s would be the predicted speed. Figure 22.

elementary quantum momentum or impulse m'c in one absolute space

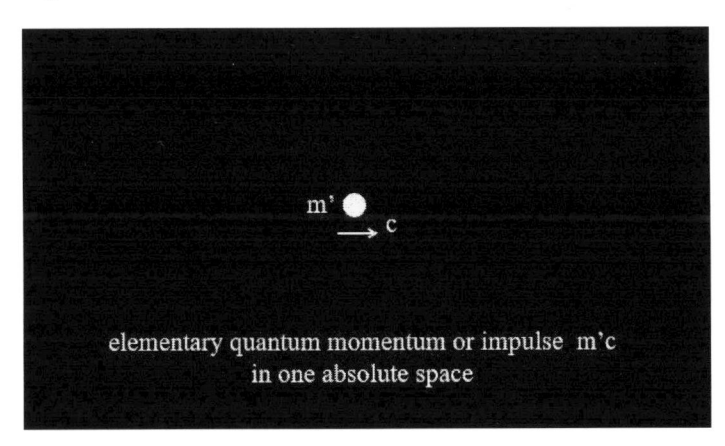

Fig. 22 – Introducing in an absolute vacuum-space an elementary mass m' with speed c there would be the momentum or elementary impulse, as Erdelyi.

And by the concept of energy of an elementary quantum, $\varepsilon = h'$ $= m'c^2$, if that quantum $m'c^2$ travel with speed c, it would be obtained $(c^2 \times c) = m'c^3$ contrary to the concept of energy. So what would travel would be to drive mc with the c speed of light, resulting then $(m'c \times c) = m'c^2$ within the concept of energy. And for the m'c impulse to travel with speed c, to result in quantum energy $\varepsilon = h' = m'c^2$, it would be necessary to have a medium of elastic mass, of extremely low density, that is, an ether of submass in the vacuum-space, that is, a referential, for the elementary impulse to go through with the speed c to have the energy m'c x c $= m'c^2$. That is, this ether in vacuum-space, referential or submass was a consequence of the existence of the elemental impulse m'c with speed c. Figure 23.

elementary quantum impulse m'c
with speed c in the ether in the espace
resulting in energy m'c x c = m'c²

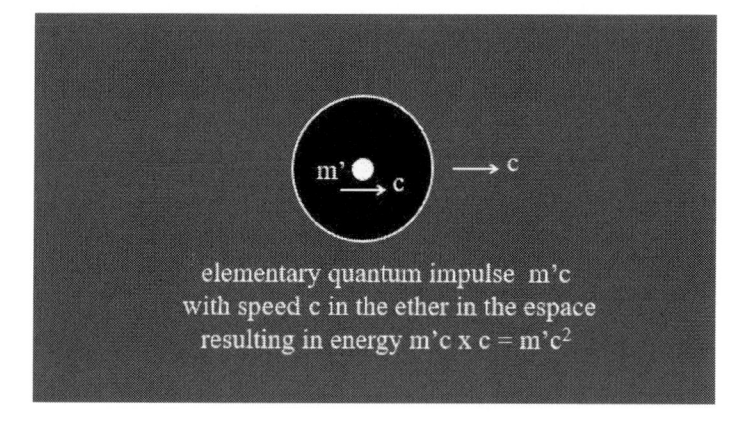

Fig. 23 – While in an absolute vacuum-space the mass m' presented speed c, forming m'c, a medium of elastic mass of extremely low density, that is, an ether of submass in the vacuum-space presented a referential for the elementary impulse m'c go with speed c, according to the quantum impulse of Erdelyi.

The Medium Ether – Around 600 BC, as Anaximander the space where the stars were rotating would be filled with a continuous medium, a special pure air, a fluid, an ether, which filled all the space. And around 350 BC, as Aristotle the light was similar to the sound vibrations, in which a luminous object would have vibration that was propagated through a transparent medium. In the 1600 for Descartes there was matter of a subtle type, of pressured globules in contact with each, acting as a rigid body. Newton, between 1664 to 1690, indicated that the light was made up of flux of particles emitted by vibrations of brilliant bodies with violent agitation of motions of the ether that triggered the incandescent substances to emit light, which reflected, refracted and difracted in other bodies, and the vibrations of these fluxes would be propagated through this ether. Between 1665 and 1678 Hooke and Huygens observed the light as a wave, in which a luminous body generated a vibration, the light, propagated by a medium, an ether. Just as the wave of the sea or sound would exist a medium like water and air, for light it would be an ether. The wave of light admitted as longitudinal should cross the planes of the crystal, similar to a sound wave that went through a horizontal or vertical opening. However, between 1817 and 1824, according to Young and Fresnel, the explanation would be that the wave of light should vibrate in all transverse planes, propagating perpendicularly to these planes, by an ether that would fill all the bodies and that it should be of a kind of rigidity that act as a rigid body for propagation with the high speed of light. In 1845, Stokes presented an ether as a incompressible fluid, which was dragged by the bodies on the surface of the Earth, but would remain undrag above the surface. In 1862, J. J. Loschmidt had formulated that the atom was enveloped by an ether, which originated the action between atoms. In 1873, Maxwell and between 1881 and 1889, J. J. Thomson and Heaviside considered for electric and magnetic phenomena a medium, an ether that permeated, entangled and or penetrated the matter and received the transmission of motion through matter, and it was the medium that propagated the electromagnetic wave, as the air propagated the sound wave, the water the wave of the sea and the ground the seismic wave. And in 1892, Lorentz through its electron theory, considered that the ether penetrated the whole matter and was the receptacle for electromagnetic energy, the means for all forces to act on the ponderous matter and the absolute referential always still. So, the energy of the electromagnetic field in the ether resulted in the ether being able to exert a force acting on other charges, that is, the ether would transmit a resulting force, ponderomotive force, subsequently named Lorentz force or total electric force . If in an ether site the energy increased by the introduction of an electric charge, the ether could transmit that energy to another site of the ether and act a ponderomotive force on another charge. So it was the state of ether with greater or lesser electromagnetic energy that would exert this ponderomotive force. This ether for being always still remains without receiving any tension, therefore no speed, acceleration and also no need to define its mass and no forces applied to it. A lake with completely flat water would remain flat if the Earth was immobile in space or in motion, but when a stone was thrown into that water the formed wave would be propagated to be the Earth immobile or in motion. That is, the speed of the wave would be independent of the velocity of the source, because it was only the water that propagated the wave. Also the seismic wave by the ground and the sound wave by air were independent air of the emitting source speed. In the Michelson-Morley experiment in 1887, to verify the existence of the ether, the light as a wave would be propagated by a medium of this immobile permeable ether, also independent of the velocity of the emitting source, but which presented the result null. So, Michelson considered there was an ether similar to Stokes'.

Elementary Impulse and Refractive Index – And in a dense medium like water, in 1850 Foucault observed that for light to be transmitted in the water, the time was greater than in the vacuum-space-ether and concluded that the speed of light was lower in the dense medium, which would prove the wave theory that was contrary to the theory corpuscular that by the attraction would increase the speed in the dense medium. In 1601 Harriot, in 1621 Snell, and in 1637 Descartes had observed that in a denser medium the light was refracted according to the relationship between the distances of sine from a medium to the other. And between 1665 and 1678 Hooke and Huygens observed the light as a wave, in which a luminous body generated a vibration, the light, propagated by a medium, an ether. Just as the wave of the sea or sound existed a medium like water and air, and for light it would be an ether. The colours were deviations, that is, refractions. As Huygens in the most refracted medium, the light as wave would be an extension, running a shorter distance at the same time from the previous medium, would have a lower speed and depending on the ratio between speed, frequency and wavelength, $v = \mathcal{V}\lambda$, the light would also have a smaller wavelength.

However, by the theory of elementary impulse, m'c would run the particles with speed c and so, in the denser medium also circumventing the atoms with the speed c. The path of light in a denser medium is greater than in the space-vacuum in a longer time, but with the same speed of the previous medium, contrary to Foucault's conclusion of lower light speed in the denser medium. And this larger path with more time, from the speed c of the elementary impulse, comes to explain the refractive index. Then, the Arago experiment of 1810 is explained, because both the light of a star focusing on the direction of the orbital motion of the Earth or in the opposite direction run the prism with equal speed c of incidence, circumventing the atoms that form the prism. And differing only in the greater path of distance from the light of the Earth or smaller by the approximation of the Earth, but with equal internal speed c in the prism and consequent equal refractive index.

Admitting as an example the light running a length L, the average free path, and half of the radius circumference R, of a molecule, πR, and a proportional route factor: $N = (L + \pi R)/(L + 2R) = (L + \pi R)/(L + 2R)$, and $L = 1/(2)^{1/2} (\pi D^2) N_m$, where D = molecule diameter, $A_{mol/cm3}$ = number of molecules in a volume of 1 cm^3. For nitrogen is obtained N = 1.000725, which is very similar to the refractive index of nitrogen.

By Einstein's theory of relativity, the Arago experiment is explained by the speed of light being always constant independent of the velocity of emitting source, and in the case the speed of light of c/n in the prism, where n is the refractive index in the middle (prism) and applying the space contraction and time dilation to keep the c speed in the space-vacuum always constant relative to the emitting source.

Also, quantum impulse considers the c speed of stellar light in the prism in the direction and in the opposite direction to the motion of the Earth, however it explains the path of light with the apparent lower velocity c/n in the middle by the impulse m'c that circumvents the atoms at the speed of light c.

Moment or Elementary Impulse – In the theory of the momentum or elementary impulse of Erdelyi, the concepts and units of momentum, impulse, force and energy are effectively observed in their real physical meanings. So these concepts applied to the quantum energy of Planck's postulate and the quantum light of Einstein shows that: momentum or elementary impulse, m'c, where m = elementary quantum mass = 7,37... x 10^{-48} g, c = speed of electromagnetic radiation, elemental force = m/t, t = second or s, and quantum elementary energy ε = (m'c/t) x |c| (1cm) = m'c x c = m'c^2 . Results when in ε = h$\mathcal{V}$ where ε = quantum energy for frequency $\mathcal{V}$ and h = Planck constant,= 6.6... x 10^{-27} [erg s]. Dividing both sides for 1 second, ε/s = |$\mathcal{V}$|h' [erg], where h' [erg] = h [erg s]/1 [s], h' = 6.6... x 10^{-27} [erg]. To $\mathcal{V}$ = 1 cycle/second, quantum elementary $\varepsilon_{(\mathcal{V}=1)}$ [erg] = ε [erg] = 1 x h' [erg] = h' [erg]. In CGS, h has the unit [erg s], "action", without greater physical significance, but h' has the unit [erg] of understandable physical energy concept. While for Einstein's relativity and quantum theory, a blue color quantum is a single large unit of condensed energy, a red-colored quantum is a single small unit of energy, but for the elementary impulse theory a blue-colored quantum is formed fo |$\mathcal{V}_a$|h' and a red-colored quantum by |$\mathcal{V}_v$| h'. There is for elementary impulse theory only an elementary quantum h' that forms all the radiations |$\mathcal{V}$| h'.

On 1952, after H. R. Hertz the discoverer of electromagnetic waves have completed that force was the consequence of absolute grandeur, of mass, velocity and time, the theory of the elementary impulse of Erdelyi showed the elemental quantum force m'c/t exactly as consequence of mass, velocity of momentum, this is speed c of m', and time.

The actual force that would apply to all physical phenomena, in mechanics, optics, electromagnetism and gravity, sought by Faraday, Maxwell, Newton, Hertz and other scientists was in reality the quantum force of Erdelyi's theory of impulse. At the time of these great scientists who sought the real cause of the physical phenomenon was nonexistent to quantum mechanics, and subsequently came to be formulated presenting the concept of quantum without mass, without dimensions, and without the actual physical concept. But, by the theory of quantum impulse, quantum has the elementary mass and its energy presents the real physical concept.

Quantum Mass – By the theory of relativity in 1905, a mass m_0 at rest increases to m by v velocity increase : $m = m_0 / \sqrt{1-(v^2/c^2)}$ and for the mass $m_{quantum}$ of quantum would be: $m_{quantum} = m_{0quantum} / \sqrt{1-(v^2/c^2)}$, and to $v_{photon} = c$ = speed of electromagnetic radiation, so $m_{0quantum} = m_{quantum}$ x 0 = 0. So for that mass $m_{quantum}$ of the quantum is $\neq \infty$, the mass $m_{0quantum}$ motionless would have to be zero by the theory of relativity.

But between 1936 to 1941, Proca had presented the quantas with mass, in which the speed of the quanta could be variable, occurring superluminary speed of electromagnetic radiation, an imaginary mass or a negative mass. And the speed of light would be a function of the frequency without being a single constant c, In that case the theory of relativity would remain valid by the quantum with mass in limit of speed.

In 1940, L. V. P. R. De Broglie also presented a quantum with mass.

And in 1952 Erdelyi, by the elementary energy m'c x c = h' in which the quantum mass m' could be in the state of tension in the constitution and field of particles and radiation, as m'c x c = h'. And the equation of the mass increase, also deduced by the theory of elementary impulse, as for the electron: $m_e = m_{0e} / \sqrt{1-(v^2/c^2)}$ was applied to determine m_e = n' x m' when the mass of the electron still m_{0e} received an additional mass of n x m' for increased mass m_{0e}, in which the velocity v of the electron would present the limit < c, i.e. $(v^2/c^2) < 1$.

By the theory of relativity the immovable electron has the mass m_{0e} and the motionless quantum has zero mass. But when the electron emits light, the quantum at speed c has energy, therefore mass, because by relativity energy is mass, and the electron then loses energy and mass. Therefore, the electron loses mass in the form of light. And so the electron should be formed of light. However, by relativity the motionless quantum if it constituted the electron would have zero mass! But by the theory of elementary impulse, the electron is formed of n quanta m'c x c = n h', i.e. the electron and all the particles are formed of light, or other electromagnetic radiations according to the mass, velocity, energy of the particles and the frequency with which these quanta elementary ε = h' be in the particles.

By the theory of the momentum or elementary impulse of Erdelyi the electrons and particles would be formed by quanta or photons m'c x c = m'c^2, in that c speed of light with relation to emitting Earth source, and also that c speed as a referential speed in principle to the ether in the vacuum-space. And that the m'c^2 could be in a state of tension in a space of configuration of elementary particles like the electron, and emitted when they received other impulses of energy m'c^2. Figure 24.

> Schematic drawing of the electron formed by the quantum energy m'c x c of mass m' with speed c rapport to the emitting source and in principle to the vacuum-space-ether
>
> n x m'c x c
> **electron by quantum impulse**

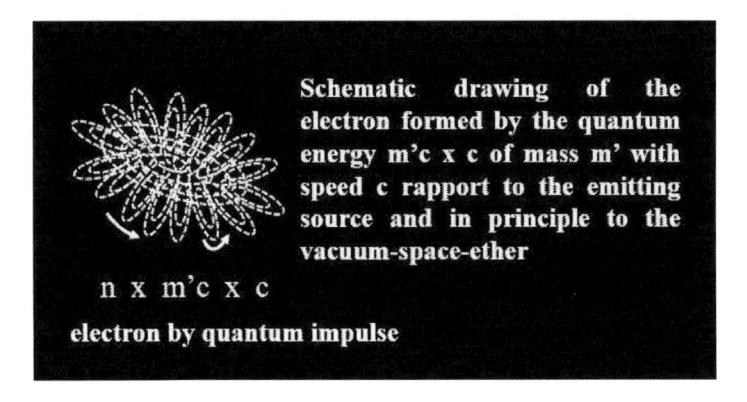

Fig. 24 – By the theory of elementary impulse of Erdelyi, the electron and other particles are formed by m'c x c = m'c², where c is the speed of light in relation to the source and in principle to the ether in the vacuum-space.

In 1955, I. Bass and Schrödinger also presented the possibility of quantum with mass.

Electron Radius by Elementary Impulse Theory – If the electron were a sphere with the classic ray of J. J. Thomson, around 10^{-13} cm, the velocity of rotation would be nine times greater than the speed of light, by the expression of angular momentum, but considered to a value found around 10^{-16} cm, the rotating velocity is less than the speed of light, while the elementary impulses would run with speed c the structure of the electron in motion, that is, m'c x c = h', which would constitute the electron. Therefore, the basic part of the electron would not be punctual or spherical, but a structure formed by quantities of m'c x c = h'.

By the elementary impulse the radius r_{0e} of the still mass electron m_{0e}, is obtained by the length of the circumference of the basic circle of the immovable electron:
$$\lambda_0 = 2\pi r_0, \quad r_{0e} = \text{electron base radius,}$$
So, $\lambda_0 = h/m_{0e}c$ and $r_{0e} = h/(2\pi m_{0e}c)$, therefore $\lambda_0 = h/m_{0e}c = 2,42... \times 10^{-10}$ cm

λ_0 = wavelength of the electron base circle = $(|c|/n) \times$ cm = $h/m_{0e}c = 2,42... \times 10^{-10}$ cm

$r_{0e} = h/(2\pi m_{0e}c) = 3,86... \times 10^{-11}$ cm with the diameter $d_e = 7.72 ... \times 10^{-11}$ cm

In 1915, Sommerfeld had determined a constant $\alpha = hc/2\pi e^2 = 137,03...$ the fine structure constant that was experimentally related between the constants c, h and the charge e of the electron with the narrow spectral lines of the atoms. Admitting that $\lambda_{electron}$ be the length of the base circle of the electron related to the energy of the electron $m_{0e}c^2 = h\, \mathcal{V}_{m'}$ where $\mathcal{V}_{m'}$ is the quantity of m' frequency that resulted in the mass m_{0e} of the electron:

$h\mathcal{V}_{m'} = h\, c/\lambda_0 = m_{0e}c^2$, so $hc = m_{0e}c^2\lambda_0$, and replacing in the expression of α is obtained:

$[m_{0e}c^2\lambda_0 / 2\pi\, e^2] \times 137{,}03...$, and $e^2 / m_{0e}c^2 = \lambda_0 / 2\pi\, 137{,}03...$

In case it is related to charge e of the electron with its total energy $m_{0e}c^2$, for the symmetrical spherically charge density within an electron radius r sphere, which is the electrostatic charge, is obtained in the CGS (electrostatic):

electrostatic electron force $f_{electrostatic} = e^2/r^2$

electrostatic potential energy = $f_{electrostatic}$ x distance = $(e^2/r^2)\, r = e^2/r$

And in balance the total energy $m_{0e}c^2$ it equates potential energy:

Then, $e^2/r = m_{0e}c^2$, and $r = e^2/m_{0e}c^2 = 2{,}82 ... \times 10^{-13}$ cm, results:

$\lambda_0 = 2\pi\, r \times 137{,}03... = 2{,}426... \times 10^{-10}$ cm and diameter $d = 7{,}722... \times 10^{-11}$ cm

So, the diameter $d_{e(\alpha)} = 7{,}722... \times 10^{-11}$ cm of the electron from the constant α of thin structure, is obtained the same diameter $d_{electron\ circle} = 7{,}72... \times 10^{-11}$ cm, obtained by the elementary impulse m'c which travels with velocity c a length λ_0 of a base circle of the electron, with a frequency $\mathcal{V}_0$ equivalent to the Compton wavelength:

Being: $c = \mathcal{V}_0\lambda_0$, $\alpha = (h/2\pi)\, \mathcal{V}_0\, \lambda_0/e^2$

$(h\mathcal{V}_0) = m_{0e}c^2 = nm'\, c^2$, in that m' = elementary mass = $7{,}37... \times 10^{-48}$ g

$\alpha = nm'\, c^2 \times \lambda_0 /2\pi e^2 = 137{,}03...$ and $\lambda_0 = (2\pi e^2 \times 137{,}03...)/m_{0e}c^2$

The base diameter of the electron $d_e = d_{e\,(\alpha)} = 7{,}722... \times 10^{-11}$ cm is greater in relation to the diameter obtained from 10^{-16} cm by the measurements of the CERN "Center and Conseil Européean de Recherche Nucléaire-Organization Européne de Recherche Nucléaire", European Organization for Nuclear Research, "Laboratoire Européen pour La Physique des Particules", European Laboratory for Particle Physics. However, by quantum impulse can indicate an extension of the electron field and present greater density in the center of the electron.

Dirac had presented in the decade of 1930 that the observed mass differed from the mass initially in the equations because there was electron interaction with the electromagnetic field, which altered the mass of the electron, when it referred to the normalization of the mass in the field equations, in the production and annihilation of particles. That is, the extent and density of the field in the electron could present a difference in obtaining the diameter of the electron according to the measurement effected.

Equation of the Mass Increase by the Elementary Impulse – Although Lorentz presents that the mass increase of a particle as the electron was due to the electromagnetic force, to explain the equation of mass increase m with v velocity of one of the particle: $m = m_0/\sqrt{1-(v^2/c^2)}$, Einstein came to explain only by coordinates of a reference system or the transformation for other reference system which in this transformation the velocity was variable and the momentum preserved, and presented that the mass was due to increased velocity.

It would have to be admitted that the motion would "condense" forming mass. But, the motion is the product of mass and speed, mv, and the speed is just a x distance by time, x/t, which cannot form mass without a cause for that speed increase. It would be an increase in speed through a "ghostly" condensation of the motion creating mass, by Einstein's relativistic theory.

By the elementary impulse is the elementary mass of the quantum, m which increases the mass of the particle. A particle like the electron that has its mass at rest m_{0e} with a portion of this mass in a base circle, receiving impulses m'c results in motion of translation, as a helicoid. Admitting that a m'c runs the circumference of the circle of the electron at rest, then in motion will travel a helicoid. Extending this helicoid into a plane results in a triangle, in which the hypotenuse is the trajectory of m'c with the speed c of light; The larger side is the projection of the trajectory of m'c of the hypotenuse in the rotate, spin, of m'c around the axle, at velocity v_s of c projection; And the smaller side is the projection of m'c of the hypotenuse in the linear direction of translation of m'c, at velocity v_e of the electron that is the linear velocity v_{linear} of the projection of c.

The hypotenuse represents: the mass m_e of the formed electron of n' quantity of m' of the impulse m'c; the speed c of m'c; the λ projection of the trajectory of m'c in the helicoid or helix.

The larger side represents the mass m_{0e} of the electron at rest formed from n m'c; spin velocity v_s of the projection of m'c; λ_s the projection of λ and the length of the circumference of the basic circle of the electron at rest traveled by m'c.

The smaller side represents the mass m_e of the electron; velocity v_{linear} linear projection of m'c, which is the linear velocity of the electron and the velocity v considered of the electron; λ_{linear} the length of λ projection in the direction of the electron motion, and it is the length traversed by the electron and the projection m'c during a spin in the electron. And that results in the increase in mass with the same expression of theory of relativity. Figure 25.

J. P. Hansen

Increased electron mass by the elementary quantum impulse

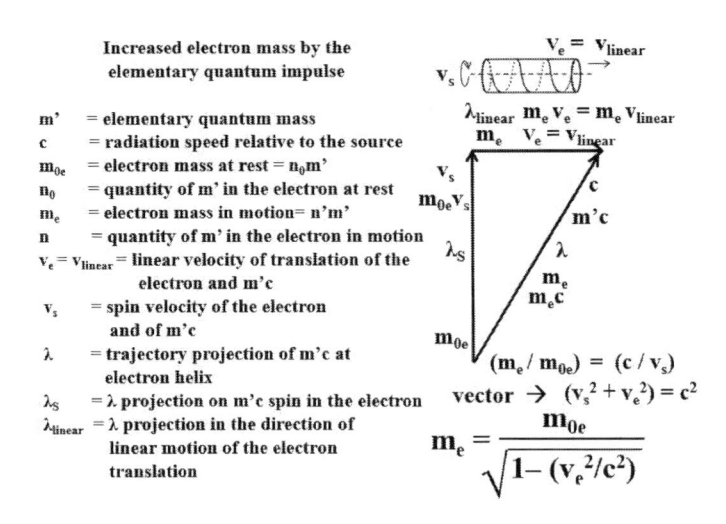

m'   = elementary quantum mass
c   = radiation speed relative to the source
m_{0e} = electron mass at rest = n_0m'
n_0 = quantity of m' in the electron at rest
m_e = electron mass in motion = n'm'
n   = quantity of m' in the electron in motion
$v_e = v_{linear}$ = linear velocity of translation of the electron and m'c
v_s = spin velocity of the electron and of m'c
λ   = trajectory projection of m'c at electron helix
$λ_s$ = λ projection on m'c spin in the electron
$λ_{linear}$ = λ projection in the direction of linear motion of the electron translation

$V_e = V_{linear}$
$λ_{linear}$ $m_e V_e = m_e V_{linear}$
m_e $V_e = V_{linear}$

$(m_e / m_{0e}) = (c / v_s)$
vector → $(v_s^2 + v_e^2) = c^2$

$$m_e = \frac{m_{0e}}{\sqrt{1 - (v_e^2/c^2)}}$$

Fig. 25 – The projections of m'c in the helicoid or helix of the electron forms a triangle that results in the equation of the increase in particle mass, according to the *O Triângulo Mágico* (*Magic Triangle*) of Erdelyi, constant in References.

By simple proportion:
$[(m_e/m_{0e}) = (c/v_s)]$, so $[(m_e v_s) = (m_{0e} c)]$

Being: $(v_g^2 + v_e^2) = c^2$, and $v_s = (c^2 - v_e^2)^{1/2}$, in which to $v_e = 0$, $v_s = c$, which is the velocity of m'c of the elementary impulses in the immovable electron. When the electron velocity is increased by n' m'c × c = n' h', the n m'c × c = nh' of the electron at rest, all (n + n') (m'c × c) continue with velocity c while the electron shows the rotating velocity (spin) v_s and the translation v_e.

And replacing: $m_e = m_{0e} / [1 - (v_e^2/c^2)]^{1/2}$, which is the equation of the mass increase and consequent increase in velocity, by the elementary impulses m'c to the electron and similarly extensive to the other particles.

Also, by the triangle of Figure 58, applying the momentums or impulses involved in the particle motion, and to the electron:
$$\text{vectored: } m_e^2 c^2 = m_e^2 v_e^2 + m_e^2 v_s^2$$
$m_e/m_{0e} = c/v_s$, this is, $m_e v_s = m_{0e} c$
And replacing: $m_e^2 c^2 = m_e^2 v_e^2 + m_{0e}^2 c^2$,
Results: $m_e = m_{e0} / [1 - (v_e^2/c^2)]^{1/2}$, which is again the equation increase in mass and consequent velocity increase for the electron and other particles, by the elementary impulses m'c.

The m' mass of the elementary quantum h' always has the speed of electromagnetic radiation c as elementary impulse m'c confined in the configuration of the particles, which is still in tension in the particles or in the electromagnetic radiation with speed c.

Particles are created from energy supplied, such as gamma rays of energy = $h\mathcal{V}_\gamma$, of quanta $h = |\mathcal{V}| h' = |\mathcal{V}| m'c^2$, which contains speed c, but also the mass m' responsible for providing this elementary masses for particle formation.

It is the quantum force of impulse m'c by time t, with the energy/t of quantum elementary $h' = (m'c \times c)/t$, which injects mass into the particles resulting in the increase of mass and consequent particle speed increase. And not a distance/time, the speed, increasing the mass by space-time, through a "ghostly condensation"!

In the case of mesons μ at high speed is due to receive more energy h' which results in mass increase and consequent higher speed. And the highest average life time is the result of greater time for loss of mass of n m'c x c = n h' in meson μ.

In the transformation of coordinates according to the theory of Lorentz electrons and Einstein's theory of relativity appears the factor squared in the equation of the mass increase, similar to the factor squared of the triangle rectangle in the equation of the electron motion by the theory of the elementary impulse that results in the same equation of the mass increase of the electron. However, by the theory of relativity it is a mathematical relationship of mass increase in function of speed, as a "ghostly" condensation of speed turning into mass! Whereas by the elementary impulse is a quantum, physically determined explanation.

And by the elementary impulse, the force m'c/t and the elementary quantum energy m'c x c = m'c² = h' applied to a body is limited in the ability to configure the particles that form this body, limit that close to the c speed of light. And it is the explanation that the speed of a body is increased when it is injected force n m'c/t or energy n m'c x c = m'c² = h', up to a limit in which the particles forming the matter have the saturated configuration. The expression of mass increase in function of speed by the theory of relativity is only mathematics, condensing the speed "phantasmagorically" in mass. However, the physical explanation is presented by the elementary impulse of force and energy.

So the equation of the mass increase as for the electron: $m_e = m_{0e}/1 - (v_e^2/c^2)]^{1/2}$ is applied to determine $m_e = n \times m'$ when the electron mass at rest m_{0e} receives the mass of n x m' for the increase of electron mass, in which the velocity v_e of the electron presents the limit < c, i.e. $(v_e^2/c^2) < 1$.

And also, the energy is different from mass, because were being matched concepts and unitys different from mass and energy.

Electron Equation by Elementary Impulse Theory – By the theory of elementary impulse, m'c, the electron mass m_{0e} immobile and with mass m_e in the linear velocity of the electron v_{linear} it consists of the quantum impulse m'c that travels at the speed c of the light, forming a helix, with the projections of the velocity v_s of the "spin" and the linear, in which λ is the projection of h' trajectory in the helix and the λ_s projections of the "spin" and λ_{linear} of the linear electron motion. Figure 26 .

electron helix

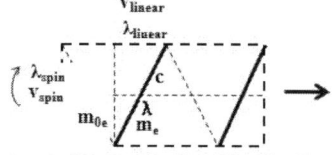

$V_{linear} = $ linear velocity of the electron
$V_{spin} = $ turn (spin) velocity of the electron in the helix
$\lambda_{linear} = $ projection of the linear trajectory of the electron in the helix
$\lambda_{spin} = $ projection of the turn (spin) trajectory of the electron in the helix
λ = projection of m'c x c = h' in the electron helix
$m_{0e} = $ mass in rest of the electron
$m_e = $ electron mass with linear velocity v_{linear}

Fig. 26 – The projections of m'c which constitutes the electron, in the helix of the electron motion, presents the dimensions of the electron as its velocity, with the deduction of Erdelyi by quantum impulse, as References.

By the equation of Planck's electromagnetic energy and Hasenöhrl or of Einstein's total energy, for the mass m_e of the electron with v velocity; h = Planck constant; c = speed of electromagnetic radiation and m'c; $\mathcal{V}$ = quantum frequency for total electron energy and h' frequency in electron helix with velocity v and mass m_e:
$$h\mathcal{V} = m_e c^2 \rightarrow \mathcal{V} = m_e c^2/h$$
$\mathcal{V}$ = frequency of h' in the electron helix with velocity v, mass m and $c = \lambda \mathcal{V}$
$$h\mathcal{V} = m_e c \, \lambda \mathcal{V} \rightarrow \lambda = h/m_e c$$
$$V_{linear} = \lambda_l \mathcal{V} \rightarrow \lambda_e = V_{linear} \, h/m_e c^2$$
$$v_s = \lambda_s \mathcal{V} \rightarrow \lambda_s = v_s \, h/m_e c^2$$

By projection of the electron helix according to the triangle of masses, velocities and momentums, with the equation of the mass increase by the theory of elementary impulse:
$m_e = m_{0e} / [1 - (v_{linear}^2/c^2)]^{1/2}$, then $v_{linear} = (c/m_e)(m_e^2 - m_{0e}^2)^{1/2}$

By simple proportion:
$[(m_e/m_{0e}) = (c/v_s)] \rightarrow [(m_e v_s) = (m_{0e} c)] \rightarrow v_s = m_{0e} c / m_e$

Results:
$$\lambda_{linear} = (h/m_e c) [(m_e^2 - m_{0e}^2)/m_e]^{1/2}$$
$$\lambda_s = (h/m_e c)(m_{0e}/m_e)$$
$$\mathcal{V}_0 \, h' = m_{0e} c^2$$
$\mathcal{V}_0 = $ frequency of h' in the electron at rest with mass m_{0e}
$$\mathcal{V} h' = m_e c^2$$
$$\mathcal{V} = (m_e/m_{0e}) \, \mathcal{V}_0$$

$$n = (m_e/m_{0e})$$
$$\mathcal{V} = n\,\mathcal{V}_0$$
$$v_{linear} = \lambda_{linear}\,\mathcal{V}$$
$$\lambda_{linear} = (v_{linear}/\mathcal{V}) = v_{linear}/n\,\mathcal{V}_0$$

$\lambda_{linear} n = v_e / \mathcal{V}$ is the length of the electron with mass m_e at the linear velocity v_{linear}
of the electron. Similarly: $\lambda_s\,n = v_s / \mathcal{V}_0$ and $\lambda\,n = c / \mathcal{V}_0$.

Table 4 it presents the increase in mass of the electron when in the particle accelerators they are accelerated by the supplied energy of $nh' = n\,m'c \times c = n\,m'c^2$ of the electric and magnetic fields, resulting in the increase of speed and the variation of dimensions of the electron structure. By the largest mass there is an increase in the frequency of $h' = m'c \times c$ of the electron and decreased the velocity of the "spin" component of the basic helix between two elementary energies h'. The diameter and wavelength of the electron spin component are diminished. And the length of the electron is increased up to around 75% of the speed of light, and after a sharp increase in the frequency of h' the wavelength λ decreases resulting in the decrease of the length of the electron.

Tab. 4 – Electron dimensions in function of its mass and velocity, extracted from *Magic Triangle* of Erdelyi in References.

c (cm/s) c = 2.99792458 × 10¹⁰ speed of light relative to the Earth source	c = 0	0,80 c
v_{linear} (cm/s) translation velocity of the electron	0,00000000	2,39839396 × 10¹⁰
v_{spin} (cm/s) spin velocity of the electron	2,99792458 × 10¹⁰	1,79868385 × 10¹⁰
m_e (g) electron mass	$m_0 = 9,109382 \times 10^{-28}$ $m_e = 0,00054\,m_{proton\ at\ rest}$	$14,23276 \times 10^{-28} = (1,56\,m_0)$ $m_e = 0,00085\,m_{proton\ at\ rest}$
$\mathcal{V}(1/s)$ m'c frequency in the electron	$\mathcal{V}_0 = 12,36 \times 10^{19}$	$19,31 \times 10^{19}$
λ_l (cm) electron length	0,00000	12,42047 × 10⁻¹¹
λ_s (cm) length of the electron base circumference	2,42550 × 10⁻¹⁰	0,93147 × 10⁻¹⁰
d_e (cm) electron diameter	7,81448 × 10⁻¹¹	2,96495 × 10⁻¹¹

c (cm/s) speed of light relative to the Earth source	0,999999849 c	0,999999996 c	0,999999999 c
v_{linear} (cm/s) translation velocity of the electron	2,9979241 x 10^{10}	2,99792456 x 10^{10}	2,9979245799 x 10^{10}
v_{spin} (cm/s) spin velocity of the electron	0,00166063 x 10^{10}	0,00034641 x 10^{10}	0,00007503 x 10^{10}
m_e (g) electron mass	16444,186 x 10^{-28} (1805,19 m_{0e}) m_e=0,98 $m_{proton\ at\ rest}$	78834,39 x 10^{-28} (8654.19 m_{0e}) m_e= 4,71 $m_{proton\ at\ rest}$	364084,1 x 10^{-28} (39968.03 m_{0e}) m_e=21,76 $m_{proton\ at\ rest}$
ν (1/s) m'c frequency in the electron	22315,35 x 10^{19}	106981,12 x 10^{19}	494075,17 x 10^{19}
λ_l (cm) electron length	1,343 x 10^{-13}	2,80 x 10^{-14}	6 x 10^{-17}
λ_π (cm) circumference length of the electron base	7 x 10^{-17}	3 x 10^{-18}	1 x 10^{-19}
d_e (cm) electron diameter	2,10^{-18}	1,10^{-19}	4,10^{-21}

Accelerators are nothing more than the injection of m'c x c of the electric fields that accelerate electric and magnetic charge particles that direct their trajectories to the final target, consisting of a large number of electricity conductors and gigantic magnets. In this acceleration the particles have their mass increase according to this energy received from h' = m'c² = by the theory of quantum impulse, and not by the speed that would condense "phantasmagorically" in mass according to the theory of relativity, or mathematical abstractions that create mass fields to give mass to particles, but that in reality would be precisely the electric field formed of m'c x c = h', which is directly injected into the particles from the beginning of its acceleration!

So, protons, atomic nuclei of lead ions, electrons and other charge particles are accelerated, where at the end they are recorded in the detectors.

They are two fundamental physical concepts that are present in the particle accelerators, which is the speed increase of the particles charged by the electric field and the magnetic field that deflects these particles. So, who increases the mass of the particles to reach close to the speed of light? If it was the energy of the electric field that increased the mass of the particles, then the energy of the field has mass and results in the increase of speed, but never the speed being increased by energy without mass of the electric field. Yes, energy is mass x acceleration x distance. And how could an increase in speed cause a "ghostly condensation", Einstein's relativity and the current Physics, to increase mass as an effect of increased speed? It's the biggest invers from the history this current Physics! And the mass of the electric field would not be produced by mathematical abstractions that created mass fields by replacing the electric field. The electric and magnetic field consists of m'c x c = m'c² = h' that comes effectively to be injected into the particles for mass increase and consequent acceleration. Figure 27.

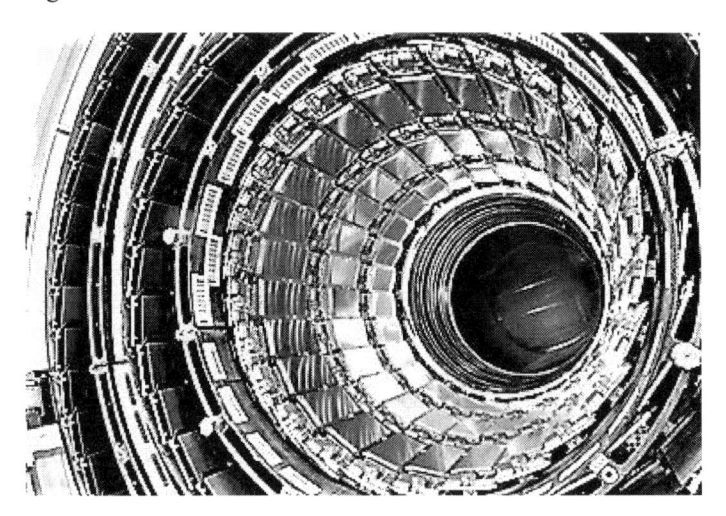

Fig. 27 – Accelerators are nothing more than m'c x c injection of electric fields that accelerate electrical and magnetic charge particles that direct their trajectories to the final target, consisting of a large number of electric conductors and gigantic magnets. In this acceleration the particles have its mass increase according to this energy received from h' = m'c² = by the theory of quantum impulse, and not by the speed that would condense "phantasmagorically" in mass according to the theory of relativity or mathematical abstractions that created mass fields to give mass to the particles, but that in reality would be precisely the electric field formed of m'c x c = h' which is directly injected into the particles from the beginning of its acceleration! Photo Authorized of CERN.

And by applying the equation of relativity to quantum would be obtained for the mass of the quantum, $m_{quantum} = m_{quantum\ rest}$ / $\sqrt{1-(v^2/c^2)}$ = ∞, because the quantum has the c speed of light . Therefore, the quantum is reduced to the rest mass $m_{quantum\ rest}$ = 0, because $m_{quantum} = m_{quantum\ rest} / \sqrt{1-(v^2/c^2)}$, so $m_{quantum}$ x 0 = $m_{quantum\ rest}$ = 0, for $m_{quantum}$ = 0/0 which is undetermined to be able to obtain radiation energy $\mathcal{E} = m_{quantum} c^2 = h\mathcal{V}$. However, by the theory of relativity the mass $m_{quantum}$ of the quantum results from the speed, because for v = 0 mass $m_{quantum\ rest}$ quantum at rest is null. That is, by the theory of relativity, the mass of the quantum is increased by the kinetics, that is, by the speed. And the higher the frequency of the electromagnetic radiation, the greater the mass of the quantum at the speed of light. The theory of relativity shows that the radiation presents mass of inertia, however the mass of inertia is zero when at rest, in contradiction with the concept of mass. Yes, any body has mass in velocity and at rest.

Structure of the Particles by the Elementary Impulse – Other particles will also have in their structures a variation of dimensions. For example, in 1942 in the University of Chicago reactor, in the same year it was obtained by Fermi and other the first chain reaction of the uranium, was effected by Fermi an experiment, later called the Fermi bottle, which was a container placed inside a pile of uranium. The neutron that formed the hydrogen atom with a proton and an electron, departing in proton, electron and antineutrino, obtaining the first measurement of the average life of the neutron. But while the protons and electrons formed inside that bottle were retained by the material of the bottle, the neutrons coming from the uranium nuclear fission in and out of the material that formed the bottle. Therefore, the neutron with higher energy, that is, greater quantity acquired from h' = m'c x c, and consequent greater mass and higher speed, resulted by the equation of the electron and the particles, in the decrease of its dimensions, by the quantum impulsion theory, thus going through the material of the bottle. The proton and the electron being formed then released h' by the nuclear equilibrium, thereby increasing its dimensions and thus getting retained in the bottle.

Creation of Electron Pair and Positron by the Elementary Impulse – By the theory of elementary impulse particles emit radiation $|\mathcal{V}|$h' and by experiments with polonium and radioactive beryllium emitting gamma rays which form pairs of electron and positron, would be admitted by the theory of elementary impulse an electron structure formed of n m'c x c = n m'c² = m_ec² = n h', where m_e = electron mass with linear velocity, c = speed of light, m'c = elementary quantum impulse. For example, if those nh' of the electron at rest were with a higher energy density in a base circle, in which the elementary impulse m'c ran with speed c a length λ_{0e} of this base circle with a frequency $\mathcal{V}_0$ equivalent to a length that could retain these n m c x c = nh', would be obtained for the immovable electron of mass m_{0e} and energy = m_{0e} c² = h' $|\mathcal{V}_0|$ = h$\mathcal{V}_0$.

And by Planck's energy radiation equation, Hasenöhrl and Einstein:

$$h\mathcal{V}_0 = m_{0e} c^2 = nm'c^2 = nh'$$
$$\mathcal{V}_0 = c/\lambda_0, \quad \lambda_0 = 2\pi r_0, \quad r_0 = \text{electron base radius;}$$

So, $\lambda_0 = h/m_{0e}c$ and $r_0 = h/(2\pi\, m_{0e}c)$
$$\lambda_0 = h/m_{0e}c = 2{,}42... \times 10^{-10}\text{ cm}$$
λ_0 = length of electron base circle = $(|c|/n \times 1\text{cm})$ = $h/m_{0e}c$ =
$$2{,}42... \times 10^{-10}\text{ cm}$$
$$r_0 = h/(2\pi\, m_{0e}c) = 3{,}86... \times 10^{-11}\text{ cm}$$
diameter $d_{\text{circle of eletron}} = 7{,}72... \times 10^{-11}$ cm.
$$\mathcal{V}_0 = c/\lambda_0 = 12{,}36... \times 10^{19}\text{ s}^{-1}$$
n = quantity of m elemental mass in the electron of mass m_{0e}
$$n = m_{0e}/m' = 12{,}36... \times 10^{19}$$
And that quantity of n is the same as h' that forms the electron and that can be obtained by the equation, $m_e c^2 = n_{h'}\, h'$, so, $n_{h'}$
$$= 12{,}36...\times 10^{19}$$
$$h = \text{constante de Planck} = 6{,}6... \times 10^{-27}\text{ erg s}$$
$$h' = h/s = 6{,}6... \times 10^{-27}\text{ erg} = m'c^2$$
m' = elementary mass = $0{,}73... \times 10^{-47}$ g, another fundamental constant.

Where c = speed of light in relation to the observer on Earth and in principle in relation to ether in vacuum-space is a fundamental constant.

The annihilation and formation of the electron pair and positron are effected through two-ray frequency range $12.36... \times 10^{19}$ which is remarkably the frequency of $\mathcal{V}_0$ that retains the nh' radiation that forms the electron.

This length of the basic circle circumference of the electron results equal to the wavelength of the electron in which the photons with wavelength λ_1 and frequency $\mathcal{V}_1$ when focusing on the electron result in a higher wavelength λ_2 and lower frequency $\mathcal{V}_2$.

In the two calculations, by Compton and by the elementary impulse, the frequencies involved result in the same value of the electron wavelength and the length of a base circle of the electron, that is, $\lambda_0 = h/m_0c$. That is, the Compton wavelength is the length of the circumference of the base circle of the immovable electron.

By the theory of quantum impulse, the frequency of m'c × c = $m'c^2 = h'$ form the total quantity of energy $\mathcal{E} = |\mathcal{V}| h'$ [erg], contrary to more than 10^{23} types of photons from the entire frequencies of vibration of the electromagnetic radiation spectrum, of condensed $\mathcal{E} = (h\mathcal{V})$ [erg s × s], in CGS, of Einstein and Planck.

The quantity of n m'c × c = n h', which form the gamma rays, the electron and the positron, also form their fields, electric and magnetic. This field consists of the n m'c × c which run lines of force in planes perpendicular to the basic circle.

By the theory of elementary impulse, therefore the electron and positron are formed of m'c × c = h', and in 1934 when in the collision of the electron pair and positron was observed the annihilation of these two particles in two photons of gamma rays of equal energies, and that by the theory of the elementary impulse these gamma rays are formed by n elementary photons h'.

For electron and positron with low speed and mass $\approx m_{0e} = m_{0p}$:
$$m_{0e} c^2 = m_{0p} c^2 = h \mathcal{V}_\gamma = n \ h'$$
$$m'c \times c = h'$$
$$c = 2{,}997925 \times 10^{10} \text{ cm s}^{-1}$$
$$h = 6{,}62517... \times 10^{-27} \text{ erg s}$$
$$m_{0e} = m_{0p} = 9{,}11 \times 10^{-28} \text{ g}$$
$$\mathcal{V}_\gamma = 12{,}358... \times 10^{19} \text{ s}^{-1}$$
$$m_{0e} c^2 = m_{0p} c^2 = h \mathcal{V}_\gamma = 81{,}86... \times 10^{-8} \text{ erg}$$
$$n = m_{0e} c^2 / h' = m_{0p} c^2 / h' = 12{,}358... \times 10^{19}, \text{ so } n_{h'} = 12{,}358... \times 10^{19} \ h'$$

To $v_e = 0{,}999 \times c$ in particle accelerators:
$$m_{e(0,999c)} = m_{p(0,999)c} = 203{,}71119... \times 10^{-28} \text{ g}$$
$$\mathcal{V}_{\gamma(0.90c)} = 290{,}0147... \times 10^{19} \text{ s}^{-1}$$
$m_{e(0,999c)} c^2 = m_{p(0,999c)} c^2 = h\mathcal{V}_{\gamma(0.999c)} = 1830{,}8653... \times 10^{-8} \text{ erg} \approx$ 22,36 times higher energy for gamma rays than for gamma rays for electron and positron at low speed.
$$N_{(0,999)} = m_{e(0,999)} c^2 / h' = m_{p(0,999)} c^2 / h' = 290{,}0147... \times 10^{19}$$
So $n_{(0,999)} = 290{,}0147... \times 10^{19} \ h'$

The formation of particles and fields seem to originate from a kind of condensation of photons. A condensed atom, at temperatures close to or just above absolute zero, forms an resonance by magnetism coupling the atoms, based on the statistical study of Bose and Einstein of 1925. There may be this grouping by action of a magnetic field of that network, as V. Ginzburg and L. Landau, in 1950. And particles like the electron can group together in the crystalline network, supplanting a Coulomb repulsion, as J. Barden, L. Cooper and R. Schrieffer, in 1957. In the case of photons, the intrinsic helix formed from the impulses m'c x c = h' may be the cause of being grouped together forming the particle. The particle creation by gamma rays were obtained in targets as atomic nuclei of larger masses, in intense magnetic field, and in vacuum nuclei the particle formation were of minimal momentums (virtual).

Fine Structure Constant Deduced by the Theory of Elementary Impulse – In 1915, Sommerfeld had determined the constant $\alpha = k \ h \ c/2\pi e^2 = 137$, the fine structure constant, where k is the electric constant, in which α is related experimentally between the constants h, c and the charge e of the electron with the narrow spectral lines of the atoms. However through the theory of elementary impulse is obtained admirably the expression of the fine structure constant, which represents an electron equation, for the electrical charge e, and the mass m_{0e} of the immovable electron through the expression of the length of the base circle of the electron $\lambda_0 = h/m_{0e}c$ with the constant $h = h' \times 1s$.

By the theory of elementary impulse, m'c travels with speed c the length λ_0 of the electron base circle with a frequency $\mathcal{V}_0 = c/\lambda_0$, equivalent to the wavelength of the electron $\lambda_0 = h/m_0c = 2{,}42\ldots \times 10^{-10}$ cm

d_e = diameter of electron circle = $\lambda_0 / \pi = 7{,}72\ldots \times 10^{-10}$ cm
$\mathcal{V}_0 = c/\lambda_0 = 12{,}3\ldots \times 10^{19}$ s^{-1}
m' = elementary mass = $0{,}73\ldots \times 10^{-47}$ g
m_{0e} = mass of the electron = $|\mathcal{V}_0|$ m' = $(12{,}3\ldots \times 10^{19})(0{,}73\ldots \times 10^{-47}g) = 9{,}10 \times 10^{-28}$ g
h' = m'c × c = h/1s
h'1s (erg) (1s) = h (erg s)
$m_{0e}c^2$ = immovable electron energy
$m_{0e}c^2 = $ (m'c × c) $|\mathcal{V}_0| = $ h' $|\mathcal{V}_0| = h\mathcal{V}_0$

And relating $m_{0e}c^2$ with h':
$m_{0e}c^2/$h' = $9{,}10 \times 10^{-28}$g × 9×10^{20}cm^2 s^{-2} /$6{,}6\ldots \times 10^{-27}$g cm^2 s^{-2} = $12{,}3\ldots \times 10^{19}$

Mesquita related $m_{0e}c^2/$h' to a mass of 1 gram, and after generalizing for a mass m_x:

$m_x c^2 /$h' = m_x $9 \times 10^{20}/ 6{,}62\ldots \times 10^{-27} = 136{,}4\ldots \times 10^{45}$
$m_x c^2/$h' = $136{,}4\ldots \times 10^{45}$ ou $m_x c^2 / $ h' $\times 10^{45} = 136{,}3$

Multiplying and dividing by the elementary impulse m'c:
$[(m_x c^2$ m'c$) / ($m'c h' $\times 10^{45})] = 136{,}3\ldots$

In the denominator c = $\mathcal{V}_0 \lambda_0 = \mathcal{V}_0 \pi d_e = \mathcal{V}_0 2\pi d_e/2$ and in the numerator m'c^2 = h', is obtained:
$\{[m_x$ h'c (g g cm^2 s^{-2} cm s^{-1})]/[m' $\mathcal{V}_0 2\pi (d_e/2)$ h'$\times 10^{45}$ (g s^{-1} cm g cm^2 s^{-2})]$\} = 136{,}3\ldots$

Multiplying and dividing by 1s and making h'1s = h,
$\{[m_x$ h c (g g cm^2 s^{-1} cm s^{-1})]/[2π m' $\mathcal{V}_0 (d_e/2)$h $\times 10^{45}$ (g s^{-1} cm g cm^2 s^{-1})]$\} = 136{,}3\ldots$

It is seen in the denominator:
m' $\mathcal{V}_0 (d_e/2)$h $\times 10^{45} = 0{,}737\ldots \times 10^{-47}$ g $\times 12{,}35\ldots \times 10^{19}$ s^{-1}$\times 3{,}861\ldots \times 10^{-10}$cm $\times 6{,}625\ldots \times 10^{-27}$ g cm^2 s^{-1} $\times 10^{45} = 23{,}3\ldots \times 10^{-20}$g s^{-1} cm g cm^2 s^{-1}, and canceling (g) de m_x (g) / m'(g), results:
$(|m_x|$ h c$)/23{,}3\ldots \times 10^{-20}$(g cm^2 s^{-2} cm), where $23{,}3\ldots \times 10^{-20}$(g cm^2 s^{-2} cm) = e^2 (erg cm) or momentum of energy.

So, m_X h c / $2\pi e^2 = 136{,}3\ldots$ in which $|m_x|= 1$.

But, $\alpha = m_x$ h c / $2\pi e^2 = 137$ it is the fine structure constant that is related experimentally between the constants c, h, and the charge e of the electron with the narrow spectral lines of the atoms.

So for $\alpha = 137$, the value of $| m_x|$ it would be 1.005, meaning a larger fraction of 0.5% of the formed mass of m'c × c = h'. This fraction of 0.5% is contained in the lower densities of h' around the base circle, which forms the field of the electron. And could be admitted $| m_x | = 1$, and obtaining:

$$\alpha = hc/2 \pi e^2 \approx 137$$

That is, through the theory of elementary impulse is obtained brilliantly the expression of the fine structure constant, which represents an electron equation, for the electric charge e, and the mass m_{0e} of the immovable electron through the expression of the length of the base circle of the electron $\lambda_0 = h/m_{0e}c$ with the constant $h = h' \times 1s$.

Meaning of the Fine Structure Constant and Electron Diameter – In case it is related to cargo e of the electron with its total energy $m_{0e}c^2$, for the symmetrical spherically charge density within an electron ray sphere, which is the electrostatic charge, is obtained that the diameter of the base circle of the electron is 137 times the diameter of the electron if considered spherical of radius obtained by the charge and total energy of the electron.

Then, in the CGS (electrostatic):
electrostatic electron force $F_{electrostatic} = e^2/r^2$
electrostatic potential energy = $f_{electrostatic} = (e^2/r^2) \times$ distance $r = e^2/r$
So $e^2/r = m_0c^2$
$r = e^2/m_0c^2 = 2{,}82 \times 10^{-13}$ cm and $D = 5{,}63... \times 10^{-13}$ cm

The diameter $d_{e(\alpha)} = 7{,}722... \times 10^{-11}$ cm of the electron from the constant of thin structure, is obtained the same diameter $d_{electron\ base\ circle} = 7.72... \times 10^{-11}$ cm, obtained by the elementary impulse m'c which runs with speed c the length λ_0 of a base circle of the electron, with a frequency $\mathcal{V}_0$ equivalent to Compton wavelength.

Relating $d_{e(\alpha)} = 7.722... \times 10^{-11}$ cm / D, where $D = 5.63... \times 10^{-13}$ cm: $(d_{e(\alpha)} D = 137.03...$
That is, the diameter of the base circle of the electron is 137 times the diameter of the electron if considered spherical of radius obtained by the charge and total energy of the electron.

Electron and Particle Charge by Elementary Impulse Theory – Since the early days of attraction and repulsion of materials that the concept of negative and positive charge of particles is followed by current Physics, such as the electron, positron and proton, without any explanation of the cause of the charge to be negative or positive, and the attraction or repulsion between charges of opposite and equal signals, it seeks to explain with infinite field lines and interaction of virtual photons between particles.

But, by the quantum impulse the charge is explained by the interpretation of Planck's law with physical significance of fundamental concepts of basic constants of Physics. By quantum impulse, the charge presents the fundamental constants m and without any square root in the description of the electricity, because e^2 is charge both negative electron and positive charges of the positron at the same time, being the electron-positron with a chirality or helicity to the right or to the left and the positron-electron with another chirality or helix on the left or right. Figure 28.

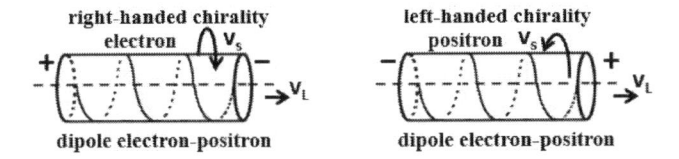

Fig. 28 – The electric charge e^2 it charges at the same time the charge of the electron and the positron, and the chirality or helix sets the electron of the positron, as Erdelyi.

And only the square of the absolute value e^2 of these charges has the meaning in energy as a momentum of energy, [erg cm] in CGS. The photon m'c x c that traverses the basic circle of the electron or positron which retains n m'c x c = nh', of the electron mass-positron forms the negative-positive charge of the electron or positron. However, according to A. E. Noether's theorem by which the laws of nature are only existing fundamentally with symmetry, shows that m'c x c running the base circle and lines of planes perpendicular to the circle that constitutes the field of the electron or positron, can present the two charges. For example, an electron in which m'c x c would be rotating in relation to a plane, in the direction of the clockwise clock Left → Right, and another electron in the direction counterclockwise Right → Left that rotates 180° on this axis of, will be in the hourly direction Left → Right equal to that of the other electron, proving to be a single charge for the electron or positron. But the negative charge e^- of the electron or e^+ positive of the positron is explained by the helix on the left or right or chirality left or right of the elementary impulse when the electron or positron rotate in the direction of the motion. Therefore, the chirality enters as a component of the "spin" intrinsic in the direction of the particle motion.

By the fine structure constant, experimentally determined and deduced by the elementary impulse: $\lambda_0 = (2\pi e^2\ 137) / (m_{0e}c^2)$
λ_0 = length of an electron base circle, often $\mathcal{V}_0$ = frequency of h' in the basic circle of the immovable electron
$$e^2 = (m_{0e}c^2\ \lambda_0)/ (2\pi 137)\ [\text{g cm}^2\ \text{s}^{-2}\ \text{cm}], [\text{erg cm}] = \text{energy momentum}$$
$$\hbar c/e^2 = h/2\pi\ c\ /e^2 \approx 137,$$
$$h = h'1s = m'c^2\ 1s$$
Replacing: $e^2 = (m'c^2 1s\ c) / (2\pi\ 137)$, in which m' and c are the fundamental constants of the charge e.

That is, the charge of the electron is related to the unit of the elementary energy h' = m'c x c = m'c² and to the length λ_0 among the h'.

$e^2 = c / (2\pi \ 137 \ h') = (3 \times 10^{10}) / (2\pi \ 137 \ h') = 3{,}48... \times 10^7 \ h'$
$\qquad\qquad\qquad\qquad$ 1cm

Or: $e^2 = (m_{0e} c^2 \lambda_0)/(2\pi \ 137) = (nm'c^2\lambda_0)/(2\pi \ 137) = (12{,}4... \times 10^{19}$ h') $\times$ (2,42... $\times 10^{-10}$) / $(2\pi 137) = 3{,}48... \times 10^7$ h' 1cm

For this equation: $e^2 = $ (m'c x c 1s c)/$(2\pi 137)$ presents the fundamental constants m' and c without square root in the description of the electricity, because e^2 it would charge both negative electron and positive charges of the positron at the same time, being the electron-positron with a chirality or helicity to the right or to the left and the positron-electron with another chirality helix on the left or right.

And if the charge of the electron e^- it is the same when the electron is with the increased mass of h', i.e. with mass m, then the length λ_m of the electron helix travelled by h' should be, by the equation of the charge in function of the mass, smaller to maintain the same charge of the still electron mass m_{0e}.

So:

For the immovable electron: $(h\ c\ /2\pi)/e^2 = 137{,}03...$
$$h = h'1s = m'c^2\ 1s$$
$$c = \lambda_0 \nu_0$$

λ_0 = length of an electron base circle which is the distance of the h' turning in the basic electron circle at rest,

ν_0 = frequency of h' in the basic circle of the immovable electron
$$[h'\ 1s\ \ \lambda_0 \nu_0 /e^2] = 137{,}03...$$
$$[h'\ \lambda_0 |\nu_0| / e^2] = 137{,}03...$$
$$|\nu_0|\ h' = m_{0e}c^2$$

$|\nu_0|$ = quantity of h' in the basic electron still
$$[\lambda_0\ m_{0e}c^2 / e^2] = 137{,}03...$$

For the motion electron with mass m:
$$[(h\ c\ /2\pi) / e^2] = 137{,}03...$$
$$h = h'1s = m'c^2 1s$$
$$c = \lambda_m \nu_m$$

λ_m = helix length in motion electron with mass m

ν_m = frequency of h' in the motion electron with mass m
$$h'\ 1s\ \ \lambda_m \nu_m /e^2 = 137{,}03...$$
$$h'\ \lambda_m |\nu_m| / e^2 = 137{,}03...$$
$$|\nu_m|\ h' = mc^2$$

$|\nu_m|$ = quantity of h' in the electron helix in motion with mass m
$$[\lambda_m\ mc^2 /\ e^2] = 137{,}03...$$

being the ratio of electron masses at velocity v linear: $m/m_{0e} = n_{m/m0e}$

equating $[\lambda_0\ m_{0e}c^2 / e^2] = 137{,}03...$ with $[\lambda_m\ mc^2 / e^2] = 137{,}03...$
$$\lambda_0\ m_{0e}c^2 = \lambda_m\ n_{m/m0e}\ m_{0e}c^2 \text{, so:}$$
$$\lambda_m = \lambda_0 / n_{m/m0e}$$

That is, the charge of the electron is related to h' and the length between the elementary energies h' , but independent of the quantity of h', because the length between each h' decreases with the increase of the quantity, keeping the charge invariable! As M. Rivas the electron charge is a simple point outside the center of the electron, and the charge is in a circular motion of radius $h/2\pi m_{0e}c/2$, moving with the speed of light c. By the elementary photon is m'c x c with speed c that traverses the electron helix.

By Einstein's relativity, the space would present contraction that would result in less length of matter formed by the particles, but would be inconsistent with the relativistic quantum mechanics in which the electron is a non-dimensional point that should also present this contraction, in which the zero contracted dimension should pass into a negative space! But, the laser light experiment showed that it is the light that traverses larger paths in space by receiving the inertia of the emitting source, therefore without space contraction. And dimensional variations of particles occur according to the electron equation of the elementary impulse theory. In 1934, Stücklberg had presented the hypothesis that the positron could be an electron with negative energy coming from the future to the past.

But by the theory of elementary impulse, the positron-electron is a dipole in which the chirality or helix determines to be the electron or the positron, without travel coming from the future or returning to the past!

In 1912, Nicholson had observed that Planck's constant h could have a physical sense, as angular momentum, for the energy levels of the atom. And by the theory of quantum impulse, it would indicate remarkably the projection of an angular momentum to h', by running circular line of particles at rest or helixs when moving.

XXXI. DIRAC QUESTIONS ON FUNDAMENTAL CONSTANTS AND THE THEORY OF QUANTUM IMPULSE – 1963-1964

In 1963, Dirac presented an article "The Evolution of the Physicist's Picture of Nature" in Scientific American, May 1963, about Planck's constant, c speed of light and charge e of the electron, with the question that what physical phenomenon would represent these three constants. Because of these constants one was the derivative of the two other fundamentals. Why should the electron charge without square root in the equations appear? And if the electron could present different shapes and dimensions by the different speeds and energies.

These questions were answered in 1964 by Open Letter of Erdelyi to Dirac and to the Scientific American, with a copy to Journals Nature and International Science & Technology.

Dirac – What is the physical phenomenon relating to h, c and the charge e of the electron?

As Erdelyi is the equation of the electron presented by the theory of impulse relating the constant of fine structure: $hc/2\pi e^2 \approx 137$.

Dirac – What are the fundamental constants?
By the quantum impulse theory the constant c speed of light and electromagnetic radiation is the fundamental quantity, as it is the speed of the elementary impulse m'c.

Dirac – What is the constant derivative?
By the elementary impulse is: $h = n\, m'c\, \lambda_e$, where $n = m_e/m' = 1{,}2364 \times 10^{20}$; m' = elementary mass; $m_e = 1{,}2364 \times 10^{20}$ m', c = speed of light, and λ_e it's the wavelength. That is, h is the derivative quantity that contains the fundamental c and m'.

Dirac – Why the charge e of the electron appears with the square root?
By quantum impulse theory the charge e^2 it is a dipole that contains the negative and positive charge in a helix, with the helix on the right or left in the direction of the motion, so it appears elevated to the square: $e^2 \approx hc/2\pi 137$.

Dirac – How could the electron present different shapes and dimensions by different velocitiess and energies?
The quantum impulse theory m'c showed the deduction of the electron equation in which the dimension and shape of the electron are depending on the mass and the consequent velocity of the electron.

XXXII. PROTON MODEL BY QUANTUM IMPULSE THEORY – <u>1952 to 1969</u>

By quantum impulse theory of Ederlyi in 1952 to 1969, for the proton-antiproton, with the mass m of 1,836.12 of the positron or electron, being:
$\lambda_{linear\,(m = 1,836.\,12e)}$ = total length of the helix that forms the proton with mass of 1,836.12 of the positron or electron.
$n_{(m = 1,836.\,12e)/me0}$ = amount of positrons or electrons with mass m_{e0} what forms the proton
$v_{linear\,(m = 1,836.12e)}$ = linear electron velocity or positron for mass of 1,836.12 x electron mass or positron (1,836.12 m_{e0}) = linear velocity of the projection of h' in the helix that forms the proton.
$\mathcal{V}$ = h' frequency in electron or positron with velocity v_{linear}
$\mathcal{V}_0$ = frequency of h' in the basic circle of the electron or positron at rest
$$n_{(m = 1,836.12)/me0} = 1{,}836.12$$
And with the equations of the electron by the triangle projection of the helix of m'c x c = h' which constitutes the electron and the equation of Planck's electromagnetic energy and Hasenöhrl or

widespread of Einstein's total energy, for the rest positron-electron, with the frequency $\mathcal{V}_0$ of h' in the basic circle of the positron-electron with mass at rest m_{e0} is obtained:

$$h\mathcal{V} = mc^2 \rightarrow \mathcal{V}_0 = m_{e0} c^2/h = 12.36... \times 10^{19}\ s^{-1}$$

e $\mathcal{V} = n\mathcal{V}_0$

velocity $v_{linear(m=1,836.12e)} = 2.997924... \times 10^{10}$ cm/s

$\lambda_{linear(m=1836.12e)} = 1.3214049... \times 10^{-13}$ cm

$\lambda_{linear(m=1,836.12e)} \cdot n_{(m=1,836.12e)/me0} = \lambda_{proton} =$ positron-electron length with mass 1,836.12 x electron mass or positron (proton mass) = total helix length forming the stationary proton = $2,426260... \times 10^{-10}$ cm, which is exactly the Compton wave length of the stationary electron, $\lambda = h/m_{e0}c$. And is deduced by the model quantum impulse of the proton. The figure 29 shows the basic model of the proton formed by positrons-electrons, which is in the course of advanced research in Physics at São Paulo University-USP, Brazil, presented by Paulo Ferraz Mesquita, Doctor and Professor in Engineering at USP.

J. P. Hansen

UNIVERSIDADE DE SÃO PAULO
ESCOLA POLITÉCNICA
ADMINISTRAÇÃO - EDIFÍCIO DO DEP. DE MINAS - CIDADE UNIVERSITÁRIA

ELetron (Positron) parado

a
Visto em perspetiva

b
Visto de topo

FIGURA 21,13

ELetron (Positron) em estado de deslocamento com velociddde v_x

$n = \frac{m_v}{m_o} = \frac{massa\ na\ velocidade\ v_x}{massa\ parado = 9,1083 \times 10^{-28}\ grs}$

$2,42626 \times 10^{-10}\ cm = \lambda_s = \lambda_{vm} \sqrt{\frac{n^2}{n^2-1}} = n \cdot \lambda_{vgn} \geq n \cdot \lambda_{vm}$

FIGURA 21,14

Proton (Anti-proton)

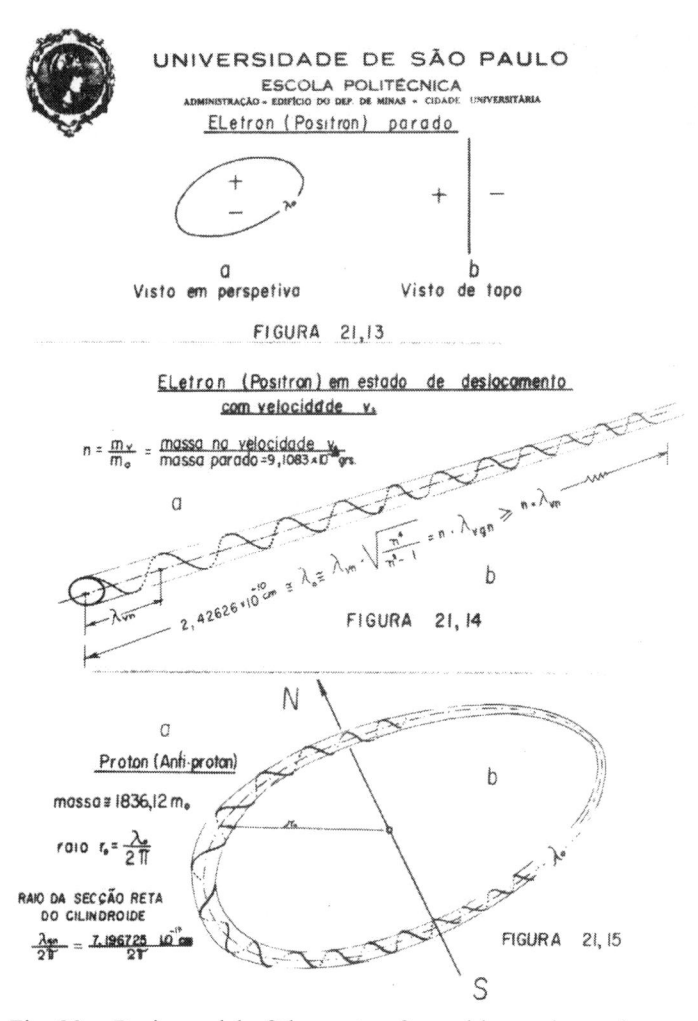

massa ≈ 1836,12 m_o

raio $r_o = \frac{\lambda_o}{2\pi}$

RAIO DA SECÇÃO RETA DO CILINDROIDE

$\frac{\lambda_{vp}}{2\pi} = \frac{7,196725 \times 10^{-14}\ cm}{2\pi}$

FIGURA 21,15

Fig. 29 – Basic model of the proton formed by positron-electron helixs by the theory of quantum impulse of Erdelyi, presented by Mesquita in Course of Research in Advanced Physics taught at the Polytechnic School of São Paulo University-USP, Brazil, as Reference.

By the theory of quantum impulse, this Compton wavelength represents exactly the length of the circumference of the electron circle at rest. And that length of the proton helix forming a circumference is exactly the area of the electron circle at rest, which indicates why the proton charge with mass 1,836.12 greater than that of the electron has the same absolute value as that of the electron.

This is one of the questions raised by the current Physics of why the mass of the proton is 1,83612 times that of the electron and not another mass! And it is significant that J. J. Thomson had presented that the hydrogen atom would be formed from around 2,000 electrons in a sphere with a positive charge fluid, and Rutherford had presented that the nucleus of the atom would be formed of a concentration of electrons next to the positive-charge fluid. It is observed by the equations of deduction of the proton model of the quantum impulse theory, that if the proton had greater mass, the frequency $\mathcal{V}$ of m'c x c = h' which constitutes the proton would be larger than usually listed in the electromagnetic spectrum and like this be at the limit of the constitution of the proton with this and not another mass!

XXXIII. MASER, LASER AND QUANTUM IMPULSE THEORY – 1916 to 2018

Between 1916 and 1917 Einstein using thermodynamics with Planck's energy density and radiation and Einstein's light quantum, and in the transition from one power state to another, obtained $h\mathcal{V} = E_2 - E_1 = hc/\lambda$ and also $E\lambda = hc$, $E_{(ev)} = hc/e$, where h = Planck's constant; E = energy; c = speed of electromagnetic radiation; λ = wavelength; $E_{(ev)}$ = energy in electron volt; e = electron charge = $1,6 \times 10^{-19}$ coulombs; 1 ev = $1,6 \times 10^{-12}$ erg. And with the probability P of the atoms in state 1 and 2 resulting in P_1 = K exp (E_1/kT) e P_2 = K exp (E_2/kT), onde K = constant, k = Boltzmann constant e T = temperature, obtained the probability of spontaneous emission when the energy level is increased. However, this probability of emission is reported to the brownian motion that only reemits light.

But in the case of a higher level of energy, it could be obtained a stimulated emission in which the wave associated with the incident photon it would originate a wave associated with the second photon, and these waves should be in phase and same frequency and in greater intensity, ie frequency coherent and intense amplification. And it occurring a diffusion of light originated from smaller frequencies in the emission stimulated.

Between 1951 and 1953 C. H. Townes presented the maser, i.e. "microwave amplification through stimulated emission of radiation". And the first stimulated emission devices were gas, with ammonia, by J. P. Gordon, H. J. Zelger and Townes and after of crystals. In 1957 G. Gould created a prototype laser. And in 1960 T. H. Maiman presented the first laser, "light amplification by stimulated emission of radiation" by ruby which presented the red light.

In 1964, N. Basov and also A. Prokhorov developed the emission stimulated for the laser.

But in 1958 a fact astonishing occurred on the discovery of the laser in 1960. In 1958, the scientific consultant F. Knoll presented information to the scientific research companies, and for the scientific and technological research company Hughes Aircraft

Company, where it was Maiman, with the following report: *"The elementary quantum theory of energy of Erdelyi that operates in Ford in Brazil and after also at the Engineering School of Mackenzie University in Saint Paul in Brazil, indicates that the meaning of Planck's law $\mathcal{E} = h\mathcal{V}$ it leads to new concepts in quantum mechanics, affecting all the theories and physical interpretations of the electromagnetic radiation currently followed. And it is expected to be beneficial for research related to its application to electromagnetic radiation, such as the spectra of ultraviolet ultra frequencies, of visible light and of infrared, with which they are researching the Hughes Aircraft Company"*. And in 1960, Maiman presented the first laser. And in 1963 and 1964 it was presented that in crystal entered a red light beam of a ruby laser and on the other side exit an ultraviolet beam filtered into the blue. By quantum mechanics that was followed always the energy went from a level superior to a lower, which was exactly the case of the laser. But it occurred that one part of the beam went from a lower level to another superior, contrary to quantum mechanics, in which by the photoelectric effect the intensity of the light when increased should maintain the same frequency, therefore the same wavelength, because the speed of light is considered constant. The Figures 30 and 31 show work on lower frequencies of light emitted resulting in higher frequencies, presented in Scientific American.

Advances in Optical Masers, by Arthur Shallow. Cover photo by Bob Terhune.

Fig. 30 – Photo of Scientific American cover of July 1963, with photo of laser emitting light of less frequency and the output obtaining a higher frequency, contrary to the photoelectric effect and explained by the higher intensity of the elemental photon h'.

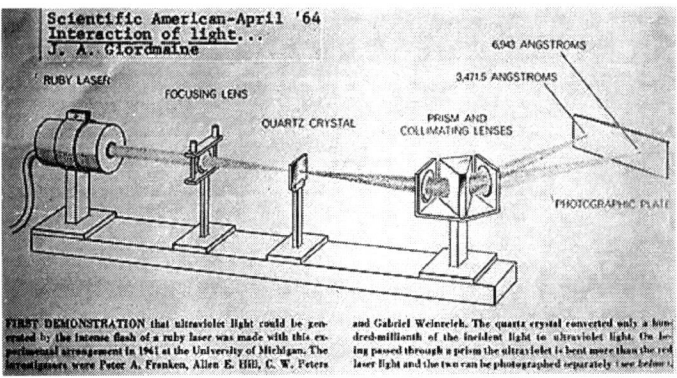

Fig. 31 – Another work presented in the Scientific American of April 1964, showing a higher frequency of light in relation to the emitted laser light, contrary to the photoelectric effect, and explained by the higher intensity of the elemental photon h'. Photo of Scientific American.

And in 1963, he founded the Korad Company which was incorporated into the Union Carbide Corporation which produced acetylene and joined in 1917 to the Linde Gas of Germany and the United States in which K. Linde introduced the first cryogenic plant in the air, nitrogen and oxygen, to produce a high temperature flame gas for metal cutting. Also, Linde Gas had made a contribution to the development of the nuclear fission atomic bomb, when the Linde Gas scientists have perfected a refining process for the treatment of uranium concentration through gaseous diffusion. Then, in 1963, the Union Carbide Corporation requested through the Ford in United States in order for Erdelyi to send his other publications, however, without the Union Carbide Corporation mentioning the name of the requester, who should be the Korad Company subsidiary who had received Knoll's technical information about the theory of the elementary impulse of Erdelyi. But at the request of the Union Carbide Corporation the publications of the Erdelyi theory were sent to that company to the occult identity requestor.

As J. Trémolières, on 1966: *"The laser despite numerous publications and applications, since 1916 and 1917 with Einstein's theory of emission spontaneous, after 50 years is still again to be investigated on the quanta, to open new chapters of Physics on the deeper constitution of matter, and the most advanced is yet to discover"*. And in 2018, almost more than 100 years of the first publications of the maser and laser, also the most advanced would still be undiscovered! But, it would be by the quantum impulse that the higher level of knowledge of the laser and quantum mechanics could be achieved.

With the maser, the applications of micro-waves have allowed the use in radio astronomy as radar, radar echoes on planets and asteroids, spectroscopy, frequency patterns, communications with satellites where signals are reflected and other possibilities as weapons for defense systems.

XXXIV. NEUTRINO MASS AND QUANTUM IMPULSE – 1930 to 1969

Pauli had presented in 1930 the existence of a minimum particle without charge, that is, neutral to preserve the conservation of charge and energy in the beta decay. This particle would have a minimum mass or even zero mass if it came out of the nucleus at the speed of light, by the theory of relativity. Subsequently, it was named neutrino.

Between 1955 and 1956 Cowan and Reines detected the neutrino near the Savannah River reactor. In the reactor the reaction is: antineutron → antiproton + positron + electron neutrino.

In 1964 R. Davis showed that it was antineutrino.

By quantum impulse theory presented in 1952, the neutrino contained mass, through $\varepsilon_{neutrino} = n\ m'c \times c = n\ h' = m_{neutrino} \times c^2$, by the correct interpretation of Planck and Einstein quantum by quantum impulse theory.

In 1956, W. H. Barkas, W. Birnbaum and F. M. Smith presented that the neutrino by the observed experiments should be without mass or minimum mass. In 1960, Pontecorvo and also M. Schwartz presented an experimental proposition for the production of neutrinos in particle accelerator to verify the neutrino masses from the electron and muon. And in 1962, Lederman, Schwartz, Steinberger, G. Danby, J. M. Gaillard, K. Goulianos and N. Mistry and in 1964 G. Bernardini, J. K. Bienlein, A. Bohm, G. Dardel, H. Faissner, F. Ferrero, Gaillard, H. J. Gerber, B. Hahn, V. Kaftanov, F. Krienen, G. Manfredotti, M. Reinharz, P. G. Seiler, A. Staue, H. J. Steiner and R. Salmeron performed this experiment and measured results without detecting mass.

And then in 1969, Pontecorvo and V. Gribv presented the hypothesis that the electron origin neutrinos from the Sun fluctuated in other types of neutrinos such as the meson μ neutrino.

Spin, Helicity and Chirality – It is mentioned by quantum mechanics that the spin of the particles, the spin + 1/2 (up) or – 1/2 (down). The spin is set to the right (right hand) or left (hand-left) with the direction of the thumb, which is its linear momentum. And in the direction of the particle motion in the thumb direction of the spin that rotates as to the right (right hand) is set to helicity + 1. For the motion of the particle in the direction of the thumb, of the spin that rotates to the left (left hand), is defined as helicity – 1.

And in these two cases, the particle being seen in the opposite direction to the motion, results in the spin rotating clockwise. And the particle being seen on the side of the motion will result in spin rotating as counterclockwise. But as the observer surpassed the motion of the particle to any other side of the motion there would be a clockwise and counterclockwise reversal of the right (right hand), left (left hand) and helicity which is defined as positive if the particle has motion towards the direction of the right spin (right hand) or left (hand-left), and negative otherwise.

However, for particles considered without mass, by Einstein's quantum mechanics and relativity as the photon, the observer could not exceed the speed of light, so helicity would be permanent in this case. According to D. Griffiths, *"The helicity of massless particles is not an artifact of the observer's reference system, but a fixed and fundamental property"*.

But, the chirality to the right (right hand) or left (left hand) of a particle is absolute, without relying on the reference system, of the particle being at rest or moving or having mass or almost without mass, that is, it is always intrinsic of the particle. However, there is considerable reference to the positive helicity particle as to the right (right hand) and to the negative helicity particle as to the left (left hand), which creates confrontation with the notion of chirality to the right and to the left. Also, the alternating particle with mass or no mass to revert its helicity as the neutrino that was without mass and after with mass.

The photon has spin $+ 1$ $\hbar$, in which $\hbar = 2/\pi$, which corresponds to transverse circular polarization on the right and $- 1$ $\hbar$ as circular cross-polarization on the left. The electron has the orbital angular momentum which is a value of the electron velocity around the nucleus and has n $\hbar$ values. And also, the angular momentum intrinsec of the electron, the spin $+ 1/2$ $\hbar$ and $- 1/2$ $\hbar$. Also, proton, neutron, neutrino, mesons, with spin $\pm 1/2$. Figure 32.

spins and helicities for the electron according to parallel motion or antiparallel to the direction of the spin momentum

helicity +1 ⟶
helicity −1 −−→

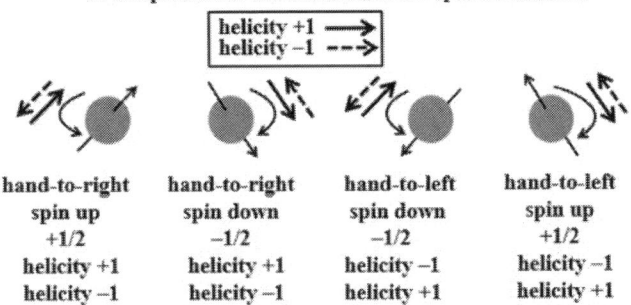

hand-to-right	hand-to-right	hand-to-left	hand-to-left
spin up	spin down	spin down	spin up
+1/2	−1/2	−1/2	+1/2
helicity +1	helicity +1	helicity −1	helicity −1
helicity −1	helicity −1	helicity +1	helicity +1

Fig. 32 – The current Physics defines particles with spins as on the right (right hand or left hand), by position $+ 1/2$ (up) and $- 1/2$ (down) and helicity $+ 1$ or $- 1$ according to the direction of the motion and the direction of the spin to the right or to the left.

However, by quantum impulse it is the helicity or chirality absolute, right or left of the direction of the motion that defines the particle or particle. Figure 33.

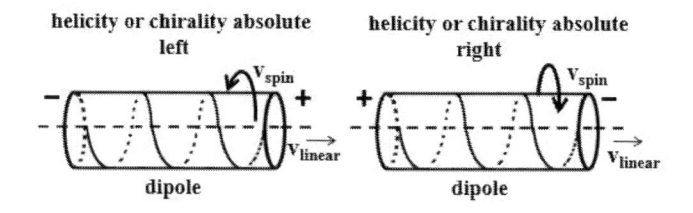

Fig. 33 – Chirality or absolute helecity defines particle and antiparticle by theory of quantum impulse of Erdelyi.

Two types of mesons with masses, spins and equal charges decay, that is, disintegrate in two and three pions, that is, different formations of pairs. Lee and Yang in 1956 proposed an experiment to observe whether there was a preference of direction when particles were created. And C.S. Wu and others, observed that in the beta decay, the cobalt 60 fell in nickel, antineutrino and electron which were emitted mostly in the opposite direction to what would be the motion of the nuclear spin.

In 1957, Lee and Yang; A. Salam and also Landau presented that the neutrino should have felt to the left (left hand) of helicity -1, and the antineutrino spin as to the right (right hand) and helicity $+1$. And in 1958, Goldhaber, L. Grodzins and A. W. Sunyar, and in 1960 A. I. Alichanov, Y. V. Galaktionov, Y. V. Gorodkov, G. L. Eliseev and V. A. Lyubinov presented experiments in that neutrinos were only left (left hand) and antineutrino were on the right (right hand).

By quantum impulse theory the particle and antiparticle is a single structure of quantum impulses m'c with speed
c forming a helix, in which the absolute, right chirality (right hand) and left (hand-left) or helicity right (hand-right) or left (left hand) is that it defines whether it is particle or particle.

PART 5 – FROM PARTICLES MASS TO ZERO MASS IN PARTICLES AND FIELDS, AND OF THE ZERO MASS TO MASS CREATED OF FIELDS AND PARTICLES

XXXV. UNIVERSAL BETA INTERACTION AND NEW PARTICLES
1938 to 1957

In 1938, O. B. Klein had presented that the beta decay would be mediated by a particle of greater mass in the atomic nucleus that named W, weak. So, neutron → proton + W⁻, W⁻ → electron + neutrino. In 1948, O. B. Klein and also G. Puppi presented the hypothesis that the neutron and meson µ decay: neutron → proton + e^- + neutrino, and $\mu^- \rightarrow e^-$ + 2 neutrinos presented a same coupling constant as weak-force interaction of the beta decay. And in 1949, this hypothesis was confirmed by Tiomno, Wheeler and also Lee, Yang and Rosenbluth, that is, there would be a universal beta interaction. In 1949, Bethe presented that the coupling constant for nuclear or strong interaction was the expression of $[40c\ (h/2\pi)]^{1/2}$. In 1950, the Seriff, Leighton, C. Hsiao, E. W. Cowan and Anderson and also L. Michel presented that the meson µ was disintegrated into electron and two types of neutrino: $\mu \rightarrow e^-$ + neutrino 1 + neutrino 2. In 1957, Schwinger presented model in which weak particles W⁺, W⁻, which would act virtually in the vacuum around 10^{-7} to 10^{-14} seconds at distances of only 10^{-16} cm, like the photon, and thus being the model of the electromagnetic and weak interaction, or electroweak. In 1958, Feynman and Gell-Mann, H. Jensen, A. H. Wapstra and F. Bochm, and also Marshak, E. C. G. Sudarshan, J. J. Sakurai and Lopes presented a theory of universality of the beta interaction, and supplemented by N. Cabibbo for experimental adjustment through a geometric angle on the axis of the decay diagram. Feynman and Gell-Mann presented that the universal weak interaction should be effected by particle creation and quantum annihilation at a minimum interval of time, i.e. virtual particles, and named weak, W⁺ and W⁻. And Lopes presented that the photons of electron-electromagnetic radiation interaction should be similar to the universal weak interaction, presenting also another neutral particle Z°. Also, Schwinger and S. L. Glashow presented these same particles W⁺ W⁻ and Z°, which were called bosons, of the term derived from condensed atoms, of Bose studies. These particles W⁺, W⁻ and Z°, should have mass, but because these particles are virtual the mass should be created by some field of mass that was unknown by the Physics.

The field that would create mass would be the quantum elementary impulse that would form all the particles. But, the theory of elementary impulse was at the beginning of his presentation and unknown to the scientific community to be admitted.

XXXVI. ELECTRIC AND MAGNETIC FIELDS WITHOUT MASS AND WITH MATHEMATICAL ABSTRACTIONS – **1963 to 1965**

On 1963, as Feynman: *"The natural interpretation of electrical interaction is that two objects attract each other simply, more against less. However, this was discovered to be an inadequate idea to represent this. A more appropriate representation of the situation is to say that the existence of the positive charge, in some sense, distends, or creates a condition in space, so that when we put a negative charge the space feels a force. This potential to produce a force is called an electric field"*. Also as Feynman: *"We can write to force F over a charge Q moving with a velocity v as $F = q(E + v \times B)$. We call and the electric field and B the magnetic field at the charge site. It is precisely because they can be specified E and B at every point in the space that is called a field. A field is any physical quantity that assumes different values at different points in space. There were several inventions to help the mind visualize the behavior of fields. The most correct is also the most abstract: we simply consider the fields as mathematical functions of position and time. The only sensible question is that the most convenient way to observe the electrical effects. Some prefer to represent the field as interaction at a distance of charges, and use a complicated law. Others prefer the field lines. The best way is to use the abstract field idea. Which is abstract, but necessary"*.

And according to A. D. Fokker, in 1965: *"The field is a non-material symbolic medium for momentum and energy. We call such a zero mass medium that in quantum theory reported to the electromagnetic field are photons, which are immaterial, without mass"*.

So, the community of Physics remained focused on particles formed in the quantum vacuum, contrary to the theory of the elementary impulse in which the quantum mass formed the other particles, as was observed in the annihilation and formation of particles by gamma-ray photons.

XXXVII. PARTICLES WITHOUT MASS AND MASSLESS FIELD THAT CREATES MASS BY MATHEMATICS: THE HIGGS FIELD – 1954 to 1964

Between 1960 to 1961 J. Goldstone and Y. Nambu had presented a particle theory, the term boson derived from condensed atoms of work of Bose, no mass. And in 1964, P. W. Higgs; F. Englert and R. Brout; G.S. Guralnik, C. R. Hagen and T. W. B. Kibble presented that the invariance obtained in particle interaction by calibration transformation ("gauge") applied by Yang and Mills in 1954, and which resulted in Lagrange's equations in no-mass bosons, it could be modified by configuring the symmetry of the Lagrange equation of the particle interaction energy for a hidden symmetry or spontaneous break. So setting in Lagrange's equation a symmetry like L= f (φ) in which φ represents particle field variables, but L also representing the classic configuration of L = $E_{kinetic\ energy} - E_{potential\ energy}$, and developing with new parameters like η, so L = f ($E_{kinetic\ energy}$) – F ($E_{Potential\ energy}$)≡ F [($E_{kinetic\ energy}$)η$_1$ + F [($E_{kinetic\ energy}$)η$_2$ – F [($E_{potential\ energy}$)]η$_3$ – F [($E_{potential\ energy}$)]η$_4$, it resulted in a hidden symmetry or a spontaneous break in symmetry, in which one of the terms contained a kinetic factor α^2 = 1/2 [(mc)/(h/2)]^2, or a particle with mass. And applying for the interaction of two particle fields with variables φ_1 and φ_2 was obtained in the development L = f (φ_1 + iφ_2) that would also appear to be a spontaneous break symmetry, in which in one of the particle fields appeared the mass, but there was another massless field along with one or more without mass particles named Goldstone bosons. So by applying a local invariance, or in that the variables are not altered, for the interaction of particles resulted in all the fields without mass, but introducing a covariant, that is, a relationship between variables, derivative of the field, ie that allows equal equations for the local field and new fields of variables φ_3 and φ_4, then it was obtained from the field of variables φ_3 a particle with mass and the field of variables φ_4 a massless particle of the Goldstone boson, and another field with mass. However, by applying a new calibration invariance ("gauge"), a "cunning" choice was eliminated this particle from the Goldstone boson without mass, that is, resulted in a mass term for particles like the W^+, W^- and Z bosons. This transformation, of symmetry to hidden symmetry or breaking of symmetry with calibration invariance ("gauge") for field and massless particle, and the application of covariant derivative and restoration of Lagrange equation with another calibration invariance ("gauge") to obtain field and particle with mass was named the Higgs mechanism.

Thus the physics extended the mathematical abstraction, presenting massless particles, involving symmetry, calibration or calibration invariance ("gauge") and covariance of Higgs mechanism, this is only mathematical artifice, to obtain from a variable without mass a relationship with another invented variable to get mass!

XXXVIII. ELECTROWEAK THEORY – 1961 to 1968

According to Glashow, Salam, Weinberg and J. C. Ward, from 1961 to 1967 there would be a particle type photon, boson, W^+, W^- and another neutral $Z°$ that should be like the photon, without mass. But they should acquire mass, be condensed, because the constant coupling of the energy mc^2 of the particles for equilibrium indicated a mass of 80 Gev (*), 87 x proton mass, for W^+ and W^- and ≈ 90 Gev (97 x mass of the proton) for the $Z°$ particle. There would also be two other particles in the interaction $W°$ and $B°$ and intensities between the particles forming a geometric angle with of the interactions $Z°$, gamma, $B°$ and $W°$, angle of Glashow, Weinberg and Cabibbo. Similar to the spin 1 of the photon, also W, Z and B was spin 1. By the Higgs field there should be four particles of that field H^- H^+, $H_1°$ and $H_2°$. And the particles W^+, W^- and Z could be each in helicities + 1 or – 1 and after interaction in helicities + 1, – 1 or zero, while the Higgs particles would have helicity zero. In 1967, Weinberg and in 1968, Salam presented new studies to those of O. B. Klein, Lopes, Schwinger and Glashow of unifying electromagnetic and weak interactions, containing the massless gamma bosons, W^+ and W^- of masses 75 x mass of the proton by Weinberg, and $Z°$ mass 95 x mass of the proton, in which was used the calibration theory ("gauge") of particles involving these interactions and the Higgs mechanism of hidden symmetry or breaking of symmetry to provide mass to the bosons W^+, W^- and $Z°$.

Thus, with the artificial Higgs mechanism was presented the interaction of formation of the universal beta decay. However, quantum impulse was the actual explanation of the particle constitution by the photon m'c x c.

* Power is defined as energy/time, and the electric current unit as the quantity of electric charges (electrons)/time. and the power/electric current set to voltage. For the power of 1 watt = 1 joule/s = 1 [kg x m/s^2 x m]/[s] = 1 [10^3 g x 10^2 cm x 10^2 cm]/[s] = 1 x 10^7 [erg]. For the electric current is defined the current of 1 ampere which corresponds to 6.24196 x 10^{19} electrons = 1 coulomb. So for voltage, 1 volt = watt/ampere = 1 x 10^7 [erg/coulomb] and for an electron charge: 1 ev (electron volt) = 1 x 10^7 [erg]/6,24196 x 10^{19} = 1.6021 x 10^{-12} [erg]. and 1Mega ev = 10^6 x 1.6021 x 10^{-12} [erg] = 1.6021 x 10^{-6} [erg] and for 1 Giga ev or 1 Gev where 1 Gev = 1.6021 x 10^{-3} [erg].

XXXIX. THE CREATION OF ELEMENTARY PARTICLES, OCTETS, ACES AND QUARKS, GLUONS OR "PASTE WITHOUT MASS" – **1947 to 2000**

Octets – From 1960 the current Physics, to explain energy spikes observed in particle accelerators, sought provisions such as octets in which at the vertices would occur these peaks, i.e. concentrated energies or possible particles.

What had been calculated by Yukawa, for a mass prediction around 1/7 of the mass of the proton, a median particle, was the meson. In 1947 observations of cosmic rays, formed of collisions between mainly protons with nuclei of atoms of the atmosphere, showed the meson π, by Lattes, Muirhead, Occhialini and Powell. This interaction occurred at a distance of 10^{-13} cm and would require a coupling constant around 1,400 times greater than the electron-photon electromagnetic. In the case of electron-photon interaction other virtual photons could interact and an additional addition of the coupling constant would allow the electron-photon balance. But in the case of meson, the probability of other mesons interacting would be elevated which would cause instability in the proton-neutron equilibrium. For a more accurate calculation for this proton-neutron equilibrium there should be another type of interaction.

In 1960/1961 Gell Mann and Yuval Ne'eman gathered the particles by masses from 0 to 3 of the mass of the proton, the strangeness, with the positive, negative and neutral charges, obtaining the observed and foreseen particles, according to experiment in 1953 by Fermi, R. Martin, D. E. Nagle and H. L. Anderson, R. H. Dalitz; in 1960 by L.W. Alvarez, M. Alston, P. Eberhard, M. L. Good, V. Graziano, H. K. Ticho, S. J. Wojcicki; in 1961 by A. Pevsner, R. Kraemer, M. Nussbaum, C. Richardson, P. Schlein, R. Strand, T. Joohig, M. Block, A. Engler, R. Gessaroli and C. Meltzer, Gell-Mann, S. Okubo, Sakurai, E. R. Erwin, R. March, W. D. Walker, E. West, Alvarez, B. C. Maglic, A. H. Rosenfeld, M. L. Stevenson, Alston, Eberhard, Good, Graziano, T. Wojcicki, Tiomno, A. L. L. Videira, N. Zagury; in 1963 by J. J. Swart, A. N. Didens, E. W. Jenkins, T. F. Kycia and K. F. Riley, Schlein, D. D. Carmony, G. M. Pjerrou, W. E. Slater, D. H. Stork, Ticho, Schlein and others, P. L. Connoly and others, in 1964 by S. Coleman and H. J. Schnitzer, V. E. Barnes and others. G. R. Kalbfleisch and others, M. Goldberg and others, with masses between 549 to 2,190 Mev. However, there were particles outside the octet and a new interpretation of how particles would be formed would be necessary.

Aces and Quarks – For a new interpretation of the particles outside the octet, in 1964 Gell-Mann and G. Zweig suggested that protons and neutrons were composed of quarks particles by Gell-Mann and Aces by Zweig. Nuclear particles and antiparticles would be formed from 3 types of quarks: u (up), d (down) and s (stranger). The charges would be split to result in the final charges of the particles: $u = +2/3$, $d = -1/3$, $s = -1/3$. And the antiquarks $u = -2/3$, $d = +1/3$, $s = +1/3$. In the case of the proton formed by the uud quarks and the neutron by ddu, would exchange in strong interaction a u by a d to form the charges of the neutron 0 and of the proton + 1. And they would have spins + 1/2 and –1/2. Also, other particles like barium and mesons would be made up of these quarks. In 1953, R. L. Hofstadter, H. R. Fechter and J. A. McIntyre had initiated research experiments on structure of nuclei, protons and neutrons, and in 1969, Friedman and H. W. Kendall and R. E. Taylor through elastic spreading of electrons observed sharp proton deviation, which evinced particles as concentrated in a part of the whole particle, in which Feynman styled partons or parts of the proton. Between 1964 and 1969 experiments for the detection of quarks was presented in 1964 by F. Gursey and L. A. Radicati; in 1965 by Salam, R. Delbourgo and J. Strathdee; in 1966 by W. Chupka, J. Schiffer, C. Stevens; in 1967 by Zweig, R. Gomez, H. Krobak, A. Moline, J. Mullins, C. Orth and J. Putten; Vine and G. H. Renninger; in 1968 by D. M. Rank; and in 1969 by C. B. A. McCusker, L. S. Peak and R. L. S. Woolcott. And it was isolated each electron dispersion with different targets in which resulted in the determination of electric charges of these dispersions with the planned fractional charges of + 2/3 and -1/3 for the quarks. As per M. Veltman on 2000: *"For all we know the particles like the electron and quarks are like dots, different from the proton, neutron, nuclei and atoms that have dimensions that can be measured".*

Paste Without Mass, the Gluons – In 1968 at the Stanford Linear Accelerator Center-SLAC and at CERN was observed constituents of the proton with the expected charges and spins, the partons of the proton or quarks, with their masses and velocities, that is, their momentums. But, the sum of these momentums and masses were inferior to the mass and momentum of the proton. The masses of the quarks were smaller than 1/3 of the proton mass and all the momentums of the quarks were 1/2 of the proton. Neutral particles have also been observed, this is without charge, along with quarks and that should effect the interaction between quarks, likewise as the massless photons when at rest (real estate) make the interaction with the electron of the particles. Then it was concluded that the quarks were made of gluons that would be without mass when at rest, similar to the photons, with time to compensate the momentums of the quarks.

Thus, it continued the basis of untouchable relativity, and the quantum mechanics in which the interactions of the particles were normalized, that is, subtractions and or additions to adjust the mathematical calculus that balanced all the particles involved in the interaction. And gravity was still outside the standard model of Physics.

PART 6 – QUANTUM IMPULSE AND GRAVITATIONAL RADIATION

XL. GRAVITATION AND QUANTUM IMPULSE – 1557 to 1965

In 1965, Mesquita presented the significance of Newton's constant universal gravitation in quantum terms by placing unified gravity with all the particles and electromagnetic radiations, within the spectrum of radiations, "The Meaning of the Constant of Isaac Newton's Universal Gravitation and Consequent Specific Discovery of Gravitational Radiation Reveal and Prove, 'data venia', Misconception of Albert Einstein, Max Planck and Contemporaries". This work was sent to the Brazilian Academy of Sciences, but that academics preferred to omit to analyze a theory that could lead to the development of gravitational control by quantum mechanics. Also, it was presented in 1965 in Congress and Conference at São Paulo University. Figures 34 and 35, and Document 8.

Fig. 34 – Mesquita presenting in Conference in 1965 on the significance of Newton's gravitational constant by the theory of quantum impulse, with the discovery of the gravitational radiations by Mesquita. Photo Authorized by Brazilian Interplanetary Society-SIB.

THE MEANING OF THE UNIVERSAL GRAVITATION CONSTANT

of Isaac Newton evinces the specific existence of

GRAVITATIONAL RADIATIONS

identified in the electromagnetic spectrum and reveals mathematical solution of the problem of the

UNIFIED FIELD ON FUNDAMENTAL PHYSICS

A conference held in 1965, December, 17th, at the Faculty of Architecture and Town Planning of the University of São Paulo, in the «COLLOQUIUM FOR THE REVISION OF SOME OF THE FUNDAMENTAL CONCEPTS OF PHYSICS»

by the Civil Engineer and Geographer
Prof. Dr. Paulo Ferraz de Mesquita
of the
University of São Paulo — Brasil

Doc. 8 – Article cover "Gravitational Radiations", of Mesquita, presented at 1965 in Conference at University of São Paulo-USP, Brazil, in References.

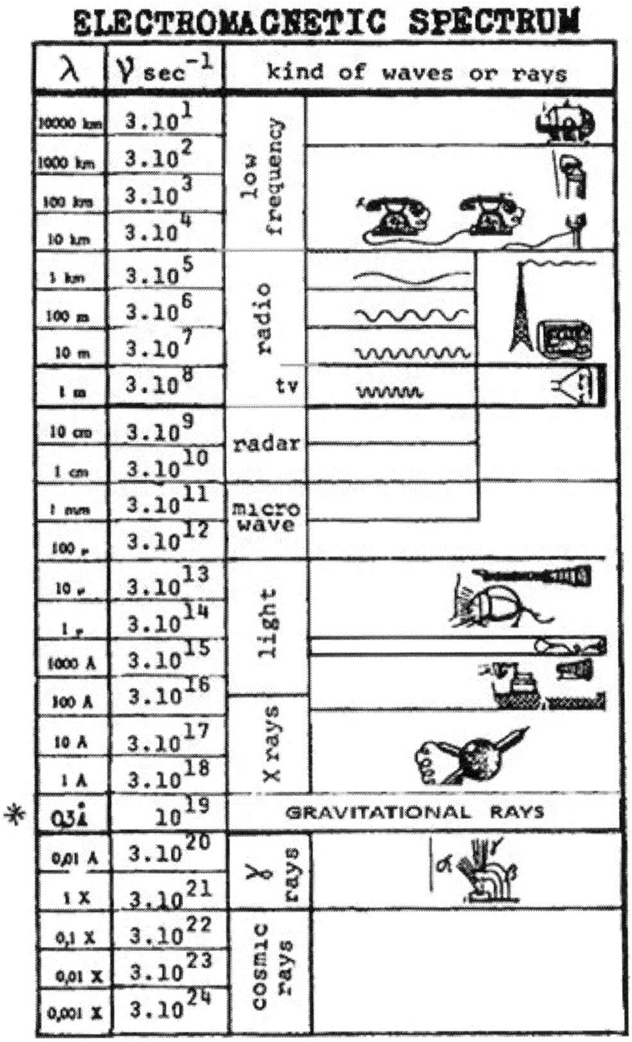

ELECTROMAGNETIC SPECTRUM

λ	ν sec^{-1}	kind of waves or rays
10000 km	3.10^{1}	low frequency
1000 km	3.10^{2}	
100 km	3.10^{3}	
10 km	3.10^{4}	
1 km	3.10^{5}	radio
100 m	3.10^{6}	
10 m	3.10^{7}	
1 m	3.10^{8}	tv
10 cm	3.10^{9}	radar
1 cm	3.10^{10}	
1 mm	3.10^{11}	micro wave
100 μ	3.10^{12}	
10 μ	3.10^{13}	light
1 μ	3.10^{14}	
1000 Å	3.10^{15}	
100 Å	3.10^{16}	
10 Å	3.10^{17}	X rays
1 Å	3.10^{18}	
* 0,1 Å	10^{19}	GRAVITATIONAL RAYS
0,01 Å	3.10^{20}	γ rays
1 X	3.10^{21}	
0,1 X	3.10^{22}	cosmic rays
0,01 X	3.10^{23}	
0,001 X	3.10^{24}	

Courtesy of S/A Phillips of Brasil—São Paulo

* The identification of these gravitational radiations introduced in the spectrum above reveals the fundamental solution of the problem of the unified Field on Physics that relates the microcosm to the macrocosm making Physics become more deductive. These radiations are responsible for the phenomena of falling bodies or universal gravitation according to a thesis presented to the Brazilian Academy of Science and to the Section of Geodesy, Astronomy, Gravimetry and Geomagnetism of the II Brazilian Congress of Cartography held at Rio de Janeiro, in July of 1965 by the Engineer Professor Dr. Paulo Ferraz de Mesquita of the University of São Paulo, São Paulo, Brasil.

Fig. 35 – The electromagnetic spectrum with the inclusion of gravitational radiation by Mesquita, according to his work presented to the University of São Paulo-USP, Brazil, as Reference.

According to the Mesquita, in the law of universal gravitation, the constant of gravitation K_G it is of the universal constants, which presents greater uncertainty of its correct value, relative to the value of $6.673 \pm 0.015 \times 10^{-8}$ [cm³/(s² g)] or: [(g cm)/s²] [cm²/g²] in which the force unit appears in the gravitational constant [(g cm)/s²] = [dina], in CGS. Being this very small constant, the sensitivity of a system is fundamental to observing the gravitational attraction between bodies.

Einstein by general relativity related the constant of universal gravitation $K_G = 6.673... \times 10^{-8}$ [cm³/(s² g)] with another constant k:

$K_G = c^2k/8\pi$, obtaining $k = 1.86... \times 10^{-27}$ [cm/g]

This constant k is in the order of magnitude, minus the numeric coefficient, of Planck's constant $h = 6.625... \times 10^{-27}$ [erg s], but without any perception of Einstein on that scale of magnitude 10^{-27} of the Planck radiation constant contained in the constant K_G.

By the theory of elementary impulse:

elementary quantum impulse = m'c [g cm/s],

where

m' = elementary mass

c = speed of light, m'c x c = h', in the vacuum measured in the Earth, that is, the speed of the m' elementary mass.

elemental quantum force = m'c/s [g cm/s²]

Then, brilliantly Mesquita introduced the impulse frequency m'c through:

force = m'c $\mathcal{V}_{m'c}$ [(g cm/s) (1/s)], where $\mathcal{V}_{m'c}$ is the frequency of m'c

And for elementary energy:

m'c x c $\mathcal{V}_{h'}$ = m'c² $\mathcal{V}_{h'}$ = h'$\mathcal{V}_{h'}$ [(g cm²/s) (1/s)], where $\mathcal{V}_{h'}$ is the frequency of h'

h' [erg] = m'c x c[erg] = h [erg s]/1[s] = $6,625... \times 10^{-27}$ [erg] = $6.625... \times 10^{-27}$ [g cm²/s²]

h = Planck's constant = $6.625... \times 10^{-27}$ [erg s] = $6.625... \times 10^{-27}$ [g cm²/s]

In Newton's gravitational law:

$F_G = m_1 m_2/r^2$ [g²/cm²] K_G

$K_G = 6.673... \times 10^{-8}$ [cm³/s²g]

To introduce Planck's constant through h' = m'c x c, would be required an ajustment of $6.673/6.625 = 1.007245$ in numeric coefficient. However, as noted by Mesquita, determinations of the constant K_G the classical methods present different values according to measuring processes, apparatus, conditions and measurement locations.

Gravitational Constant by Factors of Newton's Gravitational Force – According to the Mesquita, through calculus based on the factors of Newton's gravitational force:

$$F_G = g_\varphi = (m_a M_{Earth}/R^2_{Earth}) K_G$$, where: g_φ = acceleration of gravity according to the latitude of $\varphi = 0$ to $90°$ by the Geodesic and International Geophysics Union: $g_\varphi = 978,049 (1 + 0.0052884 sen\varphi^2 - 0.0000059 sen2\varphi^2)$ [cm/s²]

$R_{Earth\,(\varphi)}$ = Earth radius, at different latitudes, through Earth dimension patterns by the Geodesic and International Geophysics Union of the terrestrial ellipse with axes a = 637838800 cm and b = 635,669,100 cm, with Earth radius equal to the volume of the ellipsoid $R_{ellipsoid} = (a^2 b)^{1/3} = 637,122,100$ cm and volume $V_{Earth} = 4/3\pi R^3_{Earth} = 1.0833196 \times 10^{27}$ cm³; M_{Earth} = Earth mass and m_a = 1 gram, in which by Newton's law $M_{Earth} = (R^2_{Earth}\, g_\varphi / K_G)$; and for $K_G = 6.652 \pm 0.071 \times 10^{-8}$, the average of classic determinations:

$$M_{Earth} = 5.920 \times 10^{27} [g] \text{ to } 6.048 \times 10^{27} [g]$$

With these values Mesquita obtained for $R_{Earth\,(\varphi)}$ and $g_{(\varphi)}$ with φ from 0 to 90°:

$$K_{G\,(\varphi)} = (R^2_{Earth\,(\varphi)}\, g_\varphi / M_{Earth}) = 6.632 \pm 0.022 \times 10^{-8} \text{ to } 6.622 \pm 0.022 \times 10^{-8}$$

And for $R_{Earth\,(\varphi 0/90)}$ with φ from 0 to 90° and $g_{(\varphi 45)}$ with $\varphi = 45°$:

$$K_{G\,(\varphi)} = (R^2_{Earth\,(\varphi 0/90)}\, g_{(\varphi 45)} / M_{Earth}) = 6.649 \pm 0.022 \text{ to } 6.605 \pm 0.022 \times 10^{-8}$$

So it's more likely a value of $K_G = 6.627 \pm 0.022 \times 10^{-8}$, so it would have an adjustment of 1.0003, but the Planck constant $6.625... \times 10^{-27} \pm 0.00023$ presents a greater tolerance than this setting.

Newton's Gravitational Constant by Kepler's Law – Verification of the value of Newton's constant by Kepler's 3rd Law: "The time square of the period of the planets around the Sun is proportional to the cube of the distance of the half-axis of its orbits".

And by the attraction between the masses of the Sun and the planets, Kepler's third law can be related to Newton's gravitational constant K_G:

$$T^2 = a^3 / K_G (M_{Sun} + M_{planet}), \text{ where,}$$

T = time of revolution of the planet around the Sun
a = large axis of planet elliptical orbit

For the Earth-Moon system:

$$T^2 = a^3_{Earth-Moon} / K_G (M_{Sun} + M_{Earth} + M_{Moon}), \text{ where,}$$

$a_{Earth-Moon}$ = semi-axis larger of the barycenter orbit of Earth-Moon to the Sun

According to E. Rabe and A. Danjon:
$$M_{Sun} = 2 \times 10^{33} \text{ g}$$
$$M_{Earth} = 6.00763 \times 10^{27} \text{ g}$$
$$M_{Moon} = 0.08157 \times 10^{27} \text{ g}$$

Also, $K_G = n^2 a^3_{Earth-Moon} /(M_{Sun} + M_{Earth} + M_{Moon})$, where: $n = 1/T$ is the average of the angular motion of the barycenter of the Earth-Moon, and n can be replaced by the average of the angular sidereal motion of the center of the Earth, therefore to n and n':
360 ° x 60' x 60"/365.25636273 average days = 3,548.192810" average/day
n' = 3,548.192810"/206,264.806" = 0.01720212419 rad/average day
n = 0.01720212419/24 x 60 x 60 s rad/s

and $a_{Earth-Moon}$ it can also be replaced by the largest half-axis of Earth's orbit to the Sun by the Earth's equatorial radius = 637,838,800 cm and the equatorial parallax of the Sun = 8.79835" obtained by Rabe:
a'_{Earth} = 206,264.806"x 637,838,800 cm/8.79835" = 14.95322 x 10^{12} cm

With these values is obtained: $K_G = 6.62689136 \times 10^{-8}$, therefore a difference of only 0.00172×10^{-8} in relation to Planck's constant.

Newton's Gravitational Constant by the Gauss Constant – Verification of the value of Newton's constant by gravitational Gauss constant:

This constant was determined by the theory of the elliptical moton based on elements of the motio of the Earth, the mass of the Sun in relation to the mass of the Earth and the Moon, the semi-axis Earth orbit and the average solar day:
$k_{G (Gauss)} = (n^2 a'^3_{Earth})/[M_{Sun} (1 + M_{Earth-Moon-Sun})]$
n = 3,548.192810" /206,264.806" = 0.01720212419 rad/average day
$a'_{Earth} = 1$
$M_{Sun} = 1$
$M_{Earth-Moon-Sun} = (M_{Earth} + M_{Moon})/M_{Sun}$
So $K_{G (Gauss)} = n'^2 /(1 + M_{Earth-Moon-Sun}) = 0.0002959912208286$
$K_G = n^2 a'^3_{Earth}/[(M_{Sun} (1 + M_{Earth-Moon-Sun})]$
n = 0.01720212419/24 x 60 x 60 s rad/s
a'_{Earth} = 206,264.806" x 637838800 cm/8.79835" = 14.95322 x 10^{12} cm
$M_{Sun} = 2 \times 10^{33}$ g
Replacing $(1 + M_{Earth-Moon-Sun}) = n'^2 /k_{G (Gauss)}$ in:
$K_G = n^2 a'^3_{Earth} /[(M_{Sun} (1 + M_{Earth-Moon-Sun})]$,
$K_G = [n^2 a'^3_{Earth} K_{G (Gauss)}]/n'^2 M_{Sun}$,
$K_G = 6.6268921 \times 10^{-8}$

So a difference of just 0.00172 over h.

Newton's Gravitational Constant by the Danjon Equation – Verification of the value of Newton's constant by the equation of

A. Danjon of the gravitational attraction of the Earth over the Moon with the gravitational force exerted by the Sun:

$$K_G (M_{Earth} + M_{Moon}) = n''^2 \, a''^3 [1 + (1/2 \, n^2/n''^2)(1 + 3/2 \, e^2_{Moon})] = n''^2 \, a''^3 \, f_{Sun}$$

$f_{Sun} = [1 + (1/2 \, n^2/n''^2)(1 + 3/2 \, e^2_{Earth})]$
n = medium Earth sidereal motion =
$[(3,548.19281''/206,264.806'')/(24 \times 60 \times 60)'']$ rad/s
n' = medium sidereal motion of the Moon =
$[(47,434,890''/206,264.806'')/(24 \times 60 \times 60)'']$ rad/s
e_{Moon} = Moon eccentricity = 0.0549
e_{Earth} = Earth orbit eccentricity = 0.0167

So f_{Sun} = 1.002799, with the eccentricity of the Moon = 0.0549, and slope on the ecliptic = 5° 9', is obtained f'_{Sol} = 1.002723
a'' = greater semi-axis of the Moon orbit = $R_{Earth\,(\varphi)} / P_{Moon\,Parallax}$
$R_{Earth\,(\varphi)}$ = Earth equatorial radius = 637,838,800 cm
$P_{Moon\,Parallax}$ = average parallax of the Moon =
3,422.7''/206,264.80'' = 0.01659371788

And, according to E. Rabe and Danjon: M_{Earth} = 6.00763 x 10^{27} g
M_{Moon} = 0.08157 x 10^{27} g

Then, it is obtained in Newton's constant gravitation by the equation of Danjon,
K_G = 6. 62583x10^{-8}, with difference of 0.00066 less than h tolerance of 0.00023 x10^{-8}.

Constant h' within Newton's Gravitational Constant – So by several determinations is found the value of Newton's constant at 6.627 to 6.625 x 10^{-8} and therefore the constant h' [erg] = m'c x c [erg] = h [erg s]/1 [s] can be introduced into Newton's constant without any adjustment, meaning a coincidence of the values of h and K_G of the grandeur of 10^{-8}, but also that it would reveal a physical significance between the two constants by the frequency of the elementary impulse m'c x c $\mathcal{V}_{m'c}$ which is an elementary gravitational force.

Introducing h' and energy frequency $\mathcal{V}_{h'} = \mathcal{V}_G$ in Newton's constant, where $\mathcal{V}_G$ it's the gravitational frequency:

K_G [cm³/s²g] = h' [g cm²/s²] $\mathcal{V}_G$ [1/s] [cm s/g²]
K_G = 6.6... x10^{-8} [cm³/s²g] = 6.6... x 10^{-27}[gcm²/s²] x 10^{19} [1/s]
[cm s/g²]

where, 10^{19} [1/s] = $\mathcal{V}_G$ [1/s] = frequency of gravitational radiation
and, λ_G = gravitational wavelength = c/ $\mathcal{V}_G$,
λ_G = 0.2997925 x 10^{-8} cm. Figures 36 and 37.

J. P. Hansen

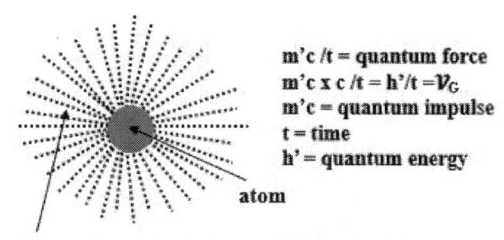

m'c /t = quantum force
m'c x c /t = h'/t = $\mathcal{V}_G$
m'c = quantum impulse
t = time
h' = quantum energy

atom

field of $\mathcal{V}_G$ = gravitational radiation frequency

The brilliant introduction of Planck's constant h in Newton's gravitational K_G constant, by the conceptualization of quantum energy frequency by Mesquita, originated from the quantum impulse theory of Erdelyi.

Newton's gravitational Law	$F_G = m_1 m_2 / r^2$ [g²/cm²] K_G
Newton's gravitational Constant	$K_G = 6.6... \times 10^{-8}$ [cm³/s²g]
quantum impulse	m'c [g cm/s]
quantum impulse energy	m'c x c [g cm² s²] = h' [erg]
quantum force	m'c / time [(g cm/s) /s]
frequency unitary	(1/s) = 1 cycle/s = $\mathcal{V}_1$ [s⁻¹]
unitary quantum force	m'c $\mathcal{V}_1$ [(g cm/s) /s]
frequency of quantum impulse	$\mathcal{V}_{m'c}$ [s⁻¹]
quantum force frequency	m'c $\mathcal{V}_{m'c}$ [(g cm/s) /s]
quantum impulse energy frequency	m'c x c $\mathcal{V}_{m'cxc}$ = h' $\mathcal{V}_{h'}$
h' and h Planck's Constant	h' [erg] = h [erg s] / 1[s]
h' = h/1s =	6.6...x 10⁻²⁷ [erg = g cm²/s²]

Introducing h' $\mathcal{V}_{h'}$ in Newton's Constant:

K_G [cm³/s²g] = h' [g cm²/s²] $\mathcal{V}_{h'}$ [1/s] [cm s/g²]
$K_G = 6.6... \times 10^{-8}$ [cm³/s²g]
$K_G = 6.6... \times 10^{-27}$ [gcm²/s²] x 10^{19} [1/s] [cm s /g²]
where, 10^{19} [1/s] = $\mathcal{V}_{h'}$ [1/s] = $\mathcal{V}_G$ [1/s] is the gravitational radiation frequency

Fig. 36 – The brilliant introduction of Planck's constant h by the concept of quantum energy frequency by Mesquita, from the quantum force of Erdelyi, in Newton's gravitational K_G constant, it shows gravitational radiation in this constant K_G.

Gravitational Radiations in the Electromagnetic Spectrum

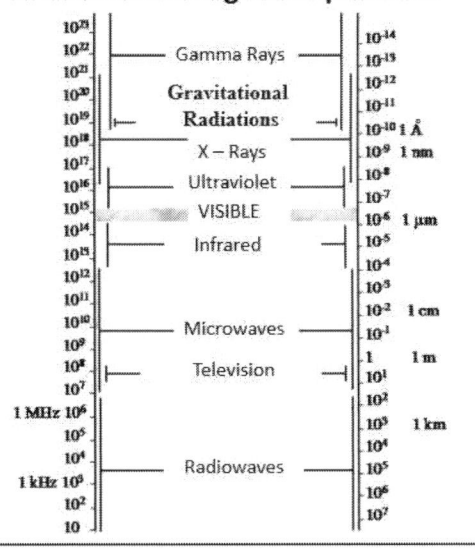

```
10²³                                    10⁻¹⁴
10²²          Gamma Rays                10⁻¹³
10²¹                                    10⁻¹²
10²⁰          Gravitational             10⁻¹¹
10¹⁹          Radiations                10⁻¹⁰  1 Å
10¹⁸                                    10⁻⁹   1 nm
10¹⁷          X – Rays                  10⁻⁸
10¹⁶          Ultraviolet               10⁻⁷
10¹⁵          VISIBLE                   10⁻⁶   1 μm
10¹⁴                                    10⁻⁵
10¹³          Infrared                  10⁻⁴
10¹²                                    10⁻³
10¹¹                                    10⁻²   1 cm
10¹⁰                                    10⁻¹
10⁹           Microwaves                1      1 m
10⁸           Television                10¹
10⁷                                     10²
1 MHz 10⁶                               10³    1 km
10⁵                                     10⁴
10⁴           Radiowaves                10⁵
1 kHz 10³                               10⁶
10²                                     10⁷
10
```

Fig. 37 – Gravitational radiation obtained by the Mesquita through the frequency of the elementary impulse (m'c)/t = m'c $\mathcal{V}_G$ where $\mathcal{V}_G$ is the gravitational frequency of 10^{19} s^{-1} within the electromagnetic spectrum in which all the radiations are also formed by m'c.

Also,
$$K_G \,[cm^3/s^2 g] = h \,[gcm]^2/s^2] \,[s] \; \mathcal{V}_G \,[1/s] \,[cm/g^2]$$
$$K_G = 6.6... \times 10^{-8} \,[cm^3/s^2 g] = 6.67... \times 10^{-27}[gcm^2/s^2] \,[s] \times 10^{19}$$
$$[1/s] \,[cm/g^2]$$
therefore: $F_G = m_1 m_2/r^2 \,[g^2/cm^2] \; 6.6... \times 10^{-27} \,[gcm^2/s^2] \,[s] \times 10^{19}$
$$[1/s] \,[cm/g^2]$$
$$n = m_1 m_2/r^2 \,[g^2/cm^2] \,[cm/g^2] = n/cm$$
$F_G = n/cm \; h\mathcal{V}_G$, which is the force between the center of masses m_1 and m_2.

Multiplying both members by cm is obtained the energy:
$\mathcal{E}_G = n \, h \, \mathcal{V}_G$, which is the potential for gravitational energy between
mass centres m_1 and m_2.

And: $P_G = n \, h' \, \mathcal{V}_G$ which is the gravitational power between the mass centres
of m_1 and m_2.

So, Newton's gravitational law follows Planck's law, em que para n = 1:
$$\mathcal{E}_G/s = h \, \mathcal{V}_G / s = h' \mathcal{V}_G \text{ , and } \mathcal{E}_G = h' \, |\mathcal{V}_G|$$

The elementary impulse shows that gravitation is caused by the impulse flows m'c x c = m'c^2 = h' on the frequency of 10^{19} hertz in the atomic nucleus, and a field formed of nh' involves all the particles, in which the electromagnetic radiation as the light in the vicinity of the bodies formed by the particles are curved by a field of h' involving that body through the particles of that body.

By the theory of elementary impulse, the fundamental constants of nature, the m' mass, and radiation with speed c relative the emitting source or c'relative to the vacuum-space-ether, is that they form the fundamental configurations of the particles and their fields, with their helix structures on the left or right, by the elementary energy h' = m'c x c. In these settings, the fields of the particles follow lines around the base of the particles, and other h' as they approach these fields enter the curved configuration of the lines of the fields. And it would be the elementary force m'c/t that would act gravitationally, without distorting the space that would cause gravitational acceleration.

Gravitational energy would be a potential energy manifestation of a radiant frequency of 10^{19} s^{-1}, in which the elemental power = 6.6... x 10^{-27} [erg] x 10^{19} [s^{-1}] = 6.6... x 10^{-8} [erg s^{-1}]. And the elementary power = 6.6... x10^{-8} [erg s^{-1}] x 10^{-7} = 6.6... x 10^{-15} [watt] shows the quantum significance of the gravitational constant.

Gravitational energy would transfer quantum energy flows h' = m'c x c from one particle to another into the bodies by ether, formed of submass, subquantum, all penetrating. So, m'c x c as radiation would be radiant free energy, different from the term found in all current literature which is that of "pure energy"! The radiant free energy consisting of m'c x c would form all elementary particles when retained in a state of tension in this ether, forming a high density field of nh' energy.

A force field exists when any variation of energy occurs, that is, of extras or decreases of h'. When in steady state the particles are in potential state.

Between 1557 and 1585, according to Piccolomini, Scalinger and Benedetti the consecutive variations of the momentum, impulses, communicating by the impressed force were the cause of the fall of the bodies. And in 1600, Gilbert conceived gravity as a magnetic emanation. And also Kepler, in the 1600, attributed to the Sun a magnetic force responsible for the attraction of the planets, while other bodies were also attracted by some magnetic effect. As Newton in his questions of "Optics", *"The bodies do not act on each other by the actions of gravity, magnetism and electricity? And the magnetic, electric and gravity attraction between the bodies, could not be by impulse or other means that is still unknown, because nature is very attuned to a common standard?"*. So, in 1862, Weber formulated that these almost imponderable particles were electrified and attracted electrically to the central part of the atom.

By the elementary quantum impulse would be impulses of m'c/t, with the energy m'c x c = h' and frequency of 10^{19} s^{-1} of low intensity that would cause the gravitational pull and these pulses of m'c/t that would form the particles and the electromagnetic field responsible for the attractions and repulsions.

The elementary impulse shows that gravitation would be caused by the pulse flows m'c x c at the frequency of 10^{19} hertz.

Between 1834 and 1845 Faraday sought to relate the gravitational motion as equivalent to the electric current generating electromagnetic field. But, the electric current produces electric field and magnetic field flows, according to electrodynamics, and that by the elementary impulse would be flows of m'c x c = m'c^2 = h'.

In 1920 according to Q. Majorana the matter itself would be the source of gravitational flux, a quantity of gravitational energy that would continually emanate from the ponderous matter.

And for decades, research teams are looking for gravitational waves through general relativity in the frequencies between 10^{12} to 10^{-18} hertz, that is, in the part of the spectrum between short and long waves, of the orbital oscillations of massive bodies that would form the gravitational wave of vacuum-space-space-time.

But, by the elementary impulse, gravity would occur by the energy flow of frequency $\mathcal{V}_G = 10^{19}$ cycles/sec with energy $\mathcal{E}_G = 10^{19}$ (m'c x c) = 10^{19} h' with low flow density and or intensity, of the atomic nucleus that forms the gravitational field. Therefore, in the gravitational motion, for example, of a body in free fall, flows of continuous m'c x c of the force field m'c/t of the gravitational mass, which would result in the acceleration of the body. And in massive bodies in collisions, nuclear transformations could occur which emitting gravitational radiations of gamma rays of 10^{19} hertz.

PART 7 – THE QUANTUM IMPULSE AND THE RETURN OF MEDIEVAL FORCES

XLI. NATIONAL AND GLOBAL DISSEMINATION OF QUANTUM IMPULSE THEORY – 1950 to 1971

Acting Press – In January 1966, one of the largest circulation newspapers in Brazil, "O Globo", presented an interview with Erdelyi of Mackenzie University and author of quantum impulse theory, reporting that this theory put in check the theory of relativity of Einstein. Erdelyi revealed that the theory of impulse was elaborated as a consequence of obtaining knowledge with detailed physical significance of the structure of matter, while the relativistic theory was based on an experiment that considered light as wave and that did not present the expected interference of waves of light paths. The theory of impulses explained the light as a monophoton (photon of one cycle by second), with a different behavior of a wave, explaining this experiment and other interpretations of relativity, but with physical significance by quantum impulse. And mentioning that the development of the theory would result in broad knowledge of the atomic structure of the atom, because the electron of the nuclear orbit had its structure explained by the theory of quantum impulse in which monophotons entered into its constitution, that is, formed of light. So with the known atomic structure would be determined depth fission and fusion processes, by the controlled transformation of the energy. But in view of the resistance from world physics to the new ideas, the effort would be to establish an international body for the development of science by dissidents of pre-establishead Physics. He considered that current Physics by continuing to admit Planck's constant h of quantum radiation without a physical significance, because energy x time was distant from significance, and insisting on following Einstein's relativity, would only lead to observing experiments on particle accelerators. But, the particle profusions formed in the collisions of these accelerators, without a real physical theory, would prevent the knowledge of what exists in the reality of the particle structure.

Cause and Effect, Without Mass and Mass Creation – What should occur in particle accelerators was that the higher the energy produced in the particle shock, the higher concentrations of energy generated more other types of particles. However, what would be obtained would be knowledge of cause and effect, without the depth of the actual constitution of the formed particles. Thus, one goal would be to obtain particles from the Higgs field and transform massless field into mass field to give mass to the particles, according to work presented in 1964 by Higgs Englert and Brout; Guralnik, Hagen and Kibble. But it would continue the lack of physical significance to the field without mass and what mass was this,
created by the particle or Higgs field that would constitute the particles with the basis of Pre-Establishead Physics.

The Brazilian Interplanetary Society-SIB, The Astronautics Journal and the International Astronautical Federation-IAF – In 1953 was founded the Brazilian Interplanetary Society-SIB, and in 1955 Flávio A. Pereira – Graduated in Science from São Paulo University, Brazil, and President of the Brazilian Institute of Aerospace and Space Sciences-IBACE – organized and was elected the first President of the Scientific Council of the SIB. Next to SIB joined the International Astronautical Federation-IAF founded in 1950 and in 1971 was with 350 Institutes in 48 Countries. Figure 38.

Fig. 38 – Werner Von Braun and L. Sedov at the IAF Annual Meeting, interconnected to the SIB, in New York in the United States, in 1968. Authorized photo of IAF Publication IAF, 15 June 1971, Paris-France.

In 1961, the SIB was consulted for the elaboration of spatial activities of the Brazil. And its with the Research National Council-CNPq presented to the Government of the Brazil the proposition of the organization Group of the National Committee of Spatial Activities-GOCNAE, created then in that same year of 1961, which subsequently in 1971 was transformed into the Space Research National Institute-INPE.

In 1962 the SIB was presided over by Octavio Magdalena, Director of the Electronic Computing Center of the Brazilian Bank-Bradesco, which directed the Observatory and Research Center Bradesco, which despite being a banking institution maintained a Center of Culture and Science. And the Scientific Council in 1962 was chaired by Erdelyi, Teacher since 1960 at Engineering School of Mackenzie University.

In November 1967, the initial edition of "Astronáutica" (Astronautics Journal) was presented as an organ of the SIB and Bradesco Observatory and Research Center, with the sponsorship of Bradesco.

Open Letter to Scientists – In July of 1966 on 18th. Meeting of the Brazilian Society for the Science Progress-SBPC, a Statement was submitted by IBACE, through the Review Committee of Physics Knowledge consisting of Mesquita-USP, Erdelyi-Mackenzie University and Flávio Pereira-SIB-IBACE. This Statement it presented that the theory of relativity and quantum mechanics resulted in divergences the that prevented the development of the knowledge of the internal structure of the matter and the physical nature of the phenomena in relation to the mathematical structures. Thus, new concepts, such as the monophoton theory of light, the structure of the electron formed from elementary impulse, the increase of mass of the electron generating the increase of the speed, contrary to the speed increase that would generate mass by theory of relativity, the discovery of the gravitational radiations dreamed of by Einstein, showed that new ideas resulted in the development of the deep knowledge of nature. And quoting Einstein's own words uttered in 1950: *"What we need are new principles. Further extensions of the ancient principles only increase the dilemma".* And that in 1965, at the Meeting of the Nobel Prize Winners in Cambridge in England, it was reported that Dirac had informed that the current quantum theory was a cause of the stagnation of atomic Physics and would present within two years a new theory. Also, that the quantum explanation for the theory of the impulse led to physical interpretation to the deflection angle of the light rays of the stars passing on the edges of the solar disk; to the slow rotating motion of the plane of the orbit of planets around the Sun and the redshift of spectral rays of light produced on the surface of the Sun, compared with the corresponding spectral rays produced on the Earth's surface.

And Mesquita at the closing session of 18th. Annual Meeting of the SBPC of 1966 mentioned that in the Symposium on Perspectives of Physical Research in Brazil of this Meeting, the Professor Engineer and Physicist M. D. Souza Santos was the rapporteur of a work presented by Doctor Professor P. S. Toledo of the Institute of Atomic Energy in which this scientist manifested that there was still need for basic conceptualization in Physics. And Mesquita added: *"Are new basics that should be transmitted to new generations of researchers, to replace preconceptions still existing".*

Erdelyi Works Abroad – Between 1952 to 1966 Erdelyi's works were sent to institutions in the Brazil and Abroad to dozens of Universities, Academies, Physics International Scientific Journal and Renowned Physicists, for publication. That Physics Journals sent out reports that they were publishing articles of research and revisions of existing theories, but not of original theories. One of these Journals was the Review of Modern Physics by the American Physical Society. And communications from the Universities were archival in them Libraries of these publications received, and without other manifestations except for a few scientists in the United States and Europe who realized the importance of a quantum theory that modified the current abstract and inconsistent Physics of Einstein's theory of relativity and quantum mechanics itself. It was found that new ideas could only be presented as long as the current ones were kept! In 1965, the IBACE had formed a Committee to Review the Postulates of Physics, in view of reports that researchers from the United States and Europe presented disagreements with the prevailing theory of Einstein's theory of relativity. And in 1966 he sought to initiate an officialization by IBACE for the submission of the works of Erdelyi to the Committee of Award of the Nobel of Physics with the purpose of a more disclosure of quantum impulse theory.

Document Without Authors! – And the IBACE sent shipping request to Abroad of the quantum impulse theory of Erdelyi to the Ministry of Foreign Affairs of Brazil in which this Ministry requested information to the Latin American Center of Physics which it submitted to the Department of Physics of the University São Paulo-USP. And this Department issued in July 1966 a document that revealed the lowest qualification of followers of dogmas and sects to maintain the obscurantism and the predetermined by the leadership domes. It was a document without a signature responsible for its contents, which was referred to the USP letterhead to Latin American Center for Physics and Ministry of Foreign Affairs of Brazil! Also, this document USP letterhead it was published in the newspaper "Folha de São Paulo", with the news: *"USP does not approve Professor Erdelyi"*, without any request after, of the rejection of this news by the Department of Physics of USP, characterizing agreement or omission with that opinion by the direction of this Department. The IBACE by its President, Flávio Pereira, considered the development of the impulse theory as:*"Of possible reflexes in the history of Science and avant-garde technology, because it is of interest to both the history of Science and of imperatives linked directly to national security"*. And that a Statement should be made to be answered to the Ministry of Foreign Affairs on the unauthorized opinion of the nonsense committed in that document without signature or name of anyone, without any responsibility for this nefarious document that mentioned:

"1-The Department of Physics of USP had carried out research in 1966 of a large number of Journals of Physics, of various nationalities, including "Physics Abstracts" that published summaries of works of a thousand Physics Journals, without locating any of Erdelyi. Only rudimentary print writings were located and some publications by IBACE.

2-The theory of the impulse was the elementary impulse of energy propagated in a medium such as ether and which would be subphotons contrary to the photons of the current theories. But, like the Michelson-Morley experiment, to verify whether the ether existed or not, it was null would result in the exclusion of the ether. So the theory of Erdelyi should also be excluded by presenting the ether. Also, as there was no experimental evidence that these subphotons existed then the Erdelyi theory did not present any proof of its validity.

3 – This theory of Erdelyi led to the placement of gravitational waves in the spectrum of electromagnetic radiation between the X-rays and gamma rays, but these radiations had never been detected experimentally. It also predicted that the constant of universal gravitation presented the numerical value of Planck's constant 6.62517 ± 0.00023 which was incorrect because the gravitational constant was 6.670 ± 0.005.

4-The theory presents the highlight of coincidences of values between physical constants, only meaning "numerology".

In the end concluding that: *"The work inspired by Professor Erdelyi is obscured by a pseudoscientific language, being evident in them the lack of coherent scientific methodology, which indicate still certain confusion regarding certain elementary knowledge of Physics modern".*

Manifestation of Erdelyi – In August 1966 Erdelyi replied by SIB to IBACE about that document without signature letterhead from the Department of Physics of USP: *"An formal opinion of a Superior Department of Physics should be of content scientific without any idea pre-established, and compatible to the level that a Physics Department of a University should present. The ether of the Michelson-Morley experiment was admitted as an ether of minimal concentration matter that penetrated the entire substance. Since the result was null, then Einstein ruled out the ether. However, this ether of minimal materiality would not exist, but there could be another type of ether that is presented by the theory of quantum impulse, that is, an ether with sub-mass, sub-quantum, but without matter, because matter involves energy that encompasses the mass. For example, the light has mass, but it's not matter. And the rapporteur forgets the Compton effect which is taught in high school, where the photon, as mentioned by the draftsman of that letterhead from the Department of Physics at USP, it was admitted as unique and indivisible, by colliding with an electron it provides kinetic energy to the electron that emits a photon of lesser energy. So, this photon doesn't subdivided into smaller photons, that is, sub-photons? Therefore, a proof of its existence and more than evidence! What exists is just a sub-photon with different frequencies".*

Open Letter of Mesquita – In August 1966 Mesquita also replied in the SIB on this letterhead of the Department of Physics USP, for sending to the Ministry of Foreign Affairs: *"It is a Pseudo-Document, for it is omitted any responsible for that opinion. And omits that the publications of the works of Erdelyi were published by the Engineering School of the Mackenzie University, Brazilian*

Interplanetary Society-SIB and by the author himself. The "numerology" mentioned shows that the rapporteur is unaware of statistical foundations of the theory of accidental errors, such as Gauss's law and criteria of its application, since it is quoted only a value of the gravitational constant, and out of which the methodology would lack coherent scientific alleged. In this way, it unaware the foundations of Geodesic and Astronomy interpreting the mathematical structure of the laws of Kepler and Newton with numerology, because the gravitational constant presents different values according to the physical analysis of gravitation. It is unaware of Gauss's law of gravitation, with its gravitational constant, recommended by the International Astronomical Union in 1938, with 15 significant digits replacing the constant of Newton's universal gravitation. and the numerical concordance, between the constant gravitational and Planck quantum, it is due to the theory of quantum impulse of the sub-photon or monophoton, which can be verified experimentally, but that the rapporteur unknown what is experimental. The rapporteur ceases to mention that the Department of Physics of USP received invitation to the presentation of the work "The Meaning of the Gravitational Constant with that of Planck", at the end of 1965 in the USP itself by the Faculty of Architecture and Urbanism, but none of the Physicists of the Department of Physics of USP showed up, and come to present opinion without due knowledge. The fact of not having found in a thousand journals the work of Erdelyi does not invalidate it, because they do not apply criticism or approval to publish, which occur in all published works. Omissions do not justify reproaching any of this work. And a beginner's argument from this rapporteur, in which the Michelson-Morley experiment ruled out the ether of Physics, when the Nobel Prize for Physics Paul Dirac stated that it is necessary to introduce a new image of ether that is in accordance with quantum theory. It is what the theory of quantum impulse presents with a new image of the ether. On the other hand, while the hidden author of this false information is unaware of Erdelyi's publications, Mackenzie University's Engineering School reissued the same in 1965".

Continuity of Posture or Dogmatic Imposture! – This opinion of the Department of Physics of USP could only be attributed to the ruling dome of the same, because it was the only one that could immediately, when being published in renowned newspapers, refute this opinion as unfounded by the Department of Physics of USP, to these newspapers, the Latin American Center that received this opinion and the IBACE and the SIB. But the omission revealed the agreement of that opinion. And the aggravating is that with the purpose of preserving interests and position in power of science educators, seeking to show the defense of the bases of Physics as an argument to defend the absolute truth in which they were supported, no longer observed that the Nobel Prize in Physics of its foundations, Dirac, just presented the need for other bases in Physics for quantum mechanics! Nothing if altered since the dawn of science, in which dogmas were followed by arguments of crumbling foundations, for this dome wanting show that represented the *"Doctors of Law and Science"* and interpreters of the utmost scientific erudition, in the right to opine what should be followed or disposed of, beyond the false summit or true farce to the issue it without a name and surname! Almost 20 centuries after Christ had faced the *"Doctors of Law and Science"* with dogmas for their interests and power, the same posture or dogmatic imposture was maintained in this new era!

"Zeal" by the Scientific Method – The leading summit of science presented that it only fulfilled the craft of caring for the scientific method, of hypothesis-proof and quality, so that theories and scientific papers could be approved. In the impossibility of direct detection of quantum impulse m'c, and the sub-photon m'c x c = h' used the argument of theories and propositions as pseudo-scientifics equal to perpetual motorcycle and quackery. However, theories and propositions that triggered new experiments and brought innovative knowledge could prove these theories and propositions formulated. Also, the photons at rest without mass of radiation, admitted by the basis of formal Physics, are only indirectly obtained by the interpretation of the theory of relativity of the Lorentz expression in which the mass of the photon at speed c: $m_{photon\ (c)} = m_{photon\ (v)}/([1 - (v^2/c^2)^{1/2}]$, where $m_{photon\ (v)}$ it would the mass of the photon at speed v. And to v = c, the denominator would be zero, which would result $m_{photon\ (v)} = m_{photon\ (c)} \times 0 = 0$. That is, the pre-establishead Physics did not weigh the photon to know their mass but it admitted as a zero mass by relativistic calculus. However, by quantum impulse, this expression of the mass increase is limited by the value of the m mass of the elementary photon with speed c, i.e. m'c x c, as deducted by quantum impulse theory. So it is not about zeal for the scientific method, but of zeal for the cover-up of scientific development! The fact that pre-establishead Physics didn't "weigh a photon at rest on a scale" didn't mean he had zero mass! Simply, it had not been measured its mass! And on the basis of current Physics, classical mechanics is considered to be limited, as it did not reach the theory of relativity and quantum mechanics. However, Bohr one of the creators of quantum mechanics presented the principle of correspondence using classical mechanics, which was considered limited by pre-establishead Physics, with quantum mechanics, while general relativity was dissociated from the quantum mechanics and special relativity presenting divergences of interpretations and inconsistencies. Several experiments in quantum mechanics showed results such that an atom could be in two places at one time, by the superposition of the high wave function squared, of mathematics, which is nonetheless the type of arrangement of the experiment and the interpretation of the result that led to that conclusion. So how could it be considered as proving the current quantum theory linked to classical theory that was limited by formal Physics and the relativistic theory dissociated from quantum and classical, as absolute truth proved?

But, the authors of this document from Physics-USP by mentioning that: *"As there was no experimental evidence that these subphotons existed, then the Erdelyi theory did not present any proof of its validity"*, they deny that an electromagnetic radiation of a frequency of one cycle does not could exist, which is precisely the subphoton, when nothing prevents in the quantum mechanics that could exist that frequency! That is, they deny the own Physics that follow!

In reality, the current quantum theory wants to be unfastened and now anchored in classical theory, and relativistic escapes from classical without even getting close to quantum. So what foundations these are that seek to bind themselves by fragile twigs like that of the principle of correspondence of pre-establishead Physics? Are these foundations that should be considered, for the zeal of the current academic, and be proclaimed as pseudo-science any theory that would be contrary to these bases in flicker, and which stagnates any scientific advancement? But, a fact that showed the universality of following dogmas, and more than stagnation, was published in the United States in the Science Journal in March of 1963 by M. K. Hubbert, who questioned presenting doubts about the development of Science in United States, whose article title summarized his thought: *"Are we Retrogressing in Science?"*.

Referral to the Swedish Academy of Sciences – And the IBACE forwarded in December 1966 the work of Erdelyi to the Swedish Academy of Sciences for Nobel Prize analysis of Physics, as a duty to inform a larger official institution to have knowledge of what was prevented by blocking the disclosure of new ideas.

Military Technology and National Security – On May 4, 1967, the newspaper *"O Globo"* had reported that Flávio Pereira would forward a memorandum in Open Letter to the President of Brazil denouncing the serious situation of national physical science, because the theory of quantum impulse of Erdelyi presented important reflexes in military technology and national security, it is their duty to send this memorandum to the head of state, what happened next.

Pronunciation to the Brazil's Leader- On May 11, 1967 the IBACE chaired by Flávio Pereira sent a memorandum in Open Letter to the Government of Brazil, Marshal Arthur Costa Silva, who was one of the leaders who delivered a military coup between May 31 to April 1, 1964 tipping the President J. Goulart which had increased the number of vacancies in all the universities and colleges of the country to broaden the aspirations of study and development of the new generations. But, with allegations that Goulart demonstrated some socialist-communist position and induced the hierarchical division in the armed forces, intervention was carried out in the name of the defence of freedom of expression, hierarchical order, education and scientific research in Country. Flávio Pereira mentioned that: *"In view of the pronunciation of your excellency and of your Government in which it is a priority for education and scientific research in the Brazil, I emphasize that in fact it is necessary to review the current intercourse in the Science in Brazil. The public universities of Brazil, which guide the application of appropriations for the law for the scientific promotion, come to release them only to those who follow the doctrines imposed by those who control their beliefs, away from new discoveries and theories, when the builders of the scientific bases present the need to revise these bases. The positioning adopted by these groups that control the Science of the Brazil is the one that Voltaire presented in which only the charlatans affirm the certainty. Thus invalidating scientific analysis. In Brazil, Engineer Teacher Istvan Erdelyi of Mackenzie University presented transcendent discoveries that are linked to the atomic nucleus and space conquest. And I come to conclude that when Lieutenant Charles De Gaulle in France had been ostracized when he defended new combat technologies as armored with great mobility to contain invaders, the invasion took place with extreme success in France! Of your Lordship depends on whether you want the ostracism of a great scientist of the Brazil".*

But, the military government was in reality with concern in the political and economic area, and the educational and scientific system were items more distant to receive an attention its. However, Officers of the Armed Forces realized the importance of Erdelyi's statements to the press in the interview of the newspaper *"O Globo"* in January of 1966 making an invitation to obtain knowledge of the theory of quantum impulse. Thus, Naval Officers such as Admiral Primo Nunes de Andrade, Superior Professor of the Navy; Emmanuel Gama de Almeida, Officer of Higher Studies of Sciences; and other Senior Officers they understood that the current bases of Physics should be reviewed.

Solid Foundations of Ruins! – And shortly after June 2, 1967, proponents of the millennial orthodoxy of the bases that were to be untouchables of Physics were manifested by the press, through

Professor Hugo Kraus, Director of the Department of Theoretical Physics of the Institute of Physics of the University Federal of Paraná-Brazil, stating that: *"Has no basis, nor foundation the affirmation of Flávio Pereira, President of IBACE, in "denunciation" to the head of the nation, according to which is serious the situation of national physical science, because Brazilian scientists refuse to discuss the theory of the impulse which puts the theory of Einstein's relativity on the ground"*.

This immediate announcement by the Director of the Department of Theoretical Physics of UFPR, the first University created in Brazil, showed that they did not consider serious that the current bases of Physics were maintained, when Nobel Prize Physicists like Yukawa and Dirac they were at the time criticizing the current bases of Physics. Previously on 1948, in his book *"The Nature of the Physical World"* of Eddington, which claimed: *"There is a radical error with the fundamental concepts of Physics and is not being seen as eliminating them"*.

Also, theories like the quantum impulse that went against Einstein's relativity would not matter that it was not analyzed by the pre-establishead Physics, because relativity should be correct and it would be a waste of time that was analyzed!

In the previous decade, in the United States the Government and American scientists had presented a fundamental appreciation to scientists who came from Abroad, like the Hungarian physicist Edward Teller who was the main articulator of the manufacture of the first nuclear fusion bomb, the hydrogen bomb, while in Brazil was this discredit against the Engineer Istvan Edelyi which came as Teller of Hungary to Abrod, to be received that way by the scientific summit of the Brazil.

XLII. THE MACKENZIE UNIVERSITY AND THE QUANTUM IMPULSE THEORY – 1942 to 1967

Scientific Conference – To officiate the support and importance of the theory of quantum impulse as a recognition for the development of science and technology, the Engineering School of the Mackenzie University in São Paulo-Brazil with the Physics, Philosophy, Sciences and Letters Faculty of Mackenzie University decided to promote a Scientific Conference of Erdelyi, Teacher of Thermal Machines at the Engineering School of Mackenzie University. And in 1966 as an engineering student at Mackenzie University, I came to participate as coordinator of this Conference that was held on November 16, 1966 at Mackenzie University. Erdelyi presented the millennial question of the constitution of light, such as wave, particle, and particle wave, to the light as the impulse of elementary mass with its determined speed c, the quantum impulse; the structure of the atomic particles of the electrons and positrons by these impulses, which initiated the last phase of the elaboration of the atomic age, that is, to obtain the remaining structures of the nuclear particles to control the disintegration of matter; and the proposition of experiment that Mackenzie University could promote so that it could determine the displacement of the Earth in space, which was impossible by the theory of relativity. With relativity there would be contraction of space itself shortening measuring rulers, and the speed of light would always be the same in any direction independent of any velocity of the emitting source of light to any direction. The auditorium was crowded showing a great appreciation to the lecturer, in which Erdelyi also presented a proposition of an experiment on the quantum impulse theory for that was effected by Mackenzie University, which by coincidence, in which a former Professor at Mackenzie University, Jaroslaw Voinoff, had presented in article in the "*Mackenzie Engineering Magazine*" in November of 1942, an equipment to effect experiment with some similarity to the proposed by Erdelyi.

Guided Experiment – And after the Conference handed down by Erdelyi in November of 1966, I asked the Academic Center of the University of Engineering of the Mackenzie University for referral to the University for the experiment with laser, called the "Space Compass" by Erdelyi, to determine the possible absolute velocity of Earth in space in different directions. Attended by the President of the Academic Center, Walter Malouf, who referred this request to the Office 304/66-67 to the Rectory of Mackenzie University. And the Journal of the Academic Center in April 16, 1967 presented the editorial by the Director of the Newspaper, Cristiano Kok, in which he mentioned that: "*Professor Erdelyi's theory was of a clear and irrefutable logic, in which this experiment replaced the Michelson-Morley experiment that was inappropriate to get the measurement of a displacement of the Earth in space. But instead of support, Professor Erdelyi had received fierce opposition from a group of USP teachers against his theory*". In fact, the logical reasoning of the theory of impulses presented an understanding with physical significance of nature, against the mathematical abstraction contained in the current bases of quantum Physics and relativistic. And the opposition to the theory of Erdelyi by a group of USP was devoid of scientific arguments against the logic and advancement of the theory of Erdelyi.

By relativity the absolute space or a medium as an ether was nonexistent, for in all the references the light would always have the same speed in any direction and it would be space and time, that would be contracted or dilated so that the c speed was always constant. In this case, any measured displacement would be incorrect, because the Earth could be moving in undetermined directions by space-time, and any measurement would be just a random measure and could vary in indeterminate periods.

However, a displacement that could be determined by the ratio between the velocity c of the light and the velocity V of the Earth in different directions, with the displacement X the Earth to present in space or ether and the displacement d of light when exiting perpendicularly from a fixed laser on Earth to a graph target, could record this displacement X, so the displacement of the Earth in a certain direction in space. That is, a simple proportionality $V/c = X/d$. Because, in this experiment, the laser would be pointed perpendicularly to different directions of space in reference to a certain initial coordinate of space and stars considered fixed. Also, by relativity the speed of light is independent of the velocity of the emitting source of light, such as the laser on Earth, is the emitting source at rest or in motion, so that the light always has the same emitting speed, measured by the observer in any reference system. While the Michelson-Morley experiment related light paths with wave interference, this experiment proposed by Erdelyi aimed at a direct measurement of the photons of the laser light to a target moving along with Earth in space.

In principle, as the speed of light is always c in relation to vacuum-space to any direction, so much for by relativity, wave theory and by quantum impulse, upon exiting the laser the light should come out at the perpendicular and reach the target that moved with the Earth in space and thus get an X shift in some direction. But, also point a result in relation to that of Michelson-Morley which had been null or different than expected.

Scientific Experiment with Authorization of the Pope! – On April 18, 1967, the Director of the Engineering School of the Mackenzie University, Professor Engineer José Justino Castilho submitted a Letter to the Academic Center on the application of Professor Erdelyi's theory, signed by the Physicists Teachers, Nelson Martins and Ronald U. Pauli, of Mackenzie University, who issued the following objections:

"1. There is the historical fact that new theories present an immediate reaction against ingrained concepts. And substitutions of theories may occur, but only after discussions and deep studies.
2. Teacher Erdelyi's theory has already been the subject of an opinion by the Physics Department of University São Paulo-USP, which manifested itself against the said theory.
3. We do not agree with his theory and it is not normal from a scientific standpoint the mode adopted by Teacher Erdelyi to present his theory.
4. Before replacing one theory with another it is necessary first of all that the theory to be replaced is wrong or that it is not experimentally verified its conclusions. Which is not the case, because the theory of relativity has been at all its points proven by

experience, including in the earliest countries including Einstein's theory in the curriculums of the middle schools.
5. As for quantum mechanics it is personal opinion that is the best the human mind has achieved until the present day for explanation of atomic and nuclear phenomena.
6. It is a universally adopted norm that a physical theory be submitted to publication in scientific journals and sent to renowned scientists to collect opinions. Otherwise, if it is published in particular, it can hardly be taken into account of a scientific work.
7. Before arrangements for conducting this experiment, only if some magazine accepts its work is because it has scientific value, because it will be judged by experts of renowned capacity.
8. And if this theory is published, highly skilled laboratories will make the experience with much more property than we would do here at our university, because without a doubt, they will have more than us, staff and apparatus. Examples of laboratories such as Berkeley-USA, CERN-Switzerland, etc.
9. We are not opposed to the realization of this experiment, but we just do not agree with the way in which we intend to accomplish it".

By these objections it would seem necessary also a papal authorisation, from the Medieval Inquisition to approve this experiment, because it had just been judged as of high heresy!

New Disagreement of Another Nobel to the Bases of Physics – A few days after the opinion of these two physicists, against the Erdelyi experiment, in which they defended *"... the best that existed in the human mind as to the theories of current Physics..."*, on April 27, 1967 Yukawa, Nobel Prize in Physics, announced that it would present in August 1967 a new theory on elementary particles, with disagreements to Einstein's theory of relativity and quantum mechanics which did not present dimensions to several elementary particles, which were the foundations of modern physics, and mentioning that: *"These fundamentals of Physics must be reviewed".*

Erdelyi's Response to the Two Physicists Teachers – On May 22, 1967, Erdelyi sent Letter to the President of the Academic Center of the University of Engineering Mackenzie, on the opinion of the two physics teachers at Mackenzie University, which is:
"1. On the opinion of the Department of Physics of the University of São Paulo-USP, or these opinions are supported in a document with a signed name (s) and in this case mentioning the name (s) of the person who issued it, or repealing that statement without foundation, for it is of an anonymous document without any signature of who issued that opinion.

2. When I disagree with my theory, anyone can do so, but it should present arguments that justify that fact.

3. A few days ago Hideky Yukawa, a Nobel Prize in Physics, which predicted the existence of the mesons, revealed the need for physical directions contrary to relativity that is considered by the issuers of this letter as in all its proven points. Or Yukawa should be taught by the two Teachers that relativity is proven and he withdraw what he said? In the equation of the electron mass increase also deduced by relativity $m_e = m_0 /[1 - (v^2/c^2)]$, where m_e is the mass of the electron in motion, m_0 the mass of the electron at rest, v the speed of the electron and c the speed of light, Einstein read the equation as that mass increases by increasing speed, concluding the relativity of the mass. But only by miracle a new increase in speed could be obtained. But, it is the force that increases speed through the impulse of time. Another serious mistake of relativity is that it considers energy = mass x c^2 which is the Hasenöhrl equation after 20 years of experiments and that Einstein has generalized for all kinds of energy. So by relativity, Energy = Mass x $1/[1 - (v^2/c^2)]$, where $1/[1 - (v^2/c^2)]$ is number without dimension and Energy = N x Mass. But as Joseph Fourier in 1822 consolidated Physics as science by showing that equal amounts should have the same units. Would the two Teachers agree that their students would write equations in which the units were different on both sides of the equation? The Michelson-Morley experiment was carried out based on the wave theory of light. But with quantum theory, in the transverse direction to the direction in which the Earth moves, light does not follow the trajectory prescribed by the wave theory. The light would come out perpendicular, while to adjust the interference in the Michelson-Morley experiment the mirrors should be moved from their perpendicular position to hit the interference. But, the two Physicists could perform this experiment without the Michelson-Morley interference model which is not appropriate to prove relativity correct. As professional Physicists would be more able to do a better analysis. And this experiment if negative would not prove that relativity is correct, because the examples cited already show the insufficiency of the theory of relativity. The other countries in following the theory of relativity do not mean that everyone should also follow it.

4. Paul Dirac, Nobel Prize in Physics, also disagreed with current quantum theory as he recently mentioned, but for the two Physicists Teachers of the current quantum theory is the best in Physics and as a personal opinion they asserted.

5. How to expect Journals from Abroad to publish theories whose specialists are indoctrinated in the present dogma. How to expect opinions from scientists mentioned whether they are developing their own work or Dirac who received work from the theory of impulse and never responded to any receipt?

6. The proposition that the experiment should be carried out should be Abroad, the two Teachers profess the insufficiency of their science or submission to what other Countries lead and teach what suits them and to be a copier of what others present".

It should be observed that as quoted by Erdelyi on the interference of the waves of light paths, which in the article of Michelson-Morley they refer to the perpendicular path of light as being necessary to be corrected by the micrometric screws, due to aberration of light, that is, the deviation of the light that originates from this motion of the Earth in the ether, and that should come out at the perpendicular but there was the adjustment in the mirror. However, the light as quantum coming out perpendicularly would reach a target moving along with the Earth and show the displacement in the ether or space, without the need for interference from waves, therefore without need to adjusting different mirrorsS of 90 ° to the interference initial null.

Real Power Source – This response from Erdelyi to the two physicists was presented to the SIB in September 1967 and added that: *"It continues an era of scientific intolerance and that scientists did not have to ask for foreign countries to perform simple experiments to verify their theories. Science is not a noble toy of some intellectuals, but the only real source of power".*

Stagnation in Physics – With the contrary opinion of the two physicists against the Engineer Teacher Erdelyi for the realization of the experiment would come to magnify more still the Mackenzie University and the Physics of all the subjects taught at the University, in fact it was an opinion contrary to show grandeur, imposition, self-proclamation of *"Doctors of Science"*, of law and of absolute truth, dominion, power and submission to the predetermined of their lords of the pre-establishead Physics.

For it was only an experiment, but whose result would allow a more thorough appreciation of what was being presented, for appreciation after the scientific community. However, the owners of knowledge demanded approval from the outside, so that an experiment of minimal cost and simple equipment could be effected, and even if it was carried out should be by countries of the outside. The indication of the CERN for an experiment of this simplicity showed the maximum point of devaluation to the Teacher Engineer Erdelyi of the School of Engineering of Mackenzie University and his students who supported this experiment, claiming high cost of the experiment and implicitly indicating the inability of engineers, physicists and technicians of the country to carry out a project of reasonable execution. Or they did not even know how to conduct an analysis of costs and ignorance to make a project achievable, and thus signing their incompetence, or else an inappropriate prank by quoting the name of CERN. The experiment would require a laser that was only produced in the United States, but at no cost would be unfeasible to use it.

This opinion of grandeur silenced the students of Physics of the Faculty of Philosophy, Sciences and Letters of Mackenzie University where the mastery of these two physicists Teachers was active, but the Engineering Mackenzie continued to support the Engineer Erdelyi, although these two physicists also were Teachers of Physics of the Engineering School of the Mackenzie University and also the author of this book where I was studying at the School of Engineering of this University.

But, the students of Mackenzie Engineering were not intimidated by this opinion of the two physicists who stated in the missive that "... *Only if some Journal from Abroad accepts its work is because it has scientific value, because it would be judged by experts of renowned capacity...* ", Thus signing total submission and inability to effect a scientific opinion within their professional area of Physics. They should annul their physical qualifications by confessing to being incapable of a technical report in their area of specialization in Physics. Or, what kind was of professionals, and who taught Physics in an Engineering School?

And the fact that there was no outward opinion would not mean that the work would be considered without validity.

But it was clear that the arguments presented by these two physicists far from any zeal for the scientific method only showed that these arguments were far from a higher level. And was noticed a broad absence of any attitude to support who showed study and dedication to the cause of Science. Thus, the University could present an experiment, which independently of the result obtained would have a repercussion throughout the Abroad by the theories involved in this experiment. But the two physicists preferred to propose the level of attitudes to stay in stagnation, to prevent any type of action that presented a value of knowledge, improvement and greater realization. In summary, the argument of this opinion against the realization of this experiment by the two physicists at Mackenzie University, requested by the engineering students at Mackenzie University, was steeped in stagnation, antiscience and behavior of submission, devoid of any greatness.

Thus, I presented in the newspaper of the Academic Center of June 15, 1967 a simple scientific demonstration showing that physicists of traditional Institute in the United States, Drexel Institute of Tecnology, as stated in the book *"Physics of the Atom"* of 1959 and published by the University of São Paulo-USP, reported accentuated deadlocks in Physics in which *"Physics is in a chaotic sieve of the structure of matter. We need to interpret the data that not only solves the enigma, but that elevates our vision to levels that so far we cannot predict".* And informing that Erdelyi's theory could unravel what these American physicists were looking for, but that physicists from the USP summit and two from Mackenzie, who indicated and published books that showed disagreements in Physics, denied the new theory as that of Erdeliy, because it should be pseudo-scientific for contradicting the theories fixed in mind them!

Experiment Postponed – And on August 28, 1967, the Lady Rector of the Mackenzie University, Dr. Esther Figueiredo Ferraz, presented to the Academic Center of the School of Engineering Mackenzie its manifestation on the application of the Erdelyi theory, emphasizing: *"This academic directory may be aware that this rectory has the utmost interest and that it will give all cooperation within its reach, in order that the studies of Teacher Erdelyi can be successfully crowned",* and presented that the opinion of the two physicists of the engineering school itself recommended that there be approval of Journals, specialists and renowned scientists from Abroad, because only then the two physicists reported that it would have scientific value the theory of Erdelyi, besides the cost they claimed would be elevated to the University. And the Rectory presented that before the realization of this experiment it would be possible to have more dissemination Abroad to receive appreciations of this theory.

Despite Erdelyi's previous response that specialists in foreign Physics Journals followed their dogmas by refusing opposing publications, and renowned scientists concerned about developing their own theories and not having their time for other theories, there seemed to be insight by the Rectory of Mackenzie University that should occur some opinion of Journals or renowned scientists Abroad on the work of Erdelyi for next so authorize the realization of this experiment. Thus, it was postponed the realization of this experiment, by an opinion of two physicists from the University itself, against the own scientific method followed by the Physics, which is the experiment to obtain the knowledge of scientific fact.

XLIII. NEUTRON AND QUANTUM IMPULSE – <u>1933 to 1967</u>

As Fermi in 1933 the neutron decay through a virtual particle action that presented a spin similar to the spin photon that acted in the electron-photon interaction.

And the neutrons when free from the atomic nucleus, presented a half-life, that is, half of them were transformed into proton, electron and neutrino, in the time of 10 minutes, and half of the remaining neutron would fall in another 10 minutes, and so on. So after 1 hour there would be around 1.5% of neutrons.

So, just like the aces and 'doughless glues', that is, quarks and gluons entered into the strong interaction of the nucleus, also other particles entered into the interaction for the neutron decay outside the nucleus.

The neutron that composed the nucleus with the proton, in the free state of the atom, presented a decay in the particles of the proton, electron, and neutrino.

And the neutron by the standard model formed from 3 fractional charges, u, d, d in the proton duu, electron and antineutrino.

Experimentally the lifetime of the free neutron was 887.4 ≈ 14.7 minutes that corresponded around 72% of most of the neutron volume decayed in the experiment, in which half or half-life was ≈ 614.9 seconds. However, if it were considered only a neutron for the decay in which all its mass was reformed into proton, electron and antineutrino, it could be obtained a theoretical time through the disintegration of the atomic nucleus that originated, according to the Mesquita, the flow of quantum energy $m'c \times c = m'c^2 = h'$ which constitutes the gravitational radiation with frequency of 10^{19} s^{-1}.

In fact, in July of 1967 it was presented by Mesquita on the 19th. Annual Meeting at SBPC, the Communication *"The Duration of the Free Neutron Life Calculated Rationally Constitutes New Crucial Test of the Specific Existence of Gravitational Radiations Identified in the Spectrum of Electromagnetism'*. Document 9.

SB CIÊNCIA E
PC CULTURA
SOCIEDADE BRASILEIRA PARA O PROGRESSO DA CIÊNCIA

ABSTRACT

The duration of the free neutron lifetime rationally computed is a new crucial "test" of the specific and experimental existence of "gravitational radiations" identified in the electromagnetic spectrum.

PAULO FERRAZ DE MESQUITA *

In a communication already presented: to the Brazilian Academy of Sciences and, simultaneously, to the Section of Geodesy, Astronomy, Gravimetry and Geomagnetism of the II Brazilian Congress of Cartography held at Rio de Janeiro, July 1965; in the "Colloquim for the Revision of some of the Fundamental Concepts of Physics" held under the auspices of the "Brazilian Institute of Astronautics and Spatial Sciences" held in

1965, December, 17th, at the Faculty of Architecture and Town Planning of the University of São Paulo; and at the XVIII Yearly Meeting of the Brazilian Society for the Progress of Science held in July, 1966 at Blumenau, State of Santa Catarina, Brasil; we have revealed the specific existence of "gravitational radiations", we entitled also "G rays", identified in the Electromagnetic Spectrum between the so called "X rays" and the "gama rays", with the wave length

$$\lambda = \frac{c}{\nu} = (0.2997925 \pm 0.0000004) \text{Å (angström)}$$

and the frequency $\nu = 10$ quintillions per second, being c the velocity of light (or of the electromagnetic waves) expressed in centimeters per second. These radiations are responsible by the phenomena of universal gravitation. We have now to communicate that we discovered since the first semester of 1965 that the duration of the free neutron lifetime, experimentally determined, confirms his rational duration we have deduced and computed from the classic equation of Poisson $\Delta\Phi = 4\pi K\varrho$ which characterizes the spherical flux of the universal gravitation formed around a point center of ponderal mass, being K the Newtonian Constant and ϱ the density of the ponderal matter. From that equation we deduced the rational expression

and the value $\Delta t = \dfrac{m_n c^2}{4\pi h'\nu} \approx (1807.9 \pm 0.3)$

seconds of time which represents the integral duration and the respective tolerance of the free neutron lifetime, that is, the interval of time, that the neutron set free from the atomic nucleus takes to vanish, reducing to zero his energy so irradiated, being $m_n = (1.67470 \pm 0.00004)10^{-24}$g its rest mass $h' = h.\sec^{-1} = (6.62517 \pm 0.00023)10^{-27}$ erg from the Planck's constant, $c = (2.997925 \pm 0.000004)\,10^{10}$ cm.sec^{-1} the velocity of light and $\nu \doteq 10^{19}$sec^{-1} the frequency of the "gravitational radiations". It is convenient to remark that $h'\nu = K$ represents, numerically, the Newtonian Constant of the universal gravitation in the C.G.S. system of physical unities. The "mean life" of the set free neutron will be, for that, the interval of time $\Delta t_m = \Delta t$

$(1 - \dfrac{1}{e}) \approx (1142.8 \pm 0.3)$ seconds for re-

ducing its energy to $\dfrac{1}{e} \approx \dfrac{1}{2.71828}$ of its

original value. Experimentally, the work entitled "Fundamental constants of Physics" by E. Richard Cohen and others — Intersciences Publishers — New York — London — 1957, reproduces two values of the free neutron "mean life" in the pages 91 and 102, accompanied by its respective standard error (root-mean square error) indicating the origin and authors: 1) (1040 ± 130) sec — P. E. Spivac, A. N. Sosnovsky, A. Y. Prokofiev and V. S. Sokolov, "International Conference for the Peaceful Use of Atomic Energy", Vol. II, July 1955 (United Nations, New York, 1956), p. 33; 2) (1120 ± 320) sec — J. M. Robson's earlier value [Phys. Rev. 83, 349 (1951)] agrees with the measurement of Spivac et al. As we see the value of the "mean life" we have deduced coincides perfectly with the experimental results within the precision that the late were obtained by the authors cited above. By another side, we have deduced also the period of semi-desintegration (half-life) of the set free neutron, that is, the interval of the uniform time $\Delta t_{0.5} = \Delta t(1 - 0.5) \approx (903.9 \pm 0.3)$ sec ≈ 15 minutes and 4 seconds, the free neutron takes to have its energy reduced to 0,5 (an half) of the original value, which (half-life) coincides very nearly with the value "15 minutes" indicated, without pointing out mean error or tolerance, by Linus Pauling to the period of spontaneous semi-desintegration of the free neutron in the page 650 of his work translated to the Spanish language under the title "Química General — Tercera edición: Diciembre, 1955 — Aguilar — Madrid". Without comments any more, these crucial tests reiterate the proofs of specific existence of GRAVITATIONAL RADIATIONS or "G Rays" we identified in the electromagnetic spectrum, confirm the meaning of the Newtonian Constant of the universal gravitation we have discovered and corroborate the solution of the problem of the unified field on fundamental Physics we have found and already exposed before.

* Civil Engineer and Geographer of the Polytechnic School and of the Department of Applied Sciences of the Faculty of Architecture and Town Planning of the University of São Paulo, Brasil.

Doc. 9 – Communication presented by Mesquita at SBPC in 1967 on the duration of the atom free neutron, as if all its mass was disintegrated by the frequency of gravitational radiation, which results in the fantastic probability of being equal to the experimental time! As Reference.

So, as a Mesquita:

Applying the concept of energy = force x distance, in Newton's law of gravitation, for a mass M and another unitary in a unitary distance, which represented a potential of the gravitational field, was obtained: $E_G = M \times 1/1^2 \times 1 \times K_G$, where K_G = Newton's gravitational constant.

And in the Poisson equation φ [erg]/[cm^2] = $4 \pi K_G \rho$ [erg]/[cm^2], which represented the spherical flux of gravitational energy of Newton around a matter density point, which originated the gravitational field, obtained from the ρ mass density of unit volume to a unitary mass over a spherical radius surface unit, then:

φ [erg]/[cm^2] = 4π 1 [g cm^2] K_G [cm^3/g s^2] ρ [g/cm^3] 1/cm^2 = $4 \pi K_G \rho$ [g cm^2/s^2]/cm^2,

that is, Φ [erg] = $4 \pi K_G \rho$ [erg]

By the rates of energy flow and density in time:

$d\Phi dt = 4\pi K_G (d\rho/dt)$, where:

$d\Phi/dt$ was the energy flow $d(mc^2/dt) = dm/dt$

being, m = mass and c = speed of light in vacuum measured in the Earth

$\int (d\varphi/dt)\, dt = \int_0^t d(mc^2/dt) dt = m_{neutron} c^2$, where $m_{neutron}$ = neutron mass

$K_G (d\rho/dt)$ represented the energy flow of m'c x c = h' with the gravitational frequency of the neutron $\mathcal{V}_{neutron} = \mathcal{V}_n = 10^{19}$ s^{-1}

$\int_0^t 4\pi K_G (d\rho/dt) = \int_0^t 4\pi h' \mathcal{V}_n\, dt = 4\pi h' \mathcal{V}_n \Delta t$

$M_{neutron} c^2 = 4\pi h' \mathcal{V}_n \Delta t$

$\Delta t = m_{neutron} C^2 /4 h' \pi \mathcal{V}_n \approx 1807.9$ s.

Fantastic Probability! – For gravitational radiation, which came out of the neutron for its complete dissolution, it would be necessary for the frequency to be 10^{19} cycles/second, which would mean an almost zero probability of being obtained. That is, the frequency would have to be just one for a sextillion of frequencies of the electromagnetic spectrum so that the dissolution time was 1807.9 seconds. And it was exactly that unique frequency of gravitational radiation that dissolve the neutron in 1807.9 seconds! It couldn't be just an unbelievable coincidence, but the reality of the frequency of gravitational radiation discovered by the Mesquita by the frequency of quantum impulse.

There are radioactive atoms in which their isotopes are transformed into fractions of a second like actin, radiation from the polonium, with 5 x 10^{-3}s and the thorium of mass 232, with billions of years, 1.39 x 10^{10} years. And particles that are free from atoms have a duration of only one millionth of a second to disintegrate into other particles. Thus, in this period one could not have an accuracy of the probability rate of the disintegration of these particles. However, out of the nucleus the free neutron is what presents the greatest time of disintegration, and so can be measured its rate of atomic decomposition, for example, in a trillion neutrons.

So, Mesquita replaced the atoms with gravitational radiation impulses m'c with the frequency of 10^{19} s^{-1}, and obtained the time around 15 minutes, equal to the experimental measurements. However, other measurements with the free neutron in the magnetic field presented half life in 10.1 minutes and other measurements in 12, 5 minutes. But under the conditions that "if"all neutron mass, fully insulated, is continuously transformed by frequency 10^{19} s^{-1}, the time would be around 15 minutes. But between 12.5 to 15 minutes of the half-life time, it continued within a fantastic probability the frequency for this disintegration by gravitational radiation, for 10^{18} s^{-1} the time would be 75 minutes and for 10^{20} s^{-1} the time would be 7, 5 minutes. That is, the frequency 10^{19} s^{-1} for decomposition if it were of the entire neutron mass, it is largely between 10.1 to 15 minutes, which is an fantastic probability that it is not a simple coincidence, but the reality of the gravitational radiations of 10^{19} s^{-1}. Figure 39.

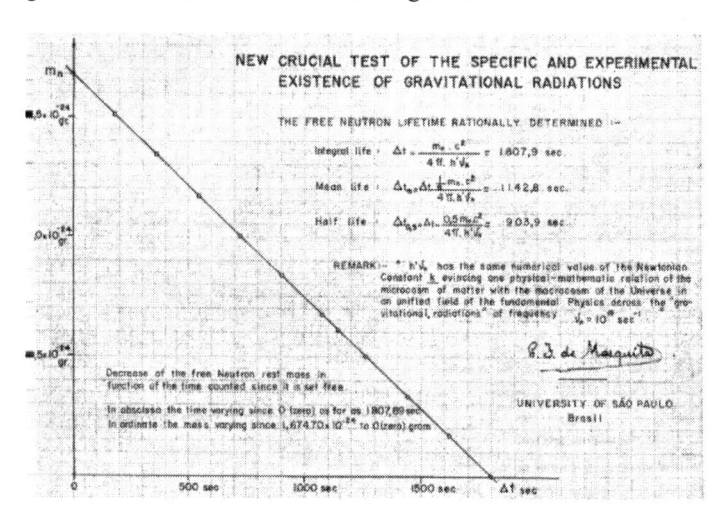

Fig. 39 – Decay of the neutron mass if it were totally by emission of gravitational radiation as Mesquita in his article presented to SBPC in 1967, as Reference.

So, the life of the neutron that would be decomposed into gravitational radiation indicated a crucial proof of the existence of these radiations deduced by Mesquita.

XLIV. UNKNOWN SOLUTIONS – 1968

On August 12, 1968 was presented in the Polytechnic Forum of the University São Paulo-USP, by Mesquita presided by Professor Dr. Wagner Vaneck Martins of the Polytechnic of USP, which was made the graphical impression of presentations on Basic Physics Research already done in the Polytechnic Forum, for its proper dissemination and consultation of researchers, without they could present allegations of ignorance of what was published and catalogued . For, despite the presentation of works in Congresses and Annual Meetings of SBPC, physicists still ignored or preferred to omit these findings, as an example, in the publication of the book "*General and Experimental Physics*", from the Director of the Physics Department of the São Paulo University-USP, Physics Professor Dr. José Goldemberg, cited by Mesquita in which Goldemberg reported that: *"The last reason for attraction universal law why do the bodies fall? – It is not answered by the current Physics: it is possible that we may discover this reason yet or there is another simpler law, from which the law above and others may be deducted. There are in nature three types of different forces that cannot be refuted in each other: gravitational, electromagnetic and nuclear, and it is impossible in the current state of Physics to say why these forces exist and we are content to describe the motions depending on them. It is possible that they are not completely independent, but the link between them has not been established satisfactorily yet. This is a very important point for the development of Physics: if we discover why a body has an electrical charge, a new avenue for progress would open. Sciama says that what we call inertia is the result of the gravitational action of the stars on us, and Einstein found out by elaborating principles of equivalence that gravity is a property of the physical space in which the matter lies and the gravitational pull is due to curvature of space".*

However, as Mesquita, in the previous works presented as "*The Meaning of Newton's Constant Universal Gravitation*" and the "*Lifetime of the Neutron*", based on the theory of quantum impulse, there were these questions sought by the pre-establishead Physics, replacing dogmas still in existing and indefinite concepts.

On August 21, 1968 in that same Polytechnic Forum chaired by Professor Dr. Giorgio E. O. Giacaglia of the Polytechnic and the Department of Astronomy of USP, Mesquita also of Politecnic Engineering School presented Memorandum on Didactics in the Teaching of Physics. And showed that failures in the base of Physics remained because questions were prevented, and due to the continuation in being repeated everything that was always taught without any modification.

And Mesquita exemplified that: *"Einstein in his work, "The Meaning of Relativity", writes that energy $E = m/[1 - (v/c)^{1/2}]$ where m = massa and $q_{vc} = v/c$, being v the speed of a body and c the speed of light, and declares that mass and energy are equal. But due to the lack of dimensional homogeneity, it becomes a counter-sense, still in the case of v = c, the energy E would be infinite what has no physical significance, soon didactically* *inappropriate, which leads the scholar or student of Physics to serious doubts. However, physicists declare that Einstein provoked a true conceptual revolution by creating a new philosophy of Physics, through new concepts of space, time and mass. And to express it they use advanced mathematics and specialized and abstract forms, which discourages him and goes on to disbelieve his own intelligence and his power of creativity, thwarting his delusions of researcher. His dismay increases when they force him to believe that the mass of a photon of light at rest is zero. And then, the hope of its creator power falls to zero and begins to decorate the mathematics of the teachers of the pre-establishead Physics, and sometimes also decorators, not to be had as little intelligent or not to be disproved, or begin to philosophize with abstract entities without the enough physical and experimental control. Teach the current Physics teachers that the mass $m = m_0/[1 - (v/c)^{1/2}]$, in which m_0 it is the mass at rest, it is mass of relativity and science fiction seriously speaks that thus an accelerated electron with close speed of light can turn into a galaxy, mistaking cause in effect. They speak in relativistic time and general relativity is admitted without observing that as Einstein was a provisional theory. The teachers of the formal Physics become so more realistic than the king, thwarting the creative power of the scholar or student for new research".*

XLV. NEW EXPERIMENTS – 1968

Laser Illuminating the Moon – On September 30, 1968, Géza L. Vadja, Ph. D. Mechanical and Eletricity Engineer, Ph. D. by the Sorbonne, Paris-France, Ph. D. in Physics by Kaiser Willelm Institute-Berlin-Germany, President, Chief Physicist of Advance Science Trends Associates Research Laboratory, California-USA, sent Letter to Erdelyi, published by the Astronautical Magazine-SIB in December 1968, mentioning that it was preferring a laser experiment to observe the displacement of the Earth in space according to the velocity indicator in different directions of Erdelyi, or "space compass", through a laser loan at the Jet Propulsion Laboratory in July 1968 in which they were studying the travel programs for the Moon of Surveyor 8 and Apollo 3. However, it had to postpone the realization of this experiment because it was more than three weeks in critica physical condition. On this experiment a greater attention was so that the flexion of the equipment didn't come to interfere in the accuracy of the result. He reported that the stain displacement caused by a laser beam, which he observed for the first time was when they illuminated the surface of the Moon, in which the illuminated area, that is, the light spot, had focused on another point other than that which was determined and aimed at by the telescope. The difference in the location achieved by the laser was considered as a result of technical imperfections.

Gravitational Control – In October 1968, the Astronautical Magazine-SIB presented a work of the Doctor in Mechanical, Engineering from the University of Rome and Professor of the Faculty of Industrial Engineering of the Catholic University in Sao Paulo-FEI, Ernesto Emanuele Enrico Geiger on experiments with magnetic, electric and gravitational field, which was patented, as it opened possibilities for possible gravitational field control.

XLVI. TRIBUTE TO ISTVAN ERDELYI – IMMORTAL CREATOR OF QUANTUM IMPULSE THEORY – 1969

Code of Physical Nature – On February 14, 1969, Erdelyi died, and on February 15, 1969 Flávio Pereira, President of the IBACE, extolling the theory of Erdelyi, came to declare: *"From these to Science the keys of the secrets which predecessors had not been able to reveal. You gave us the code of Physical Nature. Those who admired him, God granted to live with an authentic giant of Science. Istvan Erdeleyi will shine in the firmament of the History of Science as he shines on the face of the Earth some early civilization. And him name will always be alive in the grateful memory of the posters. It is not necessary to envy those that centuries ago were contemporaries of a Tales, Copernicus, Newton! Yes, we also had a big one like them, Erdelyi!*

XLVII. NEW PROPOSITIONS

Directions of Science – In May 1969, the Astronautical Journal-SIB presented a work of Professor Salvatore Salvo from the University of Industrial Engineering and the Mauá School of Engineering on modern Science and nature, in which he questioned the course of Science in the light of reality.

Gravitational Field with Lorentz Force – In June 1969, the Astronautical Journal-SIB published article of the Physicist Zsigmond Bocker involving electrons, magnetic field and gravitation, which could be correlated as interpretation by the Lorentz force between electric field and magnetic field.

XLVIII. INTERNATIONAL SYMPOSIUM ON ASTRONOMY DIRECTED BY ASTRONOMICAL MYOPICS! – 1969

Between 4 and 12 September 1969, an International Symposium on Astronomy was held in São Paulo-Brazil, organised by the Astronomical and Geophysical Institute of USP represented by Professor Dr. Jay de Moraes; With the participation of the Institute of Mathematical Research-USP represented by Professor Dr Giorgio Giacaglia; Polytechnic School-USP by Professor Dr. J. Augusto Martins; Technological Institute of Aeronautics-ITA by Professor Dr. Silvio Ferraz Mello and the National Observatory of Rio de Janeiro with Dr. Luiz Muniz Barreto.

However, although Mesquita, Professor Doctor in Sciences also Astronomical Teacher, besides Geodesic and Topography in the Polytechnic School-USP which was one of the Symposium Organizations, the Coordinator of this Symposium, Giorgio Giacaglia prevented Mesquita from presenting a work that involved the quantum impulse theory, although the inscription of the work has been carried out, such as the other 48 scientific works of the outside and other 5 of Brazil, with 1 of them of Giorgio Giacaglia himself who was the Coordinator of the Symposium, 2 of the Department of Mathematics and 2 presented by Silvio Ferraz Mello. It also prevented the personal distribution by Mesquita of this work to the congressmen. Participated 70 congressmen being 25 from Brazil.

Thus, the Coordinator Giorgio Giacaglia used a Public Institution, USP, one of the main Universities in the Brazil, for the exclusive promotion of its group, prevented that the outside could have information about the disclosure of work involving the quantum impulse theory in the Astronomy area. Once again, actions were applied to maintain the predetermined dominance against any new knowledge for the scientific bases for formal presentation.

XLIX. ADVANCEMENT IN TEACHING AND IN THE NUCLEAR AREA
1970

Advanced Researchers – Between April and November 1970 Mesquita presented a training Course for Advanced Researchers in a Unified Field of Basic Engineering Sciences in the form of free university extension for engineers, teachers, scientists and researchers or disclosers of science, making formal the quantum impulse theory by Geographical Engineering, Physics or Dynamic Geodesics and Astronomy of position within the Department of Transport Engineering of the Polytechnic School-USP.

J. P. Hansen

Planck's Constant in the Newton's Gravitational Constant in the SBPC – In July 1970 Mesquita presented at the 22nd. Annual Meeting of the SBPC a Scientific Communication *"Newton's Law Satisfies Planck's Postulate"*, published in the SBPC Yearbook, showing that the constant h Planck's was within the constant of the attraction universal gravitation, extracted from the previous work of Mesquita on gravitational radiations. Document 10.

NEWTONIAN LAW OBEYS PLANCK'S POSTULATE (ABSTRACT)

PAULO FERRAZ DE MESQUITA

(Full Professor of the University of São Paulo)SP-Brasil

(1) $F(d) = \dfrac{m_1(gr) m_2(gr)}{r^2 (cm^2)} \; 6,6\ldots \times 10^{-8} (\dfrac{cm^3}{sec^2 gr})$, Newtonian Law, CGS

(2) $1 cm = \dfrac{10^{19}}{10^{19}} \dfrac{sec}{sec} cm$

Multiplying each member of these two expressions (1) and (2), simplifying and taking F(d.cm) as E (erg) we have

(3) $E(erg) = \dfrac{m_1 m_2}{r^2} \; 6,6\ldots \times 10^{-27} (gr \dfrac{cm^2}{sec^2} sec) \, 10^{19} (sec^{-1})$, where

(4) $\dfrac{m_1 m_2}{r^2} = n$ is a pure number, a natural number (n=1,2,3,....)

(5) $6,6\ldots \times 10^{-27} (gr \dfrac{cm^2}{sec^2} sec) = h$ (erg.sec), Planck's constant

(6) $10^{19} (sec^{-1}) = \nu_g$, is the expression of an elementary gravitational radiant frequency. Consequently,

(7) $E = n.h. \nu_g$

This expression, so deduced, shows that the potential gravitational radiant energy between two ponderal masses m_1 and m_2 obeys, rigorously, the Planck's Postulate, what, "data venia", was not perceived by Einstein, by Planck himself and contemporaneous. Other works of the author confirm this conclusion.

Doc. 10 – Newton's Law and Planck's Postulate by Mesquita, SBPC Yearbook, 1970, as Reference.

Laser Radiation in Nuclear Fusion – While Erdelyi had presented emphasis for studies to be deepened to the dominance of the atomic particle structure to control the disintegration of matter and obtaining the initial structures of the electron and positron by the theory of quantum impulse, in 1970 in the Abroad was for the first time in the world performed experiment by electromagnetic radiation with laser to obtain the nuclear fusion, according to article presented by Fernand Lot, published in the Astronautics Journal-SIB, September 1970. Thus, nuclear energy was largely linked to the energy of electromagnetic radiation, the basis of the quantum impulse theory, in which the particles that constituted the atomic nucleus were formed from these radiations and applied in the first nuclear experiment with these radiation.

L. FORMAL REPORT OF CURRENT PHYSICS ON UNIDENTIFIED AERIAL OR FLYING OBJECTS – 1969 to 1971

Formal Report of the Physicist Condon to the United States Air Force – Between 11 and 12 September 1971 the Dr. Gauthier of Keating-Hart in Solid State Physics and Magnetism, of the Center for Nuclear Studies of Grenoble-France and the Institute of Atomic Energy of São Paulo presented, in a Brazilian Congress on Unidentified Aerial or Flying Objects-UFOs, Organized by the IBACE and the Center for Advanced Studies for the 21st century, a Communication on the Scientific Report published in 1969 directed by the American Physicist E. U. Condon of the University of California, which had been one of the main builders of quantum mechanics, indicated by the United States Air Force for the formal opinion of the American Government on Aerial Objects or Unidentified Flyers-OANI or UFO. In this communication, Keating-Hart reported that Condon's formal report concluded that UFOs had no scientific interest. and Keating-Hart reported that: *"The conclusion of this report contradicts a large part of what is presented as: physico-chemical effects, on supermotors, television, radio, aircraft oscillations, heat, burns, instantaneous accelerations, curves at almost straight angles with high velocity, light-beam deviation, lighting tubes looking solid, absorption or reflection of radar waves. And the study of UFOs can be of great scientific importance, by the fact that they seem to counteract our current scientific knowledge. Current Physics considers light as particle-wave and without mass in rest. And with the theory of relativity there are magnificent mathematical formulations, but we are looking for unitary equations that could bind time, matter, energy, gravitation and electromagnetism. Refuting these facts as unscientific is actually a pretext to show ignorance. These phenomena that apparently go against the laws of inertia and gravitation accompanied by electromagnetism could lead to the solution of the great modern problem of gravitation".*

It was evident by this communication from Keating-Hart of the reason for pre-establishead Physics to treat unknown phenomena as pseudocientíficas, as well as to consider the authors who observed and published reports or articles of these phenomena as paranoid unconscious! It is exactly the position to maintain their academic and power positions in their institutions, Not allowing any modification of the physical bases even in collapsing, ignoring and omitting new ideas that bring knowledge and that could lead to the explanation of these unknown phenomena.

The Engineer Alberto N. Martino presented an article on propulsion that would involve UFOs.

And researcher Adhemar Eugenio de Mello presented an article on hypothesis of control of gravitation.

LI. THE QUASARS AND GRAVITATIONAL RADIATIONS OF QUANTUM IMPULSE – **1963 to 1971**

In December 1971 the Astronautics Magazine-SIB published a scientific Letter from the American scientist Bohumil Vanco, Ph. D., on the gravitational radiations discovered by Mesquita, presenting that: *"Admitting the existence of their discovery of the frequency of gravitational radiations I checked whether they would be agreeable to the observation, which is ultimately the only means of attesting any theory. And it has been proven that her postulates correspond fully to nature, and even more so, the author's pretensions were overtaken by a very large margin, arriving in the present work to the notion of being the postulated radiation the basic element of the entire existence of the world physical".*

The work of Vanco *"The New Face of the Physical World"* dated April 1971 revealed a correlation between the gravitational radiations obtained by the Mesquita of quantum impulse with quasars, which were objects detected since 1963 by M. Schmidt as a radio source and in the optical telescope as a bright spot that presented a sharp detour to the red, looking like a nearby star, but with these variables of galaxies distant from sharp brightness and radio source, almost star galaxies, quasars. And that "if " quasars were at long distances as admitted by the pre-establishead Physics, the gravitational radiation it would emit would be by the Doppler effect received as light, and the light emitted by quasars would come by this Doppler effect as radio waves of quasars, In which quasars display exactly the noise of radio waves and the intense brightness of the light.

LII. FORMAL CONFESSION OF FORMAL PHYSICS – **1972**

The Scientia, Scientific Journal, of May-June 1972 presented article of the Physicist José Leite Lopes, which mentioned: *"We do not yet have a quantum theory of gravitation that is satisfactory. We don't even know if a new type of treatment other than that of Newton-Maxwell-Einstein-quantum mechanics would be required to well understand the subnuclear phenomenology. Perhaps the notions of space and time, although necessary as the scenario of our knowledge of the physical, macroscopic, atomic and nuclear* *world, no longer apply to areas of extension less than 10^{-13} cm or high energies. What other ideas will then replace space and time, forms Kantianas a "priori" of pure intuition?"*. But, Erdelyi's theory of quantum impulse showed other concepts that replaced the current Physics, but their physicists preferred the omission of quantum impulse and gravitational radiation.

LIII. ASTRONOMY OBSERVATORY AND THE QUANTUM IMPULSE THEORY – <u>1962 to 1972</u>

The banking institution, Brazilian Bank-Bradesco, maintained a Center of Culture and Science, with an Astronomy Observatory directed by Octavio Magdalena, Director of the Electronic Computing center of Bradesco, and in 1962 was the President of the SIB. And Bradesco came in 1967 to sponsor the Astronautics Journal-SIB as an organ of the SIB and the Observatory and Research Center Bradesco. The Astronautics Journal presented news and scientific articles presented by scientists from the Brazil and Abroad.

Missile to Astronautics Journal-SIB – However, on 16 June 1972 the Director of the Institute of Physics of the University of São Paulo-USP, the Physicist Professor Dr. José Goldemberg sent two-page Letter to the President of the Bradesco Foundation, Mr. Amador Aguiar, also Chairman of the Executive Board of the Bradesco, in which Goldemberg pronounced: *"Praising Bradesco's initiative in publishing a Journal such as Astronautics for scientific topics. But, what articles were published without having been approved by an editorial committee that was not in the publication, and were articles that departed from bizarre premises. I regret that such an important initiative would be misrepresented by such publications"*. And suggested to Mr. Amador Aguiar that*: "If you were advising on the organs of the University of Sao Paulo, through its institutes or scientific societies such as the Brazilian Society of Physics for a prior and constructive analysis of the material to be published, offering immediately the collaboration of the Institute of Physics"*, as an Open Letter issued by Mesquita that had been sent to the Editor Dr. José Reis of the *"Revista Ciência e Cultura"* of SBPC, (Science & Culture Journal).

New Missile To Astronautics Journal – And only 3 days later, on 19 June 1972, the Physicist Professor Dr. Sylvio Ferraz Mello, of the Aeronautical Technological Institute, sent a Letter of a page addressed to the Bradesco Foundation. Came to praise articles published in the sector of Newsletter aof Astronomers such as Nelson Travnik and José Manoel Luiz da Silva, and journalistic information of a certain interest. However, he considered *"appalling"* the reading of Astronautics Journal-SIB, for they were published articles of *"Diatribes anti-scientific and some form of schizophrenia, but that such facts were everyday and would not deserve greater attention. However, such facts received the endorsement of an institution of the highest reputation in the Brazil as the Brazilian Bank-Bradesco and regretted the slip-by-lure that the Bradesco Foundation should have been taken, which is a disservice to Brazilian culture. But surely the Bradesco Foundation will do everything to act with the seriousness that characterizes the name that it bears and will know how to correct that little fault. And it would be better to know the Bradesco Foundation and would suggest a contact with CNPq and the Foundation of Support to Research of the State of São Paulo-FAPESP."*, as was listed in an Open Letter of Mesquita to the Editor Dr. José Reis of Science & Culture Journal of SBPC.

Dual Warhead Missile, to Astronautics Journal-SIB and Mackenzie University! – But also 3 days after this Letter from S. Ferraz Mello, on June 22, 1972, the Physicist Pierre Kaufmann who was Director of the Center of Radio Astronomy and Astrophysics of the Mackenzie Institute sent Letter of also a page to the chairman of the Executive Board of the Bradesco Organization, Mr. Amateur Aguiar and with a copy to the Deliberative Council of Mackenzie University.

He mentioned that: *"We want to warn about inappropriate articles published in the Astronautics Journal-SIB, which involves the prestigious name of Bradesco"*. It pointed to the publication *"The New Face of the Physical World"*, published in this Journal, on the gravitational radiations discovered by Mesquita, as deposing against the Bradesco Foundation. And requested support and use of Bradesco's computers, when out of office, as a Copy of the Letter as Mackenzie University sent to Bradesco, according to the Open Letter of Mesquita to the Editor José Reis of Science & Culture Journal of SBPC.

Astronomer Praised by the Physicist Missile Launcher Praises Mesquita Article – On June 27, 1972, the Director of the Electronic Computing Center and the Bradesco Astronomical Observatory, of Astronautics Magazine and President of the SIB, Octavio Magdalena, he informed Mr. Aguiar that, on the article that was denounced by Pierre Kaufmann, he quoted the comment of the astronomer Nelson Travnik in which the other whistleblower S. Ferraz Mello praised precisely articles of Nelson Travnik published in the Astronautics. On this article involving the Mesquita work of gravitational radiations, Nelson Travnick had reported in a Letter addressed to Magdalena that: *"I read the discovery of Prof. Dr. Paulo Ferraz Mesquita. It was one of the subjects that impressed me the most. Took to the Juiz de Fora*

University-UFJF to several Teachers Professors Doctors who are reading it carefully", as being per Open Letter to SIB, from Magdalena to Mr. Aguiar President of Bradesco.

Deadly Missile to the Free Tribune of Astronautics Journal-SIB – But in early July 1972 again the whistleblower, the Physicist S. Ferraz Mello sought to contact Mr. Amador Aguiar to continue the "denounce"!

And on July 14th, 1972, Magdalena as a Letter to SIB, a Public Utility Institution, he informed that the Astronautics Journal-SIB was closed.

In July 1972 Mesquita sent in Open Letter by the Polytechnic School of the University of São Paulo-USP to the President of the Bradesco Foundation, Mr. Amador Aguiar, mentioning that the defense of pre-establishead Physics with the theory of relativity by these whistleblowers not they found solutions, while the theories presented in Astronautics Journal showed new ideas and resolutions of Fundamental Physics. And Astronautical Journal-SIB it was a free tribune, but these detractors refused to debate, to simply prevent the free expression of new ideas, and the worst is that they offered, on the pretext of scientific level, to exercise the editorial inquisition and annul any debate that they reject, as a Copy of the Open Letter from Mesquita to the President Mr. Amador Aguiar of Bradesco Foundation.

On July 18, 1972 Mesquita sent Letter to Dr. Professor José Goldemberg, Director of the Department of Physics-USP, on the Letter of Goldemberg of June 16, 1972, mentioning that he Goldemberg had the knowledge of that Mesquita was held the office of President of the Scientific Council of the Astronautics Journal-SIB, and him assertions were derogatory, frivolous and contrary to the elementary principle of professional ethics. And asked to Goldemberg to approve the *"Course of Advanced Researchers in the Unified Field"* came to be taught at the Institute of Physics as a Post-Graduate segment as requested in January 1972, as Open Letter. On December 30, 1972, Mesquita sent to the graduate coordinator of USP, Professor Henrique Fleming reiterating the approval for a *Physics Course Fundamental*.

Beyond those denouncers present demonstrations without any justification of specifying in which they would be incorrect, thus revealing an empty complaint, without content, which resulted as synonymous with frivolity of opinion against the authors of published articles in Astronautics Journal-SIB, they also came directly to the Bradesco Institution which sponsored and published this Journal, by claiming the whistleblower who *"The Bradesco Organization should have a name to watch out for there being a lack committed in this Organization"*! A complaint sought only to show a false scientific zeal as coming from a pretentious scientific erudition. The whistleblowers presented themselves self-titled from the highest knowlodge, as connoisseurs of absolute scientific truth, to oversee what should be followed and presented by any author of scientific work. And also in intending to teach to any institutions how they should be addressed and what should be published!

Unprecedented Launch Against the Missiles – While the three whistleblowers presented in July 1972 4 unprecedented pages of warning of "high scientific value" for tutorial obedience to the whistleblower under penalty of institutional relegation of the Bradesco Organization and the withdrawal of the "credit personnel" of whistleblowers to this banking institution, as stated in Letter of these whistleblowers, Mesquita presented in July of 1972 in the SBPC an article of a page linking the Laws of Coulomb and Planck, unprecedented in the scientific literature world. This work of Mesquita was presented at 24th. Annual Meeting of the SBPC and published in the Yearbook-SBPC of 1972 with the title: *"Coulomb's Law Obeys the Planck Postulate"*.

By Coulomb Law for two electrons in the vacuum: $F = e^2/r^2$, where e = electric charge and r = distance between electrons. Introducing the fine structure constant, the Compton wavelength and corresponding frequency $\mathcal{V}_0$, and the classic electron radius "**a**", in which the quantum impulse is present, and multiplying by r distance, is obtained: $F \times r = e^2/r^2 \times r$, so: $E = h\mathcal{V}_0\,\mathbf{a}/r = h\mathcal{V}_0\,n = h\mathcal{V}$. That wasn't perceived by Einstein, Planck and contemporaries. Document 11.

B

RESUMO

TÍTULO DA COMUNICAÇÃO: A LEI DE COULOMB OBEDECE AO POSTULADO DE PLANCK

NOME DO AUTOR: PAULO FERRAZ DE MESQUITA (Professor Titular da Universidade de São Paulo)

Tenhamos em vista (1) expressão envolvendo constante de fina estrutura da Física Atômica

(1) $\dfrac{hc}{2\pi e^2} \cong 137,037$ (número puro) que relaciona a constante h de PLANCK, a velocidade c da luz, a carga e do eletron e sôbre a qual conjecturam os Físicos modernos a propósito de seu significado físico.

No entanto, se nela fizermos $c = V_o' \lambda_o$ onde $\lambda_o \cong 24,2626 \times 10^{-11}$ cm é o conhecido comprimento de onda COMPTON do eletron, resulta

(2) $V_o' = \dfrac{2,99792 \times 10^{10}}{24,2626 \times 10^{-11}} \cong 1,2356 \times 10^{20}$ seg^{-1} (inteiro mais próximo); e indicando as grandezas em unidades CGS, obtem-se :

(3) $e^2 = \dfrac{h V_o' \lambda_o}{2\pi \cdot 137,037} = \dfrac{h \nu_o r_o}{137,037} = \dfrac{m'c^2 (\text{erg seg}) \mu (\text{seg}^{-1}) r_o (\text{cm})}{137,037} = x_o o^2 a$

$= h V_o' a \cong 23,06717 \times 10^{-20}$ (erg cm) onde

(4) $h = m'c^2 \text{ seg} \cong 6,62517 \times 10^{-27}$ (erg seg) [constante de PLANCK]
(5) $m' \cong 0,73715 \times 10^{-47}$ (gr) [massa quântica elementar num campo unificado pela Teoria da Impulsão]
(6) $V_o' m' = m_o \cong 9,1082 \times 10^{-28}$ (gr) [massa do eletron estacionário]
(7) $\dfrac{\lambda_o}{2\pi} = r_o \cong 3,8615 \times 10^{-11}$ (cm) [raio de um círculo girante que representa o lugar dos pontos de maior densidade de energia do eletron estacionário]
(8) $\dfrac{r_o}{137,037} = a \cong 2,81786 \times 10^{-13}$ (cm) [o conhecido e chamado raio clássico do eletron]
(9) $m_o c^2 = h V_o' \cong 8,1860 \times 10^{-7}$ (erg) [a energia do eletron estacionário]

Assim,

(10) $e = \pm \sqrt{h V_o' a} \cong 4,80283 \times 10^{-10}$ (ues q) [a carga do eletron e/ou do sitron representado por um dipolo girante como um disco].

Pela LEI DE COULOMB, a fôrça atrativa entre dois eletrons à distância r (cm) um do outro, no vácuo, será :

(11) $F = \dfrac{e^2}{r^2} = \dfrac{h(\text{erg seg}) V_o'(\text{seg}^{-1}) a(\text{cm})}{r^2 (\text{cm}^2)}$ E a energia potencial para

um eletron à distância r (cm) de um outro será :

(12) $E = \dfrac{h (\text{erg seg}) V_o' (\text{seg}^{-1}) a (\text{cm})}{r (\text{cm}^2)}$ onde fazendo $\dfrac{a(\text{cm}^2)}{r(\text{cm}^2)} = n$ (número puro)e
(13) $V_o' n = N$ número inteiro mais próximo :
(14) $E = h V_o' n = h N$ Por exemplo, com $r = 1$ cm, $E \cong 23,0672 \times 10^{-20}$(erg).

Esta expressão mostra que a LEI DE COULOMB OBEDECE AO POSTULADO DE PLANCK tam bém como a Lei de NEWTON (demonstrado antes) o que passou desapercebido de EINSTEIN, PLANCK e de contemporâneos.

Instituição: ESCOLA POLITÉCNICA DA UNIVERSIDADE DE SÃO PAULO

Doc. 11– Original Communication of Mesquita, with unprecedented work in the world scientific literature: "Coulomb's Law obeys Planck's Postulate", presented at 24th. Annual Meeting of the SBPC in 1972, in References.

Bradesco Foundation Exalts Mesquita and Does Not Deliver Astronautical Magazine-SIB to Whistleblowers – On 15 August 1972 the Bradesco Foundation, through Mr. Aguiar and Basilio Trunko Filho, sent Letter to Mesquita considering it as who contributed so much to the prestige of the Astronautics Magazine-SIB with the valuable support of Mesquita and with their respect and admiration. They reported that the Magazine was published for educational purposes without ever intending to establish controversy over the subjects. And the publication had been suspended for internal reasons of the Bradesco Foundation, as Copy of Letter to SIB.

Missile Warhead Back Against its Launcher! – And on August 18, 1972 Mesquita sent Letter to the Rector of Mackenzie University, Professor Dr. João Pedro de Carvalho Neto, questioning whether the Deliberative Council of the Mackenzie University had authorized Pierre Kaufmann to issue a missive to the President of the Bradesco Organization in name of that Council.

On September 9, 1972 the Rector of Mackenzie University reported in a Letter the Mesquita that Pierre Kaufmann did not write on behalf of the Council and was not allowed to do so to the President of Bradesco.

LIV. THE BRAZILIAN SOCIETY OF PHYSICS-SBF AND THE QUANTUM IMPULSE THEORY – <u>1966 to 1976</u>

The Brazilian Society of Physics-SBF, which became allied to SBPC, was founded during 18th. Meeting of the SBPC in 1966, in which the Founding Committee was chaired by José Golbemberg, being Mesquita as one of the founders, Oscar Sala, Jaime Tiomno, José Leite Lopes, Alceu G. Pinho Filho and Ernest W. Hamburger among others.

During the 22nd. Meeting of the SBPC in 1970, the Physicist Professor Dr. Ernest Hamburger, of the Institute of Physics-USP, had prevented the realization of a Symposium of Theme in Physics requested and that would be taught by Mesquita during this Meeting.

In 1971 the publication of the Brazilian Journal of Physics of SBF was initiated, with the Editor Physicist Dr. Jorge Leal Ferreira, sponsored by Financer of Studies and Projects of the Federal Government-FINEP.

And the 23rd. Annual Meeting of the SBPC of 1971, as Secretary General of the SBF Ernest Hamburger again prevented the realization of a *"Physics Basic Course"* during this Meeting of the SBPC that Mesquita organized and would minister.

In 24th. next Meeting of the SBPC in 1972 was the Physicist Professor Giorgio Moscatti as General Secretary of SBF who sought to veto a course *"Brief Training Course of Advanced Researchers"* prepared by Mesquita for this Meeting, but which was effected, in which it was included quantum impulse theory and its developments.

In February 1972 at the II Symposium on the Teaching of Physics, Mesquita presented manifestation on the SBF in which it presented it as dominated by a dome oligarchy that he had taken an abusive view of the SBF executive and acted as if he owned the sole owner of this society. Those who were not sympathetic to this oligarchy would have their omission or depreciation by these owners of Physics, which was exemplified by the public declaration by José Goldemberg on 31 January 1973 that: *"Every step falls a building under construction. Who has been looking for an approximation with the engineers and has not been able to"*. But according to the Mesquita, it was the opposite, because it was the Summit of SBF, in which Golbemberg was an Advisor, who continued to try to prevent the Engineer from being able to collaborate with Physics as a Science and not just as Technology.

In July of 1973 on the 25th. Meeting of the SBPC, the President of SBF, the Physicist Professor Alceu G. Pinho Filho of the Pontifical Catholic University-PUC of Rio de Janeiro, came from 1972 with evasive and omission to allow a course *"Training of Advanced Researchers"* that would be taught by Mesquita for the SBF and supporting Ernest Hamburger which was contrary to include a code of ethics in SBF to allow free manifestation of new scientific ideas. This proposition of Mesquita was voted in a Meeting at SBF receiving 23 opposing votes and 5 favorable, while the votes of more than 100 members present at that Meeting were considered as abstention.

On 14 July 1975 Mesquita presented, at the General Assembly of the Brazilian Society of Physics-SBF, demonstration against rejections to the freedom of teaching, of research and dissemination of knowledge to Teachers of Physics, which was the purpose Statutory SBF. But, that manifestation no longer transcribed into the Minutes by SBF.

On 9 July 1976 the General Assembly of the SBF Mesquita reiterated the publication of the demonstration of the previous year of 14 July 1975 in the Minutes of the Meeting.

Then it was perceived that in addition to the three whistleblowers to the Bradesco Foundation on Astronautical Journal-SIB appeared Ernest Hamburger as a fourth "zealous of pre-establishead Physics" preventing the disclosure of basic concepts of Physics and the free manifestation of scientific ideas. It was complet the "Quartet of the Vigilantes of Law and Science" with the inclusion of other doctors of "protection'of the formal Physics.

On July 14th, 1976 in 10th. Annual Meeting of the SBF during the 28th. Annual Meeting of SBPC, Mesquita presented a Communication: *"Training of Researchers in a Unified Field of Fundamental Physics"* that was published in the Annual Summary, in which presented the own current Physics in the words of Einstein mentioning the failures its: *"Cead to an attempt to find a purely algebraic theory for the description of reality. But no one knows how to get the basis of such a theory"*.

And presented the lecture on this Communication, but that was not published despite presenting in writing what had been uttered. In addition to showing the failure of Einstein's theory of relativity by Einstein's own words with the theories of pre-establishead Physics, Mesquita cited the problems its in the words of Newton, Huygens, Planck and Millikan, that is, surprising the unconditional followers of the current Physics, in which nothing should be modified!

LV. CRYSTALLIZED PHYSICS AND THE FANTASTIC PROBABILITY OF FREE NEUTRON DISINTEGRATION – 1972

In July 1972, the Physicist Paulus Aulus Pompéia, who had participated in the years 50 of experiments on mesons and other studies, with physicists of the country, sent an opinion on the work of Mesquita *"The Duration of the Free Neutron"*, at the request of Mesquita for which Pompéia could refer to the Brazilian Academy of Sciences for being a Member of this institution. So, Aulus Pompéia communicated that the atom free neutron mass disintegrating instantly, after a time that would be termed average life and half life, as a probability. And the mass of the neutron would remain invariant until its instantaneous disintegration, while by the Mesquita's work the neutron mass would decrease over time. And: *"By not monitoring the development of Physics in recent years and not being able to defend this work from any objections to the Academy, I consider it inappropriate to forward this work"*. And ended with the content: *"I reaffirm only to have the ability to reason within a scheme, possibly already crystallized, and therefore I admit that I may be wrong"*.

This opinion of Pompéia showed that the disintegration of the neutron in electron and neutrino was seen by the current Physics as the interaction or weak force of the beta decay of the neutron, as a simple probability around 50% of when leaving the nucleus after a while disintegrate instantaneously all its mass, being that time denominated of average life or half life.

Among the neutrons coming out of an atom-gram, a part of the neutrons is disintegrated in the early minutes and another part of the neutrons may even remain even after 30 minutes without disintegrating.

However, Mesquita had considered that "if " a single neutron, free from the atom and without interactions with other neutrons, was disintegrated according to the gravitational frequency of emitting its mass formed by the elementary quantum mass of the gravitational photon, the average life time and half life should be equal to the time recorded experimentally. But, as there are between 1 and more than 10^{22} frequencies in cycles per second, the probability that the gravitational frequency of 10^{19} cycles per second, discovered by Mesquita, to disintegrate the mass in the neutron formed from the elementary mass in a time that was equal to the determined experimentally, would be around to get a specific frequency in each sextillion of types different of frequencies. And it was exactly that frequency of 10^{19} cycles per second for the disintegration of the total neutron mass in full time or for mass in the average or half life time, obtained experimentally. Just a coincidence? No! But, a fantastic coincidence! And whereas the radioactivity experiments of Irene and Pierre Curie and that of Walther Kaufmann of increased electron mass, which involved the mass of the electron with the frequency of gamma rays in the annihilation and creation of electron and positron and the electromagnetic mass that constituted the photons of electromagnetic radiation, is a fantastic evidence that the neutron it would also be formed from the elementary quantum mass of the impulse theory of the frequency between gamma rays and X-rays with the frequency of 10^{19} cycles/second.

LVI. THE TRIAL OF "THE NEW FACE OF THE PHYSICAL WORLD" 1972-1973

On December 21, 1972, Mesquita sent a request for clarification to the Physicist Pierre Kaufmann of Radio Astronomy of the Mackenzie University on the article "*The New Face of the Physical World*" which had been published in the Astronautics - SIB in December Journal of 1971 by the Astronomer Ph.D. Bohumil Vanco of the United States, in which Kaufmann had criticized this article. As Kaufmann had sent Letter to the President of the Bradesco Foundation without authorization from the Rectory of Mackenzie University and without formulating justifications about the complaint as inappropriate and pseudoscientific, there was no way to avoid responding to the Mesquita's request on what it would consist of such pseudoscience.

The Judge – On 30 December 1972 Pierre Kaufmann replied to the Mesquita that: *"The evidence of gravitational waves was contradictory; the approximate formula of the Doppler effect used in the article impaired all the text and did not reflect the physical phenomenon; some z values were nonexistent from the Doppler shift; the author of this article often confused constellation with galaxy and quasar; state of ultraviolet ray emission by hydrogen; cosmic background noise ignored; confused gamma rays with cosmic rays; sectioning of the electromagnetic spectrum that was continuous, in which X-rays, gamma rays, ultraviolet, light, infrared, heat, radio, and others were just conventions of several energies; the article did not know the radio window of the atmosphere between 10 meters to millimeters; it confused telescopes with funnels ignoring interferometers; it used the Hubble constant of 100 km/s, but it was not a definitive parameter; the gravitational and intergalactic Doppler effect and the distant nature of quasars were questionable; every phenomenon involving* *the electromagnetic spectrum appeared as streaks of continuity and not just with a specific frequency of 0.3 angstrons; the increase in wavelength to situate at any location in the spectrum was not demonstrable; the Sun emits accumulated rays, so on several frequencies, just as the plasma would emit across the spectrum; theory of relativity was the best theory that explained the control and foresight of countless technological products and all the evidence to which it had been submitted confirmed its excellence; the theory of relativity to be criticized should present impropriety if there were".*

The general aspect of the Letter of Pierre Kaufmann considered that: *"The author of the article had not been aware of the latest advances in Astronomy, as it did not mention the revision of Press and Thorne of 1972; it was super simplified; it presented a heap of affirmations, numbers, propositions with no references, no order, no coherence, no brevity; ramblings, speculations, without demonstrations, on the basis of patchwork of partially used or deformed information; refused confrontation and criticism; lacked rigor; it despised those who seriously worked in science; he proposed new concepts of transcendental character showing gratuity, in which this was not science; This article could not be considered anything more than pseudoscience".*

The Replica – On 2 March 1973 Mesquita questioned the opinion of Pierre Kaufmann on the article by Vanco, stating that the revision of William H. Press and Kip Thorne of 1972 reported failure and no progress that Kaufmann indicated about the gravitational waves sought by Joseph Weber based on Einstein's general relativity. On the electromagnetic spectrum, gravitational radiation demonstrated by Mesquita in 10^{19} cycles/sec or the wavelength of 0.3 Angström between gamma rays and X-rays and that in the other Mesquita work on the disintegration of the free neutron was proof of these radiations. And the Mesquita added that in 1966 the neutron stars were discovered through X-ray observation, in that because they were so dense their matter was made so compact by the gravitational forces that they could only shine in the electromagnetic spectrum of X-rays, and that by Pierre Kaufmann acting in the area of astrophysics should understand the meaning of that remark.

Dated February 5, 1973 the scientist Ph. Dr. Bohumil Vanco sent the replica to the Physicist Pierre Kaufmann. He showed on general aspects of the article for next scientific analysis, that the article was presented in March 1971 and the quotation of Press and Thorne was 1972; It was not article of theories of gravitation, radio astronomy or astrophysics, but only of gravitational radiation discovered by Mesquita placed under the Doppler effect; then the critic should judge under this physical essence presented, but in the opinion there were then expressions of the type *"...is ignored, unknown, ignoring, despising...";* to show that the judge would know the matter, but this was inappropriate for the trial; references had been cited within the necessary, because on an original article there was a lesser quotation. The expression *"... it's not science..."* used by the judge was to denote only the mentality of a heretic.

In the technical part, the expression of the Doppler effect that the judge advocated was nothing more than that of Einstein's relativity and presented in the text, but which was no longer adopted because it did not correspond to the actual observation. And the text used was that modern astronomy used to be more in line with the observations. And another formula presented by the judge was nothing more than what was deduced in the text from another source! Therefore, the text used a correct expression validating all of it on the Doppler effect. The z values of the Doppler shift were explained. The Doppler, gravitational, and intergalactic effect showed that it was related to velocity and not with gravitational potential and intergalactic absorption. In the article it was cited that there were constellations in the distance of one billion light-years, in which in the context it was of clusters of galaxies, the "clusters" in the original English, but in the Portuguese book of the pre-establishead Physics itself was defined as constellations. And with a billion light years, what was seen were galaxies, not stars. Quasar was presented in the article as a rescendente galaxy with almost the speed of light, while smaller distances were called galaxy only. The state of hydrogen emission was not the case, but the absorption streak that was related. Background noise was not ignored, but reported as isotropic radiation in the lengths of radio-waves between 0.2 and 20 cm of weak intensity. And Vanco adds: *"Who's ignoring?"* . Therefore, the electromagnetic spectrum was treated as continuous, only that for the essence of the article, it was considered the specific gravitational radiation to be analyzed by the Doppler effect. The radio window was not unknown, but quoted as a radio-wave window! The increase in wavelength was demonstrable in the article as shown in the chart! The Hubble constant was admitted as 89 km/s, but cited that if this value were changed there would be a need to recalculate some results.

On impropriety of the theory of relativity was presented in the following communication dated May 1972 with the replica of February 5, 1973 by Vanco. And contrary to what Pierre Kaufmann reported that relativity was according to experiments, Vanco reported reports of failures in results predicted by general relativity and pointing out divergences and inconsistencies of special relativity.

By the current Physics, quasars are at the edge of the universe. However, according to H. Arp, by the observable fact that, galaxies close to the Earth with small reddening accompanied by quasars, concluded that quasars were at the same distance from the galaxies where quasars were. So the article from Vanco considering quasars if they were at the edge of the universe led to that the gravitational radiations discovered by Mesquita were seen as light on Earth, and the light emitted by the quasar as radio radiation. However, to disregard any new idea against the pre-establishead Physics, Pierre Kaufmann instinctively stated *"... parameters are being aproned and tend to make questionable the distant nature of quasars."* That is, going against your own Physics that regards quasars distant! But it was just an instant heresy, as it continued: *"They tend only because a lot of scientific work is still needed to better understand the importance of other mechanisms"*. This is, it would take a war to overthrow the pre-establishead Physics! But, thus denying that quasars could be close, it nullified the main argument against the article, which rightfully placed the quasar in the distant distance admitted by the formal Physics. The article was testing this possibility of the quasar from far away by the current Physics, in which the work of Mesquita showed magnificent success of the quasar emitting gravitational radiation. However, if in 2020 the James Webb telescope came to conclude that quasars were close, the current Physics would have to review the theory of high light emissions and radio-waves of quasars. And in this case of proximity to quasars, the gravitational radiation of the Mesquita work would remain valid. In the proximity the behavior of the quasar of emitting light and noise would maintain the gravitational radiation originated in the atomic nucleus as it occurs in the nearby galaxies, thus acting without the intense gravitational emission if it were distant and then presenting another behavior. If by the telescope James Webb were to be concluded that quasars are very distant, then the current Physics could never deny the brilliant article of Vanco and the gravitational radiations discovered by Mesquita, because they indicate how the brightness and noise mechanism emitted by quasars to elucidate a broader theory of the behavior of quasars with the observations of the teams of the formal Physics.

LVII. "SPACE COMPASS" INITIAL EXPERIMENTS – 1973

The velocity indicator or "space compass" devised by Erdelyi to obtain the Earth's displacement in space had been carried out with experiments by the Physicist Frank Nimrod of the United States, however with undetermined results.

In 1973, the Hungarian Techie Júlio Martits resident in Brazil sent a Letter to SIB reporting that it had used a velocity indicator with neon light, with the focus of light and a photographic film positioned on the horizontal plane of the Earth in different positions, to obtain the inertial velocity of the Earth in space, but the beam of light that reached a convergent lens went to the center of this film without presenting any elongation of the stain obtained, indicating that the neon light received the inertial velocity of the Earth. Although the laser was discovered in the 1960, there were still difficulties in obtaining it, which would be the ideal light for better accuracy. But, this Martits experiment presented a result for analysis of the impulse theory and indicating for laser verification of this experiment. Figure 40 shows experiment with neon light by Martits on 1973.

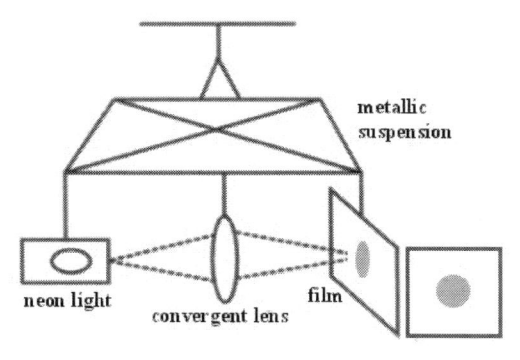

Fig. 40 – Experiment with neon light by Martits on 1973, as Reference.

LVIII. UPPER LIMIT – **1963 to 1973**

I came to participate in 1972 of an *Advanced Physics Course* promoted by the SIB, sponsored by SBPC and ministered by Mesquita, in which it was necessary to present a scientific article involving the pre-establishead Physics and new propositions of Physics. So, I presented a paper on the point electron for current Physics, and as dimensional structure the electron by the theory of quantum impulse. For the punctual electron there was an experiment in the United States conducted by Goldemberg, Director of Physics Department-USP, in which he presented a proof that was the most up-to-date electron as a point in his article *"Uper Limit of the Electric Dipole Moment of the Electron"*, which I presented in the reference of *"Electron: Interactions with Nuclei and Their Structure"*, for this course.

But with the purpose of further learning the article by Goldemberg I received his attention in 1973 on the Board of the Institute of Physics-USP in which he explained to me some questions that I presented in that attendance. The main part of the Goldemberg experiment was that the electron did not present an internal structure to the upper limit of 10^{-15} cm, that is, as the radius admitted by the electron was 10^{-13} cm, an electron with 10^{-15} cm was not detected within the electron of 10^{-13} cm. Therefore, there was no spreading when designed against a nucleus at 180 °. And another question, which was not in the article of Goldemberg, was whether the specific electron itself, in which the spin constitutes a magnetic dipole, could be constituted, for example, of photons. Goldemberg's answer was that there was no interest in it.

So there was an upper limit on which a possible small electron was not detected within the electron itself. But there was also an upper limit in not accepting new ideas, like the quantum impulse that constituted the electron and all particles.

LIX. SCIENCE & CULTURE JOURNAL OF SCIENCE PROGRESS BRAZILIAN SOCIETY-SBPC AND THE DEFENSE OF THE SCIENTIFIC INQUISITION BY THE GREATEST EXPONENT OF THE MODERN ANTI-HERESY HANDBOOK – <u>1954 to 1974</u>

In January 1974 the Science & Culture Journal of SBPC had accepted for publication an extract of the article *"Fads and Fallacies in the Name of Science"* of 1957 published in Portuguese as *"Manias and Beliefs"* in 1960, of the American author the Greatest Anti-Heretic World Exponent, Martin Gardner, an official inquisitor of the scientific pre-establishead, a book of "Inquisitor Handbook". Thus, Science & Culture Journals endorsed in science what always fought over the political freedom of opinion, that is, fought the military political censorship of the time, but justified the censorship in Science.

This extract from the book of Martin Gardner, sent by his admirer of medieval practices, indicated the procedure of how to maintain untouchable the pre-establishead Physics, justifying all the rejections of the inquisitors of scientific articles sent for publication in scientific journals controlled by the summit of *"Doctors of the Law and Sciences"* as revealed by Christ at the time. It did not present any reference to ethics or morality, to allow the referees, to continue to protect the owners of Science. He considered eccentricity or paranoia anything that contradicts the formal Physics and his ideas, denominating of perpetual motorcycle and pseudoscience, and how to make an opinion when receiving these new ideas. He defended the judges and owners of the knowledge of being with crystallized ideas and of resisting any new ideas, including the "unconscious resistance". Considered the authors of new ideas like paranoid high priests isolated in their towers.

But who isolated the formal theories of any contrary analysis was precisely that practiced by the scientific summit, its "high priests'! To be considered as normal, scientists with new ideas should accept all the appropriate arguments given by their judges and never present disagreement with the more "well-established" theories, in which "well-established" were their own ideas!

Martin Gardner had published books on the theory of relativity, condemning to the bonfire any heretic who dared to speak otherwise! But, ironically, the greatest exponent of the current scientific inquisition fell before the very inconsistency of the theory that it defended. For example, in explaining that the Michelson-Morley experiment which resulted in nullity in the displacement of fringes of interference, which by Einstein's relativity was motivated by the contraction of space and thus annul the predicted interference, Martin Gardner later he wrote that the contraction of space was "apparent"! So he annihilated what he had written before! However, Martin Gardner revealed in this extract his own incompetence in not knowing how to differentiate a perpetual motorcycle from a motorcycle, let alone a hoax of a real scientific theory. Just as the followers of the formal Physics explain the planet Venus, meteors, plasma, airplanes, satellites and atmospheric vortexes to phenomena or unidentifiable objects, which make curves almost at 90 degrees at extremely high velocity, and in fall back in ancestry, also Martin Gardner explained new ideas like witchcraft, heresy and "paranoid schizophrenia"!

By the regulation of publication of Science & Culture Journal of SBPC, two referees were consulted. So whoever sent this scientific inquisition statement found a strong appreciation of "scientific acceptance", and the publication should reach the entire network of its admirers for the "witch hunt"! From that date of January 1974, then there was an "Official Handbook" of the anti-heresy procedure within the SBPC!

LX. QUANTUM IMPULSE EXPLAINING THE KEY EXPRESSION OF THE THEORY OF RELATIVITY FOR THE PERIHELION MOTION – 1974

To explain the perihelion of the planets, the theory of general relativity showed the gravitational attraction, as acceleration effected by the curved space produced by the mass, according to a constant applied to Newton's universal gravitational attraction.

In July 1974 Mesquita presented at 26th. Annual Meeting of the SBPC the work "*Motion of the Perihelion Independent of General Relativity*" and published in the Yearbook-SBPC of 1974, in which was deducted the rotation of the elliptical orbit of the planet in its revolution around the Sun, with the formula of the transformation of Voigt-Larmor-Lorentz also obtained by relativity as its key formula, and by the theory of quantum impulse in which it presented the unprecedented introduction of Planck's constant. And it showed that it was the mass variation that increased the velocity of the planet, resulting in this variation of the perihelion, in this key formula of relativity, but interpreted with physical significance and the elementary quantum impulse within that formula, unifying gravitation with quantum mechanics, unfulfilled dream of all the scientists who seek this unification by the pre-establishead Physics.

Mesquita applied the "key" equation of the special theory of relativity, which Einstein himself did not perceive, that is, the transformation of Voigt-Larmor-Lorentz coordinates to velocity, and can relate variables with a correct interpretation of the equation of mass increase $m_v = m_0 / \sqrt{1 - (v^2/c^2)}$ by equation of the triangle of the elementary impulse theory, in which the planets approaching the star of the system would receive more mass of their gravitational field increasing their velocity. Document 12.

> MOTION OF PERIHELION INDEPENDENT OF GENERAL RELATIVITY
>
> Abstract of Communication presented at the XXVI Annual Meeting of the Brazilian Society for the Advancement of Science held in R E C I F E Pe - BRAZIL - July, 1974, by
>
> Paulo Ferraz de Mesquita
> Civil Engineer and Geographer
> Full Professor of
> The University of São Paulo
>
> Expressing, by dimensional analysis, the Newtonian Constant in terms of Planck's Constant in the third Kepler's law of motion and taking in view Lorentz's discovery which allows one to express variation of mass by the variation of velocity of a Planet in its orbit, the Author of this COM-MUNICATION derived the following formula for calculating the slow rotation $\Delta\alpha$ of the elliptic orbit of the Planet in the same sense of its sidereal revolution about the Sun:
>
> 1) $\Delta\alpha = 2\pi n \left[1 - \sqrt{1 - \frac{v^2}{c^2}}\right] s = 2\pi \sqrt{\frac{h \frac{1}{2}(m+m_1)}{a^3}} \left[1 - \sqrt{1 - \frac{v^2}{c^2}}\right] p.s$
>
> in second of nonagesimal arc per Julian century, being n=2𝜋206264,886"/T₁ = 1296000"/T₁ the mean sidereal motion of the Planet given by observation according to A. Danjon [op. cit.(1)pgs.267 and 430] in second of nonagesimal arc per mean day; T₁ the period of sidereal revolution of the Planet in mean day; v=2𝜋a/T₂ (cm/sec) the mean linear velocity of the Planet in the orbit; T₂ (sec)=86400 T₁; c=2,997925x10¹⁰ (cm/sec) the velocity of light; a(cm) the semi-major axis of the orbit; s=36525 (day)the Julian century;h=6,62517x10⁻²⁷(ergsec) Planck's Constant; v_b=10¹⁹ (sec⁻¹) the frequency of gravitational radiation; m=2x10³³(gr) and m₁ (gr) the masses of the Sun and the Planet respectively; p=206264,806"x86400 (sec).
>
> By his orthodox General Theory of Relativity, Einstein derived the following formula [op. cit. (2) pg. 97]:
>
> 2) $\Delta\alpha' = \frac{24\pi^3 a^2}{(1-e^2)c^2 T^2}$ in radian per revolution, being e the numerical eccentricity of the orbit; and T the period of revolution in seconds.
>
> Following some results of application and of "data" of observations [op. cit. (1) pg. 267] in second of arc per Julian century:
>
> | PLANETS | MERCURY"/cty | VENUS "/cty | EARTH "/cty |
> |---|---|---|---|
> | By formula 1) of the Author: | 43,28 | 9,00 | 4,01 |
> | By mean of 2) of Einstein : | 42,94± | 8,63 | 3,84± |
> | By "data" of observation : | 42,56±0,94 | | 4,6 ±2,7 |
> | | (CLEMENCE) | | (DANJON) |
>
> what confirms the coherence of both independent Theories.
>
> Op. cit. :
> (1) André Danjon -"ASTRONOMIE GÉNÉRALE"-Ed.J.& R.Sennac -Paris-1952-1953
> (2) A. EINSTEIN -"THE MEANING OF RELATIVITY"- Princeton U.Press -1955

Doc. 12 – Work summary, presented in 1974 by Mesquita in SBPC, "*The Motion of the Perihelion Independent of General Relativity*" with the correct interpretation of the mass increase generating velocity increase, by the transformations of the Larmor-Lorentz coordinate, and the equation of the triangle of the elementary impulse theory, as Reference.

LXI. AN IRREFUTABLE PROOF BY THE CURRENT PHYSICS OF THE ELEMENTARY QUANTUM ENERGY h' OF QUANTUM IMPULSE THEORY – 1974

George A. Csiky Jr., Graduated Engineer in Hungary and Master's Degree in Physics by John Carrell University in the United States, worked in electromagnetic propulsion and ionic engineering at NASA, and after in the Department of Research of the North American Army. Formulated in the United States with Erdelyi, when Erdelyi remained part of his work in Cleveland, the deduction of the expression of the mass increase of the elementary particles from its constitutions by the quantum impulses m'c and the quantum h'= m'c x c, obtaining the same expression of Voigt-Larmor-Lorentz and that of relativity, $m = m_0 /[1 - (v^2/c^2)^{1/2}$, where m = particle mass at velocity v, m_0 = particle mass at rest and c the speed of light, which Erdelyi then named Voigt-Larmor-Lorentz-Csiky equation.

On July 19, 1974 Csiky Jr. presented Scientific Conference *The Meaning of the Theory of the Impulse of Erdelyi* at the Institute of Engineering in Sao Paulo-Brazil, chaired by the Engineer Lauro de Barros Siciliano of the Institue Engineering with participation of Mesquita.

Among topics that showed details of the theory of impulse, Csiky Jr. presented the concept of restoration force in the harmonic oscillators, in which at each oscillation occurred the acceleration, therefore the electromagnetic emission, and a fundamental proof that the quantum h' obtained by Erdelyi of Planck's Law $\mathcal{E} = h\mathcal{V}$, where $\mathcal{E}$ = quantum energy in erg in CGS, h = Planck constant in erg x second and $\mathcal{V}$ the quantum frequency in cycles/second. And that Erdelyi had drawn up $\mathcal{E}$ erg/1 second) = (herg x second/1 second) x $\mathcal{V}$ cycles/second = $|\mathcal{V}|$ herg/1 second, then ($\mathcal{E}$ erg/1 second) = $|\mathcal{V}|$ h' [erg]. And to $\mathcal{V}$ = 1 cycle/second: ε [erg] = h' [erg]. That is, in each cycle of the restore force F = k*d*, where k is a constant and *d* the oscillation displacement, would be emitted the elementary quantum ε = h' = m'c² for each oscillation. That is, while the Planck quantum and Einstein was $\mathcal{E} = h\mathcal{V}$ contains all the oscillations of frequency $\mathcal{V}$ in a single quantum $\mathcal{E}$, this is, un single photon, the elementary quantum of Erdelyi was ε = h'$\mathcal{V}_1$ where $\mathcal{V}_1$ it was the frequency of 1 cycle/second. That is, by the current Physics itself, for each acceleration of an electrical charge there was the emission of electromagnetic wave, therefore it should for each charge acceleration in the oscillating circuit with its restoration force, there be the electromagnetic emission of one quantum for each oscillation.

LXII. INTERNATIONAL COMMITTEE OF SPATIAL RESEARCH – 1974

In February 1974 Mesquita had submitted the article "*The Physical Meaning of the Newtonian Constant Reveals a Way for Studying Satellite Dynamics Within a Unified Electromagnetic and Gravitational Field*" for presentation in the XVII Plenary Meeting of the Committee on Space Research, established by the International Council of Scientific Unions – COSPAR, at Symposium on Satellite Dynamics, scheduled for June 1974 in São Paulo-Brazil. It was addressed to the Executive Secretary, Z. Niemirowicz and the Chairman of the Symposium Program

Organizing Committee, Professor Giorgio Giacaglia of the Astronomical and Geophysical Institute-USP that in 1969 had prevented Mesquita from presenting Scientific Communication. And as a statement of Niemirowicz the articles would be reviewed by the Committee for eventual acceptance. And in April of 1974 Giacaglia as Chairman of the programming Committee reported that the Committee had accepted for presentation the work of Mesquita. And in June of 1974 the work was presented and published its summary in the Symposium Yearbook. Document 13.

COSPAR
COMMITTEE ON SPACE RESEARCH, ESTABLISHED BY THE INTERNATIONAL COUNCIL OF SCIENTIFIC UNIONS

PRESIDENT
PROFESSOR C. DE JAGER
SPACE RESEARCH LABORATORY
THE ASTRONOMICAL INSTITUTE
BENELUXLAAN 21, UTRECHT
THE NETHERLANDS
TELEPHONE: 762941 (A.M.)
937265 (P.M.)
TELEX: 47224

EXECUTIVE SECRETARY
Mr Z. NIEMIROWICZ

1974

SECRETARIAT
55, BOULEVARD MALESHERBES
PARIS-8°, FRANCE
TELEGRAMS: COSPACERES PARIS
TELEPHONE: 522 19-93
TELEX N° 21558 SYSTELE
PARIS EXT. 81

Symposium on Satellite Dynamics held in São Paulo, Brazil, 19-21 June 1974, in the Seventeenth Plenary Meeting of the Committee on Space Research established by the International Council of Scientific Unions and published also in Review Engineering, São Paulo, Brazil, June 1979).

THE PHYSICAL MEANING OF THE NEWTONIAN CONSTANT REVEALS A WAY FOR STUDYING SATELLITE DYNAMICS WITHIN A UNIFIED ELECTROMAGNETIC AND GRAVITATIONAL FIELD
Annexes: "Electromagnetic Spectrum"
  TABLE - "The Gravitational and Residual Fields of the Earth"
  GRAPH I - "The Gravitational Field of the Earth"

P. FERRAZ DE MESQUITA
Escola Politécnica da Universidade de São Paulo

Abstract

Expressing the Newtonian Constant in terms of Planck's Constant, through Dimensional Analysis, we obtain the following form of an elemental gravitational energy in the CGS System of Unities, based on Newtonian Law

$$dE = m\, m_1 \frac{dr}{r^2} h\, \nu_g = dn\, h\, \nu_g \text{ (erg)}$$

where dn represents a pure numerical differential factor; r the distance between the center of mass m of a Planet (the Earth, for instance) and the unitary mass m_1 of a body (one gram, for instance); h(ergsec) is Planck's Constant and $\nu_g \equiv 10^{19}(\text{sec}^{-1})$ the frequency of the here so called gravitational radiation in accordance with Planck's Postulate so that

$$h\nu_g \equiv 6,62517 \times 10^{-27} \times 10^{19} \equiv 6,62517 \times 10^{-8} \text{ (erg)}$$

would be the so looked for and called "GRAVITON" the "quantum of the gravitational energy" and dE/dr shall be the variation of energy in erg for each centimeter of variation of the distance $r = r_1 + a$ above the Planet of radius r_1. Then by integration

$$\int_{r_1}^{r} dE = n h \nu_g \quad (erg)$$

represents the necessary energy or work for elevating a body of unitary mass from the surface of the Planet as far as a distance $r = r_1 + a$ from its center, what was not perceived by Einstein, Planck and contemporaries. The scope of the present COMMUNICATION is also to put in evidence some more proofs, in Celestial Mechanics, of the importance for the Basic Physics, of the Unified Field by the Theory of Momentum, an essentially quantic theory and, consequently, independent of the orthodox Theory of Relativity.

In fact, defending the Theory of Relativity, A. Einstein ennunciated his celebrated " Postulate of Continuity" saying "ipsis litteris" : "CONTINUUM SPATII ET TEMPORIS EST ABSOLUTUM". On the contrary, for us, the experimental constants of Nature utilized in an essentially quantic "Theory of Momentum" are : the "BRAS" = m' ≡ ≡ 0,73715x10^{-47}(gr) the "Quantum absolutum of mass"; the "JED" = m'c ≡ 2,20992x10^{-37}(gr.cm/sec) the "quantum absolutum of momentum"; the "HERG" = m'c^2 ≡ 6,62517x10^{-27} (gr.cm^2/sec^2) the "quantum absolutum of energy"; as we have seen and exalted in this COMMUNICATION and in several didactic courses already ministered by us on this matter. There remains opened to the Physicists and Mathematicians the possibilities of interpreting "apparent continuities" in an Unified Field by mean of the absolutely quantic "THEORY OF MOMENTUM" here summarily suggested.

Doc. 13 – Article Summary submitted to the mediation Symposium of COSPAR by Mesquita and published its Summary in the Yearbook of this Committee, 1974, as Reference.

LXIII. THE INSTITUTE OF ENGINEERING OF SÃO PAULO-BRAZIL AND THE SCIENCE – 1974 to 1979

On 3 September 1974 Mesquita presented a proposition to the Institute of Engineering of São Paulo-Brazil so that Science also included in the statutes, with the purpose of further valuing the Institute of Engineering of São Paulo. This proposal was subscribed by more than 100 Engineers from the Engineering Institute and on 4 November 1974 there were majority in the Deliberative Council, but less than 2/3 preventing its approval. Around 16 Advisers were favorable and 12 opposed.

But in 1975 already were 800 engineers from the Institute who subscribed favourable to the inclusion of Science in Statute of the Engineering Institute.

On October 30, 1975 Mesquita presented a Lecture at the Rotary Club of São Paulo-Brazil and on December 5, 1977 at the

Engineering Institute, with the scientific theme *"The Physical Significance of the Newtonian Constant of Universal Gravitation"*, and published in the Science section, in the Engineering Magazine of the Institute at N ° 393 of Jan-Feb, 1977. And also in the Engineering Magazine, Mesquita presented *"Training of Researchers in Fundamental Physics"* in N°413 of Jan-Feb of 1979. And in N°417 of June 1979 Mesquita presented the entirety of the article *"The Physical Meaning of the Newtonian Constant Reveals Way for Studying Satellites Dynamics in a Unified Electromagnetic-Gravitational Field"*.

LXIV. SCIENCE & CULTURE JOURNAL OF SBPC AND THE INQUISITION ACTING – **1974 to 1981**

In September 1974 Mesquita had requested Science & Culture Magazine of SBPC to its Director, Biologist and Physician, Dr. José Reis, for publication in full of the work presented at the Symposium of COSPAR, whose summary had been published in his yearbook . And in October 1974 the Report of a hidden referee designated by the internal rules of the SBPC Journal, Cience & Culture, it was received.

The hidden referee mentioned that the frequency of 10^{19} cycles per second was not a universal invariant, for a "Martian" the time unit would have another numerical value other than the second that would only be worth to the Earth, so this work would be just numerology! Reported that the expression that involved together the constant K_G of Newton's universal gravitation and Planck's constant h was deducted into any text of Physics nor too advanced!

And on October 7, 1975 Mesquita sent Open Letter to the Director of the Science & Culture Journal of SBPC, José Reis, on this Report. Informed that the discovery gravitational frequency was 10^{19} cycles per second anywhere in the universe within any system of units, provided that the units represent the same physical fact, as the time when the second was the same physical value obtained from an atomic clock. On the expression that united the constants K_G and h, Mesquita mentioned that any text existed this expression of the current Physics and complemented that: *"According to contemporary physicists E. Richard Cohen, Kenneth M. Crowe and Jesse W.M. Dumond in "Fundamental Constants of Physics", 1957, is mentioned in this book that: "The only constant of Physics that is not related in any way known to the atomic constants is the Newtonian constant of gravitation, and for which there are still generally accepted theoretical formulas that relate it to any other constants or magnitudes of Physics"*. And Mesquita commented on all the other issues of this opinion and asked again for the publication in Science & Culture Journal of SBPC of this paper, as reiterated Replica of Mesquita to Science & Culture Journal of SBPC to Consultant in October of 1975.

What was more perplexing was that this occult referee not only disregarded any new idea, but still unaware of the current Physics itself, when talking about another unit of measures that altered the outcome in another place in the universe. For, a second on Earth measured by the terrestrial standard, and if the "Martian" he mentioned had another standard of measure, in which if a second, $1s_M$ "Martian" equiv1/2 second terrestrial, the "Martian" would measure for the speed of light: 300,000 km/2 s_M, that is, 150,000 km/1 s_M on Mars, which on Earth was the distance of 150,000 km/(1/2) s_E, ie 300,000 km/$1s_E$ = 300,000 km/2 s_M, because $1s_E$ = 2 S_M. So the physical phenomenon of speed of light in traverse 300.000 miles on Earth or on Mars would be equal, but the mathematics would show different numbers of 150.000 km/$1s_M$ on Mars and 150,000 km/(1/2) s_E on Earth, though with light going through the same physical distance at the same physical time. A clock in which one of them indicated half the time relative to another normal was the same physical time as the normal clock was recording.

The fact that the referee mentioned that the expression, which K_G with h in the work of Mesquita was deduced in any text of Physics, or demonstrated that unconsciously saw as logical, correct without any doubt and within the comprehension of the physical fact, or else did not understand that it was exactly the essence of work and he sought to despise scientific discovery as if it were already widely known to the whole world!

Another question of total lack of discernment to judge this work was to point out that the equation deducted from the transformation formula of Voigt-Larmor-Lorentz (Voigt-Larmor-Lorentz-Csiky) in which Mesquita introduced the constants K_G and h did not deduce Einstein's formula of general relativity from the deviation of the perihelion. But, of course, the deduction of Mesquita was part of the fundamental constants of the theory of quantum impulse and Einstein's coming from relativistic principles. But general relativity could never introduce h along with K_G for this expression because it is a macroscopic theory without achieving unification with quantum mechanics. However, only if Einstein's formula presents variables that represent bases of meaning physical, so the fundamental constants of the impulse theory could also present this Einstein formula.

And a fact that shows the total lack of ethics, is that the anonymous referee had invested as judge of the highest power, wise of the absolute truth, despising the Committee of COSPAR, including the Chairman of this Committee which belonged to SBPC itself and to USP and that he had approved this work, which was presented at the Symposium and the summary published in the Yearbook of COSPAR.

However, on 22 November 1976, José Reis informed to Mesquita that the article had been sent to another arbitrator, who also highlighted that according to the system of units the value of the frequency would be altered, so the result of the work would be invalid by considering this gravitational frequency as an universal invariant.

On December 28, 1976 again Mesquita sent Letter to José Reis refuting the claim that the gravitational frequency varied by change of unit system. and simply presented a Table with Systems of Units showing that in all of them the frequencies were equal to 10^{19} cycles per second. Table 5.

Tab. 5 – Identical frequency of gravitacional radiation for two unit systems, according to Open Letter to the Science & Culture Journal of SBPC and SIB,1976, as Reference.

CGS cm gr seg		MKS m kg seg
M_1 gr M_2 gr	MASS	$\bar{M}_1$ kg = $M_1 \cdot 10^3$ kg $\bar{M}_2$ kg = $M_2 \cdot 10^3$ kg
R^2 cm^2	(DISTANCE)^2	$\bar{R}^2$ m^2 = $R^2 \cdot 10^{-4}$ m^2
ΔL cm	DISPLACEMENT	$\overline{\Delta L}$ m = $\Delta L \cdot 10^{-2}$ m
$K = 6,6...\times 10^{-8}$ cm^3/gr.seg^2	GRAVITATION CONSTANT	$\bar{K} = 6,6...\times 10^{-11}$ m^3/kg.seg^2 $\bar{K} \neq K$
$F = \frac{M_1 M_2}{R^2} K$ dina	NEWTONIAN FORCE	$\bar{F} = \frac{\bar{M}_1 \bar{M}_2}{\bar{R}^2} \bar{K}$ newton
$\Delta E = \frac{M_1 M_2}{R^2} K \Delta L$ erg	ENERGY VARIATION	$\overline{\Delta E} = \frac{\bar{M}_1 \bar{M}_2}{\bar{R}^2} \bar{K} \overline{\Delta L}$ joule = $\frac{M_1 M_2}{R^2} \Delta L K \cdot 10^{-7}$ joule
$n = \frac{M_1 M_2}{R^2} \Delta L$	NUMBER	$\bar{n} = \frac{M_1 M_2}{R^2} \Delta L$

$\Delta E = \Delta E_x \cdot 10^{-19} \text{seg} \times 10^{19} \text{seg}^{-1}$
$\Delta E = n K \text{ erg} \times 10^{-19} \text{seg} \times 10^{19} \text{seg}^{-1}$
$= n \cdot 6,6...\times 10^{-8} \cdot 10^{-19} \text{erg.seg} \times 10^{19} \text{seg}^{-1}$
$\Delta E = n \cdot 6,6...\times 10^{-27} \text{erg.seg} \times 10^{19} \text{seg}^{-1}$

$\overline{\Delta E} = \overline{\Delta E}_x \cdot 10^{-19} \text{seg} \times 10^{19} \text{seg}^{-1}$
$\overline{\Delta E} = \bar{n} K \cdot 10^{-7} \text{joule} \times 10^{-19} \text{seg} \times 10^{19} \text{seg}^{-1}$
$= \bar{n} \cdot 6,6...\times 10^{-8} \cdot 10^{-7} \text{joule.seg} \times 10^{19} \text{seg}^{-1}$
$\overline{\Delta E} = \bar{n} \cdot 6,6...\times 10^{-34} \text{joule.seg} \times 10^{19} \text{seg}^{-1}$

$h = 6,6...\times 10^{-27}$ erg.seg   PLANCK'S CONSTANT $\bar{h} = 6,6...\times 10^{-34}$ Joule.seg
$\nu = 10^{19}$ seg^{-1} GRAVITATIONAL FREQUENCY $\bar{\nu} = 10^{19}$ seg^{-1}

$\boxed{\Delta E = nh\nu}$ erg   PLANCK'S POSTULATE $\boxed{\overline{\Delta E} = \bar{n}\bar{h}\bar{\nu}}$ Joule

P.F.de MESQUITA

It was unbelievable that another anonymous opinionist could not move from one system of units to another! And that he was still a judge of scientific articles from this Science & Culture Journal of SBPC, with this scientific culture!

On March 7, 1977 the Director of Science & Culture Journal of SBPC, José Reis, communicated the Mesquita that the Board of SBPC chaired by the Physics Professor Dr. Oscar Sala rejected to publish the article *"The Physical Meaning of the Newtonian*

Unified Constant Reveals a Way for Studing Satellite Dynamics within a Unified Electromagnetic and Gravitational Field".

On May 30, 1977 Mesquita sent Letter to José Reis and Oscar Sala reiterating the publication of the article in which as sent attachments, were the Communication Summaries accepted for presentation and publication concerning the SBPC for 29th. Annual Meeting in 1977, of the works *"Of Newton's Law to Planck's Postulate in CGS and MKS"* and *"The Graviton in CGS and MKS"*, in which in the CGS system was expressed by:

$E_{graviton} = h' |\nu_g| = (m'c^2) |\nu_g| = 6.6... \times 10^{-27} \times |10^{19}|$ erg showing that the gravitational frequencies discovered by Mesquita were invariant in the two systems of units, that is, proving the error of the two opinionists who had rejected to publish the article sent to Science & Culture Journal of SBPC.

That is, the publication in this Journal was refused, but these two Communications that presented the basis of the Mesquita's gravitational theory that was listed in that publication was approved for publication of scientific articles for the Annual Meeting in 1977 of SBPC who belonged to Science & Culture Journal!

On 8 July 1977 at the General Assembly of the SBPC of 29th. Annual Meeting, Mesquita presented also a protest to appear in Minutes on the refusal of the summit of SBPC, SBF and Academy of Sciences for publication of scientific articles in the Science & Culture Journal.

However in August 1977 the Science & Culture Journal of SBPC, through José Reis, presented Editorial on this July 1977 about Mesquita protest mentioning that it would publish this Protest, *"... as well as the contestation it deserves"*.

And in December of 1977 José Reis published in the Science & Culture Journal the *"Contesting that deserves"*, but instead of publishing the Mesquita Protest of July 1977 presented a Letter from Mesquita to the Science & Culture Journal on the non-publication of the proposed article and the position of the Jurist Prof. Washington de Barros Monteiro favorable to the publication.

But in this dispute of José Reis, who presented as "Our Clarification" in Science & Culture, Reports that *"... we take the case to the knowledge of the Board of SBPC who has decided to keep the opinions of the arbitrators chosen among Physicists of the most renowned"*.

In this case, this Board of SBPC was formed by Oscar Sala, President of SBPC, two Vice-Presidents, three Secretaries and one Treasurer, from other sectors of SBPC and José Reis.

This "clarification", by José Reis and Oscar Sala, clarified definitively the scope of the incompetence in which these "renowned" Physicists were situated, in confusing frequency with flux and with quantity of energy. And the worst is that José Reis and Oscar Sala accepted the Report despite the detailed Replica of

Mesquita showing the error made. More importantly, if these Physicists were renowned Doctors, also the Professor Engineer Mesquita was a Doctor in Physical Sciences, then Oscar Sala and José Reis should publish the work and their replicas. If the arbitrators were renowned and accountable then their Reports could be published along with the proposed article and their Replicas. Or else, this Reports were fakes to stop the publication. The first two referees did not issue any comment to the Mesquita Replica! But José Reis and Oscar Sala accepted this omission or underestimated the article sent for publication. The Mesquita arguments were ignored, or understood and omitted for not to be published. The third renowned referee had made the same mistake as the two others, and insisted on the error, so that the article and the Report of the renowned authority were published by pointing out the error of the article and the replica by the author. But beyond the unconditional protection of the renowned of the "Law, Order and Science", without their retorts to the questioning of their Reports, because they considered themselves owners of the absolute truth, this article had already been approved at the Symposium of the International os Space Research, COSPAR, presented, and published its Summary in the Yearbook of this International Symposium. As judges, José Reis, and Oscar Sala, decided by "Renowned" Physicists that the physical phenomenon, the flow of 1 gram per second in CGS was different from 0.001 kilogram per second on the MKS or a flux of 1 gram per cm^2 was different of 0,001 kilogram per 0,0001 m^2 ! For these "Renowned" the physical fact of 1 gram = 0.001 kilogram it was different, because 1 was different of 0.001, and 1 cm^2 was different of 0,0001 m^2, so they would also be different quantities of mass, energy, frequency, flow or of flux, because the numbers would be different!!! That is, the Renowned Physicists as mentioned in their point of view in Science & Culture Journal by José Reis, what they presented in reality were renowned errors! But, José Reis and Oscar Sala, also "Renowneds", ignored the COSPAR that had approved the Paper Summary that was published in the International Yearbook of COSPAR!

The refusal of Science & Culture Journal of SBPC to publish the Mesquita paper was contrary to the Guidelines of SBPC, such as freedom of research and ethics among scientists.

Society of Progress and the Retreat of Science – Another fact that was seen as an another organization by current Physics was the Brazilian Society for the Science Progress-SBPC, one of the most expressive organizations of Science in Brazil, which could only mean the progress of new ideas, directions, principles, theories, experiments and innovations. But it would seem that it was interpreted by the stagnant Physics as "Brazilian Society for the Primitivism, Protelation and Shutdown of Science"or

J. P. Hansen

"Brazilian Society for the Retreat of Science", by the vision of segments of scientific orthodoxy installed in SBPC, when that entity should represent truly by its own training Progress, Advancement and not delay and regression of science! From 1973 and up to1979, the Physicist Oscar Sala was like President of the SBPC and after until 1981 by Dr. José Reis.

LXV. ELECTRON DIFFRACTION AND QUANTUM IMPULSE THEORY
1976

In July of 1976 four students of Physics of the Catholic University, one of the best universities in the Brazil, they presented it on 28th. Meeting of the SBPC of 1976 a Communication *"Electron Diffraction According to the Impulse Theory"* in which the electron was associated to the difference of energy in the electronic beam and not to the wave associated to the matter. So, by elemental energy difference between two orbital ΔMVD = h' seg = h, where M = mass, V = velocity, D = length, h' = elementary energy, h = Planck's constant, gets the ratio of De Broglie D = h/MV. Document 14.

Doc. 14 – Communication *"Electron Diffraction According Quantum Impulse Theory"* presented by students of Catholic University, at SBPC in 1976, as Reference.

LXVI. THE REPORT OF THE PRESIDENT OF THE NUCLEAR ENERGY NATIONAL COMMISSION, PROFESSOR ENGINEER AND PHYSICIST OF SAO PAULO UNIVERSITY-USP AND CATHOLIC UNIVERSITY – 1977

In May 1977, Mesquita requested the Professor Engineer and Physicist Luís Cintra Prado so that he could make a Report on the article of gravitational radiations of frequency 10^{19} cycles/second to again send request for publication in Science & Culture Journal of SBPC.

Luís Cintra Prado graduated in 1927 in Engineering from the Polytechnic School of the University of São Paulo-USP, and Professor in General and Superior Physics of the Pontifical Catholic University of São Paulo and Applied Physics of the Faculty of Architecture and Urbanism-USP. He was Director of the Polytechnic School in 1941 and 1942, held the Rectorate of USP in 1953 and 1954. He was President of the Institute of Technological Research in São Paulo in 1956 and 1957. He held the Scientific Technical Council of the Institute of Atomic Energy from 1958 to 1961 and headed the Nuclear Engineering division of the Institute of Atomic Energy from 1960 to 1963 and in 1964 until a few years after he held the Presidency of the National Nuclear Energy Commission. In 1965 he received from the Institute of Engineering of São Paulo the title of Engineer of the Year. He was the author of a large number of works published in the Brazil and Abroad on Nuclear Physics. He was a colleague of Mesquita at the Institute of Engineering, Polytechnic School and Faculty of Architecture, both of the USP.

On May 13, 1977 Cintra Prado sent a Report contesting that the gravitational frequency could not be a universal invariant, because it would be different according to the system of units.

However, Mesquita contradicted sending Replica on June 30, 1977.

And on February 2, 1978 Cintra Prado sent new Report, but maintaining that the gravitational frequency varied from one system of units to another.

Then, Mesquita again responded on March 25, 1978 and on May 23, 1978, informing that the gravitational frequency remained the same for different unit systems.

This Report of Cintra Prado was centered on the reason K_G/h, where K_G it was the gravitational constant and h the Planck constant, which resulted not in the frequency, but in length, mass and second: in the MKS system resulted $K_G/h = [(6.6... \times 10^{-11}$ m^3/s^2 kg$)/(6.6... \times 10^{-34}$ kg m^2 s^{-2})$] = 10^{23}$ m kg^2 s^{-1}. And in the CGS, $K_G/h = 10^{19}$ cm g^2 s^{-1}. Then, Cintra Prado concluded that there were different frequencies in the two systems, because frequency is s^{-1}, and the dimensions were of different units: [m kg² s^{-1}] in MKS and {cm g² s^{-1}] in the CGS. It was a conclusion only numerical, but without physical basis, once in 1cm with [10 cycles per second] corresponds to 1 metre in 100 x [10 cycles per second]. This is, the physical phenomenon is identical, because at 0.01 meter = 1 cm there will be 10 cycles per second! An elementary drawing shows the same physical fact of 10 cycles/s:

In 1 cm: ooooooooo → 10 cycles/s in 1 cm
In 0,01m: ooooooooo → 10 cycles/s in 0,01m

Also, Cintra Prado decided to invent another proof that the frequencies were different in two systems of units, now at the beginning making the correct division between energy and the constant h for the two systems of units, in which resulted the number of the frequency in s^{-1}. But, in Newton's law of universal gravitation, Energy = K_G m²/r, where (m²/r) to m and r equal to the unit was 1 g²/cm in the CGS and 10^{-4} kg²/m on the MKS. So the energy in the MKS was 6.6 x 10^{-15} [newton x m] or 6.6 x 10^{-15} joule. And would result in energy/h in 10^{19} s^{-1} on the MKS and 10^{19} s^{-1} in the CGS. However, Cintra Prado applied in the division (Energy/h) = frequency, the gravitational energy flux by m² in MKS system, ie [(Energy/h)/m²] = [(frequency/m²)]. As in CGS the energy flux of the gravitational frequency is frequency/cm², so when moving to the MKS resulted frequency/10^{-4}m². However, Cintra Prado as he passed from CGS to MKS multiplied the CGS energy flux by 10^4 for being the unit of the MKS of 1m² to the flux. But, with this multiplication for 10,000 times it passed 10,000 times more energy than in CGS, this is 10,000 more energy in 10,000 cm²! That is, it was wrong again in the system of units, because the physical quantity should be identical in the two systems although with different numbers of the corresponding units of 1000g in 1kg, 100cm in 1m and the second that remained equal in the two systems.

The calculation that Cintra Prado presented as if it were evidence was that:

$$[(frequency/m^2)] = [(Energy/h)/m^2]$$

Then on the MKS:
$[(frequency/m^2)] = 6.6$ x 10^{-15} newton x m x $10^4/6.6$ x10^{-34} joule x s = 10^{23} s^{-1}

That is, he multiplied the energy by 10^4 in the second member of the expression, which should have also been carried out in the first member, to obtain the frequency x 10^4 corresponding to the gravitational energy that passed in 1m².

And also wrongly applied as being flux of energy the expression that wrote:

frequency = 6.6 x 10^{-11} [newton x m]/6.6 x 10^{-34} joule x s] = 10^{23} s

For 6.6 x 10^{-11} was in unity of the constant K_G in [m³/s² kg] and the correct thing would be 6.6 x 10^{-15} [newton x m] that would be in energy unit which divided by h resulted in the frequency 10^{19} [s⁻¹].

Despite the higher positions he had achieved in his professional career, Cintra Prado subsequently presented, in this Report, primary system errors for a scientific article that had already had a Report of presentation and published his Summary in the International Symposium of the COSPAR, being unacceptable errors as Engineer and Professor of General and Applied Physics!

Cintra Prado presented all the accreditation for his professional position he had reached, and that could allow the dissemination of the discovery of Mesquita internationally through Science & Culture Journal of SBPC, but made unfounded error of simple transformation of physical units. Or yet, he made a mistake that would be forged to maintain the position of a supreme being as a *"Doctor of Law and Physics"*, to fend off those who approached his pretentious absolute knowledge. Therefore, being alerted of the error committed kept the same posture, practicing as Mesquita commented on the Replica, simple tautology, when changing words and examples for the same error! Even creating an "R unit system" in the Opinion, in an attempt to prove that the frequency had another value in this system!

As a Professor of General and Applied Physics, Cintra Prado should know that the energy of a photon or quantum, Planck's law, was simply $\mathcal{E} = h\mathcal{V}$ where $\mathcal{E}$ it was the energy of the quantum, h the Planck constant and $\mathcal{V}$ the quantum frequency for that energy $\mathcal{E}$:

On CSG: $\mathcal{E}$ [erg] = 6.6 x 10^{-27} [erg s] $\mathcal{V}$ [s⁻¹]
So: $\mathcal{V}$ [s⁻¹] = [$\mathcal{E}$ /6.6 x 10^{-27}] [s⁻¹]
On MKS: $\mathcal{E}$ x 10^{-7} [joule] = 6.6 x 10^{-34} [joule s] $\mathcal{V}$ [s⁻¹],
So: $\mathcal{V}$ [s⁻¹] = [$\mathcal{E}$ /6.6 x 10^{-34} x 10^{-7}] [s⁻¹] = [$\mathcal{E}$ /6.6 x 10^{-27}] [s⁻¹]

That is, the frequencies are the same on both systems. Just like a physical distance of 1 cm in the CGS is the same distance of 10^{-2} m on the MKS, also the energy $\mathcal{E}$ erg of a photon or a quantum in the CGS is equal to the same photon or quantum of energy $\mathcal{E}$ x 10^{-7} joule in the MKS, because the unit of time is equal in both systems.

So the frequency $\mathcal{V}$ of the photon of an energy $\mathcal{E}$ it is the same for any system of units, including the "R" invented by Cintra Prado! Thus, also the energy and gravitational frequency are equal in the two reference systems, including in the system invented by Cintra Prado!

This conceptual error of Cintra Prado is what the theory of relativity and the current quantum mechanics represent the inconsistencies of the concept of physical significance. Cintra

Prado related energy that crossed 1 cm² with energy 10,000 or 10⁴ bigger than it was across 1 m², for 1 m² in the MKS system passed more flux than 1cm² in the CGS, but the energy flux was the same in any system, in which in the CGS is by cm² and the MKS is for 10^{-4} m as well as 1grama in the CGS is the same mass of 10^{-3} kilograms on the MKS! So for the standard mass of 1 kilogram, passing per second in an area of 1m², in the MKS system, also in the CSG system this mass pattern of 1,000 grams passed in 1 second in the area of 10^4 cm². That is, the frequency of the mass-pattern pass is 1 mass-pattern for 1 second, in the two unit systems. Although, in the MKS is (1kg/second)/m², and in the CGS be (1,000g/second)/10,000 cm² = (0.1 g/second)/cm², is passing on both systems the frequency of 1 standard-mass per second by the same total area. In the MKS passes kg per second and in the CGS passes grams per second, but in the two systems passes the mass pattern per second which is equivalent to the two systems of units as cycles per second for the frequency of electromagnetic and gravitational radiation. The standard metre, 1 metre, when switching from system to 100 cm is the same physical unit, continues the universal invariant of the standard-metre that is traversed in 1 second, whether 1m/second or 100 cm/second. That is, 1 standard-metre is traversed in second either in MKS or CGS!

But, Cintra Prado confused physical pattern with unit of measurement system! He presented Reports showing the superb Report of the Professor of the teaching of the Pre-Establishead Physics as if he owned the scientific truth. And the forceful analysis of the Mesquita on this type of procedure, which silenced definitively any other pronunciation of Cintra Prado that could maintain these errors of simple transformation of unit systems and confusing flux with frequency!

LXVII. EVIDENCE OF ELECTROMAGNETIC FIELD IN MASSES - 1978

In 1978, the Engineer Professor of the Industrial Engineering Faculty of São Bernardo do Sul-Brazil, Ernesto E. E. Geiger, presented new data on electromagnetic fields related to body masses, gravity, this time for appreciation of Physicists on 30th. Annual Meeting of the SBPC in 1978, with the article *"Electric, Magnetic and Gravitational Field"*.

LXVIII. THE GRAVITATIONAL INVARIANT – 1979

While three referees of Science & Culture Magazine of SBPC had refused the Mesquita article for publication, in January 1979 in the 4th. Symposium on Physics Teaching organized by SBF was approved for Communication *"Because the Frequency $\nu = 10^{19}$*

s^{-1} *it is Invariant of Universal Gravitation"* precisely an article on the elementary error of incompetence or forged of the system of units of these three referees to prevent the disclosure of that scientific article on gravitational radiation. It was a Commission of the SBF of Physicists who had in teaching the didactic as the center of their activities, therefore recognizing the precise deduction in concept in the system of units.

LXIX. DIAMETER OF THE BLACK HOLE BY THE QUANTUM IMPULSE THEORY WITH GRAVITATIONAL RADIATION OF 10^{19} CYCLES PER SECOND
1974 to 1980

Accordingly E. Parker in his book of 1974 he reported Hawking's theory of the electromagnetic radiation of black holes formed by the curvature of space according to general relativity, which followed Planck's law of the radiation spectrum. And Parker concluded: *"Since Planck's formula belongs to quantum mechanics and that the black hole is described by Einstein's theory of relativity, it seems to us to be in the presence of the first link between these two theories and perhaps to its unification"*.

In July 1979 at 31th. Annual Meeting of the SBPC Mesquita presented the Communication *"Algebraic Solution for the Diameter of Black Hole"*, in which it deduced the black hole radius from quantum impulse theory and gravitational radiation of 10^{19} s^{-1}. Document 15.

J. P. Hansen

> **31ª. Reunião SBPC - 1979**
>
> **SOCIEDADE BRASILEIRA PARA O PROGRESSO DA CIÊNCIA**
>
> FORMULÁRIO PARA A REPRODUÇÃO DE RESUMO
>
> 1. "Buraco Negro"
> 2. Astrofísica
>
> SOLUÇÃO ALGÉBRICA QUÂNTICA PARA O DIÂMETRO DE BURACO NEGRO P.F.de Mesquita (Escola Politécnica da Universidade de São Paulo)
> *[body of abstract in Portuguese with equations (1)–(5)]*
>
> * CONCISE ENCYCLOPEDIA OF ASTRONOMY-Weigert and Zimmermann-Ed.Adam Hilger - England
> ** THE KEY OF THE UNIVERSE-Nigel Calder-Ed.British Broadcasting Corporation- England
>
> PAULO FERRAZ DE MESQUITA

Doc. 15 – *"Algebraic Solution for Diameter of Black Hole"* Communication presented by Mesquita in 1979 at SBPC, as Reference.

And then in 1980, paper of Mesquita at "Revista de Engenharia", Brazil: "Buraco Negro – O que é?" *(Black hole – What is it?),* where it shows in true grandeur the diameter of the black hole that would have the same mass of the Earth, as Nigel Calder in *"The Key of the Universe"* in References. Figure 41.

●

Fig. 41 – Mass black hole diameter in true grandeur, equal to Earth diameter, calculated by quantum impulse, similar to that calculated by the current Physics.

LXX. THE PHYSICS IN THE DECADE OF 70 WITH MORE RENORMALIZATION, QUARKS AND GLUONS WITHOUT MASS, SPECIAL CHARGES OR "COLORED" AND STRING PARTICLES
1970 Decade

New Renormalization – In 1972, G. W. Hooft and M. Veltman; C. G. Bollini and J. J. Giambiagi had presented renormalization this time to the W^+, W^- and $Z°$ bosons.

Special Charges: "Colored" and "Anticolour"- In the atom, each electron can only occupy a quantum state. Thus, two electrons at the same energy level and at the same orbital momentum should have spins $+ 1/2$ and $- 1/2$ according to the Pauli principle based on the electron radiation emissions of the atomic spectra. For the formed quarks and antiquark particles it would also be necessary to validate this quantum state, and in 1965 O. W. Greenberg, T. Han and Y. Nambu proposed a new quantum number for this differentiation of quarks and antiquarks. And in 1970 H. Fritzsch, H. Leutwyler and Gell-Mann, proposed that each quark and antiquark in the particles should occupy only one quantum state, with its orbital angular momentums and spins $+ 1/2$ and $- 1/2$. But they should contain another type of charge beyond positive and negative, something that would influence the attraction and repulsion of quarks and so M. Gell-Mann, Fritzsch and J. Deena was named the charge-color. In the case of a particle composed of 3 quarks equal to uuu, ddd and sss, the spins would be identical, for example $+ 1/2$, but with different charge-color. And the gluons without mass, virtual with an interaction within their almost lifeless with a life time of 10^{-23} seconds, bearing charge-color. Thus, the proton formed by 3 uud quarks and the neutron by udd, would exchange a u for a d through the incredible properties of the gluon almost without existence, this is virtual, without mass and with special charge-color! Gell-Mann presented life calculation average meson $\pi°$ on that the higher the number of charges colors in the meson the faster it would be its disintegration, ie from 0.75×10^{-15} s to 8.3×10^{-17} s, experimentally confirmed in 1971. And between 1970 and 1973, it was observed that the quantities of particles formed were proportionate to three charges colors for each quark u, d and s.

In 1973, H. D. Politzer; D. Gross; and F. Wilczek presented the quantum dynamic chromium CDQ in which the gluons were massless particles of 8 types of charge-color-anticolor, in which in the emission or absorption of the gluon by the quark there was change and color of the quark. And the quarks the farther they drifted between them in the proton, the more the force was increased between them, which was explained by the so-called asymptotic confinement! And also in 1974, A. Xodos, R. L. Jaffe,

K. Johnson, C. B. Thorn and Weisskopf presented containment models, and in 1975, H. J. W. Muller-Kirsten; T. Goldman and S. Yankielowicz; and J. F. Gunion and L. F. Li presented potential confinement. Thus, an explanation was sought for the confinement of quarks through models, such as a space bubble exerting internal pressure. Also, a field of gluons around the quarks was imagined to form narrow pipes of charge-color between the quarks as they separated or approached, in which the resistance between them increased the greater the separation.

More Types of Quarks – In 1964 the neutrino of the muon was detected, which constituted 4 leptons particles: electron, electron neutrino, muon and muon neutrino; And there was the hypothesis for the interaction of the other particles by 4 quarks, which would be the "charm", by D. Amati; H. Abiona, J. Nuyts and J. Prentki; B. J. Bjorken and Glashow; Z. Maki and Ohnuki; and Y. Hara. In 1970, Glashow, J. Iliopoulos and L. Maiani presented the interactions between leptons and particles of hadron, in which quarks in the strong interactions and boson mediators W^+, W^- and $Z^°$ were present. In 1973, at CERN, P. Musset and others through scattering neutrinos with hadrons obtained evidence of these interactions and in 1974 at the Fermi National Accelerator Laboratory-FNAL or FERMILAB, near Chicago in the United States by C. Rubbia and others. In 1974, B. Richter and others, at the Stanford Positron-Electron Asymmetric Ring-SPEAR from the Stanford Linear Accelerator Center Integrated Safety Management-National Accelerator Laboratory-SLAC, Stanford in the United States; E S. C. C. Ting and others, in the alternating Gradient Synchrotron Brookhaven National Laboratory-BNL AGS, and confirmed in 1974 by C. B and others, in Anello d'Accumulazione-AdA in Frascati, in Italy, They discovered a new particle that would follow the strong interaction. In 1974 the quark "charm" predicted in 1964 was detected.

And in 1975, T. Appelquist, Politzer, E. Eichten. K. Gottfried. T. Kinoshita, J. Kogut. K. D. Lane and T. M. Yan; Okubo, V. S. Mathur and S. Borchardt presented the hypothesis that the particle "charm" and others that had also been discovered, as between 1963 and 1981 by Lattes and S. Hasegawa; in 1975 by Musset and others at CERN, J. Kroch and others, in the FNAL; Richter and others, in the SPEAR; Palmer, Samis and others; B. Knapp and others, would be quark states "charm". But in 1975, M. L. Perl, Richter and others, in the SPEAR, and at 1977 H. Meyer and others, in the Deutsches Elektronen-Synchrotron-DESY in Hamburg in Germany, in the collision electron-positron obtained muons, electron of the muon, another particle U that after was called tau and its electron, with mass of 1.784 Gev and life time-average of 4.6 $\times 10^{-13}$ s obtained in 1982 by G. J. Feldman and others. Then, in 1974, M. Barnett, Gursey, P. Sikivie, P. Ramond; in 1975, H.

Hariri; and in 1976, Eichten and Gottfried presented the hypothesis that there were 6 leptons there should also be 6 quarks to establish a symmetry of 6 particles. In 1977, Lederman and others, of the Fermilab they discovered a new upsilon particle of 9.46 Gev, which would be formed by a fifth quark, the "bottom", which was confirmed by W. R. Innes and others. And other similar particles of 10, 10.3 and 10.6 Gev were detected in the DESY and Cornell Electron Storage Ring-CESR From Cornell University in the state of New York in the United States.

In 1978, R. Taylor and others; L. M. Barkov and M. S. Zolotorev; C. Baltay, R. Palmer and N. Samis and others observed the neutrino + proton interaction → neutrino + neutron + meson π^+, which would be mediated by the particles W^+, W^- and Z°.

Leptons, Preons, Barium and Quarks – In 1973, J. C. Pati and Salam presented hypothesis that leptons would be formed of preons particles similar to the barium formed of quarks. In 1973, R. Gatto presented that high energy electrons could present strong force interaction, and in 1975 Lopes presented that the leptons would also be formed of quarks such as barium and other types of quarks according to Lopes, C. E. M. Aguiar, S. Fleury, M. B. G. Ducaty, A. L. B. Lopes, J. A. S. Martins and D. Spehler. In 1979, H. Harari; M. A. Shupe; and in 1981, Fritzsch and G. Mandelbaum presented a hypothesis that also quarks such as electrons and neutrinos would be formed from other particles.

Quarks and Gluons Jets – In 1975, in storage SPEAR, in electron collision and positron with energy < 2 x 18 Gev was observed the evidence of quarks jets, and at 1979 in the DESY in electron collision and positron was observed the existence of quarks and antiquark jets, which accelerated originated the interactions with energies appearing as a third radiation jet that would be that of gluon.

Free Quarks – In 1977, W. M. Fairbank, G. S. La Rue and A. F. Hebard presented an experiment for detecting charges in magnetic spheres and obtaining fractional charges. However, in 1980, G. Mopurgo and M. Marinelli considered these fractional charges as produced by electromagnetic forces. In 1972, Patie Salam and in 1974 as H. Georgi and Glashow sought to unite the strong, weak and electromagnetic interactions, but the coupling constant that involved the proton would result in average life of 10^{31} years that vetoed this model. But in high energies, densities and temperatures the protons would be annihilated in quarks.

Supersymmetry – In 1971, Y. A. Gold'fond and E. P. Likhtman; J. Gervais and B. Dokita presented the hyposymmetry hypothesis, in which all particles would have antiparticles with differential spins in relation to the particles, and the other properties of supersymmetric particles and antiparticles would be identical and they could be transformed through the modification of spins, also covering gravity with the interaction by the graviton without mass, and spin 2, which should be predicted for eventual unification of quantum mechanics with general relativity.

And in 1976, D. Z. Freedman, P. Nieuwenhuizen and S. Ferrara; S. Defection and B. Zumino observed that in a local particle transformation moving particles from one point to another would result in gravitational interaction, so there would also be the unification of all interactions: strong, weak, electromagnetic and gravitational. However, infinitesimal between forces would result in unstable unification, where normalization would also result in instability.

Strings and Superstrings – G. Veneziano in 1968 had observed, that in the strong, nuclear force, in high-energy particle collisions, a result of surprising connection to the expression of an indeterminate mathematical function of first-species integral that had been discovered by L. Euler. Then, in 1970, L. Susskind, H. Nielsen and Nambu showed that the spreading formula at particle collision angles and L. Euler was due to a slender elastic string, with different models of vibration, in which the energy was characteristic of its type of vibration, without mass, with charge, presenting 26 dimensions of space-time, and particles with superluminary velocity or tachyons. In 1971, Ramond; A. Never; and J. Schwartz, they presented the string of supersymmetry, that is, the superstring with 10 spatial dimensions and a dimension without tachyons. In 1972, J. Schwarz, A. Neveu and J. Scherk presented that the string could vibrate in up to 10 dimensions of space-time. And in 1974, J. Schwarz and Scherk presented in the development of string theory the interaction of gravity by graviton without mass and spin 2, exactly what had been predicted for eventual unification of quantum mechanics with general relativity. Also, F. Gliozzi, Scherk and D. Olive expanded the development to the vibrations for the spins of the particles. Also, string researchers created the vibrating particles, the strings with and without mass, with 10 and up to 11 dimensions that they admitted as "rolled up"!

LXXI. MASSLESS NEUTRINO BY PRE-ESTABLISHEAD PHYSICS – **1979**

On August 28, 1979 was presented a Conference *"Elementary Particle Physics"* at the Engineering Institute in São Paulo-Brazil, by the Physical Engineer Professor Roberto Salmeron, who was active at CERN, through invitation to this Conference conducted by Mesquita which had been his Teacher at São Paulo University-USP.

His initial comment was to reveal that Physics presented two major problems, which were the smallest constituents of matter and the structure and evolution of the universe. Searches sought knowledge of the structure of matter and technology. He then said that in particle shock new particles were created from the theory of relativity in which energy was disappearing to create energy. In particle acceleration it was the electric field that created a force that accelerated the particle. The particles exerted actions between them, interacting them. There were four types of interaction: gravitational by the masses, but that between particles was not detectable; electromagnetic by the fact that the particle has charge; weak, in which the particles stayed close for a time of 10^{-10} to 10^{-11} seconds; strong, in which the proximity time was only 10^{-22} to 10^{-23} seconds. The fundamental problem of Physics was to know the laws of these four interactions, which would explain everything in the universe, that is, because the bodies would be made up of a certain structure. And that between 1947 and 1967 had been obtained a major modification in the structure of the matter that passed from the particles of the photon of mass zero, electron, proton, neutron, meson and neutrino of zero mass for hundreds of particles with strange properties, called strangeness, charm and beauty, formed of quarks which were formed of fractional charges, and also particles of interaction such as photons, particles W and Z, gluons and gravitons in gravitational interaction.

Mesquita questioned: *"What was charge? And how to measure the fractional charge?"*.

Salmeron replied: *"A neutron had no charge, but had spin caused by internal charges. Charge was a concept that we do not know, nor do we know why it exists"*.

Mesquita: *"However, the concept of energy and impulse was obtained $\mathcal{E} = h\mathcal{V} = m'c \times c$, so getting m'. And the impulse theory also gets the formula $\mathcal{E} = mc^2$. It was mentioned that the neutrino had no mass, but why not assign to each energy a mass in the different types of neutrino?"*.

Salmeron: *"Because experimentally when measuring neutrino mass is found to be zero value"*.

For the power owners, the theory of relativity and quantum mechanics had all been settled in current Physics, so Salmeron's responses were an astonishing recognition of the limitations this Physics. However, as an active at CERN it was possible to prove as Salmeron presented his lecture, that the world position of formal Physics had its bases in the theory of relativity, in which experiments were admitted as 100% favorable to the theory, as the expression of energy and mass. But the energy and mass also had their relationship obtained by the theory of quantum impulse that explained the constitution of the particles, sought by Salmeron and its Physics. For example, the mass-growth formula of Einstein's relativity particles came from Voigt, Larmor, Lorentz, but also deduced by Erdelyi and Csiky Jr. By the quantum impulse that explained the macro-mechanics and micro-quantum-mechanics phenomena. Also, relativity showed the relationship of variables in a superficial way, macroscopic, and with a lot of arrangement and supplies of values to conform to the relativistic forecast. At this conference, Salmeron stated that he received dozens of articles daily and could not analyze the theory of quantum impulse when Mesquita in two opportunities in previous years delivered it personally. Thus, with the refusal of international magazines to publish new ideas, and of physicists even receiving in personal submission ignore the receipt, so what remained were the problems of Physics to be solved, but that remained unresolved, with particles of zero mass, inconsistent theory such as that of relativity taken as absolute truth and that it would have 100% of proof, and the formal confession by Salmeron that: _"... we don't know what charge is"_. However, the concept of charge was explained by the theory of impulse, in which quantum elementary energy h' = m'c x c constituted the particles with their right or left chirality turning which was the positive or negative charge.

The mass of the neutrino was derived from the correct interpretation of the expression of mass increase of electrons and other particles, in which the speed v_η of the neutrino was limited to $v_\eta \neq c$, because the speed c was that of the elementary mass m' that constituted the elementary impulse m'c. And the energy of the neutrino ε_η = n m'c x c = n m'c^2 resulted in mass m_η = n m'c^2. However, as Salmeron had presented at this conference, up to 1979 it was not possible to measure the neutrino mass, and so the Pre-Establishead Physics considered the neutrino without mass because it could not directly measure its mass! By the own theory of relativity, the mass of the neutrino could never be zero, for in this case the speed of the neutrino would be equal to the speed of light, and no particle could reach such speed, for the mass would be infinite! But for not can detect directly concluded that there was no mass!

LXXII. THE SWEDISH ACADEMY OF SCIENCES AND THE NOBEL PRIZE IN PHYSICS FOR THE UNIFICATION OF FORCES – <u>1978 to 1980</u>

In October 1978, the Physicist A. Salam, Director of the International Center for Theoretical Physics of Triste in Italy, of the International Atomic Energy Agency-IAEA of the United Nations Educational, Scientific and Cultural Organization had presented an article in the *IAEA Bulletin* mentioning that the electromagnetic force, the nuclear weak of radioactivity, the strong nuclear force of the atomic nucleus and the gravitational were identical. And that the unification of the electromagnetic and weak force was contained in the theory of Glashow, Weinberg and Salam as an experiment, of June 1978, at Stanford University in California, which in proton-based electron collisions showed some parity difference between left and right in the behavior of electrons after shock with the protons, being according to prediction of this electro-weak theory. So all forces should be aspects of one force.

In October 1979 the Swedish Academy of Sciences presented the Nobel Prize in Physics to Salam, Weinberg, and Glashow by the electro-weak theory, through the Communication of Bengt Nagel informing that this theory could open possibilities in the future of energy and was an important step to understanding the structure and origin of the universe.

And on January 21, 1980 Mesquita presented a Communication in the Division of Alternatives Energy at the Engineering Institute in SãoPaulo-Brazil on nuclear energy in relation to the theories of the Pre-Establishead Physics, such as the electro-weak theory that despite obtaining the Nobel of Physics it presented gravity and strong force outside the unification of the forces of Physics. However, this Physics continued to ignore the quantum impulse that defined the elementary mass, the elementary energy and the gravitational radiation formed by the elementary energy h' and the frequency of 10^{19} cycles per second, that is, gravitational energy $E_G / 1s = [(h\ 10^{19}\ s^{-1})/1s] = 10^{19}\ s^{-1}\ h'$.

As the dissemination of theories of quantum and gravitational impulses continued to be prevented from publication in the Brazil and Abroad, there was an opportunity for them to be sent directly to the holders of the highest level of the scientific summit of current Physics, such as the Nobel Prize Committee and Awarded by this Organization.

Thus, on February 8, 1980, I presented information to the Nobel Prize Committee on the news that this Committee reported that the award-winning theory of Physics of 1979 sought the basic principles of the constitution of the universe and the possibility of establish the unit with the nuclear force. Then, I mentioned that gravity was out of this connection and that articles already published, by the COSPAR and Congresses of SBPC revealed physical-mathematical solution of the Unified Field in Fundamental Physics, and sending the quantum impulse theory. This information was sent to the Committee representative, to three Nobel laureates, Salam, Glashow and Weinberg, as well as other researchers.

However, no receipt of any comment, as expected, because the Manual of Pre-Establishead Physics prevented such an attitude!

LXXIII. "MODERN LESSONS" FROM CURRENT PHYSICS – 1944 to 1980

As extracted from the article *"Culters of Mediocrity"* in *"De Alma para Alma"* ("From Soul for Soul") of 1944 by H. Rohden, it was appropriate to the behavior that the Pre-Establishead Physics exercised on new ideas, in which it could constitute *"Modern Lessons"* of the Teaching of Physics:

1-Ignore new ideas as it is the creation of heretics.
2-Only know what is published in the Journals of the Institution of Pre-Establishead Physics. Out that's high-danger contraband.
3-Stay between the grids of the pre-establishead saber and move away from stratospheric flights.
4-Only open doors opened.
5-Scroll only previously aligned rails.
6-Walk like sheep of the flock and never move away from the margin of the pre-establishead route.
7 Just as the greatest idealist in history, Christ, was crucified, also to move away, to silence and crucify other idealists who should have the same fate.
8-To address research and publications authorized exclusively by the "Officers" this Physics.

LXXIV. NEW EXPERIMENT ON EARTH VELOCITY – 1980

In May 1980 the University of Campinas-Brazil-UNICAMP, through the Physicist Lattes presented a result of experiment with light of mercury, in which there seemed to be indication of the motion of the Earth in space. Lattes had discovered the meson π of the interaction of the atomic nucleus along with Muirhead, Ochialini and Powell, which resulted in the Nobel of Physics only for Powell by the photographic technique. However, in July 1980 Lattes withdrew the article for publication of this experiment, informing that it would be reanalyzed. Through a diffraction grid, Lattes had observed differences in light paths during the Earth's motion in space, but was attributed to the difference in temperature between day and night. But, Lattes considered that the theory of relativity erroneously equated measurements with physical magnitude and was then incorrect.

LXXV. THE SCIENCE PROGRESS BRAZLIAN SOCIETY AND FREEDOM OF EXPRESSION – 1949 to 1980

In 1980 the Presidency of the SBPC was exercised by the Professor of Physics-USP, José Goldemberg, Vice-President of SBPC, in which in the 32th. Annual Meeting was discussed the democratic options for Science and Education, in which the press reported on the Meeting that: *"Who goes to SBPC is an immense majority of young students and researchers, just to the cata of what they cannot have in the University: the free debate around the cutting edge research"*.

And so it was expressed Professor Herman Lent one of the founders of SBPC and participant of the First Meeting in 1949: *"It is fundamental to freedom of research, freedom of chair, lack of prejudice against ideas and freedom to teach and to learn"*.

And in Physics, when analyzing nuclear energy, the Physicist Luís Pinguelli Rosa commented: *"Thermonuclear fusion is still a utopia. However, within a realistic perspective, it will be possible to think about the use of this method in many decades"*.

In 1980 the Nobel Prize for Physics Emilio Segré stated that particle Physics was the only area where the largest number of unsolved problems existed. Some of them were almost a hundred years old and the tendency of physicists was to forget them because there was no indication of a solution.

So it was seen the need to consider new ideas, for despite the theories of this nuclear Physics, the distance to results was admitted as many decades. Yes, there were new principles that were lacking for knowledge and then new developments.

And Mesquita presented the Communication *"Notion of the Impulse Theory in Fundamental Physics"*. And also, the *"The Physical Significance of Loschmidt's Constant in the Impulse Theory"*, in the SBF in its 14th. Annual Meeting with the 32th. of SBPC, in 1980, in which it was related this constant with the quantum h and the gravitational frequency of 10^{19} s^{-1}.

During the General Assembly of the 32th. Annual Meeting of the SBPC in 1980, Mesquita returned to protest on the unacceptable refusal to publish new ideas by Science & Culture Journal of SBPC, that distorted the view of SBPC what was the main theme of SBPC in this year of 1980 and the reason for SBPC, which was the freedom of teaching and research. And Mesquita reiterated in 1980 to the SBPC that the paper should be published.

By ample demonstration of approval by the Sovereign Assembly should the President of the Assembly chaired by José Goldemberg to recognize this sovereignty and approve the Mesquita paper for publication. However, Goldemberg did not consider it a Vote of Confidence, but only one Intervention, and ignoring the broad majority favorable to the publication.

On 11 November 1980 Mesquita sent a Letter to Goldemberg requesting the publication as a large favorable majority in the plenary demonstration.

LXXVI. NUCLEAR ENERGY AT THE INSTITUTE OF ENGINEERING-BRAZIL – **1981 to 2070!**

In February 1981 I sent to the President of the Institute of Engineering in São Paulo-Brazil, Eng. Luiz Alfredo Falcon Bauer, information on a predicted nuclear energy control and warning about the resistance of Pre-Establishead Physics of prevent new information. The Physicist Dr. Marcelo Damy de Souza Santos, who was Director of the Institute of Atomic Energy in Brazil, gave a Conference at the Institute of Eninnering stating that it would take 50 years of research to obtain control of nuclear fusion, this is in 2030. But, the Physicist Nobel Prize in Physics, Abdul Salam, predicted for the year 2000, and other physicists predicted for 2070 that control. However, in 1980, England interrupted the search with the ZETA reactor (Zero Energy Thermonuclear Assembly) after 20 years of operation, without any success, with the current laws of Physics.

Then, with new physical principles, could engineering act for the reach of this nuclear control, in that Mesquita, of the Council of the Institute of Engineering, faced reaction to new basic and fundamental concepts of Physics and would be necessary the constant performance of the Institute of Engineering.

There were expressive replies of agreement by official statement and members of the Institute of Engineering with full approval to the presentation of new ideas and responses of appreciation by the national industry. But the Pre-Establishead Physiccs did not to new ideas.

LXXVII. ATTESTATION OF THE INVARIANT GRAVITATIONAL FREQUENCY OF MESQUITA AND APPROVAL OF SCIENCE PROGRESS BRAZILIAN SOCIETY-SBPC AGAINST THE OWN SBPC – **1981**

In July of 1981 in the 33th. Meeting of the SBPC Mesquita presented a Communication *"Existence and Physical Uniqueness Mathematical of the Gravitational Invariant $V = 10^{19}$ s^{-1}"*, which once again proved the errors of the three judges of Pre-Establishead Physics, in the negative publication of the work of Mesquita for the Science & Culture Journal of SBPC, in which the SBPC itself approved this Communication to be presented at the Annual Meeting of 1981 and publication of the summary of this article in the Yearbook-SBPC in 1981.

LXXVIII. A MATTER NEVER CLOSED! – 1981

And at the General Assembly of the SBPC of 1981 Mesquita once again presented a Vote of Confidence for publication of him paper in Science & Culture Journal, mainly with the proof of the Gravitational Invariant and approved and presented in SBPC's Communications. However, the majority of participants in this Assembly were of the Political and Social area and were not duly aware of the existing backlog on the publication of a scientific work of Physics and preferred not to manifest, while few participants linked to the Pre Establishead Physics Summit followed the *"Modern Lessons"* of this Inquisition by rejecting the motion. The inconsistency was visible, in which the SBPC Committee approved precisely the
Communication that correctly pointed to the Gravitational Invariant, while the Assembly by a Minority approved the rejection between a Majority that there was abstained this Vote of Confidence that treated exactly of the gravitational invariant for publication in Science & Culture of SBPC! There was fifty approvals on Politics and Social and the only one of Physics was rejected by a minority with an abstention from the majority.

Despite this vote aimed at a minority approving, Science & Culture published in November 1981 in the words of José Reis, Director of the Journal, came to declare that: *"Now, with the rejection of the Vote of Confidence for the 33th. Annual Meeting, Science & Culture Journal, aware of this sovereign deliberation, is understood that the subject is definitively closed".*

But this matter, practiced with without any ethics could never be closed, in which the authors of this refusal following the *"Modern Lessons"* of the Pre-Establishead Physics Inquisition deliberately practiced everything against their own Objectives of SBPC and Vote of Confidences of freedom that had been passed hitherto by SBPC, to keep their untouched positions of *"Owners of the Power of Law and Sciences"*. What was closed was the confession of incompetence, of disagreement, superb and cover-up of primary errors of the renowned judges.

LXXIX. DENSITY AND AGE OF THE UNIVERSE 13.7 BILLION YEARS to 1982

On the 34th. Annual Meeting of the SBPC in 1982, Mesquita presented the Communications *"Age of the Universe by Quantum Impulse in the Big Bang Hypothesis"* and *"Average Density of the Physical Universe by the Quantum Impulse of Erdelyi"*. These Communications show that for a density of 3.23846×10^{27} g/cm^3 obtained from quantum impulse, of an original sphere of compressed matter in the minimum radius of the conventional black hole of the Big Bang, is obtained the age of the universe in 3.4 billion years, and from ~13.7 billion years to a density four times greater. For the average density of the universe, from 3.23846 x 10^{27} g/cm^3 is obtained 3.088 x 10^{-28} g/cm^3, while by relativity is 3.5 x 10^{-28} g/cm^3, and by estimates of astronomers of 10^{-27} g/cm^3 to 10^{-29} g/cm^3.

LXXX. THE INSTITUTE OF ENGINEERING AND THE NOBEL PRIZE IN PHYSICS – **1982 to 1984**

On 19 April and 23 June 1982, as a member of Energy Alternatives of the Institute of Engineering in Brazil, I sent to the President of the Institute of Engineering-IE, Engineer Lauro Rios a statement focusing on the nuclear energy in which it was necessary the scientific and technological development with the Physics and Engineering. And the works of the Mesquita presented the proper importance for Physics and Engineering, thus justifying it to be referred to the Nobel Prize Committee by IE.

However, on 15 August 1982 IE declined to send this referral, claiming that advanced theoretical physics was incompatible with the functions of the IE in engineering progress!

On September 7, 1982 the author replied to IE that in engineering teaching was the chair of Advanced Physics or Modern Physics, therefore an unacceptable argument to decline that request.

On October 4, 1982 Mesquita sent Letter to Lauro Rios, informing that the Councillor Engineer Braz Juliano sent him an opinion on gravitational radiations in which Braz Juliano endorsed what was written by his colleague Aron Kuppermann, Professor of Chemical Physics of the California Intitute of Technology, that mentioned: *"The rest masses of the graviton, gravitational field, photon and quantum of electromagnetic field need to be zero by firm experimental fundamentals. and the adoption of a fixed frequency of 10^{19} sec^{-1} for the graviton is arbitrary and should be by experimental facts"*.

However, Kuppermann complemented that he was not an expert in this field! What denoted gratuity and levity of opinion, besides presenting the continued fallacies of the theory of relativity and quantum mechanics of the current Physics, on particles devoid of mass when at rest and the field itself without mass.

On October 5, 1982 the Engineer Fernando A. Ramos Gonçalves of the deliberative Council of IE informed in Letter that his vote and of another adviser were favorable to the submission of the works of Mesquita to the Nobel Prize Committee by the IE, but another eighteen they voted against it.

However, CERN in Europe and FERMILAB in the United States planned the start of building bigger machines for particle acceleration experiments, which would involve the most advanced engineering in all segments. But, without the slightest insight and competence of that Advanced Theoretical Physics was Modern Physics Linked to Engineering, these 18 Advisers in 1982 voted, instead to the aggrandizement of IE, to relegate themselves. For, in 1975 more than 800 IE Engineers had subscribed to what Science should be in the Statutes of IE, and so these 18 negative votes did not represent the true majority of Engineers to elevate the grandeur of IE. Although these 18 Advisers represented the members by election by the statutes, they were far from the other attitudes of the non-voting partners of IE Engineers. So this 18 votes nothing represented, but only the absence of high values!

And in January 1983, in the CERN machines built by Engineers and Physicists, Rubbia and S. Meer, through the collision of proton and antiproton discovered the particles W^+ and W^- with 80 times the mass of the proton and in June the particle $Z°$ with 90 times the mass of the proton, considered a great scientific and technological advancement. In 1984 Rubbia and Meer received the Nobel Prize in Physics for this scientific and technological advancement!

LXXXI. UNIFIED FIELD – 1983

Between January and June of 1983 had been announced by CERN the discovery of energy spikes formed in the energy range of the particle accelerator that would promote the interaction of the weak force, the particles W and Z. Many other peaks were also observed in this range energy, but the peaks of energy more according to the masses that were supposed to effect the weak interaction of the nucleus were named W and Z that unified the theory of the weak force and the electromagnetic.

However, the energy supplied to accelerate electrons or protons for collision with the target was explained by the theory of quantum impulse, not only by the equipment of coils and magnets that produced the electromagnetic field for acceleration of the particles, such as the own constitution of the field and of the particles by n m'c x c = n h'. In 1983 N. S. Lockyer and others, and E. Fernandez and others, presented the mass of ≈ 5 Gev for the bottom quark, and the average life-time of 1.4×10^{-12} s for particulates called Ypsilon.

And to disclose the unification of the physical field among all the forces, I presented a Communication *"Rational Physical Interpretation of Planck's Constant h Leads to the Rational Solution of the Unified Field of Physics"* on the theory of quantum impulse and radiation gravitational in 35th. Annual Meeting of SBPC in 1983.

LXXXII. NEW PROOF OF GRAVITATIONAL RADIATION BY THE CONICAL MOTION – 1983

On June 7, 1983 other direct online information was sent by me author to the Abroad, to the Director of the Academy of Sciences of Sweden, Bengt Nagel, with copy to Harvard University to Glashow and Weinberg, California University for Reines, Michigan University to John Vander Velde, Nippon Butsuri Gatukai Institute in Japan and CERN for Herwing Shepper, on a unitary mass describing a conical relationship with the constant G of universal gravitation with Planck's constant h, through the theory of quantum impulse, by Communication *"Invariance Physical-Mathematics of the Conics in the Quantum Impulse Theory"* by Mesquita sent for presentation in July at 35th. Annual Meeting SBPC.

However, the Pre-Establishead Physics preferred to omit any manifestation of what was contrary to its untouchables bases, in which there was only signature of receipt by assistants, of the correspondence that had been sent.

LXXXIII. SCIENCE AT THE INSTITUTE OF ENGINEERING-BRAZIL 1983-1984

In September 1983, Science was inserted into the Statutes of the Institute of São Paulo Engineering Institute-IE, after 9 years of Mesquita propositions to magnify IE.

And on October 8, 1983 I reiterated in Letter To Engineear Plínio Oswaldo Assmann, President of IE, to be officially sent Mesquita works to the Nobel Prize Committee.

On October 17, 1983, the Engineear Hermes Ferraz, Assesssor of the Board of IE, sent to the President of IE a brilliant Report of the highest ethical and scientific consideration, endorsing the referral from the works of Mesquita to the Academy of Sciences of Sweden, because the Science of Advanced Physics was no longer a reason for IE should not be absent, since Science was the subject of the statutes of IE.

And on June 29, 1984 as requested by IE I sent again grounds on the request for referral to the Swedish Academy of Sciences of Mesquita works.

LXXXIV. SYMMETRY DETECTED – 1984

In June 1984, Rubbia detected at CERN symmetrical quarks, that is, of equal masses and opposite charges by the theory of symmetry of matter. However, in addition to explaining the origin of the mass and the explanation of what is the particle charge, the theory of quantum impulse also explained the absolute chirality or helicity of the independent particle if the observer is in the direction of meeting or moving away from the particle.

And in 1984, at CERN, there was evidence of detection of the sixth quark with mass, of particle decay W^+, W^- and $Z°$.

LXXXV. THE FIRST THEORETICAL PHYSICIST OF BRAZIL AND HIS CONCLUSIONS OF THE PHYSICS IN COMMEMORATION OF HIS BIRTHDAY OF 70 YEARS – 1964 to 1984

Unification, Particles and Fields – On July 2, 1984 was celebrated the 70th. Anniversary of the first theoretical physicist of the Brazil, the Doctor Engineer and Physicist Mário Schenberg. At the Symposium on Perspectives in Theoretical Physics in 1984 Schenberg presented that for 20 years was dealing with the unification of gravity with electromagnetism. And it had united the partial derivatives forming tensor of electromagnetism, with the four dimensions of the universe by the tensors of the Weil partial derivatives of the gravitational field, which was resulting from the curved space of the Riemann geometry and partial derivatives, that is, Riemann-Christoffel-Ricci tensors by the distribution of mass and energy generating the gravitational field by general relativity. He then united the gravitational field with the electromagnetic field, but within mathematical fields, in which the gravitational field was the distortion of space-time and the electromagnetic field as the relationship that involved charges and forces.

It was a unification without the actual quantum constitution of the gravitational or electromagnetic field.

In the book "*Thinking Physics*" Schenberg commented that the unification of the weak and electromagnetic force with the strong and gravitational force did not exist until that time.

Scheberg sought to unite quantum mechanics by Planck's constant h with the geometry of space and time. And he understood Planck's constant as unexplained by quantum mechanics, which allowed only to relate energy, time, and spatial coordinates by mathematical operations.

But, the constant h could only be explained by the correct meaning of h [(energy x time)/time] = h' (energy), by the theory of quantum impulse, because (energy x time) escaped comprehension as Eddington.

On the physical field, Schenberg reported that the concepts of particle and field were inseparable, and possibly there will be a quantum unit field theory that should explain the existence of the various types of particles and fields.

However, the theory of quantum impulse showed the constitution of the particles and their surrounding field formed of m'c x c = h', i.e. the elementary mass m' with the speed c of the light in vacuum-space measured in the Earth, forming the h' energy of the elementary quantum.

Schenberg commented, that according to Einstein the unification of electromagnetism with the gravitational field would occur extraordinary events with every reformulation of the theory of quanta.

However, it was the reformulation of quantum theory with the theory of impulse that provided the unification of the gravitational field by the constant K_G of Newton's universal gravitation with Planck's constant h, interpreted with h/second.

Schenberg continued in his conclusions by saying that Einstein considered the fine structure constant $\alpha = 1/137 = k\ e^2/hc$ a riddle in Physics, and when one understood what it represented would explain Planck's constant h. But, this ideal of explaining the fine structure constant, as Schenberg was not yet realized.

However, the theory of the impulse had already explained that this constant was an equation of the electron with the high charge squared that was the helix formed of $n\ m'c \times c = n\ h'$ in turn forming what resulted to be the charge. In other words, in addition to explaining what was the fine structure constant, it also explained what the particle charge was!

Schenberg also commented on the quarks, in which he considered complex structures, but that were not fundamental structures, and that it was quite obscure this situation of particle structure.

But by the theory of quantum impulse the quarks were also formed by the elementary quantum h' and it would have its constitution as it was deduced to the electron and to the proton by the quantum impulse theory.

Current and Future Physics, and the Freedom for New Theories – As Schenberg: *"Many studies of Physics that are assumed to be understood in reality are not, as in the case of quantum mechanics. And one cannot live from the past, for all theories will be modified and with a truly new idea"*.

Also, Schenberg said that Physics would only find another type of unification with the approximation of the development of biological processes and psychology, as seen by Heisenberg and Schrödinger.

Schenberg commented: *"General relativity itself was limited by Einstein, in which space-time presented measurements of length in geometry, but could geometry be based without the metric. And so other physical facts should be free from constraints and impositions, and the scientist would need to have freedom of creation"*.

Conclusions of Schenberg and the Terror and Obscurantism:
"An idea is original if it is a rebuttal of what is admitted as correct.

That is, they are new ideas because they are opposed to the traditionally accepted ideas. And this originality will be as much more radically innovative as the new ideas. And it takes a high determination of intellectuality to confront those of ancient visions who see it as wrong ideas to what is actually a new fact. It is the infringing of what is considered right that the new is born, and there is creation. Yes, every new theory breaks the old laws. What greater infraction was there in the laws of classical Physics than the one introduced with the theory of relativity, and then with quantum mechanics?".

Schenberg and dozens of USP teachers had been persecuted by the military dictatorship in the "Coup d'État" of 1964 which ousted President João Goulart two years before the end of his term. And he mentioned: *"That Brazil redeem these years of terror and obscurantism"*.

They were present at this Symposium, Physicists of SBF and SBPC, and among them, Goldemberg, Hamburger, Sala and Tiomno, who had criticized the Military Government for thwart freedom of thought while exercising censorship at the University. However, these renowned freedom proclaimers were the ones who had hampered the freedom of Mesquita chair at USP itself, objecting without proper scientific justification or preventing the disclosure of its request for publication of article with new ideas, exactly what Schenberg mentioned in this presentation!

The Physicist Tiomno commented that: *"In the field of particle Physics, there are more opportunities for developing ideas that lead to renewal, modification of fundamental concepts. Not only gauge theories (calibration), but also theories of supersymmetry are configured as viable paths".*

Tiomno mentioned new ideas for fundamental concepts, but by completing the phrase he kept the same ideas adopted from abstract theories of calibration and supersymmetry, ignoring the theory of quantum impulse.

The Physicist Giambiagi presented that: *"It was possible to unify electromagnetism with weak interaction, and this provides us with an indication of the path by which we must insist to complete this unification with gravitational and nuclear interactions. And with the theories of supersymmetry and great unification for nuclear, weak and electromagnetic interactions, it was the disintegration of the proton that was considered to be of infinite average life and the existence of magnetic monopoles. And the magnetic monopole would be the cornerstone of the constitution of matter".*

However, Giambiagi himself concluded that: *"With respect to the magnetic monopole, that B. Cabrera said to have observed, with machines 50 times better around the world the result was negative, even they were saying that Cabrera had detected the only magnetic monopole that was passing through the universe. As for the disintegration of the proton, also the result was negative. And over the supersymmetry statistical results showed that it didn't work. Facts that led to a certain dismay. Supersymmetrical representation is difficult, and there are complicated groups and representations, very complicated diagrams".*

But, Giambiagi kept the hope in monopoles linked to the modern concepts of mathematics, such as groups of forms and indepedentes positions of dimensions, the topology, of transformations of fields and particles in space, where the constant h entered together with the electron charge in the quantization of the electromagnetic field. And that supersymmetrical particles could be obtained experimentalally in the accelerators, in the course of the unification of the forces.

However, the constant h and charge of the particles, which were only related without the physical significance, would only be a mathematical unification of spaces and transformed fields. As for the symmetrical particles, with the quantity of quantum elementary h' energy of the impulse theory that propels protons or electrons to the collision on the target or with another proton or electron in the accelerators, peaks are continuously observed in the tracks of energy and could indicate many particles with symmetrical helicity. But the basic constitution will be in the theory of the impulse of Erdelyi and the gravitational radiation of the Mesquita for the unification of forces. Even the interaction or force itself is effected by m'c/time that results in the energy m'c x c = h'.

And Physicist Guido Beck commented that: *"In the past as a student in Europe in Vienna it was prevalent the kinetic theory of gases that initiated the structure of matter, and the molecule was admitted as a point. But then the molecule would be formed of dots representing the atoms. And at the present moment the nucleus of the atom is composed of a proton which is composed of quarks and gluons. And the future of Physics was between a greater understanding of the laws as Einstein, or just the accuracy of describing the phenomena without ever being understood as Bohr. And the tendency of Physicists is for the belief in Bohr. However, quantum theory is using redundant concept by quantifying Planck's constant h, the electric charge "e", and the constant c of the speed of light. This prevents you from getting a real image of a system. And only a theory that would eliminate the redundant concept could show the properties of the atomic system. Therefore, new Physicists should not rely too much on current concepts, as they may be provisional concepts as Einstein asserted".*

In fact, according to the theory of quantum impulse of Erdelyi which he had mentioned in a Letter to Dirac with a copy of the Scientific American, in 1964 it was the c speed of light and quantum's elementary mass the fundamental constants. And the constant h and the charge *e* they were the derivatives of c and m'.

And in the Science & Culture Journal of SBPC of July 1984, Lauro de Oliveira Lima, of the Educational and Experimental Center Jean Piaget, commented in his article *"Neutrality of Science"* that: *"It is inevitable to understand and discover, what is Science, which does not progress by accumulation, but by paradigm shifts, that is, new mental instruments of observation. So, at a time when condemning a kind of creativity of science hurts the essence of human knowledge. Why should all of your ability highlight a sector to condemn?".*

In other words, to increase their power, the "Officers" of the Pre-Established Physics seek to impose their authority to subdue opposing ideas.

Also, in the Science & Culture of SBPC of July 1984, published by José Reis, presented an article of Nature Journal of 1983 by S. Hetherington, in which he criticizes members of the scientific community who silenced or justified errors, with serious damage to the authority of science. Typical case is that they have seen, not what really existed, but what they wanted to see. And the peer review, the judging couple of articles, seemed more efficient at silencing theories, so that it would not become public in order to tarnish the reputation of scientists, hiding the truth even from the scientific community itself.

However, José Reis with his renowned members of the SBPC accepted silence for new theories like that of Mesquita, preventing his publication by following judges who could not sustain a mistake they had wrought.

So, what is the point of publishing this article from Hethrington in the Journal of SBPC which had been published in Nature Journal? Would it be just to disguise compliance with freedom of publication?

LXXXVI. NEW RADIATION DETECTED PROVES QUANTUM IMPULSE THEORY – **1984**

In February 1984, Nature published that the physicist at Oxford University, Henry Rose, had detected subatomic particles and high-frequency electromagnetic waves from the chemical radium element, in addition to the known alpha particles, beta and gamma.

However, according to the theory of quantum impulse, the matter is made up of electromagnetic radiation. So this radium experiment was a fundamental proof of the theory of impulse.

LXXXVII. NEW GENERATION OF GRAVITATIONAL WAVE DETECTION – **1984**

In July of 1984 in the 36th. Meeting of the SBPC and 18th. Annual Meeting of the SBF, the author presented a Communication *"New Physical and Mathematical Foundations Applicable to the*

Detection of Gravitational Radiation and Consequent Development of the New Structure of the Cosmos" showing that they failed to 1th. and 2th. generation of gravitational wave detection and that the 3th. generation should be based on high frequency.

Physicists were looking for gravitational field waves, this is, space-time distorted waves, which would be originated by massive mass shock disturbance. This disturbance would be propagated in that space-time distorted in the orbital oscillating frequency of these bodies, where by the distance traveled this waves would be received with some attenuated frequency between 10^{12} to 10^{-18} hertz, of which was produced of the initial energy wave of the shock of masses.

But, by the gravitational radiation of Mesquita, the flux of gravitational electromagnetic radiation energy, formed of quantum impulse energy h' = m'c x c, from the atomic nucleus was in the frequency of 10^{19} hertz in the nuclear transformation these massive masses, whose energy would cause dimensional variation for the detection interferometry of this mass shock. Document 16.

PHASES	METHODS	GRAVITY FREQUENCY	WAVE LENGTH	RESULTS
First Generation	Metal Bar Detectors	10^{-3} to 10^{5} hertz	3.10^{13} to 3.10^{5} cm	Notable Failures
Second Generation	Laser Space Probe Doppler Bars 10^{-3} °K	10^{-3} to 10^{5} hertz	3.10^{13} to 3.10^{5} cm	—
Third Generation	News Calculus Quantum Impulse Space Probe	10^{19} hertz	~ 3.10^{-8} cm	New Structure of the Cosmos

Doc. 16 – Communication drawing *"New Generation of Gravitational Wave Detection"* presented by the author in the SBPC of 1984, focusing on gamma rays as a 3th. generation in the detection of gravitational waves at the frequency of 10^{19} hertz, over the previous with metallic bar and laser interferometer between 10^{-3} to 10^{5} hertz, as Reference.

LXXXVIII. NEW INFORMATION – SCIENCE AT THE INSTITUTE OF ENGINEERING-BRAZIL – <u>1985</u>

On February 27, 1985 Mesquita communicated to the Physical Engineer Aulus Pompéia of the Brazilian Academy of Sciences on request for routing work by IE involving gravitational radiations and the life of the free neutron. Mesquita in contact with Aulus Pompéia understood that Pompéia would be fully favorable to this request for referral by IE to the Swedish Academy of Sciences, and on March 5, 1985 Mesquita asked ie that Aulus Pompéia be consulted. On April 3, 1985 was sent the formal request of IE to Aulus Pompeii to him position.

On April 15, 1985 Aulus Pompéia sent a Letter to the Mesquita and on 16 April 1985 to the President of IE, in which Aulus Pompéia communicated that it would be favorable to send Mesquita work to the Swedish Academy of Sciences, but only if that work were to confirm theoretically or experimentally. And presented two objections. The first was that the numerical value of the constants should follow the International Convention and that this work was written differently.

However, the work of the Mesquita followed exactly the value of constants according to the International Convention. For example, in the case of speed of light, the convention said that the estimation of the error is effected in the 3 standard deviations applied in the last digits, in case $c = 2.997925 \times 10^{10}$ cm/sec ± 4, which was applied in the last digit, ie 5 ± 4, while the Mesquita presented at the beginning of the work (29979250000 ± 4000) cm/sec in which the printing should be ± 40000 to apply in the last digits. However, then the print was correct when printed 2.997925 ± 0.000004 in which it is applied in the last digit 0.000005! Therefore, another "renowned" consultant, Aulus Pompéia, who found a graphical error scanning to invalidate an original work for Physics, and who added an exclamation point to extol a simple dactylographic error! Is that a "zealous" referee?

The second objection was that the neutron was disintegrating after a time instantaneously and by statistics, not by continuous loss of mass.

The free neutron by Pre-Established Physics it decay instantly after a few seconds, others up to 30 minutes disintegrate instantly, and some free neutrons may not even disintegrate. But from the Mesquita article, "if "all the mass of a free neutron, and without any interaction with other atoms, was recombined internally with the gravitational energy flow of frequency 10^{19} s^{-1}, to form the proton and emit the electron and the neutrino, the time needed for this disintegration would be 30.13... minutes and the half life of 15.06... minutes, which was the time it would be for the neutron to initiate this internal disintegration until the instant that its mass was half blown away to form the proton and emit the electron and the neutrino. But, also 15.06... minutes is half the quantity of neutrons that it disintegrates obtained experimentally, in which some neutrons disintegrate in a few seconds and others in a longer time.

Admitting that between two neutrons, one of them presented the disintegration after 1 second and another after 15 minutes, in both cases the neutron that by the theory of the impulse would be formed of n x m'c would have its constitution of around 1800 electrons as a ring in torsion and other three twists of the ring. But by quantum impulse, the basic constitution of the neutron was the proton that would have its helix closed by the ring, but that in the neutron an electron with a counter helix would neutralize the charge of the proton. By quantum impulse, the proton with the helicity advancing in one direction generated the charge as positive and in the direction that was moving away contained the negative charge, however it was the forward direction that manifested the positive charge. And the electron with the opposite helicity had the negative charge in the direction advancing the m'c motion that formed the helix. Then, the exchange of gravitational energy flux between the free neutron internally, and the probability of external interaction with the field of atoms, resulted in a time of 1 second or a longer time of 15 minutes, to form the proton, electron and the neutrino.

That is, Aulus Pompéia accepted that: *"The Atom does not age, but dies in a given instant"*, as he mentioned. He would prefer the free neutron to be disintegrated at any instant as a magic pass! But, preferring not to know that it was the internal gravitational flow of the atomic nucleus that reorganized the particle, resulting in the emission and formation of new particles.

And the fundamental fact that he did not realize or didn't want to perceive while reading the work was the essence of the work, the title itself, of what time it would be necessary for the disintegration of the neutron, if all its mass were transformed continuously by gravitational frequency to form the proton, electron, and neutrino. Since the proton was the constituent of almost all neutron mass it was the fact that in a few seconds the neutral formed along with the electron and the neutrino. However, "if "all the neutron mass was needed to transform into proton, electron, and neutrino, the time would be greater, but it should be in accordance with the time obtained experimentally. And the probability of that specific frequency to transform the whole mass of the neutron would be one in around 10^{23}, to coincide with the experimental time of 15 minutes for the half life of the neutron, and was obtained 15 minutes and 3, 6 seconds! That is, not a fantastic coincidence, but a proven fact of nature!

However, on 10 June 1985 Eng. Mário C. de Oliveira Pinto, administrative superintendent of IE, came to inform me that another consultant could analyze the work of Mesquita, and I sent request to the Physicist César Lattes of CampinasUniversity-UNICAMP.

However, Lattes refer to the Physicist, A. Rodrigues Jr., to effect this opinion, but shortly after the time A. Rodrigues Jr., he reported that he had many work scheduled without the ability to analyze this request.

LXXXIX. CREATION OF NEW IDEAS AND IMMORTAL CREATION
1985

In June 1985 Niels Jerne, Nobel of Medicine of 1984, mentioned in a presentation that: *"Great ideas arise when dogmas are buried, and you should not rely too much on current ideas"*.

And on June 17, 1985 within these ideals as Niels Jerne, Mesquita delivered as thanks to the Doctor who attended him one of his works for his knowledge, and ended on that day his existence on Earth, but leaving his immortal creation of the Theory of Gravitational Radiations of 10^{19} cycles per second, the greatest aspiration of the greatest physicists of all time in unifying all forces or interactions of nature, based on the theory of the quantum impulse of Erdelyi's immortal creation. Weinberg's weak theory, Salam and Glashow left the strong force and gravity out of the unification of Physics, and the theory of supersymmetry that sought to unite gravity with strong and weak force was not according to experimental facts. However, the work of the Mesquita presented theoretical and experimental evidence uniting Newton's gravitation with Planck's electromagnetism, containing the electromagnetic, nuclear and gravitational forces, that is, all forces of nature.

XC. HOW DISCOVERIES OCCUR IN PHYSICS –
1985-1986

In July of 1985 in the 37 ª. Annual Meeting of SBPC, the Engineer and Professor of Physics José Maria Filardo Bassalo presented a Communication on "Discoveries in Physics" in which the creative process was originated by studies, hypotheses, immediate understanding (intuition) and verification, in which the immediate comprehension was effected basically by three types: rational, arising from works aware of their authors and occurred, however, outside the existing "commands' such as Newton; of isolated ideas, outside the current scientific paradigm, such as Gilbert and Ampère; and of accidental facts such as radioactivity, as of Becquerel and Röntgen.

This Communication it was also published in the Science & Culture Journal of SBPC of November 1986.

But for those renowned Commanders in Chief and their "Officers" of the "Standard" Physics, the discoveries should only be within the formality permitted, following the current paradigm or at random if within these cases!

With the discovery of gravitational lenses by diverting light at the edge of galaxies, in January 1985, I presented a Communication *"Proof of the Deviation of Radiation, by*

Computer, Presents Fundamental Proof of the Theory of Mathematical-Physics (Basic Principles in Engineering) of the Unified Rational Field of Physics", in the 37th. Annual Meeting from the SBPC of 1985. It was mentioned the deviation of radiation electromagnetics by galactic clusters, a gravitational "lens", was predicted by general relativity based on the constant of Einsten of gravitation which is the function of Newton's gravitational constant. But, by quantum impulse the deviation of the radiation is by circumventing the atoms, as analyzed by the quantum impulse of Erdelyi in the refraction of dense means and the gravitational radiation of the Mesquita, which explain the constant gravitation of Newton and the Unification Gravitational-Electromagnetic.

XCI. THE COMPLEXITY THAT WAS LACKING – 1967 to 1985

Mathematical abstraction entered quantum mechanics, where imaginary complex numbers appeared from $i = \sqrt{-1}$, and in general relativity there was mathematical abstraction of deformations of Riemann space but with real numbers. However, in 1967 the Mathematician and Physicist Roger Penrose decided to add complex numbers to relativity, introducing spaces with four imaginary dimensions, which were the spaces called "twistors" formed of points and space that would constitute all the particles, with the purpose of uniting quantum mechanics with relativity. Thus, in addition to photons and particles, both massless, Penrose presented particles formed of dots and space! However, also this theory of Standard Physics-Mathematics did not result in any unification of nature! In other words, this theory resulted in nothing discovery, even because it is out of the causes of the scientific discoveries presented in the paper of Bassalos of 1985!

XCII. "MASSES, WHAT ARE THEY AND WHERE DO THEY COME FROM"? – 1985

Although the Higgs mechanism gave mass to the particles by the field created by mathematical artifices, in 1985 Feynman observed that the masses of the particles, m, in all theories were used as numbers which did not understand them. *"The masses of the particles, what they are and where they come from?"*.

However, by the theory of quantum impulse, the elemental quantum mass m' with the speed of light c in vacuum measured on Earth forms the m'c quantum impulse that constitutes all the particles of matter. While matter is the particle gathering, the mass by quantum impulse is a complex concept in which the elementary mass m' is a fundamental constant of nature.

XCIII. SUPERSYMMETRY AND SUPERSTRING FOR 50 YEARS
1984 to 2035!

Supersymmetry – Between 1985 and 1991, J. W. F. Valle, J. Ellis, G. Gelmini, C. Jarlskog, G. C. Ross, G. D. Coughlan, R. Holman, R. Ramond, M. Ruiz-Altaba, R. N. Mohapatra, S. Nussinov, J. J. Simpson, A. Hime and N.A. Jelley presented prediction and or evidence of supersymmetry particles.

Superstring – Schwarz and M. Green in 1984 had presented normalization in string theory, in which particles such as electrons, quarks, considered as punctuals would be a filament of 10^{-33} cm, and with vibrating energy of the strings producing mass, through $E_{vibration} = mc^2$ what form $m = E_{vibration}/c^2$. The greater the vibration of the string, it is greater the mass. For the photon, graviton and gluon, all without mass, the vibration is $\approx$ zero. And the charge, as the spin and other properties of the particles would be defined by the vibration pattern and consisted of nine dimensions and one of the time. Also, another standardization was effected by Gross, J. Harvey, E. Martinec, and R. Rohn.

And in 1985, P. Canherss, G. Horowitz, A. Strominger and E. Witten presented the explanation of the vibratory patterns of the additional string dimensions by the abstract mathematics of six dimensions by E. Kolobe in 1957 and S. Yau in 1978. And by string theory, the Higgs field would be formed from these strings of 10^{-33} cm. As Witten the string theory would dominate the Physics in the next 50 years, this is, until 2035!

But, a higher development of mathematical abstraction for the strings in these next few years in what could possibly result? Particles in the form of strings in 11 dimensions? And that could be hidden or "rolled up" as accepted by their proponents? It would be the development of metaphysics with these dimensions, keeping the theory of relativity as untouchable with all inconsistency and accumulated errors? And after this 50 years, would continue the same current problems of the bases of Physics?

XCIV. GENERALIZED FIELD OF ALL PARTICLES, STANDARD MODEL, GRAVITATIONAL FIELD – 1985

By the experiments carried out in particle accelerators up to 1985, it was confirmed that the universe was composed of fields of particle matter of the fermion type as the electron and quarks, and of the hadron type, such as proton, meson and neutron. And intermediate fields of fundamental interactions from boson particles, such as photons and gluons. These fields of these particles presented quantum fluctuations of the zero point, now with the creation or annihilation of particles at least times, and fields that were with and without mass.

However, by the theory of quantum impulse, all fields were formed by the m'c impulse with the speed of light c in vacuum measured on Earth, that formed the elementary quantum h', which would constitute all the particles.

The standard particle model comprises the interactions of the particles and fields of the weak and strong electromagnetic force, including the particles and the Higgs field, but gravity is outside this standard structure.

By general relativity, the space-time in which space is contracted and elongated time according to the velocity of the frame, in the curved Riemann space, shows the curvature of the light in the proximity of a mass. But, if the cause of the curvature of space is mass, how would the mass originate that curvature? In the case of a cubicly shaped mass the spatial distortion would be cubic, with precise angles of 90° ? Also, astronomical observations should present complex curvatures of light by the action of interconnected masses, but only deviations from light in localized bodies are seen.

The space-time distortion of general relativity only relates the accelerated motion to mass, unable to explain how mass modifies space-time.

And the eletromagnetic field that makes a part of this mass energy field that would be the fundamental cause of distorting space-time, and that emits radiation, relativity and current quantum mechanics conceives the photon as without mass, because by its equation of mass increase the photon would have infinite mass at the speed of electromagnetic radiation.

For elementary impulse theory, a ray of light that curves when entering a gravitation field follows the resulting configuration of the fields curved by the particle configuration. When the light focuses on an atom, it propagates around the atom which results in a longer time to travel a distance between two points of a line in the atom, that is, an apparent lower speed of a medium, and the consequent refractive index.

XCV – FORMAL OMISSION – 1986

For the purpose of obtaining an Opinion of Scientific Paper, in April 1986 was sent to the Academy of Sciences of Sweden the article of the *"Neutron Free"* of Mesquita, guided by the Physics Professor Zillah de Mesquita, licensed from the University of São Paulo-USP, and the author as associate of SBPC. This article was also written in the Swedish version for a special consideration of the request for this Opinion.

However, the Academy of Sweden came to maintain the formal omission, of other previous referrals such as that of the IBACE with the works of Erdelyi, and by the author as associate of the Engineering Institute of Mesquita work.

It was proven that even in the maximum stronghold of judges of Physics that grants the highest prize annually, there was no difference in the procedure of treating a scientific article that was contrary to that of "standard" Physics, failing to effect a simple forwarding of this article, which was duly requested, so that some Teacher, Physicist or an Aide could issue a simple opinion. This was not a request for a trial of the article for awards, as this was the case for the chosen one when in life.

However, not even a simple Communication of impossibility of attendance was issued, attesting the attitude of behavior of the highest award authority of Pre-Establishead Physics. Document 17 – Early pages in the Swedish language of work *"Neutron's Life Free by Gravitational Radiation"*.

J. P. Hansen

To Mr. PROF.ALF JOHNELS São Paulo, April 1986
The Royal Academy of Science Sweden
Kungl Vetenskapsakademien
Box 50005 - 10405 Stockholm

Re: Confirmation of existence of gravitational radiations ($10^{19} sec^{-1}$) through the rational determination of the medium life of neutrons by the impulson theory by I.ERDELY and gravitational theory by P.F.MESQUITA.

Dear Sirs,

In attention to the Royal Academy of Science of Sweden, due to the acknowledgement of discoveries of the Science, we submit to your approval the following work:

"LIVSLÄNGDEBEAV EN FRI NEUTRON, RATIONELLT BERÄKNAT, UTGÖR EN NY VIKTIG TEST AV DEN SPECIFIKA OCH EXPERIMENTELLA EXISTENSEN AV "GRAVITATIONSBESTRÅLNINGAR", IDENTIFIERADE I ELEKTROMAGNETISMENS SPEKTRUM".

by P.F.MESQUITA, whose previous works were sent already in 1980 to the Royal Academy of Science Sweden, to the attention of Mr.PROF.NAGEL BENGT.

The present work is a translation into Swedish (especially done for the Royal Academy of Science of Sweden) of the original thesis in Portuguese, of which only the summary is translated into English.

We are awaiting your pronouncement of this work, with the purpose of a better knowledge of the Universe.

Yours sincerely,

Prof. ZILLAH B. de MESQUITA Eng. PAULO HANSEN
Physicist Licentiated - São Member of Brazilien Society
Paulo University Science Progress

Copy: Mr.PROF.NAGEL BENGT
 Mr.STIG LUNDQVIST

LIVSLÄNGDEN AV EN FRI NEUTRON, RATIONELLT
BERÄKNAT, UTGÖR EN NY VIKTIG TEST AV DEN SPECIFIKA OCH
EXPERIMENTELLA EXISTENSEN AV "GRAVITATIONSBESTRÅLNINGAR",
IDENTIFIERADE I ELEKTROMAGNETISMENS SPEKTRUM.

Avhandling överlämnad till den Brasilianska
Vetenskapsakedemin

från

Civilingenjör och geograf, prof. Dr. Paulo
Ferraz de Mesquita, från São Paulos Universitet,
arkitektur- och urbanism-fakulteten

i brev adresserat till akademi-ingenjören,
Prof. Dr. Paulus Aulus Pompeia.

São Paulo, december 1966.

Innehåll

	sid.
Förord	III
Den fria neutronens liv rationellt fastställt (grafik)	IV
Historisk introduktion	1
"Gravitationsbestrålningar i elektromagnetismens spektrum	4
Vetenskapens entydiga begrepp	4
Påminnelse om nödvändiga och suggestiva kunskaper . .	5
Rationell fastställning av en fri neutrons livslängd såsom enhet	12
Rationell beräkning av "medellivslängden" och "halva livslängden" (eller period av semi-desintegration) av en fri neutron	17
Noggrann undersökning genom experimentella resultat	22
Slutsats	23
Biografiska referenser (op.cit.)	24

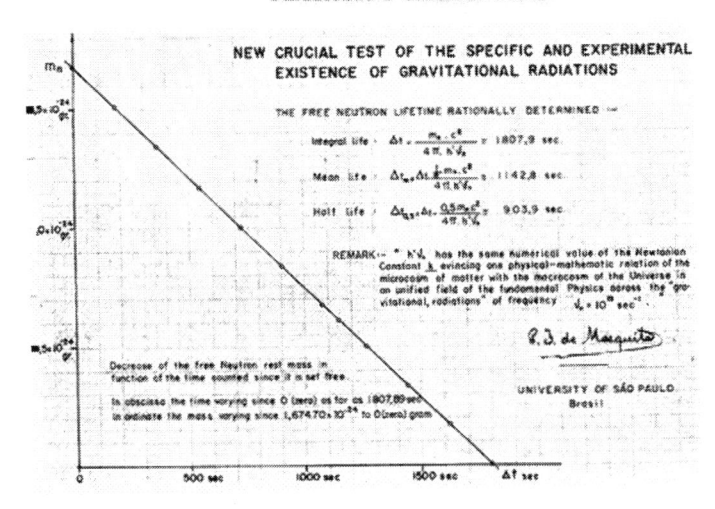

Doc. 17 – Initial pages, in the Swedish language, of work "*Neutron's Life Free by Gravitational Radiations*".

XCVI. NEUTRINO MASS – **1952 to 1987**

In 1987 Salmeron and others presented a particle collision experiment of oxygen and uranium nuclei at CERN, in which they observed evidence of breaking the asymptotic attraction of quarks.

And also in 1987 was published by the New Scientist Magazine of 20 August 1987 and 10-16 October 1987 by the Newsletter of SBPC-SBF, number 109, which: *"Data obtained from the recent supernova observation suggests that neutrinos have a rest mass of the order of ten electron-volt equivalent energy. And this estimate agrees with results obtained based on string theories and also predictions of the neutrino model for the universe".*

However, Salmeron who worked at CERN, had mentioned in 1979 in the Conference at the Engineering Institute in Brazil that experimentally was found the value zero by measuring the mass of the neutrino, ignoring the prediction of the impulse theory since 1952 that the neutrino had mass.

So it was more of a proof that the formal Physics sought to ignore theories that were far ahead of the teachings and failures of this dogmatic sect this Physics.

XCVII. LEARNING OF THE MILITARY DICTATORSHIP BY SCIENCE PROGRESS BRAZILIAN SOCIETY-SBPC – 1964 to 1987

Years of fighting the military dictatorship was what he said to SBPC since 1964 when the coup was struck against President João Goulart on the pretext of seeking to go against freedom, but adopting censorship to allow freedom of opinion. However, in September 1987 the SBPC introduced a new statute, applying exactly what it had fought, i.e. the elimination of the right of free chair exercise, and imposing strict "reins" on the scientific community to eliminate opposing ideas from summits of each entity of SBPC and the development of science. Thus, preventing or hindering cooperation between researchers, suppressing the ethical term in the new norms, and the freedom of research opinion and preventing the diffusion of science. Then, the two decades learning of the military government by the Board of SBPC was magnificently assimilated and imposed from August 1987 with the presidency of the SBPC exercised by Psychologist Carolina Bori and Board by the Physicist Ennio Candotti and Sociologist José Albertino Rodrigues, to meet what was required of Members of each sector of SBPC for the purpose of imposing their ideas.

XCVIII. IN THE NAME OF SCIENCE! 1988-1989

In 1988 the SBPC began to expand the rules for the publication of the Communications summaries of the Annual Meetings, with activities by consultants not identified to the presenters of these Scientific Communications, justifying as necessary the quality of the work. That is, a pretext for imposing control of what should be published, contrary to everything that SBPC promulgated in the previous time of the government of military dictatorship.

In September 1989 Astronomer Ronaldo Rogério de Freitas Mourão, of the National Observatory in Rio de Janeiro-Brazil, Doctor in Astronomy from the University of Sorbonne-France, author of more than 40 books on Astronomy, mentioned that: *"The*

J. P. Hansen

National Research Council–CNPq maintains in its paintings the worst in the authoritarian system. Those who fought against the political system within the SBPC are now in the positions of power keeping the same authoritarian rubbles from the time of the dictatorship".

XCIX. "A BRIEF HISTORY OF TIME" OF DR. HAWKING AND THE THEORY OF QUANTUM IMPULSE – <u>1987 to 1989</u>

In the book in "*A Brief History of Time*" of Physicist Dr. Stephen Hawking was mentioned that there was still no theory that unified quantum mechanics with gravitation. Then, for the opportunity of direct communication to Hawking about his book, because the Journals or "Magazines" of the Pre-Establishead Physics did not accept to publish new ideas, I sent a Letter addressed to Dr. Hawking informing about the quantum impulse which united all the forces of nature. Document 18.

Curitiba, Brazil, February 1989

To
DR. STEPHEN HAWKING
CAMBRIDGE UNIVERSITY
ENGLAND

Dear DR. HAWKING,

Congratulations for your work "A Brief History of Time: From The Big Bang To Black Holes", in special for yours fantastics calculus for discovery of the black holes' radiation.

I take this opportunity for to comment about the your appearance on unified theory of physics with regard to your first, fifth and tenth chapters of the work above.

Agreeable to your description, still it's not a theory that unified quantum mechanics and the gravitation – a quantic theory of gravity – the base of the unified theory of physics.

But indeed, is verified by following formulas, remarking the dimensions with its terms (by units C.G.S.system) in parenthesis, that:

-1) According "Momentum Theory" by I.L. ERDELYI, (Anexe 3):

a) $h(erg.sec) = $ constant de PLANCK $= 6,62... .10^{-27}(erg.sec)$

$h(erg.sec) / 1(sec) = h'(erg) = 6,62... .10^{-27}(erg)$

b) $E_\nu(erg) = h(erg.sec) \times \nu(sec^{-1}) = h\nu(erg)$

$E_\nu(erg) = $ quantic energy of frequency $\nu(sec^{-1})$

$E_\nu(erg) / 1(sec) = h'(erg) \times \nu(sec^{-1})$

and for $\nu(sec^{-1}) = 1(sec^{-1})$:

$E_\nu(erg) / 1(sec) = h'(erg) \times 1(sec^{-1})$, $\therefore E_\nu(erg) = h'(erg)$

$\therefore h'(erg) = $ elementary quantum of energy

-2) According "The Meaning of The Universal Gravitation Constant of NEWTON..." by P.F.MESQUITA, (Anexes 1 and 2):

f) F(dyne) = Momentum/time = Momentum X Frequency
This is: F(dyne) = Frequency of Momentum (g.cm/sec X 1/sec)

and for m'c (g.cm/sec):

$F_{m'}$(dyne) = m'c(g.cm/sec) X ν(sec^{-1})

g) By the NEWTON's law of gravitation:

F_G(dyne) = $m_1 m_2$(g^2)/r^2(cm^2) X K_G(cm^{3}sec^{-2}g^{-1})

F_G(dyne) = gravitational force

K_G(cm^{3}sec^{-2}g^{-1})=constant of gravitation of NEWTON
K_G(cm^{3}sec^{-2}g^{-1})=(6,627±0,022) .10^{-8} (cm^{3}sec^{-2}g^{-1})

r(cm)= distance between m_1 and m_2

h) Employing h' in the law above, by fact that in K_G the term (6,627±0,022) to be similar to (6,62...) of h':

F_G(dyne)=$m_1 m_2$(g^2)/r^2(cm^2) X 6,62... .10^{-27}(erg) X10^{19}.(cm.g^{-2})

h'(erg)=6,62... .10^{-27}(erg)=6,63... 10^{-27}(g.cm^2.sec^{-2})

h'(erg)=m'c(g.cm/sec).c(cm/sec)

i) Considering the concept of force like frequency of momentum, in the last formula above of the gravitational force F_G:

F_G(dyne)=$m_1 m_2$(g^2)/r^2(cm^2) X m'c(g.cm/sec).10^{19}(sec^{-1})Xc(cm/sec)X(cm.g^{-2})X(sec)

being 10^{19}(sec^{-1})=ν_G(sec^{-1}), the gravitational frequency (gravitational radiation) of elementary momentum m'c(g.cm/sec) with velocity of light c(cm/sec), this is the frequency of elementary quantum of energy h'(erg):

F_G(dyne)=$m_1 m_2$(g^2)/r^2(cm^2) X h'(erg).10^{19}(sec^{-1})X(cm.g^{-2})X(sec)

Multiplying each member by 1(cm), and being:

$m_1 m_2$/r^2 (g^2/cm^2)X(cm.g^{-2})X(cm) = n , a pure factor, it results in the latest formula of F_G:

F_G(dyne)X1(cm)=E_G(erg)= n.h'(erg).ν_G(sec^{-1}).(sec)

or: E_G(erg)= n.h(erg.sec).ν_G(sec^{-1})

and for n=1,

E_G(erg) = h(erg.sec).ν_G(sec^{-1}) = hν_G (erg)

This is, E_G(erg) = hν_G (erg) = gravitational quantum = 1 graviton
Therefore,"the quantum mechanics is inside of the gravitational law of NEWTON",according formula above with the only term of the quantum hν_G.

j) Being: h(erg.sec)/1(sec) = h'(erg) ,

h(erg.sec) = h'(erg) X 1(sec)

∴E_G(erg) = 1 graviton = h'(erg)X1(sec) X ν_G(sec^{-1})

or: E_G(erg) = h'(erg)X1(sec) X 10^{19}(sec^{-1})

∴ E_G(erg) = 10^{19}h'(erg)

k) Dividing by 1(sec) the lasts formulas de E_G(erg) above:

E_G(erg)/1(sec) = P_G = gravitational powder = h'(erg).ν_G(sec^{-1})

This is, only the term h'ν_G(erg) being present in the gravitational law of NEWTON.

J. P. Hansen

◊) In the gravitational law of NEWTON, with these calculus presenting the term $E_G(erg)$ or $E_G(erg)/(sec)=P_G(powder)=P_G(erg/sec)$:

$$\frac{F_G(dyne).1(cm)}{1(sec)} = \frac{E_G(erg)}{1(sec)} = \frac{m_1 m_2 (g^2)}{r^2 (cm^2)} \times K_G(cm^3 sec^{-2} g^{-1}) \times \frac{1(cm)}{1(sec)}$$

$$\frac{F_G(dyne).1(cm)}{1(sec)} = \frac{E_G(erg)}{1(sec)} = n.h'(erg).\nu_G(sec^{-1})$$

$$\therefore \frac{m_1 m_2 (g^2)}{r^2 (cm^2)} \times K_G(cm^3 sec^{-2} g^{-1}) \times \frac{1(cm)}{1(sec)} = n.h'(g.cm^2.sec^{-2}).\nu_G(sec^{-1})$$

$K_G(cm^3 sec^{-2} g^{-1}) = n.h'(g.cm^2.sec^{-2}).\nu_G(sec^{-1}) \times \frac{r^2(cm^2).1(sec)}{m_1(g) m_2(g).1(cm)}$

$K_G(cm^3 sec^{-2} g^{-1}) = n.h'\nu_G.1/n \ (cm^3 sec^{-2} g^{-1}) = h'\nu_G(cm^3 sec^{-2} g^{-1})$

and being, $\frac{r^2(cm^2).1(sec)}{m_1(g)m_2(g).1(cm)} = 1/n \frac{(cm).1(sec)}{(g^2)}$, a factor that, then

$K_G(cm^3 sec^{-2} g^{-1})$ is placed in the gravitational law of NEWTON with that presenting the term $E_G(erg)/(sec)=P_G(power)$, it results the dimensions of this term $P_G(power)$:

$$\frac{F_G(dyne).1(cm)}{1(sec)} = \frac{E_G(erg)}{1(sec)} = P_G \text{ (power)}$$

$$P_G(power) = \frac{m_3 m_4 (g^2)}{r_1^2(cm^2)} \times 1/n \frac{(cm).(sec)}{(g^2)} \cdot \frac{1(cm)}{1(sec)} \cdot n.h'(erg).\nu_G(sec^{-1})$$

$$\therefore P_G(power) = n_1.h'.\nu_G \ (erg/sec)$$

Also, for $F_G(dyne).1(cm)=E_G(erg)$, it obtains with the factor above without the term (sec):

$$K_G(cm^3 sec^{-2} g^{-1}) = h\nu_G \ (cm^3 sec^{-2} g^{-1})$$

So, the gravitational constant of NEWTON it represents the gravitational radiation of quantum E_G, or of $h(erg.sec)$ and of the elementary quantum $h'(erg)$

m) -3) Summary:

In summary, these calculus above reveals the discovery of the meaning for the gravitational constant of NEWTON, this is, it represents the gravitational radiation of 10^{19}hertz of the elementary quantum of energy h'(erg) or of the h(erg.sec) quantum action de PLANCK, with h'(erg) explaining that quantum action of incomplete concept.

Therefore, this theory by P.F.MESQUITA - **based** in the existence of one only quantum h'(erg) for alls radiations and the concept de force of GALILEO (momemtum/time), by I.L. ERDELYI, on the contrary to divers quantums according EINSTEIN, and considering the concept of force like frequency of momentum -, it comes to obtain a consistent quantic theory of gravitation, a theory that unifies the quantic mechanics and the gravitation, it allowing the actualization of the quantum mechanics for its larger use to the technique and technology.

With congratulation, in special, for your wonderful work about black holes's radiations, the gratitudes of to receive this letter and your possibility of one commentary of these remarks about gravitation contained inside of a quantum theory.

Yours sincerely,

Paulo Hansen

Eng. Paulo HANSEN
Member of Engineering Institute
and Brazilien Society Science
/Progress

Anexes:
(1): "The Meaning of the Universal Gravitation Constant of NEWTON..."by P.F.MESQUITA - 1965.
(2): "The Physical Meaning of the Newtonian Constant reveals a way for studying Satellite Dynamics..."by P.F.MESQUITA - 1974.
(3): "Open Letter to DR.DIRAC about number 137with the Momentum Theory"by I.L.ERDELYI - 1964.
(4): "Quantic Algebric Solution for the Diametre of the Black Hole" by P.F.MESQUITA - 1979.
(5): Abstracts of Communications presenteds at the Brazilien Society Science Progress - 1983/1984/1985/1989, by P.HANSEN

Eng. Paulo HANSEN

COMPLEMENTS

The employment, in the theory of the meaning of the gravitation constant of NEWTON by P.F.MESQUITA, of the constant of PLANCK h or h' in the gravitational law of NEWTON - by fact of h or h' to be similar to gravitational constant of NEWTON(*)-, it not appears to demonstrate a simple coincidence, by motive of that procedure to come at: to confirm the concepts quantics, of momentum theory of I.L.ERDELYI; to reveal the discovery of the gravitational radiation and its frequency; to unify the quantic theory of gravitation. (*) Anexe.

Beyond, that theory of MESQUITA, to come to present explanation about another physics concepts, as to demonstrate that the COULOMB's law of the magnetisme and electricity also it presents the quantum h or h' in itself.

As well, the theory of MESQUITA it permits to calculate rationally the motion of perihelion, independently of general relativity theory of EINSTEIN. In that case, MESQUITA utilized, in the final of that demonstration, a LORENTZ transformations formula inside a proceeding of calculus non relativist all that its interpretation! This is, a of the sames formulas of LORENTZ that they were bases of relativity theory of EINSTEIN, but not utilized by himself in that case of the perihelion.

Although, those theorys (momentum and meaning of the gravitation constant of NEWTON) of ERDELYI and MESQUITA, they to be in basic disagreement with the relativity theory of EINSTEIN, they explain about principles, conclusions and problems not solveds by relativity theory. Thus, by facts as the maxim velocity existent (radiation, light) according ERDELYI, MESQUITA and EINSTEIN, is possible to say that "God not plays dice" according mention of EINSTEIN, and therefore it not having a simple coincidence of constants in the nature, but only a physical reality.

Also, in the MAXWELL's theory of the electromagnetism, in the last part of the calculus for determination of the velocity of the electromagnetics variables, it was observed by MAXWELL that:

$v = \sqrt{1/\mu_o \cdot \varepsilon_o}$, being μ_o =permeability constant in the vacuum = $1,25... .10^{-6}$ (N/A^2) ; ε_o =dielectric constant in the vacuum = $8,85... .10^{-12}$ $(A^2.sec^2 / N.m^2)$, resulting in :

$v = 2,99... .10^8$ (m/sec) = velocity of the light!

This is, resulting the velocity of the light as a "coincidence" that MAXWELL observed with great enthusiasm, concluding that the light it was a electromagnetic radiation.

Therefore, the theory of MESQUITA - with base in the momentum theory of ERDELYI, utilizing the concept of force as frequency of momentum and employing the constant of PLANCK inside the gravitational law of NEWTON -, reveals not a coincidence between those constants. But, that theory reveals a physical reality for a quantic gravitation theory. HAWKING comments about the possibility of the discovery of a consistent quantic gravitation theory, for to be understanded amply. The theory of MESQUITA, with base in the work of ERDELYI, it come to present the discovery of a ample, consistent and unified theory of a quantic gravitation theory, for better comprehension and development of a structure of the cosmos.

(*):

Numerical values:

$h = (6{,}62517 \pm 0{,}00023) \cdot 10^{-27}$, with standard error
$h = (6{,}62517 \pm 0{,}00076) \cdot 10^{-27}$, with relative certainty
$h = (6{,}62517 \pm 0{,}00066) \cdot 10^{-27}$, with relative certainty
$K_G = (6{,}754 \pm 0{,}041) \cdot 10^{-8}$, determination of 1798
$K_G = (6{,}634) \cdot 10^{-8}$, determination of 1878 to 1898
$K_G = (6{,}655) \cdot 10^{-8}$, determination of 1901 to 1930
$K_G = (6{,}670 \pm 0{,}005) \cdot 10^{-8}$, determination of 1942
$K_G = (6{,}627 \pm 0{,}022) \cdot 10^{-8}$, on function of the Earth's latitudes and gravitational acceleration by the latitudes
$K_G = (6{,}62689) \cdot 10^{-8}$, on function of the KEPLER's law and GAUSS' universal gravitation constant by datas of the solar system
$K_G = (6{,}62583) \cdot 10^{-8}$, on function of the movement of the moon about the Earth

$h \cdot 10^{19} = (6{,}62517 \pm 0{,}00066) \cdot 10^{-27} \cdot 10^{19}$
$h \cdot 10^{19} = (6{,}62451) \cdot 10^{-8}$
$h \cdot 10^{19} = (6{,}62583) \cdot 10^{-8}$

Therefore:
$K_G = (6{,}62583) \cdot 10^{-8}$ (on function of the movement of the moon about the Earth)

$h \cdot 10^{19} = (6{,}62583) \cdot 10^{-8}$ to $(6{,}62451) \cdot 10^{-8}$, justifying the employment of the constant of PLANCK h or h' in the gravitational law of NEWTON.

References:

1 - I.L.ERDELYI - Momentum Theory - Conference at Tecnological Institute of Aeronautics - São José dos Campos (Brazil) - 1955.
2 - I.L.ERDELYI - Correct Interpretation of Planck's Constant - Conference at Engineering Institute - São Paulo (Brazil) - 1954.
3 - P.F.MESQUITA - The Meaning of the Universal Gravitation Constant of Newton evinces the specific existence of Gravitational Radiations identified in the electromagnetic spectrum and reveals mathematical solution of the problem of the Unified Field on Fundamental Phisics-University São Paulo and Brazilian Academy of Sciences - São Paulo (Brazil) - 1965.
4 - P.F.MESQUITA - The Physical Meaning of the Newtonian Constant reveals away for Studyng Satellite Dynamics within a Unified Electromagnetic and Gravitational Field - Symposium on Satellite Dynamics of the International Council of Scientific Unios of the Committee on Space Research and Review Engineering - São Paulo (Brazil) - 1974/1979.
5 - I.L.ERDELYI - Open Letter to DR.P.A.M.DIRAC, about number 137 - Copies in Nature (England), Scientific American (U.S.A.) and International Science and Technology (U.S.A.) - 1964.
6 - P.F.MESQUITA - Coulomb's Law obeys Planck's Principle - Brazilian Society Science Progress - 1972.
7 - A.EINSTEIN, L.INFELD - The Evolution of Relativity - Simon & Schuster - New York - 1938/1960.
8 - A.EINSTEIN, H.A.LORENTZ, H.WEYL, H.MINKOWSKI - The Principle of Relativity - Dover Publications - New York - 1952.
9 - I.L.ERDELYI - Michelson & Morley Experiment - Brazilien Interpl.Society São Paulo (Brazil) - 1964.
10 - A.EINSTEIN - Autobiographical Notes - Library of Living Philosophers - New York - 1978.
11 - A.EINSTEIN - The World as I see it - Philosophical Library - New York - 1949.
12 - B.PARKER - Einstein's Dream - Edições 70, Portugal - 1986.
13 - P.HANSEN - Confirmation of the Radiations' Detour by Computer presents Fundamental Proof of the Physical Mathematical Calculus Theory of Rational Unified Field of the Physics - Brazilien Society Science Progress - 1985.
14 - F.SELLERI - El Debate de la Teoría Cuántica - Alianza Editorial, Madrid - 1986.
15 - P.A.TIPLER - Physics - Worth Publishers - New York - 1976.
16 - S.W.HAWKING - A Brief History of Time: From the Big Bang to Black Holes - Bantam Books (E.U.A.) - 1987.
17 - P.HANSEN - News Basis Physical Mathematical applicable for Detection of the Gravitationals Radiations and Consequent Development of a New Structure of the Cosmos - Brazilien Society Science Progress - 1984.

Doc. 18 – Letter to Hawking informing about the quantum impulse which united all the forces of nature.

However, his Adviser, Andrew M. Dunn, came to inform you that by weakness of his health, Dr. Hawking could not read the manuscript. Also, his Adviser should not be allowed to read the manuscript or send it to some other scholar who might or may have wanted to read it. Thus, no communication with new ideas can reach the standard Physics, which with its shield of imperviousness continues to control the entire Academic Community to maintain its positions, getting the contrary ideas restricted to a minimum of scientists. Document 19.

UNIVERSITY OF CAMBRIDGE
Department of Applied Mathematics and Theoretical Physics
Silver Street, Cambridge, England CB3 9EW

Telephone: Cambridge (0223) 337900
Telex: 81240 CAMSPL G. FAX 33 4748

S W Hawking, C B E, F R S
Lucasian Professor of Mathematics

13th April 1989

Engº Paulo Hansen,
Curitiba (PR)
Brazil

Dear Paulo Hansen,

Thank you for sending your letter of the . However I do regret, that due to his severe disabilities, Professor Hawking has a very limited working day. This means that he has not been able to read your manuscripts.

I am sorry that I can not be of further help.

Yours sincerely

Andrew M. Dunn (Graduate Assistant)
On behalf of Professor S. W. Hawking

Doc. 19 – The answer from Dr. Hawking's Adviser.

In July of 1989 I presented the Communication *"Unified Theory In Physics: Dr. Hawking's Opinion and a Commentary on Unified – Translation/Letter extract to Dr. Hawking"*, on 41 [a]. Annual Meeting of the SBPC. However, despite the Communication being carried out in English, some renowned SBF consultant who judged the article submitted for submission at SBPC required it to be carried out in Portuguese, although it was not in the regulation that requirement to be in Portuguese, and otherwise would not be accepted!

C. "MATTER CLOSED" – **1981 to 1989**

In November 1989 the Science & Culture Journal presented an article on the Nature Journal which had published a work where it was subsequently observed that contained forged data to obtain a certain conclusion. And Editor J. Maddox had explained that the Journal had procedures, however they were insufficient to prove that the data were dubious and would rather err because it could exclude discoveries. But in the face of continued criticism Maddox replied that the matter was closed. However, in this article in Science and Culture presented by the Physicist Sylvio Ferraz Mello who then a Professor in the Astronomy Department of the Astronomical and Geophysical Institute of USP, he mentioned that: *"The answer to this tide of unfavorable comments was another authoritarian act by the editor of Nature, which unilaterally considered the matter closed".*

However, in November 1981 in Science and Culture it was precisely what Sylvio Ferraz Mello with José Reis and the whole summit of SBPC, who always denied the right of publication of Mesquita article in the Science and Culture, through the judgement of consultants indicating errors totally refuted, was that they had considered the question as: *"Matter Closed"*!

CI. ASSEMBLY OF PERPETUAL OPPRESSION – **1989**

In the Editorial Science & Culture Journal in December 1989, André Luiz Padillas Perondini, Maria Isaura Pereira de Queiroz and Sylvio Ferraz Mello who mentioned: *"We attend the call of the President of SBPC, Carolina Bori, and the Board of Directors who appointed an Editorial Committee formed by experienced and renowned scientists in their areas, whose task was to radically transform the Magazine. A Magazine that no article is accepted or rejected without the prior analysis of at least one expert consultant. A Journal that deals with science seriously, without the demagoguery of other vehicles. Science & Culture have been zealous in these four years for the greater interests of the Brazilian* university, publishing the thought and scientific production of its teachers, whenever it results from a serious and competent work".

That is, the scientific work of a Professor of São Paulo University-Brazil as Paulo Ferraz Mesquita, Engineer, Professor at Polytechnic School, Doctor in Science, would not be considered serious and competent to be accepted for publication in Science & Culture for this editorial microboard? It was what? It was simply a high-level job in which this so-called council was too small to be able to see any level above the mediocrity of its height! Among the members of the editorial board of Science & Culture in 1989 was the Physicist Pierre Kauffmann, who previously had issued a collection of statements of errors of editorial council without justification and when justified were totally refuted, but prestigious by the editors of Science & Culture.

Not only the Jounal of SBPC, but in the SBPC itself by the Presidency of the Psychologist Carolina Bori and the Board in 1989 with the Physicist Ennio Candotti and the Sociologist José Abbaraju Rodrigues, was applied radical transformation that resulted in perpetual oppression to the scientists who dared to disagree with the ideas of those who wanted to exercise their domain actions in all sectors of SBPC, required by the direction of these sectors, in which the Science practiced that this power exercised was not questioned, investigated or could evolve against the paths exclusively traced by these dominating Science in SBPC's inquisition sectors.

CII. INTENSE ACTIVITY OF THE HERETIC HUNTERS! – 1990

In 1990 the I sent the Scientific Communication *"Coincidental Values of Physical Constants and Consequent Fundamental Discoveries in Science"* to the 42th. Annual Meeting of the SBPC, in which Maxwell's discovery was presented that electromagnetic waves showed the same speed of propagation as the constant speed of light c, uniting Electromagnetism with Optics. And reporting that the constant of Newton's Universal Gravitation was equal to the factor 10^{19} of the Planck constant of the energy quantum, resulting in the Unification of Gravity with Electromagnetism, Optics and all the Forces of Nature, by the theories of Erdelyi and Mesquita. Document 20.

Coincidental Values of Physical Constants and Consequent Fundamental Discoveries in Science

J. P. Hansen – Institute of Engineering

Summary. Relationships between physical constants in which coincidental values come to result in fundamental discoveries are mentioned. Are related electromagnetic, quantum and gravitational constants – permeability (μ), dielectric (ε), speed of light (c) and gravitational of Newton (G) – presenting discoveries between electromagnetic radiation, light and gravitational radiation, and consequent rational unification of optics-electromagnetism-gravitation.

ELECTROMAGNETIC CONSTANTS

In Maxwell's theory of electromagnetism, in the latter part of his calculations for determining the speed of electromagnetic variables, he came to observe that: $E_{Electric}$ = electrical field component and $M_{Magnetic}$ = magnetic field component, presented the speeds:

$$V_{E(Electric)} = V_{M(Magnetic)} = \sqrt{1/\mu\varepsilon}$$, and in CGS units:

μ (dyne /A^2) = permeability constant for vacuum = 1.25... 10^{-1} (dyne/A^2)

ε (A^2 s^2 /dyne cm^2) = dielectric constant for vacuum = 8.85... 10^{-21} (A^2 's^2 /dyne cm^2)

Therefore, $V_{E(Electric)} = V_{M(Magnetic)} = 2.99... 10^{10}$ (cm/s), the "coincidental" value of the speed of light! That is, resulting in the speed of light as a "coincidence" that Maxwell found with great enthusiasm, concluding that light was an electromagnetic radiation. Thus, revealing the unification between optics and electromagnetism. (1).

QUANTUM AND GRAVITATIONAL CONSTANTS

In a similar way, the Planck constant is observed:

h (erg s) = 6.62... 10^{-27} (erg s), and as (2):

h (erg s)/1 (s) = h' (erg) = 6.62... 10^{-27} (erg) = energy quantum

h' (erg) = m'c (g/s) X c (cm/s) = m'c^2 (erg), where

m' = quantum mass = 0.73715 x 10^{-47} g

m'c = quantum impulse

By the basic concept of force, of Galileo, where: "The change of motion, in time, is proportional to force (external action), and the amount of motion being the joint measure of speed and quantity of matter", then:

force = F (dyne) = impulse/time = m'c/t' (g cm s^{-1} /s^{-1}) = quantum force

(t' ≤ 1) and m'c (g cm/s) in whole number

As (3): F (dyne) = m'c/t' (g cm/s/s) = m'c (g cm/s) x 1/t' (s^{-1}), this is:

F (dyne) = quantum impulse frequency

Therefore: F (dyne) = m'c (g cm/s) x $\mathcal{V}$ (s^{-1})

$\mathcal{V}$ (s^{-1}) = frequency of impulse m'c, being $\mathcal{V}$ (s^{-1}) in whole number

The gravitational constant G of Newton, as (3):

G (cm^3 s^{-2} g^{-1}) = F$_G$ (dyne) (cm^3 s^{-2} g^{-1})

By Newton's Law of gravitation:

F$_G$ (dyne) = m$_1$m$_2$ /r^2 (g^2/cm^2) x 6.62...10^{-8} (cm^3 s^{-2} g^{-1})

F$_G$ (dyne) = gravitational force

r (cm) = distance between masses m$_1$ (g) and m$_2$ (g)

Introducing h' into that law, and using the concept fundamental of quantum impulse frequency:

F$_G$ dyne) = m$_1$m$_2$ /r^2 (g^2/cm^2) x [m'c (g cm/s) x 10^{19} (s^{-1})] x c (cm/s) x (cm g^{-2}) X (s)

Multiplying both members by 1 (cm) and reordering:

E$_G$ = gravitational energy

n = m$_1$m$_2$ /r^2 (g^2 cm^2) x (cm g^{-2}) x (cm) = dimensionless factor

$\mathcal{V}_G$ (s^{-1}) = 10^{19} (s^{-1}) = 10^{19} (hertz) = gravitational frequency

This is, as (2) and (3) the revelation of quantum h' (erg) of gravitational radiation with the frequency of 10^{19} Hertz. Thus, unifying quantum theory and gravitation, sought continuously, but unsuccessful by Einstein and other scientists. (4).

CONSTANTS: ELECTROMAGNETIC, QUANTUM and GRAVITATIONAL
Then, light, thermal, and other electromagnetic radiation, as Planck:
E (erg) = hν (erg); The radiation of light and electromagnetic, with speed c (cm/s) discovered by Maxwell; the radiation of the quantum h' (erg) = h (erg s)/s; h' = m'c X c (erg), as Erdelyi; and the gravitational radiation of energy E_G = h' | 10^{19} | (erg) , coming unify the constants h, c, and G.

CONCLUSION:
The relations between Planck's constants h, G of Newton, with h' of Erdelyi and frequency of the impulse m'c of ν_G (s^{-1}) = 10^{19} (s^{-1}) = 10^{19} (hertz) of gravitational frequency by Mesquita, its have come to present not a simple "coincidence" between the factors of these constants. But, to reveal the fundamental discovery of physical reality – gravitational electromagnetic radiations and its frequency – widely sought by several generations of detectors of these radiations (5). This discovery of Mesquita by the introduction of fundamental concept of frequency of the quantum impulse of Erdelyi, in a similar way of Maxwell by relations and compared between constants with values or coincident factors, its came to result in the rational unification of optics-electromagnetism-gravitation.

Thus, allowing to connect concepts and laws to discoveries of physical reality. By: Galileo-concepts of movement, force, quantity of motion and impulse; Newton-Laws of the motion and gravitation (2, 3, 6); Oersted and Faraday – electricity and magnetism (1); Maxwell – electromagnetism and optics (1); Planck – quantum energy (2); Erdelyi – quantum impulse, elementary quantum energy h' = m'c^2 (2); Mesquita – frequency of impulse m'c, electromagnetism-gravitation (3). Then, being achieved the quantum-gravitational unification, in which Einstein came to admit that it could only be achieved through the search for new principles (4, 7); and this unification sought by Hawking and many other scientists, because it is gravity the determining element of the evolution of the Universe and other knowledge (8, 9), which came to be effectively obtained through Erdelyi and Mesquita (2, 3).

Numerical values:

$h = (6{,}62517 \pm 0{,}00023) \cdot 10^{-27}$, with standard error
$h = (6{,}62517 \pm 0{,}00076) \cdot 10^{-27}$, with relative certainty
$h = (6{,}62517 \pm 0{,}00066) \cdot 10^{-27}$, with relative certainty
$K_G = (6{,}754 \pm 0{,}041) \cdot 10^{-8}$, determination of 1798
$K_G = (6{,}634) \cdot 10^{-8}$, determination of 1878 to 1898
$K_G = (6{,}655) \cdot 10^{-8}$, determination of 1901 to 1930
$K_G = (6{,}670 \pm 0{,}005) \cdot 10^{-8}$, determination of 1942
$K_G = (6{,}627 \pm 0{,}022) \cdot 10^{-8}$, on function of the Earth's latitudes and gravitational acceleration by the latitudes
$K_G = (6{,}62689) \cdot 10^{-8}$, on function of the KEPLER's law and GAUSS' universal gravitation constant by datas of the solar system
$K_G = (6{,}62583) \cdot 10^{-8}$, on function of the movement of the moon about the Earth

$h \cdot 10^{19} = (6{,}62517 \pm 0{,}00066) \cdot 10^{-27} \cdot 10^{19}$
$h \cdot 10^{19} = (6{,}62451) \cdot 10^{-8}$
$h \cdot 10^{19} = (6{,}62583) \cdot 10^{-8}$

Therefore:
$K_G = (6{,}62583) \cdot 10^{-8}$ (on function of the movement of the moon about the Earth)

$h \cdot 10^{19} = (6{,}62583) \cdot 10^{-8}$ to $(6{,}62451) \cdot 10^{-8}$, justifying the employment of the constant of PLANCK h or h' in the gravitational law of NEWTON.

REFERENCES

1. TIPLER, P.A. – Physics – Worth Publishers, New York, 1976.
2. ERDÉLYI, I.L. – The Examination of Planck's Constant – Escola de Engenharia Mackenzie, 1963, e Interpretação do Conceito de Força – Instituto de Engenharia, 1954.
3. MESQUITA, P.F. – The Meaning of the Universal Gravitational Constant..., Universidade São Paulo, 1965.
4. PARKER, B. – Einstein's Dream, 1986.
5. HANSEN, J.P. ...Detecção de Radiações Gravitacionais – SBPC, 1984.
6. NUSSENZVEIG, M., CARNEIRO, F.L., ROSA, L.P. – 300 Anos dos "Principie" de Newton – COPPE, 1988.
7. EINSTEIN, A. – The Meaning of Relativity – Princeton University, 1955.
8. HAWKING, S.W. – A Brief History of Time... – Bantam Books, 1988.
9. HANSEN, J.P. – Teoria Unificada da Física, Opinião de Dr. Hawking e Comentários sobre a mesma... – SBPC, 1989.

Doc. 20 – "*Coincidental Values of Physical Constants and Consequent Discoveries Fundamental in Science*", submitted to SBPC in 1990 by the author.

However, with the pretext of quality, any report that did not follow what was taught by the pre-establishead Physics would be a cause of rejection by the Committees of the Societies that were part of the SBPC. Thus, this Communication was considered as not accepted by be the gravitational radiations of Mesquita that would have the basis of accidental coincidence of the arbitrary standard of time of 1 second, because this coincidence could not be compared with the case of Maxwell in which the concordance of values was independent of the system of units adopted, according to Letter dated April 30, 1990 from SBPC. However, SBPC's consultants had long since been received by Mesquita that the time pattern of the gravitational frequency was only the same time as the time of all the variables in all the unit systems. And that this text would not be within the history of science.

On May 18, 1990 I sent an Open Letter to the President of SBPC, Physicist Ennio Candotti with a copy to the Secretary General of SBPC, Dr. Fernando Galembeck, refuting this primitivism of allegations to refuse a simple Scientific Communication. Document 21.

Missive to the Report of the Scientific Communication

1-Referee: "An arbitrary pattern of time of 1 second was used to obtain the coincidence between the quantum and gravitational constants. And the works of Erdelyi and Mesquita are devoid of physical fundamentals by having that standard of time. "

1-Author: It is unfounded because the "arbitrary pattern" could be n seconds and the result would be identical:

$E \text{ (erg)} = h \text{ (erg s)} \: \mathcal{V} \: (s^{-1})$

$E \text{ (erg)} /n \: s = [h \text{ (erg s)} / n \: s] \: \mathcal{V} \: (s^{-1})$

$[E \text{ (erg)}/n \: s] \times n = h' \text{ (erg)} \: \mathcal{V} \: (s^{-1})$

$E \text{ (erg)}/s = h' \text{ (erg)} \: \mathcal{V} \: (s^{-1})$

If a time pattern in the Centauri galaxy were 0.25 seconds = 1 second on Earth, the speed of light there would be 1,196... 10^9 m/s $_{Centaurus}$). However, the physical phenomenon would be the same, for the light would travel the same distance in second on Earth as in 0.25 seconds of Centaurs, for 1 second of Earth = 0.25 seconds of Centaurs. It is a primary level ignorance of the Referee!

2-Referee: "There cannot be compared the coincidence of constants by Mesquita, with the case of Maxwell in which the agreement of values is independent of the system of units adopted".

2-Author: Both with Maxwell and Mesquita, the values or factors of the constants come independent of the system of units:

Constants	CGS	MKS (SI)
ε	$8,85\ldots10^{-21} A^2.s^2/dyn.cm^2$	$8,85\ldots10^{-12} A^2.s^2/N.m^2$
μ	$1,25\ldots10^{-1} dyn/A^2$	$1,25\ldots10^{-7} N/A^2$
c	$2,99\ldots10^{10} cm/s$	$2,99\ldots10^{8} m/s$
h	$6,62\ldots10^{-27} erg.s$	$6,62\ldots10^{-34} J.s$
h'	$6,62\ldots10^{-27} erg$	$6,62\ldots10^{-34} J$
G	$6,62\ldots10^{-8} dyn.cm^2.g^{-2}$	$6,62\ldots10^{-11} N.m^2 Kg^{-2}$
$E_G = G.m^2/r$	$6,62\ldots10^{-8} erg$	$6,62\ldots10^{-15} J$
$\vartheta_G = (E_G/s)/(h')$	$10^{19}.s^{-1}$	$10^{19}.s^{-1}$
$\vartheta_G = E_G/h$	$10^{19}.s^{-1}$	$10^{19}.s^{-1}$

To prevent new ideas, the Referee commits error elementary in System of Units!

3-Referee: "It is not a work of History or Philosophy of Science".
3-Author: The work involves Galileo, Newton, Oersted, Faraday, Maxwell, Planck, Hawking, Erdelyi and Mesquita. It's the work that Story? Of Scientific Confiscation or of Modern Inquisition?

4-Referee: "It is a work with nothing to add again".
4-Author: Then they would be refused all work, because when referencing Einstein and other scientists "nothing would add again". The refusal to publish this work by SBPC, is by disagreeing with the dogmas of physics and prevent the free disclosure of the work of Erdelyi and Mesquita. And it is frontally contrary to the fundamental objectives of SBPC: freedom of research; of opinion, of the right to carry out their work; of respect for the works and its authors; to communicate and discuss their work; to oversee the ethics of scientists.

Doc. 21 – Missive refuting the primitivism of allegations of a Scientific Communication.

CIII. ELECTROMAGNETIC PHENOMENA UNKNOWN BY CUURENT PHYSICS – 1990

In September 1990, Engineer E. E. Geiger, presented article "Electromagnetic Phenomena Unknown by Physics" at Research & Technology Magazine-FEI In which a different phenomenon of attraction was observed between an electromagnetic wave flow and the matter density.

This attraction would seem to result from the field of elementary energies h' which interacted between the flow of energies of the electromagnetic wave and the electrons of the atoms that formed the matter.

CIV. SCIENTIFIC AMMUNITION OF THE BRAZILIAN NAVY
1989 to 1991

The interest for new ideas by Senior Studies Officers of the Brazilian Navy continued with information between the author and the Commander of the Navy Gama Almeida and Admiral Nunes Andrade. And on January 14th, 1991 I received a large amount "ammunition" for the confrontation of "warmongers actions"! Gama Almeida, sent two scientific articles *"Basic Concepts, Fundamental Laws and General Principles Involved in a Rational and Objective Cosmological Vision of the Physical Universe"* and *"The Principles of Conservation and Evolution as Natural Foundations of Unification of the Theories of Physics"*, presented by Nunes Andrade, published in the Naval Research Journal, October 1989. Based on conservation matter-energy Nunes Andrade concluded that the theory of relativity and quantum mechanics presented concepts contrary to the phenomena of nature.

And on February 8, 1991 I came to issue Letter to Nunes Andrade about his proposition of matter-energy conservation and commenting on the Michelson-Morley experiment in the interference of light rays by different observations Cited in his article and that an interpretation of experiments involving Michelson-Morley could be performed.

On February 15, 1991 I received Letter of Nunes Andrade in which he presented detailed article *"A Cosmological View of the Physical World based on the Concept of Substantiality of Energy – The Objective Interpretation of the Experiments of Silvertooth and Marinov"* on the Michelson-Morley, which presented a null result of absolute motion, and those of E. W. Silvertooth and S. Marinov which showed deviation of light indicating the absolute motion, for the continuity of studies on the bases of Physics. He also sent three numbers of the Naval Club Magazine with the articles: *"The Subjectivism and the Objectivism in a Cosmological*

View of the Physical Universe"; "The Conceit Fundamentals of Mathematics and the Modern Languages Formalized", and *"A Statement about the Mystification of Scientific Thought in the Technological Age".*

In the article *"A Testimonial on the Mystification of Scientific Thought in the Technological Age"* of 1989, Nunes Andrade commented that: *"According to the dominant pragmatists tendencies in the technological era, it is argued that, if competent scientists dedicated to physical-mathematical research are already accustomed to using the sophisticated formalisms of the theory of relativity in the description of certain experimentally observable results, dispensing with the rational and objective interpretations of the said results, such behavior is of their exclusive agency, not fitting any objection or questioning about on the part of researchers not included in their narrow circle. This is the doctrine followed by the Brazilian Center for Physical Research – CBPF, an organism maintained by the Brazilian society for the essential purpose of promoting free research in the field that is its own; instead of empowering yourself that the scientific research project that is to guide you through the duty of office, the illustrious Director of the CBPF, in fact a competent Physicist of high feedback, prefers to align unconditionally with the external matrix ordering of standard doctrine and of which is faithful and efficient collaborator. Having personally verified the impossibility of contesting the ideas proclaimed in the above theses, it has disregarded its free debate and public judgement in the institution that it directs; on the other hand, the most competent researcher of CBPF in the sector in question has spontaneously assumed the responsibility of requesting copies of the disapproved works so that they would be part of the library of the institution, where they are awaiting a time of unconditional and unrestricted academic freedom, in which they can be debated and judged publicly in the light of reason and common sense, free from the constraints imposed by bigotry that inhibits the free manifestation of thought scientific".*

In 1989 the Director of the CBPF was the Physicist Leite Lopes, who before Glashow, Salam and Weiberg, had discovered particles of interaction of the weak force, W and Z, of mass around 1/3 of the mass at rest of the proton, but which considered contradictory with the massless photon in rest of *electromagnetic* interaction. That is, despite that Physicist itself perceive the contradiction of particles without mass and another of immense mass, remained as an effective collaborator of pre-established Physics, and so Weinberg came to quote, in the Nobel Prize for Physics by Glashow, Salam and Weinberg, the initial work of Leite Lopes.

CV. NOTHING SHOULD BE ADDED TO SCIENTIFIC KNOWLEDGE!
1933 to 1991

On March 13, 1991 the Scientific Communication "*Coincidental Values of Physical Constants and Consequent Fundamental Discoveries in Science*" was resent to SBPC for the 43 [a]. Annual Meeting of the SBPC, directed to the Committee of Physics-Mathematics, because the previous Committee of History of Science had rejected.

But, on May 17, 1991 through Letter from the Secretary General of SBPC, Dr. Fernando Galembeck, informed that this Communication was refused by this Physics-Mathematics Committee by: "*... to add nothing to scientific knowledge.*" Therefore, any article that reports any of the Physics such as the theory of relativity and quantum, "*should not add anything to scientific knowledge*"!

No theory of Pre-Establishead Physics had achieved the unification of all the forces of nature, and this Communication reported that Newton's Constant Universal Gravitation was equal to the factor 10^{19} of Planck's constant of energy quantum, of the theory of Erdelyi and Mesquita, resulting in the Unification of Gravity with Electromagnetism, Optics and of all Forces of Nature, with an analogy in which Maxwell had detected that the constant speed of electromagnetic waves was equal to that of the speed of light. But, the theory of Erdelyi and Mesquita were systematically prevented from its presentation and disclosure by its own SBPC in the Science & Culture Journal. In reality what these two Committees of Physics wanted is that "*nothing was added to scientific knowledge*"! And those two theories had already been present thousands of publications!

And on May 28, 1991 the I sent Open Letter to Dr. Fernando Galembeck, Secretary General of SBPC with copy to the President of SBPC, Ennio Candotti, reporting manifestations of noble personalities of Science against pseudo-judges. Thus, showing that there was in the SBPC only the right of a point of the statute that protected these oppressors considered not anonymous, because they were identified internally, but hidden, cloaked, hidden by the rules of the SBPC, but that only covered a point within the infinite radius of Morality and Ethics. Document 22.

Doc.22 – Manifestation again th Judges of Science, as Letter to General Secretary and Presidency of SBPC in 1991.

Demonstrations Against the Judges of Science

According to M.S. and M. Giambiagi – Center for Physics Research – in the Science and Culture Magazine of the Brazilian Society for the Progress of Science-SBPC, 10/1983: "A referee can reject a job (especially those that come out of the routine and be written in an accessible and direct manner) for its" irrelevance ", a word widely used by them, for some futile reason".

"May also reject for subjective reasons, in the name of "high scientific quality" (as Proceedings of the National Academy os Sciences, U.S.A.)".

Quoting S. Vila Nova, Science and Culture Magazine, 1980: "The supposed control of the quality of scientific production is unsatisfactory or even suspicious".

He reports that by frivolous or subjective criteria of Physical Review Letters, the Physics Letters appeared at 1961.

"The pretension to standardize the way to produce Science is frustrating and claims the space for different styles: a plurality beyond possible and desirable, to avoid inhibiting the release of potentiality of scholars. "

Frota Pessoa – Professor of São Paulo University-Brazil, in Science and Culture Magazine, 07/1984: "Presenting useful, curious, sensational facts, should be informed to the scientific community, to expose the logical relationships between facts and ideas that are typical of science".

Isaac Asimov – Scientist, in Science and Culture Magazine, 07/1984: "The teaching of science must be the broadest, most widespread imaginable."

"One of the forces that attract young people to science is their presentation."

"Science ceased to be the task of the ancient lone wolves."

Lauro Oliveira Lima – Educator, in Science and Culture Magazine, 07/1984: "The progress of science is not done by accumulation, but by paradigm shifts (the invention of new values to observe)".

"Stopping science would cut the bird's wings so it wouldn't fly anymore."

"The moment you condemn a kind of creativity is wounded the essence of man."

"You have to end the micro dictatorships that exist in school and research centres".

N.S. Hetherington – Scientist, published in Nature, 1983: "It seems more efficient to silence disagreement, so that it does not become public".

"The silencer conspiracy goes so far that it sometimes hides the truth even from the scientific community, aiming to protect another researcher's reputation."

F. W. Lima – Professor at São Paulo University-Brazil, in Science and Culture Magazine, 01/1985: "New ideas in basic research are fundamental for further development and new applications, chained adamantly to these initial studies.

A. Petroiani – Professor at the Federal University of Minas Gerais-Brazil, in Science and Culture Magazine, 03/1985, in the article "Publication of Scientific Work": "Human progress is based on the dissemination of ideas, discoveries and inventions within the most diverse areas of knowledge. The search for truth has promoted research to clarify situations. "

"The prudent thing is to present a scientific work in powerful scientific societies with ethical and moral dignity."

Ronaldo Mourão – Doctor in astronomy by the Sorbonne University in Gazette Journal of Paraná, 09/1989: "Those who fought against the system within the SBPC, are now occupying positions and maintaining authoritarian rubble. "

"There are still bodies of power making pressure so that science does not question, investigate or evolve, power to the scholastic moulds, as it existed in the time of Galileo."

John Maddox – physicist, At the time Editor of Nature magazine, says his biggest concern It was to reject work in presenting discoveries, such as the rejection by Nature in 1933, of Enrico Fermi's work on the disintegration of radioactive substance (beta decay).

Bertrand Russel-Philosopher and Mathematician, Nobel laureate, creator of International Tribunal for war crimes, he said: "The world needs frank hearts and brains, and it is not through rigid systems, whether old or new, that it can be achieved."

"I would like to see a world where education was aimed at mental freedom rather than the incarceration of spirits in a rigid armor of dogma."

J. A. Giannotti – Professor at São Paulo University-Brazil, in the Folha de São Paulo newspaper, 03/1990, mentioned: "There's a class of little professors, all of them autonomous, not having to account to anyone. "

"Commanded by their laws was associated with a kind of cultural narcissism, when the teacher and the researcher are locked in the circle of their own ideas, as if dialogue and controversy were pitfalls to threaten the continuity and advancement of thought itself, Instead of constituting, as has always been admitted, the soil from which it feeds, is fed. It destroyed the nerve that eletriza the couple tradition-innovation, provoking the sclerosis of the academic life ".

"In Brazil, in general, they want the reality to agree with the imported books. "

"For scientific activity is a process of circulating people and ideas."

"For modernization, the development of basic science is necessary."

"What matters is the ability to form new types and research groups."

"It is to create conditions for the intellectual to be recognized, by his work, by his word, by the sharpness of his thought."

Fred Allen – American Consultant, on the commission of Judges: "Commission is a group of technicians who alone can do nothing, but that together they decide that nothing can be done".

William Shakespeare, one of the greatest Writers of humanity, about when there are new ideas: "Brave New World, which has such people". As scientific inquisitors seek to keep these people away.

Primo Nunes Andrade – Deputy Admiral, Professor of the Superior Magisterium of the Navy, in Magazine Naval Research-Supplement of the Brazilian Maritime Magazine, 10/1989 and Magazine of the Naval Club number 280/1989: "According to inquisitorial norms imposed by the" intelligentsia "which commands with intolerance the modern scientific establishment, only what is published in scientific journals officially accepted, It can be considered and evaluated by the international scientific community. "

"The experiment of the American researcher E. W. Silvertooth, although it was sponsored by suitable organisms (Rome Development, Griffith AFB, Air Force Systems and Defense Research Agency), do not have achieved to obtain the "nihil obstat" that provide the disclosure of his work in some scientific magazine officially accepted. The reason of such discrimination is because this work is in contradiction with religiously consecrated relativistic myths, thus making it prohibited to free examination and to the impersonal criticisml and constructive. "

"By the current inquisitorial standard, if this Silvertooth experiment does not achieve disclosure in scientific journal accredited by the establishment, it does not merit official recognition of the international scientific community, all passing as if this experiment had never been accomplished, "

"Thus, maintaining paralysed scientific thinking, now subordinate to idols, myths and taboos of the technological age, fruits of interests and dominant prejudices."

"And there are almost all that by inertia are conformed to so many deformations of scientific thought, surrendering to cultural terrorism that stigmatizes as symptoms of intellectual disability or inadequacy to progress, preventing the "imprimatur" of the dominant scientific establishment, by the almighty control of the media.

Flávio Pereira-Graduated in Science from the University of São Paulo, Director of the Brazilian Institute of Space and Spatial Sciences-IBACE, in 1967 wrote in the Bulletin of the Brazilian Interplanetary Society-SIB: "The Professional Specialists of Physics insist on ignoring science innovations that present ideas contrary to the prevailing dogma of relativity theory. Ballistics, living in an icy climate in a regime of academic. When Einstein himself authored this theory, just before his death in 1955, came to say: "The time is propitious to new utterances, and extensions of ancient postulates will only serve to increase the dilemma; That that you don't find yourself prone the creation process will not pass of a dead man". "

"Hitler thought of building the third Empire of Germany, lasting at least a thousand years. The supporters of the theory of relativity they do not have in mind less time for the duration of their beloved theory. "

Otávio Madalena – President of the Observatory and Research Center Bradesco, 1968, reports: "_If you wish to create something that brings everyone a chance to improve their knowledge, find all sorts of free opponents to clutter the evolution of ideas. "

Istvan Erdelyi – Graduated Engineer from the University of Budapest-Hungary, Professor of Engineering at Mackenzie University, President of the Scientific Council of the Brazilian Interplanetary Society-SIB from 1962 to 1969, immortal creator of the theory of quantum impulse, had reported in 1957 in a scientific conference at the Institute of Engineering: "The behavior of professional specialists in science for new ideas does not differ in essence from those who played Giordano Bruno at the bonfire. The time of Galileo repeats itself, it is repeated. "

In 1967, at Mackenzie University, "would any priest, bishop or cardinal approve some polygamist theory in a Catholic country! Similarly, ideas contrary to those of professional physicists, they will be rejected when subjected to their specialists. "

In 1969, in the Revista Astronáutica (Astro Magazine): "There is a kind of contemporary inquisition fuelled by relativists, who all do to maintain their dogmas."

Paulo Ferraz Mesquita – Engineer from the University of São Paulo-USP, Dr. In Physical and Mathematical Sciences from Brazil University of Rio de Janeiro, Professor of Geodesy and Astronomy of Polytechnic Engineering of USP-Brazil, President of the Scientific Council of the Brazilian Society of Astros, from 1969 to 1972. At 1971 he mentioned at the General Assembly of SBPC: "At the bottom of all the causes of resistance to the progress of science is found the lack of ethics. Dominates in the Brazilian society of physics, sheltered from SBPC, the medieval criterion of promoting impetuous prepotency against the freedom of manifestation of scientific thought of associate, hindering the freedom of teaching and research".

In the Science and Culture Magazine, 11/1981, about the work of Mesquita – "Meaning of Newton's Gravitational Constant and Consequent Specific Discovery of Gravitational Radiation Reveal and Prove Misconception of Einstein, Planck and Contemporaries" – in which it was refused for publication in the Science and Culture Magazine and the Brazilian Academy of Sciences. "This work was not published by reason, as the expression of its own President: "It would be a great responsibility to publish the misconception of Einstein, Planck and contemporaries". And so, the threat and delay of the original research in the field of fundamental physics in the country has been continued for more than 15 years. Attached to the past, the professional physicists of the dome, deny taking notice of the possibility of progress offered by new hypothesis of work coherent with known experimental data ".

"Prepotency does not promote the progress of science, coming through the publication to stimulate scientific truth research through free debate."

CVI. VOTE OF CONFIDENCE FOR SCIENTIFIC ANTIOPPRESSION
1991-1992

The alteration of the Standards of SBPC was considered by its Board as necessary for the improvement of the quality of the scientific works. If that was the case, an opinion of the consultant could be issued alongside the proposed article, thus having freedom of the presentation of new ideas, and the reader is also free to decide his opinion. Also, the article could be withdrawn if the author was in agreement with the opinionist of his article. But in reality the new regulation of SBPC had the purpose of annihilate with any new ideas that would be contrary to the interests of consultants, controllers, owners of committees and summits of sectors of SBPC.

The Pre-Established Physics and the Quantum Impulse

Drawing, of "Horace Dinosaur" by artist Maurício de Sousa, published by the newspaper "O Estado de São Paulo" (The State of São Paulo), Brazil, on June 9, 1991 illustrates the essence of the mindset that was hidden by apparent progress or regulations. Figure 42.

Fig. 42 – "Horace's Dinosaur" by artist Maurício de Sousa, published by the newspaper "O Estado of São Paulo", Brazil, on June 9, 1991 illustrates the essence of the mindset that was hidden by apparent progress or regulations.

Thus, in July of 1992 I presented a Vote of Confidence to the General Assembly of the 44 [a]. Annual Meeting of the SBPC for Communications to be published with the opinions of consultants and replicas, or withdrawn from the work if they agree with the authors.

However, the President of SBPC Ennio Candotti accepted the manifestation of one participant in the SBPC, in which he considered that only some possible high-relevance article could be free of judgment and publishable.

It is clear that this relevance was subjective and without any possibility of applicability by the tax regulation.

Ennio Candotti considered that it would be to generalize all the articles as of quality to be public, so the quality considered that the motion should not be applied. As Candotti, an paper declined although it should have been accepted, one could not generalize that all other refused would be acceptable, so the quality would not matter if one or the other were rejected.

And that's what happened, because two participants were opposed to this Vote of Confidence, in an Assembly with around 40 participants who did not pronounce, and were already around 2 hours of the morning of July 17th, 1992. But even if it was approved in this General Assembly, this Vote would not be effected, as it would be contrary to the Statute and regiment of SBPC. This Vote should be addressed by the Board of Directors and the Council of SBPC, which did not occur by the objections of SBPC Sectors.

It is ironic that the next day, 18 July 1992, Candotti mentioned in an interview to the press that the Minister of Science and Technology in 1992, Hélio Jaguaribe, showed inability to tolerate the diversity of opinions, which was the essence of democracy, but that and kind of authoritarianism was coming to an end! And the Historian Maria Victoria Baku of USP, after greeting Candotti, commented that a Country without diversity did not reach modernity or democracy!

But the Vote of Confidence presented was precisely about intolerance to the diversity of ideas or opinions, under the justification of quality! Quality could not suppress diversity, but to allow the two opinions issued.

It is also ironic that in this year of 1992 no Scientific Communication involving the theory of quantum impulse or gravitational radiation was presented, and Ennio Candotti at the end of the Meeting mentioned that the public had been scarce and that no resounding research had been presented! In the case of Physics were not Scientific Communications that would be refused by the Committees in the name of supposed quality. It was the Scientific Communications permitted by the current Physics, which by the words of Ennio Candotti would appear to be irrelevant! This was the quality obtained with the changes in the statutes, regulations, laws and norms, carried out by the sectors of SBPC, for the supervision of preventing the diversity of opinions and new ideas! As the Biochemist Rogério Meneghine of USP, the visibility of science, that is, its disclosure was fundamental to scientific progress. With minimal scientific disclosure, what was coming out was a minimum of scientific results. And what mattered most to science was Scientific Communication.

According to Philip Hilts of the newspaper New York Times, the United States in 1992, had 1 million of researchers at a cost of 25 billion dollars, and that made 50% of science on the planet. And the results were to be published or else they would perish.

However, in the Brazil, SBPC amended its regulation to prevent publications that could go against the interests of power maintainers, with the quality claim! If it were for quality it would be simple, the publication would be effected with the opinion attached to the judge and with the replica of the author. But, this Vote of Confidence had been rejected or not appreciated by the President of SBPC in 1992.

CVII. NUCLEAR FUSION PORTABLE – 1895 to 1989

After the discovery of the X-rays in 1895, and scaling of diffraction waves by W. H. Bragg and W. L. Bragg in 1913, it was determined the structure of crystals in different compounds, in which properties as hydrogen absorption were observed in metal crystals as palladium and titanium.

Thus, in 1926, physicists Fritz Paneth and Kurt Peters attempted to obtain hydrogen fusion through the crystalline network of palladium which would approximate the hydrogen atoms to form helium. However, some helium obtained was attributed to the air that penetrated the vial where were the palladium and hydrogen.

In 1932, with the discovery of the neutron, John Tandberg, Torsten Wilner and Bertil Wilner experiment with deuterium to try to obtain deuterium (1 proton + 1 neutron) (Helium (2 protons + 1 neutron) + 1 neutron or (2 protons + 2 neutrals). That is, H12 (He_2^3 or He_2^4. And they obtained some helium and neutrons.

In 1947, Frank F.C., A. Sakharov, Y. Zeldovich and others theorized that the Meson μ – (muon) could involve two cores of deuterium or tritium (1 proton + 2 neutrons) i.e., H_1^3, that by the negative charge there would be attraction with the 2 nuclei of positive charge and diminishing its distance to the fusion, despite having a minimum timelife, but with mass 207 times greater than the electron.

In 1956, Luís Alvarez and Edward Teller sought the merger by the muon, but the fusion measurement by muon presented a minimum value.

In 1977, Soviet researchers at the Joint Institute for Nuclear Research in Dubna indicated that energy spikes occurred in a mixture of deuterium and tritium, which increased the fusion by muon.

In 1982, Steve E. Jones of Idaho National Engineering Laboratory in Idaho Falls used the muon beam at the Los Alamos National Laboratory to achieve cold fusion. Some of his results were promising, but the number of fusions each muon can catalyze was still too low.

And in 1989 Martin Fleischmann and B. Stanley Pons announced that from electrolytic cell of palladium, platinum and heavy water (deuterium), they obtained intense quantity of heat, tritium, neutrons and gamma rays. This heat in some cases, according to the type of material (asbets) that involved palladium, resulted in the fusion of palladium at 1554°C. However, the quantity of neutrons that was measured was much lower to reach these temperatures. Experiments with other researchers indicated contradictions between neutron detections, tritium, resulting in new theories about the heat obtained. One of them, according to Linus Pauling, was just the formation of the compound $Pd2H_1{}^2$ that releases heat from the chemical reaction. But, after there were measurements of helium that would indicate the fusion. However, other tests did not detect helium. Other explanations have been presented, such as electrons that acquire greater mass, greater acceleration, greater electric field, when deuterium is in conjunction with palladium, resulting in approximation of deuterium atoms and consequent heat emission, but without the nuclear fusion.

New experiments with deeper theories could achieve smaller equipment for nuclear power, destined for diverse applications such as spaceships. The theory of quantum impulse involves particles, with its dimensions, shapes, charges and fields, formed of elementary energy, which would be of valuable calculations for that purpose.

CVIII. NOTHING NEW! – **1993**

On the 45th. Annual Meeting of SBPC in 1993, nor has any Scientific Communication been presented about quantum impulse or gravitational radiation.

According to the news of the press after contact with the organizers of the Meeting, it was published that: *"As for the scientific part, as they say, it was not to be expected great news."*

SBPC's Regulations on 1993 control to refuse "heretical communications" were still active. And what was left *"It was not to be expected big news"*.

CIX. SCIENCE PROGRESS BRAZILIAN SOCIETY- SBPC ON TRIAL
1992 to 1994

In July 1994 the SBPC received criticism in which the newspaper "O Estado de São Paulo", Brazil, put repairs to the work presented by SBPC in those recent times.

By the new statutes existing in 1994 in the SBPC once judged a scientific work, and if it were refused there would be no replica by the author of that work. And when I had presented in 1992 a Vote of Confidence in which an article, a work, a scientific work presented to SBPC should be the replica for others to have knowledge, this Vote was rejected or not appreciated by the Board. However, the new President of SBPC, the Geographer Aziz Ab'Saber, has exercised the right of reply, i.e. exactly the replica, so that both the criticism presented and the SBPC position and or justification could be known.

But, the position and justifications presented by Aziz Ab'Saber was that SBPC was the bulwark of free expression of thought and the largest society of stimulus to science in the Southern Hemisphere.

However, as "bulwark of the free expression of thought" the replica of SBPC, in the case of the newspaper "O Estado de São Paulo" in Brazil, it was radically contrary to its Statutes against the freedom to present a simple Scientific Communication, without the right of its publication even if it had a contrary opinion of the inquisitor of the SBPC. On 23 September 1994 to emphasize to the President of SBPC the right of reply, to those who refused works contrary to the doctrines of their judges, and that were published these works along with opinions contrary, I sent Open Letter to Dr. Aziz Ab'Saber, President of SBPC.

On October 31, 1994 received a Letter from Secretary General of SBPC, Adhemar Freire-Maia, Doctor in Sciences, in which in a dignified attitude mentioned that the Board of SBPC had decided that the Board itself would indicate the Judging Commission of scientific work, and no longer the Committees of Sectors of SBPC. And also, a change of Regulation so that there was a criterion of what should be quality for your refusal. The Board of SBPC since July 1994 was chaired by Aziz Ab'Saber, as Vice-Presidents, Francisco Salzano, Doctor in Biological Sciences; Jacob Palis Junior, Engineer and Doctor in Mathematics, Member of the Brazilian Academy of Sciences; and as General Secretary, Adhemar Freire-Maia, Doctor in Science. It was a Historical Letter from SBPC informing that the Judging Committees of the works indicated by the Scientific Societies of the SBPC as Physics Brazilian Society-SBF were extinct, being replaced by a Judging Committee indicated by the SBPC itself from 1994.

CX. SPECIAL MEETING OF SBPC WITH FREIRE-MAIA AND WITH "BRAKES" ON THE HUNTERS OF HERETICS! 1994-1995

In 1994 the SBPC decided to present the Special Meetings in which a local theme of the region, where the Meeting was held, would be focused and other Scientific Communications. So in 1994 was held the First Special Meeting – The Cerrado and the XXI Century: Man, Earth and Science. And in 1995 the 2th.

Special Meeting – New Paths: Environment and Social-Cultural Diversity.

It was evident that after the resolution of the new Board of Directors of SBPC, Adhemar Freire-Maia putting brakes on the hunters of heretics, it would then be possible that Scientific Communications could be accepted. In fact, Scientific Communication "*Experiment with Laser Beams on Determination of Absolute Motion of Inertial System in Space*" that I submitted was accepted by the Judging Commission of SBPC and published in the Annals of the 2t. Special Meeting of 1995.

This Communication was fundamental to Science, especially in the case of Physics, because presented an experiment idealized by the quantum impulse of Erdelyi, in which a beam of light coming out of a laser installed at 90 ° of the direction of the Earth's motion in space for the constellation of Leo should come out at 90 °, or diagonal if it receives the inertial velocity of the system Earth-Sun-Laser. According to the theory of relativity, light always has the same speed c in space and should come out at 90 °, because even Einstein's relativity the speed of light c is independent of source speed, that is, of the system Earth-Sun-Laser, and would not receive the inertial speed of this system. However, the experiment showed that the light came out diagonally, that is, the photon of light receiving the inertial velocity of the Earth-Sun-Laser in the motion for the constellation of Leo. The conclusion was evident: the theory of relativity was demonstrate wrong, for it was inertia that carried the light varying its velocity in relation to the emitting source, contrary to the contraction of space and time dilation to keep c always constant by relativity! Figure 43.

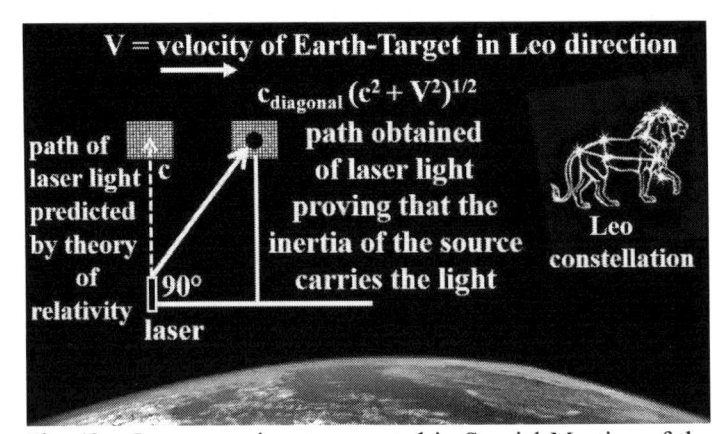

Fig. 43 – Laser experiment presented in Special Meeting of the Science Progress Brazilian Society, in 1995, it shows that the laser at 90° emits the light on the diagonal proving that the source carries the inertia of the light. Earth and Leo Clipart from *Editora Europa*.

E a velocidade da luz é sempre constante em relação ao espaço-vácuo-éter, mas é variável em relação à velocidade da fonte emitente, pois a luz é carregada pela fonte, sem contração de espaço ou de dilatação de tempo. Figura 44.

> **The source emits the light with c' velocity always constant relative to the vacuum-space-ether**
> **By laser experiment in that the laser light comes out diagonally, the source carries the light, then the c velocity is variable in relation to the source velocity and without space contraction and or time dilation**
>
> c' velocity of light in relation to vacuum-space-ether
>
> c velocity of light in relation to emitting source
>
> v source velocity in relation to vacuum-space-ether

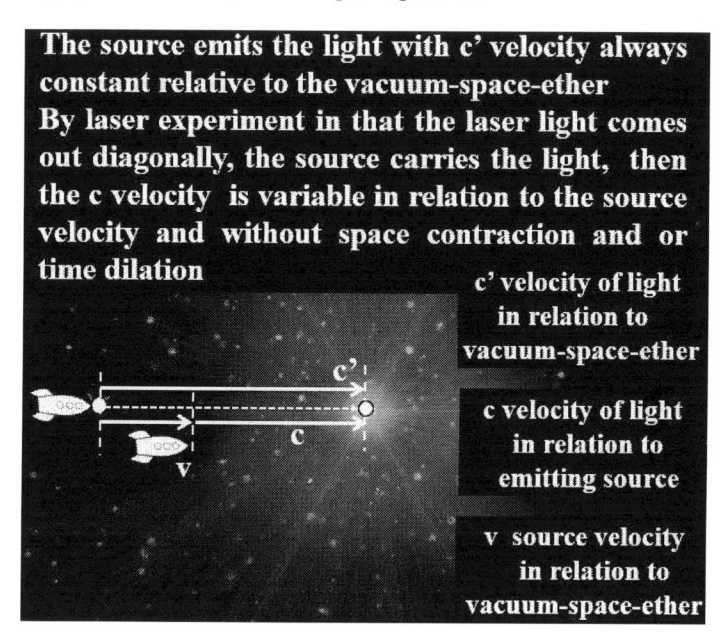

Fig. 44 – This laser experiment it shows that the speed of light c' is always constant in relation to the vacuum-space-ether, but by the emitting source carries the light, because the laser at 90° the light comes out diagonally, the speed of light c is variable relative to the source velocity. Adapted from the clipart of *Editora Europa*.

There were more than 800 Scientific Communications sent between all sectors of SBPC and more than 10,000 participants were in auditoriums and classrooms, surpassing the expectation of participation and presentation of Scientific Communications. The new Regulations against the hunters of heretics allowed for a more active participation and not copyist for the Scientific Communications.

However, the Physics Brazilian Society-SBF preferred to ignore the experiment reported in the Scientific Communication, which was in the publication of the Annual Meeting in the section C2-Physics, of Exact Sciences, because there would be no way to prove the opposite of the result obtained, then silencing to keep what they continue teaching as the theory of relativity and current quantum theory.

CXI. STANDARD MODEL WITHOUT GRAVITY AND UNIFICATION OF THEORIES OF RELATIVITY AND QUANTUM WITH THE STRINGS OF DIMENSIONS "COILED" – 1995

In 1995, in Fermilab, by the work of 800 physicists, was announced the discovery of the sixth quark named top quark, which comprised the standard model of the pre-establishead
Physics of particles and fields, by the interactions of the electromagnetic forces, weak, strong, including particles and field of Higgs.

But gravity was still outside that formal structure.

But what were quarks formed? This imposed Physics did not admit that the photons at rest had mass and were the constituents of all elementary particles, which by the impulse theory were the photons $h' = m \, c \times c = m'c^2$. This Physics had only the mathematical abstract field of Higgs to invent mass for the particles!

Also in 1995, through C. Hull Studies, P. Townsend, A. Sen, M. Duff, J. Schwarz, P. Horava, and others, E. Witten has unified all string theories, introducing 11 dimensions that would unify the theory of relativity with quantum mechanics and admitting as the theory of everything. But these 11 dimensions were admitted as "coiled" between them through mathematical abstractions distant from the physical meaning.

CXII. TRIPLE MEETING – 1996

With the success that the disclosure of Science without the exterminators of the future of Science, when performing Scientific Communications at SBPC, in 1996 the Board of SBPC came to emphasise the national and regional dissemination of Science produced in the country, scheduling three Meetings, one Annual and two other Specials.

Thus, the author sent a Scientific Communication *"Coincidental Values of Physical Constants and Consequent Fundamental Discoveries in Science"*, which was the Communication that had been refused in two consecutive years by the Committees of SBF-SBPC, for there was the dissemination of the theory of quantum and gravitational impulse interconnected to electromagnetism and light, contrary to the hunters of extermination of new ideas. And in May 1996, with the new Regulations on presenting Scientific Communications to SBPC, this Communication sent was accepted for presentation and published in the 3rd. Annual Special Meeting of the SBPC. At the opening was the Scientific Conference, *"The origin of the Universe"* of the Astrophysicist Dr. John Steiner. That by the reviewers, consultants, anonymous inquisitors of the Imposed Science by chance would be that *"...nothing would add to scientific knowledge..."* because it is already of ample knowledge? It is clear that this Lecture would be a great success for this high-level Physicist, far above the surface of creeping beings!

And the Communication of the author presented and published in the Yearbook-SBPC of 1996, after several years of being prevented by the occult reviewers of the antiprogress of Science!

With the ban on hunting the work of new ideas that was exercised by heretic hunters, it was presented in this 3rd. Meeting Special more than 500 Scientific Communications, around 2,400 subscribers and more than 7,000 participants, being again a success. This Communication by the author who had previously been refused by reviewers, with a primary claim *"not to add anything to scientific knowledge"*, merely showed the inability to add any scientific knowledge to mediocrity of their minds! The Scientific Conference conducted by Astrophysicist John Steiner on the universe, big bang and Hubble law, by these reviewers nothing should add to scientific knowledge, because thousands of times had already been quoted previously. However, the plenary was full in his presentation!

But, in May, in the Annual Special Meeting of 1996 it was a success, in the 48th. Annual Meeting of the SBPC in July 1996 there were no Scientific Communications presented on the theories of quantum impulse and gravitational radiation, and there was the presentation of Scientific Communications already presented in other periodicals and others that were considered of low quality by various sectors of SBPC. In other words, SBPC itself observed the lack of originality, that is, new ideas in all sectors of SBPC. Without new ideas, the researchers of the Imposed Science sent the same articles already published or teachings that they received for the counting of works produced and thus fulfill their academic goals. But, new ideas were prevented from being taught, and with the rejection of similar works already published in other places, little would remain, ie, according to what the Organizers of the Annual Meeting of the SBPC of 1993 had commented with the Press was only that: *"There was no big news to be expected"*! What happened again in 1996 with works copy and re-copies, because the leadership summit continued to prevent the creation of new ideas.

However, in other warm Meetings of the SBPC, hundreds of works on the basis of knowledge previously presented by scientists, continued to be disseminated with ample success, without constituting a nothing to scientific knowledge. Except for the inability of these reviewers to absorb any that was seen from new angles to the scientific knowledge presented in these works. by SBPC.

J. P. Hansen

CXIII. THE JOURNAL NATURE AND THE PAPER SUBMITTED OF LASER EXPERIMENT, WHICH PROVED THE FALLACY OF RELATIVITY – 1997

On 17 March 1997 I drew up the full article resulting from the Scientific Communication in which had been published its Summary in the Yearbook-SBPC Special Meeting of 1995, sending to the Nature, considered the most prestigious Scientific Journal worldwide, for submission of the article *"Laser Light Experiment about Inertial System Determination"*, to the Editor Mr. Philip Campbell, in which this article checked the inertia of light. That is, the photon of light received the velocity of the Earth moving in space. Thus, the speed of light c depended on the inertial velocity of the emitting source which was the Earth, i.e. c' diagonally = $(c^2 + V_{Earth})^{1/2}$. Document 23.

LASER LIGHT EXPERIMENT ABOUT INERTIAL SYSTEM DETERMINATION
J.P.Hansen - Universidade Mackenzie/Instituto de Engenharia, Brasil (a)

It is known that the Michelson-Morley and equivalent experiments, since 1887, they came nothing to detect about velocity (v) of inertial system, like Earth-Sun-Milky Way Galaxy (ESG). However, from 1951 to 1987, works involving measures of: redshifts, temperature in the cosmic blackbody radiation and light stationary phase, related to direction of motion of the ESG system, they came to present velocity v = ~ 350 km/s for the Earth-Sun (ES) component of that motion through space or relative to the cosmic blackbody radiation, in the direction towards the proximity of the constellation Leo. The present experiment, utilizing laser projected in one millimetre screen (to distance d = 3m), in the direction of that constellation and in then to 90°, admitting in principle the velocity of the laser light (c = ~ 300.000 km/s) independent of v, with basis in a device spacial displacement pointer, by radiation quantum momentum (impulse) theory,constituent of the light, it would come to point in the screen one displacement (x), by the relation of the times of trajectories of laser light (t') and of ESG (t'') ; t'=d/c , t''=x/v , t'=t'' , then x=3.5mm. Results of laser light in those directions presented x=0, conforming Martits experiment, in 1973, with neon light and convergent lens of the rays of that light, with basis also in that device spacial displacement pointer. Despite of the present experiment, as well of the Martits, they not register that displacement, they came to point notable evidence of the photons of the laser light or neon light they to be with inertia of the ESG, requiring to the analysis of the experiments which not registered that inertia. Also, the present as well as of Martits experiments not being in agreement with variations of the space-time, by the Einstein relativity theory, because to be those projections of the laser or neon light (that they would point x) in the transverse direction to motion; and in case of light curvature, for explain x=0 , it would must to be of α = arc tg (v/c) = 206", not calculated by the general relativity.

INTRODUCTION - It is known that the Michelson-Morley and equivalent experiments, since 1887,they came nothing to detect about absolute velocity (v) of inertial system, like Earth-Sun-Milky Way Galaxy (ESG) [1 to 3]. These null results could be explained by : dragging of hypothetical ether [1] ; matter contraction in the direction of itself motion [3 to 5] ; Einstein relativity theory, with the variable time and the own space for constancy of the light velocity (c= ~300.000 km/s) in void for all inertial systems, that is, light without inertia in any reference system [6] ; light with inertia of the emitting source, in permanently stationary phase [7]; light without inertia by oblique rays presenting trajectories in equal times in the two transverse directions of the Michelson-Morley interferometer arms, by an adjustement ≠ 90° of the interferometer arms for interference of the light rays that results in permanently stationary phase [8 and 9].

About the dragging of hypothetical ether, the stellar light aberration existence in conjunction with the Fizeau experiment of 1851 came to confirm not to have that dragment [10 and 11]. The matter contraction in the Lorentz electronic theory, it came to be contradicted in various experiments [3, 12 to 14] , despite of favourable propositions to itself [15]. But, by fault of other results of that theory, and by other results in accordance with the relativity theory — variable space-time, stellar light aberration, Doppler effect, light velocity by binary star and solar source, mass increase, light deflection, perihelion motion , light velocity in diffracted, reflected or transmitted moving sources, particles scattering in high velocity, time dilation of electromagnetic sources and of fast meson, waves velocity (χ and γ rays) by particle in high velocity, electron and muon g-factor, equivalence principle, gravitational waves and recession of galaxies — came to have primacy by this theory [2, 3, 16 to 23]. Although, results explained by the relativity, also they have been explained by not relativistic propositions [2, 7 to 9, 15, 24 to 42] . The light, in the case of to be in permanently stationary phase, it would come in the Michelson-Morley and equivalent experiments, for any direction of motion of the reference system, in principle to present absence of interferences, like confirmed in those experiments [42].

370

The Pre-Established Physics and the Quantum Impulse

From 1951 to 1992, detection about displacement of ESG system, through space or relative to the cosmic blackbody radiation, across brightness galaxies clusters and their consequents recession redshifts, they came to register velocities values of ~ 350 to 700 km/s for Milky Way Galaxy-Local Group (GG), to direction between Virgo and Orion, this is, in the proximity of the constellation Leo [43 and 44].

From 1967 to 1977, detections about displacement of the ESG system in relation to cosmic blackbody radiation, across of radiometers, they came to indicate a difference of temperature of that radiation, that is, different frequencies and different energies received by ESG, in direction of that displacement and opposite to itself, with: ~355 km/s for the component Earth-Sun (ES), declination (dec.) = ~ −13°, right ascension (r.a.) = ~ 12h ; and of ~ 600 km/s for Milky Way Galaxy (G), dec. = −18°, r.a. = +10h, towards the proximity of the constellation Leo [45].

And since 1974, experiments involving equipments with toothed wheels (accelerated solid bodies rotating), forming stationary phase light, but permiting obtention of a phase difference of that stationary phase, in the light trajectories, they came to relate the variation of that phase with the displacement of ESG in space, with the velocity of ~303 km/s, dec.= ~ −21°, r.a. = ~13h, also towards the proximity of the constellation Leo [46], despite of objections to those experiments [3].

Posteriormente, from 1983 to 1987, experiments realized without utilization of accelerated solid bodies rotating, but also with basis in the light in initial stationary phase (but not permanently, across displacement of a movable platform), they came to present detections of that phase shifts related with directions of motion and opposite to itself of ESG, being obtained for the ES velocity through space of ~ 378 km/s, dec. = ~ −20° , r.a. = ~ 11h, newly towards the proximity of the constellation Leo [47].

By the experiments above, since 1951, is verified that in the ESG system the ES component velocity of ~ 350 km/s, it come to be related through space or to the cosmic blackbody radiation towards the proximity of the constellation Leo. In these experiments, of detections of phase variations, the light in initial stationary phase, it come to present velocity c (measured in Earth or ESG reference system). So, the light would come to present the inertia of the system, in relation to the space or cosmic blackbody radiation, with the velocity $c'=c+v$ in the direction of the source (ESG) motion, but always equal to the c when measured in any reference system of the type ESG (in that the light would be with the inertia of the reference systems). Also, in these experiments, to the to have fixation of initial stationary phase (but not permanently, across resources in that equipments for displacement of itself), they come to permit detections of interferences to the set out of the stationary phase. Because, in case of the permanently phase, all system equipment-light phase would be in motion, and so not coming to register any interference. It being that in these experiments above, with that initial phase fixation, themselves come to dispense of to be or not the light with inertia of the source (ESG).

EXPERIMENT - With the objective of to verify that displacement of ESG with the ES component v =~350 km/s, dec. = ~ ± 20°, r.a. = ~ 11 to 12h, towards the proximity of the constellation Leo, or then a detection of an inertia of light, this experiment came to utilize laser light projected on one millimetre screen, being the laser and the millimetre screen fixed in opposite extremities of the a bar, over one support with rotatory axes. The laser light came to be projected in that millimetre screen to distance d = 3m, in that direction towards the proximity of the constellation Leo, and in sequence to 90°, being admited in principle the velocity of laser light c independent of source (ESG) velocity, with basis in a device spacial displacement pointer, by radiation quantum impulse theory, constituent of light, in that: h (in unity of erg.s) / 1s = h' (in unity of erg)=elemental quantum energy= elemental photon= $m'c \times c = m'c^2$; $(m'c^2)/1s = h'v'$; where h = Planck constant = $6.6...\times 10^{-27}$ [erg.s], h'= $6.6...\times 10^{-27}$ [erg], v'= radiation unitary frequency, m'=elemental quantum mass, m'c = elemental radiation quantum impulse, (m'c)/1s = quantum force; and v= radiation frequency, h'v = (energy, of all photons h' with frequency v , in 1s) [9, 26, 27, 41]. Where for the laser light, $h'v_{laser} = (4,285...\times 10^{11}$ h')/1s . So to the 90° of the direction of the ESG motion, the laser light would come to point, in the millimetre screen, one displacement (x), by the relation of the trajectories times of laser light (t') and of ESG (t''); t'=d/c ,t''=x/v , t' = t'', so x = 3.5mm. And, in the direction of the ESG motion, the laser light coming to attain exactly in the millimetre screen centre ,by to follow the same direction of the ESG displacement (Fig.1).

371

RESULTS - It not came to be registered, in the millimetre screen, by laser light, any displacement. So, attaining the laser light always in the milimetre screen centre, in the directions of the ESG motion and in the transverse of the motion. It coming to confirm the Martits experiment [48], in 1973, with neon light and utilization of converging lens for themselves rays until one receiver film, across a metallic suspension and directed to different directions in the space, also with basis in the device spacial displacement pointer of m'c radiation quantum impulse theory, (Fig.2). Photo 1 shows exemple of observed result in all registers of Martits, in different directions and dates, where the none elongation of the little circle come to be obtained, of the neon light continuously projected in different directions for each position in the film, that is, it not presenting any displacement in relation to the milimetre screen centre. Photos 2 and 3 show exemple of the present laser light experiment result, observed in all these registers, in direction of ESG motion and transverse itself, also in different dates.

CONCLUSIONS - Despite of the present experiment, like that of Martits, they not register that motion of ESG, they came to point, by experiment characteristic, notable evidence of inertia of the photons — $m'c^2$ of velocity c in relation to ESG; or $m'c'^2$ with velocity c' in relation to the space or cosmic blackbody radiation, modifying the initial principle of velocity c of the radiation quantic impulse of that theory — of the laser light or neon light, in the ESG inertial system, where so noted for the photons the velocities $c'= c + v$ in the direction of the motion and $c' = (c^2 + v^2)^{1/2}$ in the transverse direction of itself, in relation to the space or cosmic blackbody radiation, but always obtained the velocity c when in relation to the source (ESG). So, the observation of absence of the any shift, in the millimetred screen, of those photons, in the transverse direction of that motion, it came to point total coincidence with the trajectory of the inertia of themselves, justifying also the Michelson-Morley and equivalent experiments.

Also, the results of that present experiment, like that of Martits, they come to be contraries to the space-time variations in the motion direction of the special relativity. So, in that present experiment, like that of Martits, the Michelson-Morley and equivalent null results come to be explained by the trajectories of the laser light or neon light in the transverse direction (that would point the displacement of ESG). And, by the raport of Einstein, in 1931, in that the Michelson-Morley experiment came to open the path for the special relativity and this to the general [49] — in that by general relativity, the light would come to present curvatures by the space-time variations —, in that if had presented a curvature, by experiment above for x=0, it would must to be of α=arc tg (v/c) = 206", value not determined by relativistic theory.

Experiments about not validity of inertia by the light, involving particles (electrons, mesons and positrons) in high velocity [2 and 19], in view of the Martits and present experiments, they come to require analysis to the behaviour of themselves as much as : braking in the magnets of equipments of the experiments and emissions of the χ and γ electromagnetic radiations, with basis in an energy saturation (as high frequency of the photons in the emission). These analysis involving radiation quantum impulse theory (and also gravitational radiation quantum impulse theory in that $m'c \times v$ = impulse frequency = force ; $m'c \times c \times v = m'c^2 \times v = h'v$, $h'= h/1s$; and of the relatement of h , h' ,and impulse frequency, with the expression of the Newton's law of the gravitational force $F_G = (m_1 m_2 / r^2) K_G$, where K_G the Newton gravitational constant, coming to be obtained the gravitational radiation frequency $v_G = 10^{19}$ hertz, a radiation of high frequency and of low intensity, in that $h'v_G = 10^{19}$ h' / 1s —) where like exemples of these theories, the mass increase by the radiation quantic impulse theory deduced by itself, and the other aspects involving these theories [9, 26 to 28, 30, 31, 34 to 39, 41, 50]. So allowing to come confirm the photons inertia, alike obtained by Martits experiment with neon light and the present with laser light, where in those experiments the photon inertia come to be comproved without necessity of the Fitzgerald-Lorentz matter contraction or Einstein space-time variations, where these contraction or space-time variations come to follow that coincident neon photon or laser photon inertia.

REFERENCES

1- A.A.Michelson,E.W.Morley. American Journal of Science. 31, 385 (1886) / 34, 333 (1887).
2- J.G.Fox. American Journal of Physics. 33, 1 (1965).
3- W.A.Rodrigues,J.Tiomno. Foundations of Physics. 15, 945 (1985).
4- G.F.Fitzgerald. Nature. 46, 164 (1892)
5- H.A.Lorentz. Verhandelingen der Konink Akademie van Wetenschappen,Amsterdam,1, 74 (1892).
6- A.Einstein. Annalen der Physik. 17, 891 (1905) ; Preussische Akademie der Wissenschaften. 668 (1916).
7- W.Ritz. Annales de Chimie et Physique. 13, 145 (1908) ; Oeuvres, Gauthier Villars, Paris. (1911).
8- I.Voinoff. Revista de Engenharia Mackenzie. November, 5, Mackenzie University Engineering Scholl,São Paulo,Brasil., (1942).
9- I.L.Erdélyi. Momentum theory,copyright I.L.Erdélyi,São Paulo,Brasil. (1954) ; Brazilian Interplanetary Society SIB,São Paulo,Brasil, (1963) ; Mackenzie University Engineering Scholl,São Paulo,Brasil. (1965) ; Revista Astronáutica. (Brazilian Interplanetary Society SIB), 24, 2 (1969) / 28, 3/ 29, 1/ 30, 1 (1970).
10- J.Bradley. Philosophical Transactions Royal Society. 35, 637 (1728).
11- H.L.Fizeau. Comptes Rendues Academie Sciences. Paris. 33, 351 (1851).
12- F.T.Trouton,H.R.Noble. Proceedings Royal Society. 72, 132 (1903).
13- D.C.Brace. Philosophical Magazine. 6,7,p-317 (1904).
14- R.J.Kennedy,E.M.Thorndike. Physical Review. 42, 400 (1932).
15- L.Janossy. Theory of relativity based on physical reality,Akadémiai Kiadó,Budapest. (1971).
16- F.W.Dyson,A.S.Eddington. Philosophical Transactions. 220, 291 (1920).
17- I.Shapiro. Physical Review Letters. 13, 789 (1964).
18- J.C.Hafele,R.E.Keating. Science. 177, 166 (1972).
19- T.Alvager,F.J.M.Farley,J.Kjellman,I.Wallin. Physics Letters. 12, 260 (1964).
20- D.Newman,G.W.Ford,A.Rich,E.Sweetman. Physical Review Letters. 40, 1355 (1978).
21- R.Eötvös. Annalen der Physik. 59, 354 (1896).
22- J.H.Taylor,A.Wolszczan,T.Damour,J.M.Weisberg. Nature. 355, 132 (1992).
23- A.Friedman. Zeitschrift für Physik. 10, 377 (1922).
24- B.Vanco. Revista Astronáutica (Brazilian Interplanetary Society SIB). 50, 15 (1971).
25- C.Lugo. Los rayos simultáneos,Editorial Américale,Buenos Aires. (1967).
26- I.L.Erdélyi. The new model of the electron and the positron,Mackenzie University Engineering Scholl,São Paulo,Brasil. (1964), and edited by F.Ninrod,California,San Francisco,U.S.A. (1965).
27- I.L.Erdélyi. Open letter to P.A.M.Dirac-University of Cambridge,copy to Nature and Scientific American, about article in Scientific American of May,1963 by P.A.M.Dirac-The evolution of the physicists picture of nature. (1964) ; Revista Astronáutica SIB. 39/40/41, 1. (1971).
28- P.F.Mesquita. Mass variation by quantic momentum theory, Yearbook of the 33th.Annual Meeting Brazilian Society Advancement Sciences SBPC and 15th.Annual Meeting Brazilian Society Physics SBF. (1981).
29- E.N.Parker. Scientific American. 4. (1964).
30- P.F.Mesquita. The meaning of the universal gravitation constant of Newton evinces the specific existence of gravitational radiations,São Paulo University USP; 2th.Congress Brazilian Cartography , presented to Brazilian Academy Sciences ABC, (1965); Revista de Engenharia (Engineering Institute,São Paulo,Brasil). 393. (1977)
31- P.F.Mesquita. Motion of perihelion independent of general relativity, Yearbook of the 26th.Annual Meeting Brazilian Society Advanced Sciences SBPC and 8th.Annual Meeting Brazilian Society Physics SBF. (1974);Revista de Engenharia (Engineering Institute, São Paulo, Brasil). 417, 23. (1979).
32- H.Dingle,W.H.McCrea. Nature. 177, 298 ; 178, 299. (1956).
33- P.N.Andrade. The objective interpretation of the Hafele-Keating experiments with clocks, cit.in Revista Pesquisa Naval (Rio de Janeiro, Brasil). 3, 1 (1990).
34- P.F.Mesquita. Planck constant in the celestial mechanics, Yearbook of the 26th.Annual Meeting Brazilian Society Advanced Sciences SBPC and 8th.Annual Meeting Brazilian Society Physics SBF. (1974).
35- P.F.Mesquita. The physical meaning of the newtonian constant reveals a way for studying satellite dynamics within a unified electromagnetic and gravitational field,Symposium on Satellite Dynamics of the 17th.Plenary Meeting of the Committee on Space Research by the International Council of Scientific Unions. (1974).
36- J.P.Hansen. News basis physical mathematical applicable for detection of the gravitationals radiations and consequent development of a new structure of the cosmos, Yearbook of the 36th.Annual Meeting of the Brazilian Society Advanced Sciences SBPC and 18th.Annual Meeting Brazilian Society Physics SBF,p-301 (1984).
37- J.P.Hansen. Unified theory in Physics by impulse theory and quantum gravitational radiations, Yearbook of the 41th. Annual Meeting Brazilian Society SBPC and 23th. Annual Meeting Brazilian Society Physics SBF,p-315 (1989).
38- J.P.Hansen. Coincidents values of physicals constants and consequents discoveries in Science, Yearbook of the 3th.Special Meeting Brazilian Society Advanced Sciences SBPC,p-377 (1996).
39- I.L.Erdélyi. Revista Astronáutica (Brazilian Interplanetary Society SIB). 12, 7 (1968).
40- H.Arp. Quasars,redshifts and controversies,copyright H.Arp,U.S.A. (1987).
41- P.F.Mesquita. Process for determining apparent place of the point on the celestial sphere to which should be directed the absolute movement of the Earth,23th.Annual Meeting Brazilian Society Advanced Sciences SBPC and 2th.Annual Meeting Brazilian Society Physics SBF. (1971).
42- E.B.Geiger. Papers presented in the 21th.and 23th.Annual Meeting Brazilian Society Advanced Sciences SBPC. (1969/1971).
43- P.J.E.Peebles.Physics Today. January,p-18 (1978).
44- S.Flamsteed,citing the "seven samurai" (D.Burstein,R.Davies,A.Dressler,S.Faber,D.Lynden-Beel,R.Terlevich,G.Wegner) and D.Weinberg,W.Freedman,J.Primack,M.Pierce,R.Kirshner,J.Tonry,W.Press,P.Demarque,M.Postman,T.Lauer,J.Ostriker,M.Strauss, B.Paczynski. Discover. 3, 66 (1995).
45- G.F.Smoot,M.V.Gorenstein,R.A.Muller. Physical Review Letters. 39, 898 (1977).
46- S.Marinov. Czechoslovakia Journal Physics. B24, 965 (1974); "Eppur si Muova",CBDS,Brussel. (1977); General Relativity Gravitation. 12, 57. (1980).
47- E.W.Silvertooth,S.F.Jacobs. Applied Optics. 22, 1274 (1983).
48- J.Martits. Letter to Scientif Counsel of Brazilian Interplanetary Society SIB,and private communication to author. (1991).
49- G.Holton. Thematic origins of scientific thought, Harvard University Press. (1973).
50- J.P.Hansen. Electron: Interactions with nucleus and its structure; paper for receivement of certificate of approvation of Brieff Course of Basic Advanced Physics,by Brazilian Interplanetary Society SIB in Brazilian Society Advanced Sciences SBPC. (1972).

a) Correspondence adress: J.P.Hansen-R.Dep.Mário de Barros 1.130-CEP 80530-280-Curitiba-PR-Brasil

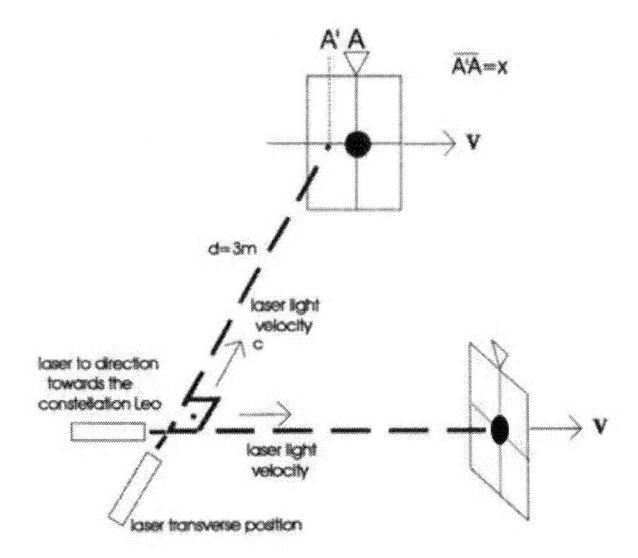

Fig.1 - Scheme with laser light experiment, for device displacement pointer by velocity v of the Earth-Sun-Galaxie system in space.

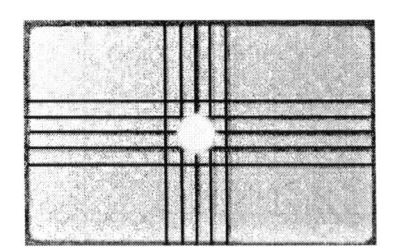

Fig.2 – Laser light experiment. Projections of laser light on the milimetred screen, towards the constellation Leo and in transversal direction, to the Erth-Sun-Galaxie motion, so has been indicated the inertia of the photons laser light.

Doc. 23 – Paper from the laser experiment sent to Nature Journal.

On 2 April 1997, Nature responded by the Letter of Dr. Leslie Sage, Assistant Editor, in which: "...*despite considering without any doubt the quality, from the work submitted..., Nature chose to publish articles that would be receptive and of great interest to a wide audience...*". However, despite the quality of the article submitted, mentioned by Nature, against those who sought to refuse articles in the name of "quality", the article sent should not attract a wide audience. In fact, the article should attract a broad refusal to go against the established authorities of the dogmas of Pre-Establishead Physics. Document 24.

nature

In reply please quote:
H03304 LS/rjl

968 National Press Building
529 14th Street NW
Washington DC 20045-1938
Tel: 202 737 2355
Fax: 202 628 1609
email: nature@naturedc.com

2 April 1997

Dr J P Hansen
Engineering Institute
Mackenzie University
Brazil

Dear Dr Hansen,

Thank you for submitting your manuscript, "Laser light experiment about inertial system determination", which we are regretfully returning with this letter.

It is *Nature*'s policy to return a substantial proportion of manuscripts without sending them to referees, so that they may be sent elsewhere without delay. Decisions of this kind are made by the editorial staff, often on the advice of regular advisers, when it appears that papers are unlikely to succeed in the competition for limited space.

Among the considerations that arise at this stage are the length of a manuscript, its likely interest to a general readership, the pressure on space in the various fields of *Nature*'s interest and the likelihood that a manuscript would seem of great topical interest to those working in the same or related areas of science.

In the present case, while your results will undoubtedly be of interest to fellow specialists, I regret that we are unable to conclude that the paper provides the sort of advance in understanding that would excite the immediate interest of a wide, general audience. We therefore feel that the paper would find a more suitable outlet in a specialist journal, rather than *Nature*.

I am sorry that we cannot respond more positively, and I hope you will understand that our decision in no way reflects any doubts about the quality of the work reported. The unfortunate fact is that we receive many more papers than we can publish, and we must attempt to select those that will be of greatest interest to a wide audience. I hope that you will rapidly receive a more favourable response elsewhere.

Yours sincerely,

Dr Leslie Sage
Assistant Editor

Nature America Inc.

Doc. 24 – Response from Nature.

CXIV. OTHER INTERNATIONAL PHYSICS JOURNALS – 1997

On April 15, 1997 was posted the submission of the article "*Laser Light Experiment about Inertial System Determination*" for submission to Physical Review, the most important Physics Journal in the world. It was the complete article of "*Experiment with Laser-Rays on Determination of Absolute Motion of Inertial System in Space*" in which the abstract had been presented at the Meeting of SBPC in 1995.

However, on 12 May 1997 the Associate Editor, Lorenzo Narducci, of Physical Review reported that the articles they would like to accept should be of importance and advances to the field that the Physics followed.

Thus, this article was not suitable for publication in Physical Review, because it was contrary to the area of work relativistic of Physical Review! The answer should mention the result, the focus of the article in which the speed of light c by relativity is by the experiment obtained $c' = (c^2 + V_{Earth}^2)^{1/2}$ contrary to relativistic theory. Regarded as the main Journal of Physics in the world, it presented only evasive words to cover up the result contrary to relativity!

CXV. PHYSICAL REVIEW LETTERS JOURNAL – 1997

On June 9, 1997 being posted this article of *Laser Experiment* for Physical Review Letters the assistant Editor, Jerome Malefant, reported that no trial was made about the work being correct or not, but only if it would be suitable for publication. And that was not the case, because this article was not motivating to its readers in that the abstract, the introduction and the conclusions were crucial to this criterion to be appreciated by a large number of its readers.

In fact, from the abstract to the conclusions the result of the experiment did not leave any motivation for its followers in support of the sect of the theory of relativity!

CXVI. AMERICAN JOURNAL OF PHYSICS – 1997

On July 28, 1997 this article of the *Laser Experiment* was sent to the American Journal of Physics, but on September 15, 1997 the Assistant Editor, Kannan Jagannathan, replied that the article was not appropriate for that Journal, without presenting any additional information.

Probably, if published the Journal would lose audience, because it was not appropriate for anyone who was read that article contrary dogma theory of relativity!

CXVII. FOUNDATIONS OF PHYSICS JOURNAL – 1997

In 1st. October 1997 this article of the *Laser Experiment* was sent to the Foundations of Physics Journal, however the Editor, the Physicist Alwin Van der Merwe reported on 6 November 1997 that this article was not of sufficient fundamental importance for the purposes of this Journal.

Although the Foundations of Physics was devoted to conceptual bases and the fundamental theories of Modern Physics, Biophysics and Cosmology, the Editor Van der Merwe did not consider that the conclusion of the article collapsing the current fundamental foundations of Modern Physics as the theory of relativity, in which any researcher could repeat the experiment and prove his veracity of that the photon presented the inertial velocity of the Earth motion. The result of the experiment it was contrary to the theory of relativity, in which the speed of light should come out of the laser at 90°, without composition with the velocity of the Earth in the direction of the constellation of Leo. However, if the purpose of this Journal was devoted to maintaining the pillars of this Modern Physics, then this article was of sufficient fundamental importance not to be published!

CXVIII. IL NUOVO CIMENTO JOURNAL – 1997-1998

On November 27, 1997 the article of the *Laser Experiment* was sent to the "Il Nuovo Cimento" Journal, and on 21 March 1998 the Deputy Director, Remo Ruffini, reported that the article was not published as expressed by the consultant, who considered that the conclusions of the article were contrary to the basic principles of the theory of special relativity, which had been widely investigated and established beyond any doubt.

On April 29, 1998 the author sent Letter to the Deputy Directors Drs. Remo Ruffini and Renato A. Ricci mentioning facts that were against to what this consultant wrote, like J. G. Fox's article on experiments that pointed to results of being more contrary than favorable to the theory of relativity. And refute the primary arguments of this consultant. And for lack of any other argument that justifies the theory of relativity, no response came to be issued by the consultant of Il *"Nuovo"* Cimento. Document 25.

Fax 00.39.51-581340
April 04 1998

To : IL NUOVO CIMENTO
Att. : Dr. Remo Ruffini – Vice Director

Copy: Dr. Renato A. Ricci,

Dear Dr. Ruffini,

I have understood your letter above, but despite unclear mentioned by your referee, to the paper *"Prove of the photon inertia…"*, he has understood very well that paper! Because, he understands that the paper presents topics contrary to the old principles of relativity, as soon as himself report!

To does the paper experiment is very simple, but your referee he not to wish to see himself the experiment result: if he to does, he would see the photon inertia, contrary to those old principles relativity, and himself old principles!

About "...*old basic principles of relativity which have been widely investigated and settled long time ago*...", mentioned by your referee, in the American Journal of Physics, vol. 33, January 1965, p. 1 to 16, J. G. Fox writes that: "...*contrary to what has been believed for several decades, that the data as on binary stars, do not offer any evidence more favorable to the relativity..., ...agree more with the contrary theory relativity*...".

By the Éditions Grass Fasquelle, 1991, "*Dieu et la Science*", the Physicist Igor Bogdanov writes: "The conclusion about Foucault's experiment is startling: the oscillation plan of the pendulum is determined by the Universe", this is, it follows the inertia! Also, Umberto Eco writes of similar mode, in the "*Il Pendolo di Foucault*", Ed. Fabbri, 1988. And of similar mode, the photon presents the inertia by mentioned in the 7309 NCBR paper, that all students can to do.

I understand that the review procedure of "Il Nuovo Cimento" presents the impossibility of that paper publication, in accordant your referee, and the "heresy" in to be contrary to the "old" principles of relativity and of your himself referee, and him as soon as similar other never mistake! But, despite your referee, by the one anterior "heretic": "Eppur Si Muova"! In the 7309 NCBR paper, contrary the mind of your referee, the photon inertia presents the inertia!

Cordially,

Paulo Hansen

J. P. Hansen
Engineering Institute
Doc. 25 – Letter-Reply to Il "Nuovo" Cimento.

CIXX. PHYSICS LETTERS A JOURNAL – 1998

On September 14, 1998 the article of the *Laser Experiment* was sent to the Physical Letters A to the Editor, Ph. D., P. R. Holland, which on 24 October 1998 reported that the consultant had refused to understand what the author was saying.

However, the consultants of the other Journals had understood very well what the author was saying! But this consultant tried not to understand the article, because the conclusion was not to his liking! For, just by drawing it was seen that the laser light reached the center of the lattice leading the inertia of the displacement of the Earth in space or the cosmic background radiation (CBR), and not leaving at 90 degrees with the simple speed of light c as absolute constant by Einstein's relativity., ie , the light coming out of the laser had the composition of speed c and the velocity V of the Earth in space or CBR towards the constellation of Leo, that is, was greater than c. The reviewer did not want to understand that the diagonal velocity was the composition of the c speed of light with the V velocity of the Earth, contrary to Einstein's relativity which is always c in any frame. Then, in the Earth's frame the measured speed was c, however in the space-vacuum frame or CBR the velocity of light was the composition of speed c with the velocity V of the Earth. Therefore, the speed of light was different when measured by the Earth's referential and the vacuum-space reference or CBR, which invalidated the theory of relativity, in which in all references the speed of light was always c.

CXX. ASTRONOMY AND ASTROPHYSICS JOURNAL – 1998

On November 23, 1998 this article of the *Laser Experiment* was sent to Astronomy and Astrophysics to the Editor, Dr. J. Lequeux, who reported that this article was more appropriate for Physics Journals for presenting a physical experiment than for Astrophysics, but presented suggestions for better understanding of the article. I replied that it was very difficult to approve an article contrary to Einstein's relativity, and in astronomy the colleague of J. Lequeux, H. Arp, found many problems with consultants who did not acepted the smallest distances of bodies that emitted intense light and radio-waves, quasars, it pointed by Arp.

CXXI. THE ASTRONOMICAL JOURNAL – 1999

By the fact this article *Laser Experiment* present the motion in the space of the Earth-Sun System, the author sent the article to the Astronomical Journal to the Editor, Dr. Paul Hodge, February 1999. And on March 2, 1999 Paul Hodge replied that by the article involving a basic issue of a basic question of fundamental Physics should be subjected to an appropriate publication of Physics.

The Editor had the realization that the experiment involved a deep fundamental question of Physics.

CXXII. THE ASTROPHYSICAL JOURNAL – 1999

On March 15, 1999 this *Laser* article was sent to the Astrophysical Journal to the Editor Dr. Helmut A. Abt and in 1st. April 1999 the Scientific Editor, Dr. Robert C. Kennicutt Jr. replied that the Journal published original observations, significant new theories and interpretations, in which the article disapproving the theory of relativity would certainly constitute such new interpretations. But, however, according to Robert Kennicutt, the results presented in the article were consistent with the predictions of the theory of relativity.

What Robert Kennicutt noted was that the speed of light in relation to vacuum-space or cosmic microwave radiation (CMR), (which had been detected in 1965 by radio telescope), was always the same as the emitting source of light, that is, the Earth stopped in vacuum-space or at any velocity in the direction of the motion of the Earth. Thus, by not depending on the velocity of the source in relation to the vacuum-space, the speed of light c was always constant in the vacuum-space in the direction of the motion of the source, which was in accordance with the theory of relativity. But, the velocity of light c_v was less in relation to the Earth, the greater the velocity of the Earth in space-vacuum or CMR, which was contrary to the theory of relativity where in any reference the speed of light c was always the same. And the meson with the velocity near the light emitting the photon at the speed of light c, that is, the velocity c_v of the photon in relation to the meson was close to zero, but by the theory of relativity the speed of the photon in its reference was the speed c in relation to the meson and not to zero speed.

The experiment showed that the velocity $c_{diagonal}$ of light when diagonally at the velocity V of the emitting source of the Earth-Sun-Laser System was variable in relation to the emitting source, contrary to the theory of relativity. Therefore, the laser light when focusing on the center of the lattice showed the inertia of the velocity of the photon of light by the source, contrary to the relativity in which the light did not present any inertia with the velocity of the source. Thus, the experiment reported in the article did not show any space contraction or time dilation for that the speed of light c was always constant relative to the emitting of light source. But, it is the inertial velocity $c_{diagonal} = (c_v^2 + V^2)^{1/2}$ and $c' = c_v + V$ of the photon of light received from the source that is the physical fact and non-contraction of space and dilation of time.

So the author sent email to Scientific Editor of The Astrophysical Journal clarifying that the predictions of the theory of relativity were contradicted by the result of the experiment, because the variables of space-time, that is, the contraction of space and time dilation were non-existent. The inertia of the velocity of the source received by the photon varied the velocity of light c_v relative to the source, and broadly explaining the Michelson-Morley experiment. And the explanation of the photon of light with the correct interpretation of Planck's constant and the other consequences showed new interpretations for Physics, Astronomy and Astrophysics.

However, on 5 April 1999, the Scientific Editor Robert Kennicut replied that he should have mentioned in the previous email that articles reporting bases of the theory of relativity were not usually published in The Astrophysical Journal. The Scientific Editor showed in his reply that he had no basis to reaffirm that this article was consistent with the theory of relativity, for it was not his area of expertise.

CXXIII. GENERAL RELATIVITY AND GRAVITATION JOURNAL – 1999

On April 28, 1999 this article of the *Laser Experiment* was sent to the Editor, Dr. David G. Blair, who reported on May 5, 1999 that this article had been sent to the consultants. And on August 2, 1999, the Editor reported that according to his consultants this article was not appropriate for this Journal.

As information, the author sent on August 3, 1999 email to the Editor showing that this article was a consequence of the article of the Marinov's Experiment published in 1980 in General Relativity and Gravitation which was contrary to the theory of relativity. Therefore, the fact that the article submitted is contrary to the theory of relativity should not be the cause of its non-publication. Or, was it not published by the fact of validating the Marinov experiment?

On August 30, 1999 the Editor responded by preferring only to inform that it did not present "enough quality" for its publication, without clarifying that "quality" was required by the consultants for that article. It was clear that there was not enough quality that the result of this experiment was favorable to the theory of relativity!

For the fact that the article of Marinov had presented many criticisms by the imposed Physics, with the allegation that there was torsion on the axis that resulted in deviation of light and consequent determination of the absolute motion of the Earth in space, contrary to the theory of relativity, it could be the case of General Relativity and Gravitation only to publish articles of the general theory of relativity and gravitation that did not refute the theory of relativity, to avoid further criticism, despite the alleged torsion there is not been proven.

PART 8 – EXPERIMENT PUBLISHED IN INTERNATIONAL JOURNAL OF FUNDAMENTALS OF PHYSICS PROVES BY QUANTUM IMPULSE THE FALLACY OF BASES OF CURRENT PHYSICS

CXXIV. PHYSICS ESSAYS JOURNAL – **2000 to 2002**

On May 17, 2000 the author sent the article of the *Laser Experiment* to Physics Essays an International Journal Dedicated to Fundamental Questions in Physics, headquartered in Canada, to the Editor, Dr. Emilio Panarella.

In a Letter dated August 11, 2000, Physics Essays sent a review of two consultants so that the author could perform corrections or justify the objections of the Reviewers. This Physics Essays procedure, to present the content of the revision and to allow the authors of articles to clarify or accept the objections formulated, resulted in a deepening of the article submitted for publication and its viability within a broad debate of opinions issued by both the authors and the Reviewers.

In the case of this 1st. Revision, one of the Reviewers asked for further clarification of the topics cited. And the other Reviewer presented among other objections that the light would always have the natural aberration, that is, the deviation of light by the velocity of the Earth, therefore, would always reach the center of the target, and recommended that the article was not published. However, on September 25, 2000 the author reported to Physics Essays that the photon of light having the inertial velocity of the Earth, that is, the aberration alleged by the Reviewer, then explained the null result of the Michelson-Morley experiment where the speed of light c with plus the velocity of the aberration of the Earth's motion resulted in the null interference, without any contraction of space or time dilation as alleged by Einstein's theory of relativity. But, also the speed of light by the theory of relativity should be velocity c at the perpendicular to the motion of the Earth in space, because the laser were placed at 90° to this motion, however in the case there was the composition with the inertial velocity of the Earth. Only this objection of the Reviewer showed the widespread lack of bases of current Physics in its own principles of this Physics.

On December 21, 2000 Physics Essays sent the second Revision by one of the Reviewers, in which he claimed that *"There was only one possibility of the light coming out of the laser"*, that is, would always present the deviation of the velocity of the Earth where the laser is fixed, and then it was against the publication of the article.

On March 8, 2001 the author replied that for the laser placed in 90° of the direction of the motion of the Earth in space, the light could come out exactly at 90° with its speed c in relation to the Earth, if it did not present the inertia of the motion of the Earth in space. Yes, the target placed in the same line of 90° of the laser was moving in space while the light going straight at 90° would reach the target with a distance X of its center of the. But, if it reached the center of the target in motion with the V velocity of the Earth, then the velocity of light diagonally could only be the vector composition of the speed c of the light in relation to the Earth if it came out at 90°, with the velocity V of the Earth. That is, the velocity of light would be greater than c, ie $c'_{diagonal} = (c^2 + V^2)^{1/2}$. If the Earth were standing in space, the speed of light would come out of the laser with its speed relative to the Earth. This is, if the light had has always the inertia of the Earth, then the velocity of light would depend on this Earth velocity, as Robert Resnick says in its Physics Textbooks. Then, the velocity of light would be differents, as opposed to relativity. On March 8, 2001, new clarifications were sent on the second Review.

On August 23, 2001, email was sent to the Editor and the Editor's Assistant emphasizing the result of the experiment presented in the article submitted for publication.

In a Letter posted March 14, 2002 Physics Essays sent the 3rd. Review of the two consultants, in which one of the Reviewers argued that: *"No matter what Robert Resnick says"*. But, Resnick is considered one of the best Physics Professors in the United States and author of Physics Textbooks adopted worldwide. That way, getting just subjective your Review. And again this Review mentioned the crucial factor of the article, in which: "*The inertia of the photon was nonexistent*". But, he did not understand that if the Earth-Laser was to be standing in space the photon would not receive this velocity V from the Earth. It was evident that if the photon came out of the laser fixed on the Earth also stopped in space, it could not go off diagonally as the laser experiment had shown with the moving Earth in space in the direction of the constellation of Leo. Also, he didn't understand how larger was the velocity of Earth-Laser, the greater the velocity of the photon diagonally to reach the center of the target, by the vector composition, that is, the photon would have greater Earth-Laser velocity inertia. What if the Earth-Laser be standing in space, the photon of light would come out at the perpendicular, because the inertial velocity of the photon would be with the inertial velocity of the Earth equal to zero. This, is, the inertia of the photon was existent! But, he continued to recommend against the publication of the article. The other Reviewer found doubt about how the photon was accelerated with its
speed c to achieve greater diagonal velocity.

On May 21, 2002 the author sent a response to the Physics Essays of this 3rd. Revision, presenting to the Reviewer that

Jordan, one of the best ball-to-basket players could err the ball far away from the basket, if it were not at the same velocity as the Earth, the ball and the basket when throwing the ball into the basket. In the case, that Jordan and the ball in his hands stopped in the air for a second, while the velocity of the Earth, that is, the court and the basket continued to move, Jordan missed the ball in the basket! For, the time that he aimed at the basket, the ball and Jordan were standing in space, that is, without inertia of the velocity of the court-basket, while the court-basket was still moving. To the Reviewer B was mentioned that the electron, when emitting the photon of light, it was already with the V velocity of the Earth and also the photons were emitted with the V velocity of the electron by the inertial velocity of the Earth, with more its speed c emitted by the electron. By Einstein's theory of relativity the light always has speed c, but diagonally the velocity of light was $c'_{diagonal}$ by the laser experiment.

On July 2, 2002, Dr. Emilio Panarella, Editor of Physics Essays sent Letter informing that the article was accepted for publication in issue Number 1-Volume 15-March 2002. The Editor Emilio Panarella was a Doctor at the University of Naples and Post Doctorate by the Polytechnic Institute of Brooklyn in the United States. He was founder President and Technical Director of Advanced Laser and Fusion Technology Inc. – ALFT, Member of the National Research Council of Canada, American Physical Society, Institute of Electrical and Electronics Engineers, European and Italian Physical Societies, Canadian Nuclear Society, author of more than 100 articles on Science and Technology and Lectures in the United States, Canada, Italy, France, England, and Editor of Physics Essays. In 2001 was the Chairman of the International Symposium on Nuclear Fusion Research in which on the Board of the Symposium was Edward Teller who was one of the main responsible for the construction project of the hydrogen nuclear weapon of the United States, detonated in 1952 in the presidency of Harry S. Truman, and between 1981 and 1989 from the Project Strategic Defense Initiative-SDI, with Laser to Nuclear Shield, also known as project "Star Wars" in the time of President of the United States Ronald Reagan. Document 26 presents the Constitution of the Board of Physics Essays, and Document 27 the Abstract of Laser Experiment paper.

Physics Essays

An International Journal Dedicated To Fundamental Questions In Physics

EDITOR
E. Panarella

ASSISTANT TO THE EDITOR
A. McLean

ASSISTANT EDITOR
X. Guo

ASSOCIATE EDITORS
V. Arunasalam, M.H. Brill, A. Hirose, M. Laroussi, R.B. Mann, and M. Vijay

EDITORIAL BOARD

A.D. Allen - CytoDyn, Inc., California	L. Qiu - Academia Sinica
P. Amendt - Lawrence Livermore National Laboratory	R.P. Rau - Louisiana State University
E. Arimondo - University of Pisa	P. Raychaudhuri - Calcutta University
E.I. Bitsakis - University of Athens	M.L.G. Redhead - Cambridge University
D. Censor - Ben-Gurion University of the Negev	K.L. Sala - Communication Research Centre of Canada
E.S. Hanff - National Research Council of Canada	M.O. Scully - University of New Mexico
J. Hilgevoord - Rijksuniversiteit, The Netherlands	I.P. Shkarofsky - MPB Technologies, Inc., Montreal
W.M. Honig - Curtin University, Australia	E. Sindoni - University of Milan
M. Inokuti - Argonne National Laboratory	L. Sobrino - University of British Columbia
L. Kostro - University of Gdansk	K.H. Spatschek - University of Dusseldorf
D. Kruskal - Rutgers University	G. Vahala - College of William and Mary, Williamsburg
T.E. Phipps, Jr., - Urbana, Illinois	L. Vahala - Old Dominion University, Norfolk

---

Physics Essays has been established as an international journal dedicated to theoreti and experimental aspects of fundamental problems in Physics and, generally, to the advancement of basic knowledge of Physics. The Journal's mandate is to publish rigorous and methodological examinations of past, current, and advanced concepts, methods and results in physics research. *Physics Essays* dedicates itself to the publication of stimulating exploratory, and original papers in a variety of physics disciplines, such as spectroscopy, quantum mechanics, particle physics, electromagnetic theory, astrophysics, space physics, mathematical methods in physic plasma physics, philosophical aspects of physics, chemical physics, and relativity.

The Journal will endeavour to reflect the environment in which best research is carrie out by providing a stimulating publication outlet for both the expression of ideas and reporting of results, within the rigour of the scientific discipline with which the Journa is concerned, namely Physics. As a dynamic new journal, *Physics Essays* combines rigorous scientific reporting with freedom to express ideas based on logically sound a well balanced points of view.

Physics Essays, an international, peer-reviewed journal of impeccable quality, supported and advised by a renowned Editorial Board, has been established as the sc journal to act as the voice of the international physics community in a truly interdisciplinary fashion.

Doc. 26 – Constitution of the Editorial Representation of *Physics Essays*.

Verification of Photon Inertia by a Laser Light Photon Radiation Experiment

José Paulo Hansen

Abstract

The Michelson–Morley experiment and those like it have detected no results on the absolute velocity of inertial systems like Earth–Sun. The null results of these experiments have been explained by the space-time variable of Einstein's relativity theory, which maintains that the velocity of light and other electromagnetic radiation is constant ($c \approx 3 \times 10^{10}$ cm/s) in vacuum, measured by an observer in a reference system for any velocity (v_s) of the emitting source of the radiation. That is, this radiation would be emitted with a velocity c independent of the velocity (v_s) of the emitting source. Contrary to Einstein's relativity theory, this implies that the final velocity (c') of this radiation depends on the inertial velocity (v_s) of the emitting source, with respect to space or to the cosmic background radiation, and then the final velocity (c') of the light radiation or any other electromagnetic radiation would be given by $c' = (c + v_s)$, in the direction of the emitting source, with respect to either space or the cosmic background radiation. That is, c' would be variable according to the source velocity (v_s), according to the Ritz emission theory. From 1951 to 1987 works involving redshifts, temperatures of the cosmic background radiation, and standing phase shifts of light presented a velocity $v \approx 350 \times 10^5$ cm/s for the Earth–Sun system motion toward the constellation Leo, with respect to either space or the cosmic background radiation. The present laser light photon radiation experiment, applied to this motion toward the constellation Leo and transverse to this motion, based on new aspects of the quantum theory, has shown notable evidence that the velocity c' of this laser light photon radiation has the component of the source inertial velocity v of this Earth–Sun motion, explaining the Michelson–Morley experiment and those like it. Also, the present paper explains the velocity (c_r) of the γ photon radiation emitted by particles at high velocity ($v_{s0} \approx c_r$). The result of the present experiment is consistent with the prediction of a constant velocity (c') of light and any other electromagnetic radiation, but only in the direction of motion of the emitting source of the radiation. And there is not a constant velocity (c) of this radiation with respect to the observer in a reference system, contrary to Einstein's relativity, which admits a constant velocity (c) of this radiation in all reference systems and in each direction of this radiation in space. But there is a constant velocity ($c' = c_r + v_s$) of this radiation with respect to space or the cosmic background radiation, because the inertial velocity (c_r) with respect to its emitting source varies according to the inertial velocity (v_s) of the emitting source of this photon radiation, which is also contrary to the Ritz emission theory. For other directions of the photon radiation with respect to the direction of motion of the emitting source, the present experiment proves that the photon radiation does not have a constant velocity (c), but is affected by the inertial velocity (v_s) of the emitting source of the radiation.

Key words: Michelson–Morley experiment, Fitzgerald–Lorentz contraction, Einstein's space-time variable, elementary quantum impulse, elementary photon, photon inertia

Doc. 27 – Abstrac of the paper published in full in Physics Essays.

CXXV. ONE YEAR AFTER: MARCH 2003

In March 2003, a year after the publication of the article showing the inertia of the photons by the *Laser Light Experiment*, contrary to the theory of relativity in which light presents only a single velocity in any reference and any emitting source, the Editor of Physics Essays, Emilio Panarella, in the Editorial that is presented once annually, concludes at the end that the theory of relativity should re-analyzed, as well as the Michelson-Morley experiment that is precisely this experiment involving the theory of relativity and that it treated the laser experiment. Document 28.

> **Physics Essays**   Volume 16, Number 1, 2003
>
> ## Editorial
>
> ### Fifteen Years Later
>
> *Physics Essays* was officially launched 15 years ago, in April 1988, with its first issue. The process of establishing the journal, however, started many years before, in 1971, with discussions with prominent members of the physics community on the need for a journal of this nature and scope and the contribution it could make to the advance of physics. The turning point was a discussion that I had with Gene Wells, editor at that time of *Physical Review Letters*. I asked him about the criteria for acceptance of papers in that journal, and he replied, "We reject the bottom 10 percent of the papers submitted to us. We publish the middle 80 percent, and reject the top 10 percent." To my question, "Why do you reject the latter group?," he replied, "We do not know if they are right or wrong."
>
> The mission of *Physics Essays* is to recover this top 10 percent of articles that would otherwise be lost. If the price to pay is to err with the publication of a few papers that might be wrong, it is worth it for the recovery of those many articles that are right. This mission has been carried out now for 15 years, and, as we embark on another 5 years of publication, I would like to share some thoughts on how this journal is fulfilling its mission.
>
> The Statement of Purpose and Editorial Policy, as printed in the masthead for 15 years, includes the following statement:
>
> > Different points of view will be accepted as long as they are logically sound and well balanced in their exposition, until the process of truth searching naturally reaches a stage of a convincing argument in favor of one point of view or the other.
>
> Many subjects of physics are debated in *Physics Essays*: cosmology, quantum mechanics, the standard model, classical electrodynamics, astrophysics, gravitational radiation, probability analysis, electromagnetic theory, and so on. The journal offers provocative thoughts on all these subjects.
>
> The debate is always professional and accommodating. There is, however, a subject of physics that is fiercely debated. I refer to the special theory of relativity (STR). Here the debate is adversarial, with the large majority of authors claiming openly that STR is wrong, and the other side claiming that the antirelativists just do not understand STR. The Editor, being in the middle, has taken note of the confusion and has decided that it is time to set the terms for the search for truth, as mandated by the Statement of Purpose mentioned above.
>
> My reasoning goes as follows. If STR is right, clearly the relativists have not done a sufficiently thorough job of explaining it. On the other hand, if the antirelativists are right, clearly they are not doing a sufficiently thorough job of explaining their arguments to the relativists. I believe that STR therefore needs to be dissected piece by piece, in a methodical way, leaving no stone unturned. Ultimately, the truth will prevail. Moreover, I believe that the historical circumstances that led Albert Einstein in 1905 to formulate the STR should also be reviewed in order to have a complete picture of the background in which this young patent office clerk found himself when he came forward with such extraordinary ideas as the constancy of the velocity of light in all reference frames and the principle of relativity itself. The Lorentz transformations, as derived or postulated by Lorentz himself, should also be reconsidered, because they form one of the components of the historical background. The Michelson–Morley experiment and the theory of the ether should also be reviewed.
>
> So *Physics Essays* wants to shed conclusive light on this important subject. More explicitly, I want the truth to prevail. Irrespective of where the truth is found, *Physics Essays*, by facilitating its discovery, is fulfilling its mission as set forth 15 years ago in this journey of its life.
>
>   *Emilio Panarella*

Doc. 28 – Editorial of Physics Essays in which the Editor Dr. Emilio Panarella comments that the Michelson-Morley experiment and the theory of relativity should be reviewed, which was what treated the laser experiment article.

CXXVI. DISCOVERY OF THE MASS OF THE NEUTRINO THAT HAD NO MASS BY IMPOSED PHYSICS! **1998 to 2018**

Between 1998 and 2002 experiments proved the oscillation of the neutrino which showed that they had mass. Also in 2010 the Gran Sasso Laboratory and the Oscillation Project Emulsion Racking Apparatus-Opera at CERN showed with laser experiment that the neutrino electron neutrino oscillation of the meson µ, proving its mass!

That is, 50 years before these experiments that showed mass of the neutrino, it already had mass by the quantum impulse theory, which contradicted the mass neutrino by relativity applied in quantum theory!

The neutrino oscillations have come to prove that they have masses for all types formed by the electron, meson µ, tau and the probable sterile neutrino so termed as being inactive in the strong, weak and electromagnetic interactions. In 2018, The Fermi National Accelerator Laboratory with Booster Neutrino Experiment-MiniBooNE detected this probable sterile neutrino with ≈ 1 eletron-volt, this is, $\approx 0.178...\times 10^{-32}$g.

By the theory of elementary impulse, the particles around the speed of light have the reduced dimensions, according to the equations that relate the rotating velocity, spin, linear velocity, wavelength and frequency of h' which forms the motion helix with its mass, as observed for the electron and applicable to other particles with other constitutions and dimensions.

So, for the neutrino, with speed close to the speed of light, with estimated mass equivalent to an energy of < 30 electron-volt quite inferior to that of the electron, would present a decrease of its dimensions by the concentration of h' in relation to the mass of rest, according to the equation similar to that of the electron by the theory of elementary impulse.

By the elementary impulse the neutrino always had mass, because with an energy that was measured at the speed v that was, would be obtained its mass m_η :

$$E_\eta \text{ total} = m_\eta c^2 = m_{0\eta} c^2 / \sqrt{1 - (v^2/c^2)}$$

Also, in 2018 the Telescope IceCube Neutrino Laboratory-ICECUBE announced detection of neutrinos (with impressive, as they) of $\approx$ 290 trillion electron-volt, this is, $\approx 5.16947...\times 10^{-17}$g. That is, after more than 50 years of the creation of the quantum impulse theory which indicated mass of the neutrino and that the formal Physics left it without mass, now this Physics considers the neutrino mass as impressive. Worse, the ICECUBE team with this result, although it is proven that the neutrino is particle with mass, presented this discovery disregarding its own scientific method in which they presented the announcement of the discovery to the press as the Physicist Darren Grant of Alberta University to BBC New Brazil: *"There is no indication that these neutrino particles can be divided into smaller points, where they are emitted by stellar explosions and moving at the speed of light"*. Therefore, they do not recognize their own experiments of annihilation of particles, such as the electron and positron that result in gamma rays, that is, formed of quantum energy m'c x c = h' of quantum impulse theory, and so the neutrino also formed by elementary energy h'.

CXXVII. QUANTUM MASS – **1938 to 2011**

In 1963, Octávio A. Ribeiro da Cunha had presented the mass quantum m_0 extracted from $E = h\mathcal{V} = h \times 1 = m_0 c^2$ getting to the mass quantum m_0 the same value as m' = 7.37... x 10^{-48} g of quantum impulse theory. And other quanta with mass in 1972 by S. Deser, P. Broglie and J.P. Vigier who presented the quantum of 10^{-65} g. Also in 1972, V. G. Hobson had presented the quantum h equal to h', with mass M_Q just like m' of Erdelyi. The quantum with mass continued to be presented in 1974 by M. Moles and Vigier, in 1990 and 1992 by Vigier, and at 1991 J.P. Narlikar and other. Between 2000 and 2010 A. Worsley had presented the quintessence quantum of the oscillator $E = nh$ where n represented the number of quintessence of Planck's quanta contained within a quantum per unit of time, which was the elementary quantum h' of Erdelyi and of Hobson. And Worsley also presented the mass m_q of the quantum quintessence with grams unit x second, with m_q = 7.37... x 10^{-48} [g s], in which by Erdelyi m' has the mass dimension, that is, in [grams]. Such was the realization that this quantum of mass m_q it was fundamental in the emission of radiation that Worsley presented in 2001 a world patent that the electron was formed by this mass quantum emitted in radiation, which Erdelyi had published in the decade 60 by the electron equation consisting of m'c x c = h'. And in 2011 M. Sato and H. Sato presented the quantum that should also have mass.

Up to 2005, measurements of pulsars, radio frequency and gamma rays resulted that if there was mass for the quantum would be less than 10^{-30} to 10^{-43} g ; and for measurements with torsional scales, Coulomb's law and concentric spheres would be less than 10^{-36} to 10^{-43} g; and in plasma, external and magnetic fields would be less than 10^{-37} to 10^{-57} g. In 2010, measurements presented for pulsars, radio frequency, gamma rays, torsional balance, Coulomb's law, concentric spheres, plasma, external and magnetic fields, that quantum mass if there were to be less than 10^{-47} to 10^{-51} g and for the severity of gravity, the gravitons, in plasma measurements, external and magnetic fields would be less than 10^{-52} to 10^{-66} g. However, the external, magnetic and plasma fields could present extensions that would alter balances, densities and concentrations, thus presenting differences in the value of the mass of these gravitons. In 1938, Kramers had presented that the theoretical mass of the added electron with a mass of electron interaction with its own field would be obtained the actual mass of the electron. Also, it is remarkable that the limit of the measurements for the quanta is exactly in the value of the elementary quantum of mass 10^{-48}g, as a limit to determining the minimum elementary mass. It would be the principle of Heisenberg's indetermination where h' is the limit of indetermination. And in the case of the elementary impulse, in which also the graviton is the elementary quantum, the difference in measurement between intergalactic outer fields and other measurements appears widely contradictory.

CXXVIII. "SPACE COMPASS": THE VELOCITY INDICATOR
1810 to 2002

An experiment with light to obtain eventual displacement in different positions in space had been devised by Erdelyi in 1954, by the theory of the impulse as a "space compass", an indicator of Earth velocity for several directions. And that in principle the light composed of the elementary mass m' with the speed c in relation to the source, with the energy of the elementary quantum h' = m'c x c, should exit the emitting source with speed c equal to Einstein's theory of relativity and wave theory, to obtain deviations in a target that along with the Earth motion in space.

The Michelson-Morley Experiment – In the Michelson-Morley experiment, the light was admitted by wave theory with the speed c measured in the Earth. And as the inclination of the telescope on Earth, that is, the stellar aberration, showed that it was due to the motion of the Earth in the vacuum-space-ether, and if the light reached the mirror exactly at 45° the light should go through the path D of the arm of the interferometer at 90°, and in that case the light would return also in 90° without interfering in the interferometer in the motion along with the Earth. Then the mirror was set, with micrometric screw, so that the light had a aberration angle and hit the initial interference with the other light paths. And in the rotation of the interferometer this interference would be expected to present variation, fringes of interference, When they were changed directions of light paths. So, the light with speed c was set to go off diagonally to compensate for the velocity V_{Esl} of the Earth in different directions in the vacuum-space-ether, but keeping the speed c. So, in this experience the speed c of the light coming out diagonally, but without the inertia of the velocity V_{ESL} of the Earth-Sun-Interferometer system in vacuum-space-ether, this is, keeping speed c diagonally. Figure 44.

Michelson-Morley Experiment

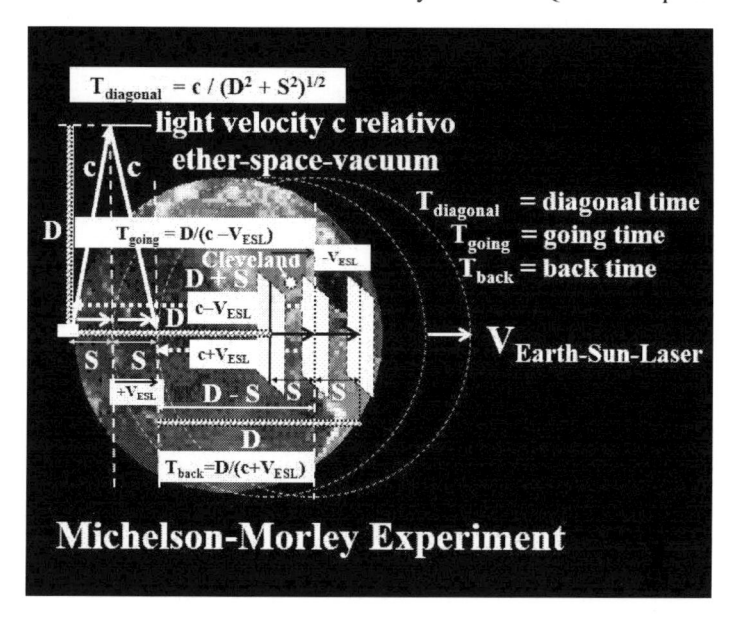

Fig. 44 – Schematic of the Michelson-Morley experiment which admitted the speed of light c diagonally without the inertia of the Earth-Sun-Interferometer in the ether, which would result in fringe of interference. The three paths, round-trip, and cross-section should be different from the Michelson-Morley projected experiment. Adapted from the Earth clipart of *Editora Europa*.

Waves, Ether and the Three-Times of the Michelson-Morley Experiment – A lake with completely flat water remains flat, it was the Earth still or moving in the vacuum-space, but when a stone is thrown into that water the formed wave is propagated, it was the Earth still or in motion. That is, the speed of the wave is independent of the speed of the source, because it is only the water that propagates the wave. Also the seismic wave by the ground and the sound wave by air are independents of the speed of the emitting source.

And in the case of an airplane when it exceeds the sound barrier, the plane does not carry the air, but it ceases to receive the vibration of the air, that is, it exceeds the wave of the sound that the airplane itself generated. Therefore, the speed of the wave is independent of the speed from the emitting source. Figure 45 .

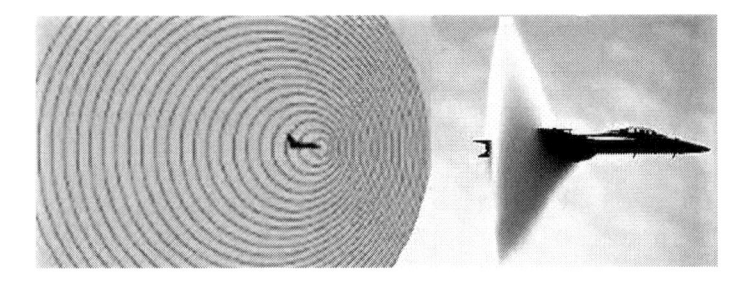

Fig. 45 – The speed of the wave is independent of the speed of the emitting source. While a supersonic plane exceeds the speed of sound, the sonic wave shows the same speed that is independent of the speed of the aircraft's emitting source. Design by the composition from cliparts of *Editora Europa*.

In the case of the Michelson-Morley experiment, the light as wave would be propagated by a medium of a still permeable ether, also independent of the speed of the emitting source. Then, as Michelson-Morley, for light with speed c go through the arm distance D of the interferometer in the direction of the Earth-Sun-Laser System with velocity V_{ESL} that traverses the distance S, at the same time that traverse D, as seen from the space the path would be D' = (D + S) for the time to go T_{going}, so D = (D' – S) = [(T_{going} c) – (T_{going} V_{TSL})], so T_{going} = D/(c – V_{TSL}) in which the Earth "flees" from the light. On the back, D' = (D – S), D = (D' + S) = [(T_{Back}c) – ($T_{back}V_{ESL}$)], then T_{back} = D/(c + V_{ESL}) in which the Earth goes against the light. On transverse path, as Michelson-Morley the light with speed c at a time $T_{diagonal}$ follow the path $(D^2 + S^2)^{1/2}$, that is, c = $(D^2 + S^2)^{1/2}$ $T_{diagonal}$. Therefore, the three paths and the three times of the wave of light in space would be different, and would cause fringe interference. But, the result of the interference was null, as Figure 44.

The Model of Einstein's Theory of Relativity – The special relativity had its confirmation in the Michelson-Morley experiment which had presented a null result. And the own origin of the theory of relativity had its first model in the Michelson-Morley experiment, according to Einstein himself: *"This experiment of Michelson-Morley was the first model that led me to special relativity"*, in his lecture *"As I Created the Theory of Relativity"* at the University of Kyoto in 1922. Einstein's explanation of the null result was by the contraction of space and time dilation, transforming the physical meaning of space and time by a mathematical and metaphysical space-time.

Results Conflicting – According to Einstein's theory of relativity in 1905, for all reference systems the speed of light is always

299,792.458 km/s or c ≈ 3 x 10^{10} cm/s what would prevent it being possible to determine any displacement of the reference system in relation to the vacuum-space or some medium as a fluid, a subtle species, an ether, because there would be contraction of vacuum-space and dilation of time so that the speed of light remain c independent of any speed of the emitting source of light. Also, by relativity in which there is no absolute reference, the displacement of the Earth in the vacuum-space could not be detected. And the light would not compose the speed c with the inertial velocity of the Earth or the Earth-Sun system, that is, it would not have an additional speed resulting larger than c. Experiments with electromagnetic radiation and times such as clocks, pendulums and global positioning system, GPS, to obtain a possible displacement in different equipments and or positions in the middle like ether, space-vacuum or subsequently in cosmic microwaves radiation, they had been presented after the Arago experiments in 1810; Airy in 1871; E. Macart on 1872; E. Ketteler in 1872; Michelson on 1881; Michelson and Morley on 1886 and 1887. The results obtained they were between null displacements, detected deviations and conclusions of theoretical contradictions, experimental or indications of inconsistencies.

Explanation of the Null Result in the Rotation of the Michelson-Morley Experiment – In 1971, L. Janossi showed that in rotation from the interferometer of Michelson-Morley the paths of light are kept in equal times, thus preventing the fringes of interference. So, by initial adjusting the light diagonally, the Michelson-Morley experiment nullified the Earth's displacement in the vacuum-space-ether, explaining the null result of this experiment.

First "Space Compass" Experiments – In 1954 Erdelyi had presented by the theory of the momentum or impulse a "space compass", an indicator of Earth velocity for several directions. And that in principle, the light with speed c should come out of the emitting source with that speed c, which is also required by Einstein's theory of relativity and the wave theory in the ether, to obtain deviations in a target that along with the Earth shifts in space.

And from 1955 to 1973 some previous experiments on this "space compass" were proposed and or performed by different researchers, using different types of light and lenses converging to the target, in which different results were obtained, and it was found that for the different positions in space it was necessary a suitable design with mechanical precision and optics for the measuring equipment. In 1971, Mesquita presented in the 23rd. Annual Meeting of the SBPC, published in this Year"s Yearbook, a procedure for emitting laser light in the laboratory with projection for the celestial dome with the purpose of obtaining a displacement of the Earth in vacuum-space. Figure 46.

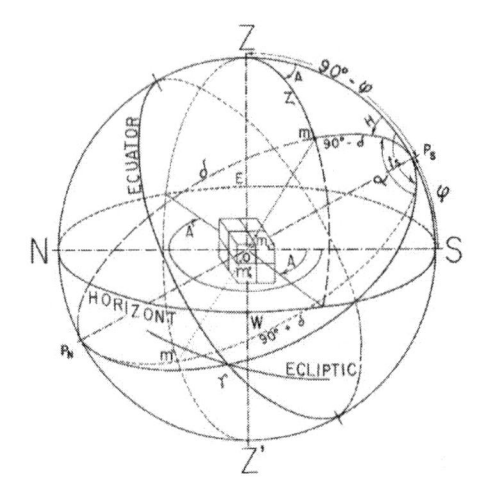

Fig. 46 – Experiment schematic in 1971 according to the Mesquita to determine the displacement of the Earth in the vacuum-space, by the "space compass" or velocity indicator. The O point is the center of rotation, of the velocity indicator, in the laboratory; M' is a point of the celestial sphere; OM'$_1$M' is the presumed sense of the absolute motion of the Earth; Z is the Zenith; α = straight ascension; δ = declination. Presented at SBPC in 1971.

Luz Neon – In the 1973 experiment of J. Martits with neon light showed that narrow rays of light presented the inertia of the velocity of the source on Earth in different directions of the vacuum-space. Figure 47.

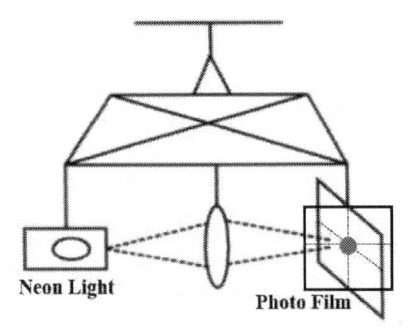

Fig. 47 – Experiment with light neon of Julius Martits for possible detection of the motion of the Earth in various directions. Design as Communication to the Brazilian Interplanetary Society-SIB and to author, as Reference.

Laser Experiment and The Michelson-Morley Experiment –
And in 1995 at SBPC and 2002 in Physics Essays, I was presented a laser experiment to verify the inertial velocity of the laser light radiation at 90° and in the direction of the constellation of Leo, it showed that by the laser positioned to emit the light radiation at 90° of the direction of the motion of the Earth-Sun-Laser system with velocity V_{ESL} = 350 km/s (obtained between 1951 and 1992 in the direction of the constellation of Leo), should be left in 90° in relation to the cosmic microwaves radiation or to the vacuum-space-ether, if it had the speed c constant of the light measured in the Earth-Sun-Laser source. But, the laser light came out diagonally in relation to cosmic microwaves radiation or space-vacuum, so carrying the inertial V_{ESL} velocity of the source. Figure 48.

With the laser in 90° of the motion of the Earth-Sun-Laser towards the Lion, the light should come out of the laser at 90° with the speed c of the laser light, reaching the target moving with the velocity V_{ESL} should present the deviation $X \approx 3.5$ mm

$c \approx 300.000$ km/s

But, the laser light presented the X = zero deviation that indicated that the laser light received the inertial V_{ESL} velocity of the Earth-Sun-Laser, with the velocity c'_d diagonally, explaining the Michelson-Morley experiment and contrary to the contraction of space and dilation of the time of the theory of relativity

$(d/c) = (X/V_{ESL})$

$V_{Earth-Sun-Laser}$ = 350 km/s · in the direction of Leo

Lion

d=3m
90°

Laser in the Earth-Sun-Laser system installed at 90° of the Lion direction

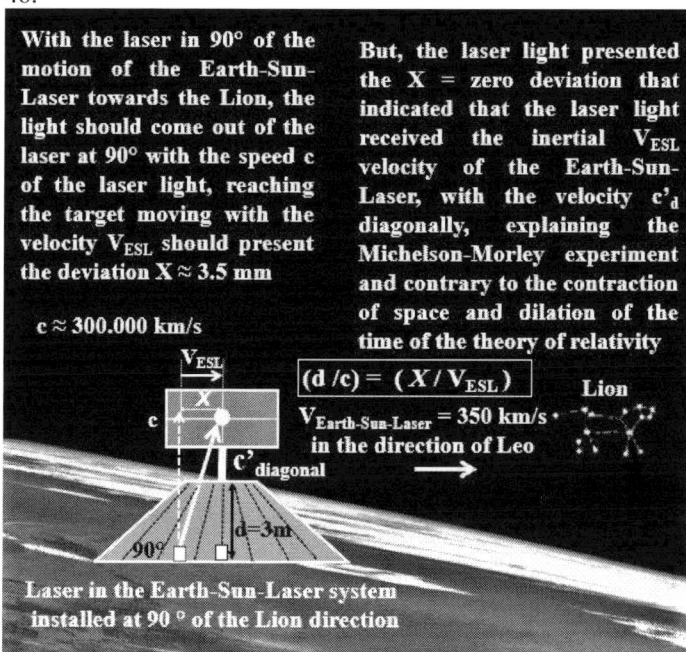

Fig. 48 – The laser positioned at 90° of the direction of the Earth-Sun-Laser with V_{ESL} velocity for the constellation of Leo in which the laser light should come out at 90° with speed c presented the velocity c_d' diagonally when receiving the velocity of the source, contrary to the theory of relativity in which the speed of light would be c and should come out at 90° to the motion of the Earth-Sun-Laser system to the constellation of Leo. Design of the Earth adapted from photo NASA and constellation Lion clipart of *Editora Europa*.

In the Michelson-Morley experiment the light was intentionally removed from the path of 90° of the Earth-Sun-Laser motion in the vacuum-space-ether, and placed at an angle of inclination of the aberration of light, for a predicted interference with the other light paths. But in this laser experiment, the light was intentionally placed to come out at 90° from the direction of the Earth-Sun-Laser in the vacuum-space-ether, precisely to get the displacement in the tarjet of the Earth-Sun-Laser in the vacuum-space-ether, or the speed of light composing with the velocity Earth-Sun-Laser if it goes diagonal. Then, in the Michelson-Morley experiment the light was already going diagonally by the inertia of the velocity of the Earth-Sun-Laser, without the perception of Michelson-Morley, and the adjustment was then solely for initial zero interference.

And in the direction of the motion of the Earth-Sun-Laser source to the constellation of Leo, the laser radiation is also carried by the source with the velocity c' = (c + V_{ESL}) in relation to cosmic microwaves radiation or vacuum-space-ether, where c is the speed of laser radiation in relation to the Earth-Sun-Laser system, without any contraction of space and time dilation, contrary to theory of relativity. Figure 49.

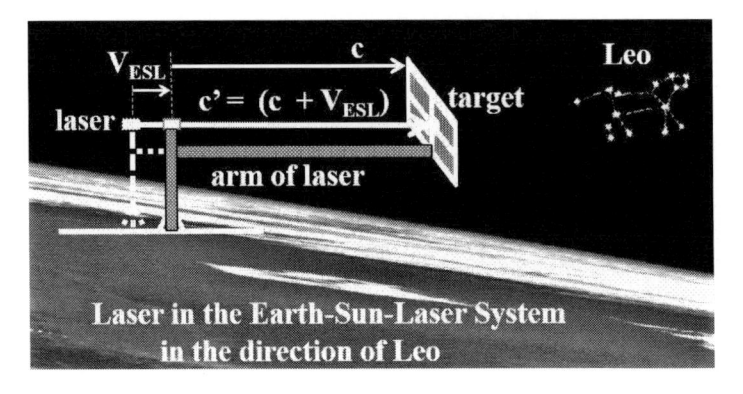

Fig. 49 – The V_{ESL} = 350 km/s is the velocity of the Earth-Sun-Laser system in relation to the vacuum-space-ether or cosmic microwaves radiation in the direction of the constellation of Leo; c ≈ 300,000 km/s is the speed of electromagnetic radiation with respect to emitting source Earth-Sun-Laser; c' ≈ 300,350 km/s is the speed of electromagnetic radiation in relation to vacuum-space-ether or cosmic microwaves radiation in direction to constellation Leo. Design of the Earth adapted from photo NASA and constellation Leo clipart of *Editora Europa*.

The laser experiment showed that the electromagnetic radiation comes out in the diagonal of the Earth-Sun-Laser system, so in the direction of the motion the electromagnetic radiation has the constant velocity of c' = (c + V_{ESL}) in relation to vacuum-space-ether or cosmic microwaves radiation, without space contraction or time dilation. Therefore, contrary to Einstein's relativity in that the speed of electromagnetic radiation is c in relation to vacuum-space or cosmic microwaves radiation and also to the emitting source, where for the constant c speed the space is contracted and the time is dilated. And in the decay of mesons µ at high velocities was observed larger times and highter distances travelled attributed to the theory of relativity. But, in the decay of mesons µ in which at high velocities run higher distances and larger times for decay, could be explained by the mass increase of the mesons µ that increased its velocity by receiving higher energy, therefore a longer time for its decay. And contrary to the special relativity of a ghostly condensation of the speed-space-time that would form and increase the mass of the meson µ, or shell, or apparent mass, without any physical significance

Other results of optical experiments, such as the binary stars are attributed to Einstein's relativity. As W. de Sitter, in 1913, in which a star advances with velocity v in the direction of the Earth if the light is emitted with velocity (c + v), while the other one that moves away with also speed v emitted the light with velocity (c − v), should presenting images with eccentricity, but no eccentricity was observed. However, in 1965, Fox had showed that the images of binary stars to present eccentricities should result in high velocities and a short period of revolution of the binary stars. And the results obtained, contrary to what has been believed for several decades, were more in line with the additions or subtractions of light velocities and binary stars than with special relativity.

In 1977, Kenneth Brecher had noted that pulsars showed the emission of X-ray electromagnetic radiation with one constant speed c, independent of the speed of the emitting source in any direction, to all the positions from the pulsar's orbit, detected on Earth. So, being considered as the greatest proof of the constant speed of light regardless of the speed of the source, in which the space is contracted, and the time is dilated for this speed c to be constant by the theory of relativity. However, the speed of light c is also independent of the speed of the source by wave theory. But, according to the laser experiment, it is the speed of light $c' = c_v + V_{source}$ in relation to the space-vacuum-ether that is independent of the speed of the source, in which the light is originated from the quantum impulse with the speed c' in the space-vacuum-ether. In case of the V_{Earth} Earth velocity in direction to constellation Leo, c_v is the c speed of light in relation to the source Earth, and V_{Earth} is the source velocity. And as the source carries light as seen by the trajectory of laser light diagonally placed at 90 °, the velocity of light c_v is variable relative to the velocity of the source, so without a space contraction and or time dilation relativistic for that the speed of light c it is always constant as theory of relativity. If the velocity of the source is null, the light comes out of the source, without any energy field saturation, with velocity c' in relation to space-vacuum-ether. And greater is the velocity of the source, as in the case of mesons that emit gamma rays, it is lower the velocity

c_V of the light and other electromagnetic radiation in relation that source. In this case, with higher velocity, the source is saturated of quantum energy and the light comes out "clingier" to the source, although it is constant in relation to the space-vacuum-ether, because the light m'c x c' is originated with the speed c' in the space-vacuum-ether. Figure 50 showed the physics process of emission of light in vacuum-space-ether.

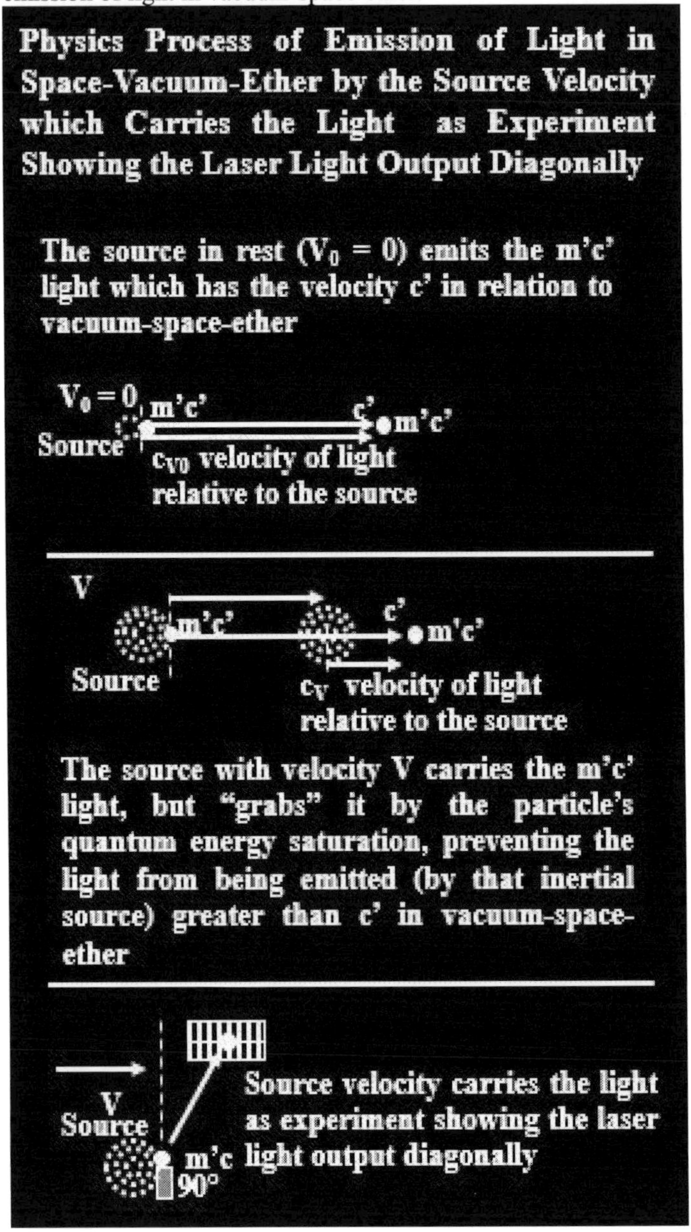

Physics Process of Emission of Light in Space-Vacuum-Ether by the Source Velocity which Carries the Light as Experiment Showing the Laser Light Output Diagonally

The source in rest ($V_0 = 0$) emits the m'c' light which has the velocity c' in relation to vacuum-space-ether

$V_0 = 0$, m'c'   c' • m'c'
Source c_{V0} velocity of light relative to the source

V
m'c'   c' • m'c'
Source c_V velocity of light relative to the source

The source with velocity V carries the m'c' light, but "grabs" it by the particle's quantum energy saturation, preventing the light from being emitted (by that inertial source) greater than c' in vacuum-space-ether

V
Source   Source velocity carries the light as experiment showing the laser m'c light output diagonally
90°

Fig. 50 – The physics process of emission of light in space-vacuum-ether by the source velocity which carries the light as experiment of the laser light output diagonally.

In the case of binary stars, both by relativity and by wave theory as Michelson-Morley experiment, the speed of light c is always constant in relation to the space-vacuum-ether. However, the c_v velocity of light is variable in relation to the emitting source, as seen in the laser experiment in which the light comes out diagonally, proving that the light is carries by the source. Figure 51.

The speed of light emitted by the binary star is always c' in relation to the space-vacuum-ether, and as the laser experiment in that the light coming out diagonally proves that the emitting source carries the light, this is , the c_v velocity of light being variable (c_1 , c_2 , c_3) relative to the v velocity of the emitting source

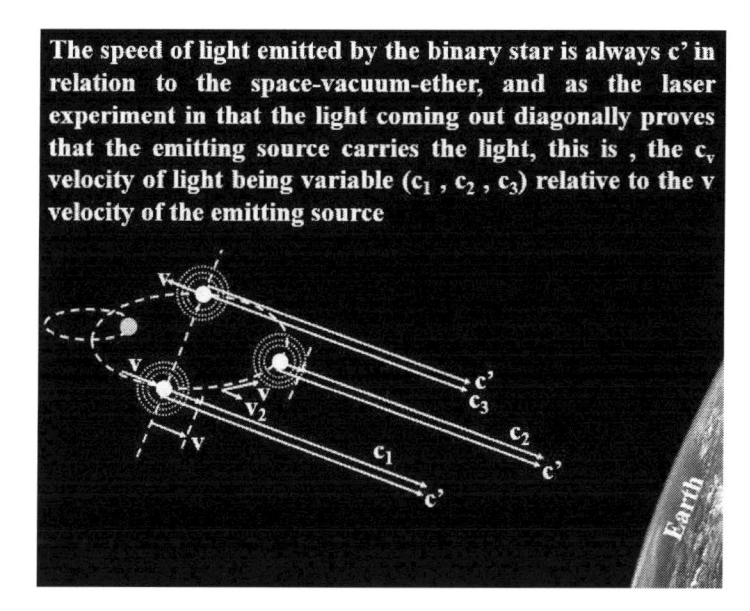

Fig. 51 – In any position the binary star emits the light with constant speed c' in relation to the vacuum-space-ether, while by the theory of relativity and by the wave theory as in the Michelson-Morley experiment it is constant the speed of light c. However, the c_v (c_1, c_2, c_3) velocity of light is variable relative to the binary star in which the c_v velocity of light depends on the velocity of the emitting source, as seen in the laser experiment in which the source carries the light diagonally. Earth clipart from *Editora Europa*.

Special relativity had its confirmation in the Michelson-Morley experiment which had presented a null result, in which in this experiment the speed of the light c coming out diagonally, but without the inertia of the velocity of the Earth-Sun-Interferometer system in vaccum-space, only for the inclination of light aberration by the motion of the Earth, to adjust the interference of the light paths, in which in the rotation of the arms of the interferometer should occur the fringes of interference. However, the light with the laser at 90 ° of the direction of the Leo motion came out diagonally with the velocity c'$_{diagonal}$ which was the cause of this null result.

CXXIX. INERTIAL VELOCITY OF LASER LIGHT AND CERN-MINOS-FERMILAB, OPERA-ICARUS-CERN – 1907 to 2014

Laser Light Radiation Inertial Velocity – In 1995 a Scientific Communication had been presented in the SBPC-Brazil in which the Summary was in its Yearbook-1995-SBPC, about the experiment for checking the inertia of light when the light is emitted from the laser, and in 2002 the detailed article was submitted and published in Physics Essays. By the Figure 48, the laser light is emitted at a distance d = 3 m for a target at 90° to the motion of the Earth-Sun-Laser system of $V_{ESL} \approx 350$ km/s velocity in relation to cosmic microwaves radiation or vacuum-space for the constellation of Leo. Being admitted the speed of light c ≈ 300,000 km/s of the light that would come out at 90° of the laser, measured by the observer on Earth, the laser light when traversing a distance d should reach a target near the laser fixed on the equipment on the ground moving in the cosmic microwaves radiation or the vacuum-space-ether in the distance X with velocity V_{ESL}. Therefore, for the same time t that the light goes through d, the target traverses X by the relationship $d/c = X/V_{ESL}$ where X it should present an displacement of 3.5 mm, but the result was $X = 0$, evincing the inertial velocity of laser light radiation emitted by Earth-Sun-Laser system source. This experiment was carried out based on the theory of the momentum or quantum elementary impulse of Erdelyi in 1952, originated by applying concepts with physical significance to Planck's quantum law, and based on the CERN experiment of 1964 gamma ray radiation by the mesons $\pi°$.

In the motion of the Earth-Sun-Laser system in the direction of the constellation Leo the laser light was directed to the center of the target. And when emitted in 90° of the Earth-Sun-Laser system motion, this laser experiment showed that the electromagnetic radiation has the deviation diagonally, therefore dependent on the velocity of the source. Then, in the direction of the motion of the Earth-Sun-Laser system the laser light is also dependent on the velocity of the source, with the speed c' = (c + V_{ESL}) which is always constant in relation to cosmic microwaves radiation or vacuum-space-ether, without any contraction of space or time dilation. According to special relativity, the speed of light is c in relation to cosmic microwaves radiation or vacuum-space measured in a reference of coordinates and time (x, y, z, t) and also c in relation to cosmic microwaves radiation or vacuum-space measured in another frame (x', y', z', t'), as a consequence of the transformation of coordinates and time, which resulted from the two postulates of special relativity in which according to Einstein: *"1. The laws by which the states of physical systems subjected to variations are not affected, if these state modifications are referred to one or the other of two coordinate systems in uniform motion of translation. 2. Any beam of light moves in a stationary system of coordinates with a certain speed c, whether the radius emitted by a stationary or moving source".*

Therefore, according to Einstein's relativity the speed c of electromagnetic radiation is always constant in relation to cosmic microwaves radiation or vacuum-space and independent of the speed of the source, and to keep c constant space is contracted and time is dilated. By the electromagnetic wave theory and also by the sound, seismic and maritime waves, the velocity of these waves is independent of the source. But by this laser light experiment the c_v velocity of light relative to the source depends on the velocity of that source V, where the velocity $c' = (c_v + V_{ESL})$ is constant in relation to cosmic microwaves radiation or vacuum-space-ether, but without contraction of space or time dilatation. Figure 49.

Many works have been presented in which the velocity of light is observed differently, in which it is dependent on the velocity of the source and of the observer in another references:
in 1996 by F. Sellers; in 2010 by M. Sato, S. J. G. Gift, M. Rasul; in 2011 by D. Anderson, G. Cao, D. Crossley; in 2012 by Gift, C. Ranzan; on 2013 by Ranzan; in 2014 by S. Das. Also, it was observed the variable velocity of light, emitted by the source, in the diagonal direction, in 2014 by R. D. Sauerheber.

CERN Experiment of 1964 and Laser Light Experiment – The CERN experiment of 1964 by T. Alvager, F. J. M. Farley, J. Kjelman and I. Walin proved that velocity c_{gamma} of the gamma rays emitted by the mesons $\pi°$ with velocity $c_{meson} \approx c$ of light measured on Earth, in the direction of the motion of the mesons $\pi°$ was $c_{gamma} = c$ as defined for the ratio of distance measurement required in the reference apparatus in the light time of the mesons $\pi°$ and gamma rays, indicating that the velocity of the gamma rays was independent of the velocity of the source. So this CERN experiment would be the best confirmation of Einstein's theory of relativity about the independence of the speed of light c in relation to the speed of the source. However, it was necessary to note that the mesons $\pi°$ having almost the speed of light c and gamma rays, i.e. $c_{meson} \approx c$ and $c_{meson} \approx c_{gamma}$, in which $c_{gamma} = c$, then the speed of gamma rays in relation to mesons $\pi°$ is $c_{gamma/meson} \approx 0$ km/s. And uniting this laser experiment with this one at CERN, the conclusion was that the velocity of the laser light in the direction of the motion is always constant in relation to the cosmic microwaves radiation or the vacuum-space-ether, but in that the higher the velocity of the source, the lower the velocity of the radiation in relation to the source, without space contraction or time dilation. So for a higher V velocity than V_{ESL} of the emitting source of electromagnetic radiation in relation to cosmic microwaves radiation or space-vacuum-ether, smaller would be velocity c_V electromagnetic radiation in relation to the motion of the emitting source, ie $c' = (c_V + V)$ or $c_V = (c' - V)$ in which c_V it is the velocity of light relative to the source and V is the velocity of the source, and c' is the always constant velocity of light in relation to cosmic microwaves radiation or vacuum-space-ether, but without contraction of space or dilation of time. And to $V_0 = 0$, the velocity c' would be $c' = (c_0 + V_0) = c_0$. So, this light laser experiment showed that in relation to cosmic microwaves radiation or vacuum-space-ether there is a c' constant speed of electromagnetic radiation, but without contraction or expansion of space, and the velocity of electromagnetic radiation is variable in relation to the source when it is correlated with this 1964 CERN experiment. The conclusion of the laser light experiment it is also contrary to the mechanics of Galileo and Newton in which the speed c'_{final} of light is an addition of c with any speed of the source, and also contrary to the Ritz in which the speed of light c is always the same in relation to the source. Figure 52.

Laser in the Earth-Sun-Laser System in the direction to Leo and the CERN Experiment

[Figure showing Laser Experiment with equations: V_V, c_V; V_{ESL}, c, $c' = (c_V + V_V)$; $V_0=0$, c_0, $c' = (c + V_{ESL})$; $c' = (c_0 + 0)$; pointing toward Leo constellation. CERN Experiment showing velocity of the $\pi^°$ meson, velocity of the gamma rays, velocity of the gamma rays relative the $\pi^°$ meson ≈ 0 km/s.]

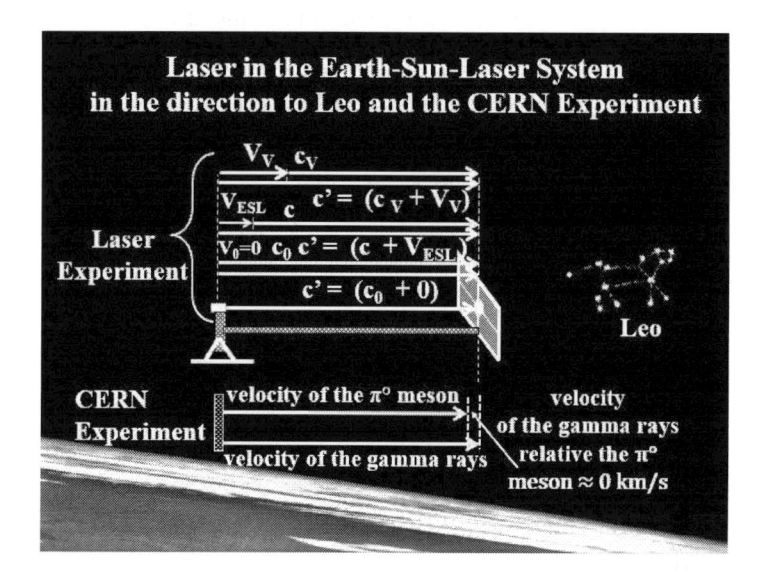

Fig. 52 – The $V_{ESL} = 350$ km/s is the velocity of the Earth-Sun-Laser system in relation to the cosmic microwave radiation or vacuum-space-ether in the direction of the constellation Leo; $c \approx 300{,}000$ km/s is the speed of electromagnetic radiation in relation to the Earth-Sun-Laser source (ESL); c_V it is the velocity of the electromagnetic radiation in relation to the ESL source with velocity V and c_0 it is the velocity of the electromagnetic radiation in relation to the ESL source with velocity V = 0. In the CERN experiment, the velocity of gamma rays is almost zero relative to the meson which has the velocity close to the gamma rays. Design of the Earth adapted from photo NASA and clipart constellation Leo from *Editora Europa*.

The Figure 53 shows a bolide-car which with its inertial velocity carries the light, but according to the speed of the car, where the higher the speed of the source, smaller will be the speed of light relative to the car.

A bolide-car with speed of 200 km/s, at Position 0, emits light with imaginary-speed of 300 km/s in space, with the Earth standing in space.

1- By the laws of the motion of Galileo and Newton, if the light received all the inertia of the speed of the car, the speed of light would be (300 km/s + 200 km/s) = 500 km/s and would reach the target at Position 500.

2 – If the light is not carries by the speed of the car, then the light reaches the target at Position 300 with the speed of 300 km/s in relation to point 0. But, in relation to the bolide-car the pilot measures the speed of light at (300 km/s – 200 km/s)/time = 300 km/s by the theory of relativity, because on his watch at 200 km/s marked around 0.3 seconds. That is, the time was slower and the space was on contraction due to the motion of the bolide-car.

3 – But by the laser experiment, the light is carries by the source, where the higher the speed of the source, smaller will be the speed of light relative to the source. Then, the car-bolide carries the light from Position 0 to the Position 200 with 200 km/s in relation to space, and the light with its speed of 300 km/s in relation to the space reaches the Position 300. Therefore, the speed of light relative to the car-bolide is from (300 km/s – 200 km/s) = 100 km/s, without the fallacy of space contraction and or time dilation of theory of relativity.

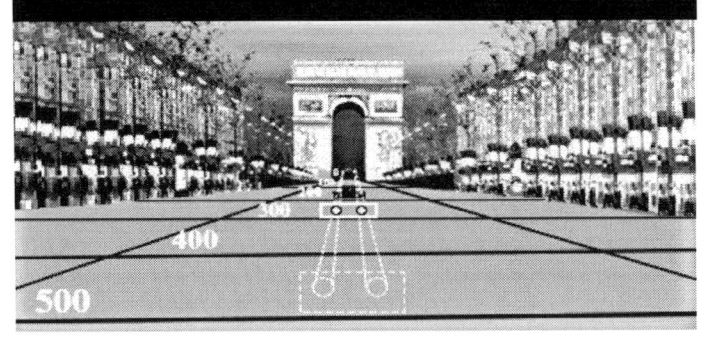

Fig. 53 – With the Earth standing in space, a car bolide-car to 200km/s carries the light from 0 to position 200. Where the higher the speed of the car, the lower the speed of light relative to the car. And the light reaches the position 300 with its specific velocity of 300 km/s in space, measured on Earth, therefore without the fallacy of space contraction and time dilation of the theory of relativity. Adapted from photo www.pinterest.com.

The laser experiment showed that in the direction of the motion, the electromagnetic radiation has the constant speed c' = (c + V_{ESL}) in relation to cosmic microwaves radiation or vacuum-space. And uniting with the CERN experiment of 1964, the velocity of gamma rays with respect to mesons $\pi°$ was ≈ 0 km/s. The bottom line is that the higher the velocity of the V source, the lower the velocity of the electromagnetic radiation relative to the source. If the Earth-Sun-Laser system had the velocity V = 0, then c' = (c_0 + 0) = c_0. That is, the velocity of electromagnetic radiation is variable relative to the velocity of the emitting source. Therefore, contrary to Einstein's relativity, which the speed of electromagnetic radiation is always constant in relation to any speed of the emitting source, through the contraction of space and time dilation, in relation to any system of reference.

By the relativity of Galileo's motion, two ships in space with control off, no astronaut inside the ship could know which ship would be at rest or which in motion. The astronaut of the ship 1 sees the ship 2 move away. But, the astronaut of ship 2 sees that ship 1 is moving away. But according to Einstein, if the astronaut of the ship 1 put a mirror in front of his face, he would see his face in the mirror. But if the ship were to move at the speed of light, the light inside the ship reflecting on its face should never reach the mirror which was also at the speed of light. So the astronaut would know if he was on the move or at rest. But the actions of the astronauts, the laws of Physics, should be equal for the astronauts at rest or on motion. So, according to Einstein, the speed of light should be constant in relation to the astronaut, and the astronaut with the speed of light would see the light coming out of his face with the speed of light that would hit the mirror and return with the speed of light, so he would see his face in the mirror. But by the CERN experiment of 1964 and the laser of 2002, if the velocity of a spacecraft was equal to the speed of light, the light would not reach the mirror in front of the astronaut inside the ship, because the mirror would always be escaping the light at the same speed.

MINOS-FERMILAB, OPERA-ICARUS-CERN – In the nucleus beta decay, in the initial experiments, the electron and the proton were detected. But, the final energy measured was observed less than the initial energy. In 1930, Pauli had presented the existence of a minimal wythout charge particle, that is, neutral to preserve the conservation of charge and energy in the beta decay. This particle would have a minimum mass or even zero mass if it came out of the nucleus at the speed of light, by the theory of relativity. Subsequently, it was named neutrino by Fermi. MINOS-FERMILAB had obtained in 2007, in reference to the speed of light admitted as c ≈ 300,000 km/s measured in any reference, the speed of the neutrino in 300,015.30 ± 8.7 km/s, and at 2011

OPERA-CERN obtained 300,007.44 ± 0.84 km/s, while in 2012 ICARUS-LNGS obtained 300,000.03 ± 0,49 km/s. Although ICARUS-LNGS shows the speed of the neutrino equal to the light, the relative inaccuracy ± 0.49 Km/s placed both the neutrino and the smallest or as higher speed than that of the light, while MINOS-FERMILAB with ± 8.7 km/s of inaccuracy and OPERA-CERN with ± 0.84 km/s of inaccuracy, the speed of light was maintained above speed c. The measurement by ICARUS-LNGS far from showing a speed to the neutrino, only showed that by the imprecision of the measure can be equal, smaller or greater than that of the light! For these three measurements of the speed of the neutrino, of MINOS-FERMILAB, OPERA-CERN and ICARUS-LNGS, where it was considered the neutrino without mass or mass, because it was above or equal to the speed of light, it was to be contrary to the relativistic theory. For, by Einstein's relativity, a mass m_0 presents by the increase of mass relativistic $m = m_0 / \sqrt{1 - (v^2/c^2)}$, and for light with speed c would be m x 0 = m_0, that is, $m_0 = 0$, and case $m_0 \neq 0$ then m would be infinite. And with speed c for the neutrino, according to ICARUS-LNGS, it would be without mass. But, data from neutrino transformations indicated that it would have mass, and by these three neutrino measurements by MINOS-FERMILAB, OPERA-CERN and ICARUS-LNGS this by relative inaccuracy, with speed > c would indicate infinite mass. So while the observations proved that the neutrino had mass, all three of the neutrino speed measurements showed that it would have zero mass or infinite mass if it were to follow relativity. But MINOS-FERMILAB, OPERA-ICARUS-LNGS-CERN and NASA were observing this experiment of the inertial speed of laser light in relation to cosmic microwaves radiation or vacuum-space and the CERN experiment of 1964 would see explained the speed and mass of the neutrino without the incompatibility of relativistic theory.

Three Equal Times in the Michelson-Morley Experiment – By the laser experiment presented in summary at SBPC in 1995 and articles on 2002 and 2016 in Physics Essays, the result showed that, seen by the reference of the cosmic microwave radiation or vacuum-space-ether, the laser light is with the speed c', which is the speed c of light in relation to the Earth-Sun-Laser System plus the velocity V_{ESL} of the Earth-Sun-Laser system in relation to cosmic microwave radiation or vacuum-space-ether in the direction of Leo. By this laser experiment and the interpretation of the CERN experiment with mesons near the speed of light emitting gamma rays, the light and all other electromagnetic radiations its have in the direction of the motion the velocity c' = (c_v + V_{ESL}) relative to cosmic microwave radiation or vacuum-space, where the higher the velocity of the source, the lower the velocity of light in relation to the emitting source. In the case, if

the Earth-Sun-Laser system decreases its velocity to zero, the light would have the speed c' = $(c_v + V_{ESL}) = [c_0 + (V_{TSL\,(0)} = 0)] = c_0$.
That is, c_v is variable according to the velocity of the source. In this case of $V_{TSL\,(0)} = 0$, the light in all directions of cosmic microwave radiation or vacuum-space would have the velocity of c' = $(c_v + V_{ESL}) = c_0$. However, in the opposite direction to the source motion, with the velocity of light c' = $(c + V_{ESL})$ does not have inertial velocity V_{ESL} of this motion, therefore c' = c. And c in the field of the Earth-Equipment-Assembly has the inertial velocity in the opposite direction of the motion, ie $- V_{ESL}$, then (c $- V_{ESL}$). The velocity of the light's inertia by the velocity of the Earth-Sun-Laser source would be similar to a skater with his specific velocity V_{skater} in relation to Earth, and the V_{wind} velocity of the wind on going path and back. And then finds a sharp curve, the skater to back with his specific velocity, he has in the contrary direction inertia V_{wind} velocity. The theories of the emission of light are known in that their velocity is always c relative to the source is fixed or in motion. However, in one of these theories, as that of the new source, in which the velocity of light is c, but relative to the image of the source that is obtained in the mirror, it there are different interpretations existing on the behavior of light in reflection by mirrors. Figures 54 and 55.

The quantum impulse (m'c) always has the velocity c and it can be dragged by the source velocity

$c'_{motion\;direction} = (c + V_{source}) =$ **velocity of (m'c) relative to the fixed mirror**

$c'_{motion\;direction}$

(m'c)

c = velocity of m'c relative to the source

fixed mirror

(m'c)

V_{source}

V inertial of the source

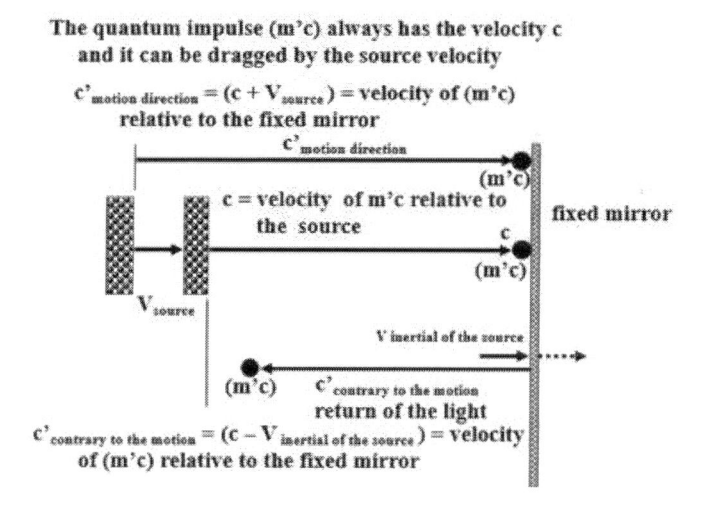

(m'c) $c'_{contrary\;to\;the\;motion}$
return of the light
$c'_{contrary\;to\;the\;motion} = (c - V_{inertial\;of\;the\;source}) =$ **velocity of (m'c) relative to the fixed mirror**

Fig. 54 – The quantum impulse m'c (c relative to the source) or m'c' (c' relative to the vacuum-space-ether) always has the same velocity and forms the particles by running helix, and in electromagnetic and gravitational fields. And in the motion of the source, the particles with the impulses are dragged with the velocity of the source.

Own Velocity and Drag or Inertia Velocity

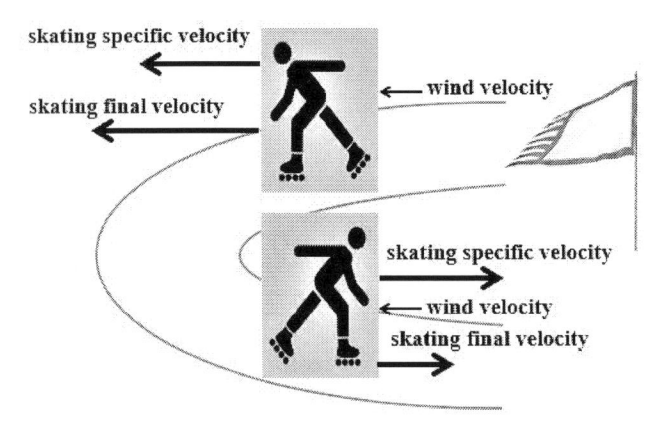

Fig. 55 – The skater has its specific velocity and plus the wind velocity on going, and the back has its specific velocity minus the wind velocity. In both cases, in going and in the back, the skating specific velocity is the same. Adapted from clipart of *Editora Europa*.

The laser experiment shows that in the Michelson-Morley experiment, as Figure 56, in goig path, to traverse the distance D of the interferometer and S of the Earth-Sun-Laser moving in cosmic microwave radiation or vacuum-space-ether, this is, (D + S) in time t_{going}, $c' = (c + V_{ESL})$ containing the inertial velocity $(+V_{ESL})$. So $[(t_{going}\ c_{going}) = (D + S)] = (D + V_{ESL}\ t_{going})$. So, $t_{going} = D/(c_{going} - V_{ESL}) = D/(c + V_{ESL} - V_{ESL}) = D/c$.

For the distance back of the light on that $+V_{ESL}$ it goes against the direction of light, the light when reaching the mirror at the end of D back with the speed $c' = c$, because there's no V_{ESL} in this direction, in relation to the Earth-Sun-Laser system, but it has the inertia of V_{ESL} in direction of the Earth-Sun-Laser motion, ie $c'_{back} = (c - V_{ESL})$, in time t_{back} where light travels (D – S) while the Earth-Sun-Laser source traverses S, $[t_{back}\ c'_{back} = (D - S)] = (D - V_{ESL}\ t_{beck})$. So, $t_{back} = D/(c'_{back} + V_{ESL})$, is obtained $t_{back} = D/(c - V_{ESL} + V_{ESL}) = D/c$.

As in the direction of the source motion the speed of light in relation to the vacuum-space-ether is $c' = (c + V_{ESL})$, the light if it came out exactly at 90° of the direction of the source to Leo would be without inertial velocity V_{ESL} which is in the direction of the motion. So, if the light came out exactly at 90° it would have the speed $c' = (c + 0), = c$. Then, in the transverse direction to the motion, the velocity of light has the vector composition of c and V_{ESL}. And with velocity c at 90° it would travel D, that is, $t = (D/c)$, that it's the same time t that velocity V_{ESL} of the Earth-Sun-Laser would have to traverse the distance S in the cosmic microwave radiation or vacuum-space-ether, that is, $t = (S/V_{ESL})$ to Leo. Therefore, by the composition vector velocities c and V_{ESL} is obtained $c'_{diagonal} = (c^2 + V^2_{ESL})^{1/2}$ and $t_{diagonal} = D'/c'_{diagonal}$, for the time $t_{diagonal}$ that the laser light travels on the diagonal is the same time t that the light would do in the perpendicular direction when traversing D, that is, $t = (D/c)$, and it is the same time t that the velocity V_{ESL} of the Earth-Sun-Laser would have to traverse the distance S in the cosmic microwave radiation or vacuum-space-ether, that is, $t = (S/V_{ESL})$. Therefore, the time traveled to a distance D of the arm of the interferometer on the going and back is then $t_{going} = t_{back} = (D/c)$, which is equal to the time of the diagonal $t_{diagonal} = (D'/c'_{diagonal}) = (D/c) = t$. Thus, with the equal three times the result could only be what was obtained, that is, zero interference!

$$t_{diagonal} = D'/c'_{diagonal} = D'/(c^2 + V^2_{ESL})^{1/2} = t = S/V_{ESL} = D/c$$

$$t_{going} = D/(c'_{going} - V_{ESL}) = D/(c + V_{ESL} - V_{ESL}) = D/c$$

$$t_{back} = D/(c'_{back} + V_{ESL}) = D/(c - V_{ESL} + V_{ESL}) = D/c$$

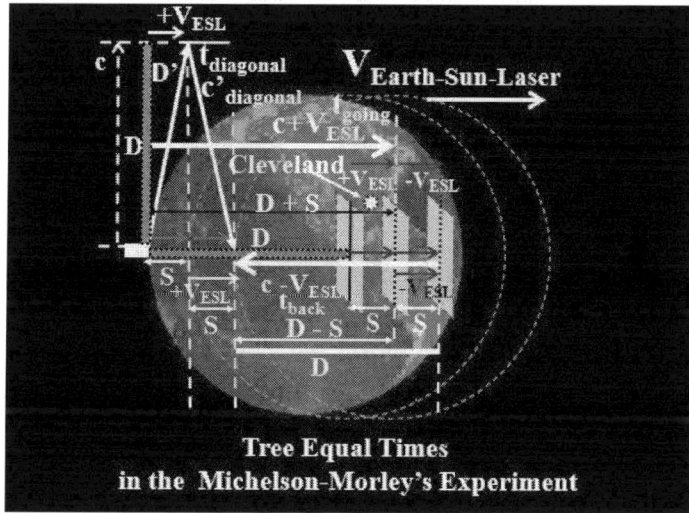

Fig. 56 – Three equal times in the Michelson-Morley experiment. Earth clipart from *Editora Europa*.

This analysis of the path times of light could find also, that in the case of light with velocity $c' = (c + V_{ESL})$, in relation to vacuum-space-ether, in the direction of the motion, the back from the mirror without V_{ESL} inertia, that is, $c'_{back} = c$ and plus the velocity at which the Earth-Sun-Laser system goes against the light, that is, $c'_{back} + V_{(ESL)going} = (c + V_{ESL})$. In that case, the path of going-back would be $2D = t [(c + V_{ESL}) + (c + V_{ESL})] = t 2(c + V_{ESL})$. And transverse $2D' = t'\ 2[c'_{diagonal} = (c^2 + V^2_{ESL})^{1/2}]$, therefore there would then be a time difference between the two paths, in which to zero in on the interference which would be detected in the rotation of the interferometer, the mirrors were adjusted. And as Janossy, in the rotation of the interferometer the time difference remains the same for any rotation, therefore without any light interference fringes for the paths. So, what resulted in equal times for the light paths of the Michelson-Morley equipment. Also, other analysis in which the going light $(c + V_{ESL})$ hit the mirror and return with that same speed $(c + V_{ESL})$, in that V_{ESL} goes in the opposite direction of light, in vector reversal $(-c - V_{ESL}) + V_{ESL}$, what would result $t_{return} = -(-D/-c) = D/c$.

The laser light equipment has been constructed with all rigidly fastened parts, including parts within the laser, preventing flexion or torsion, to the accuracy of the position of the laser light received on the target.

This laser experiment used the emission of light with the laser at 90° of the direction of the Earth-Sun-Laser system motion in the direction of the constellation of Leo. Then, the laser light would keep the perpendicular to the direction of Leo, by Einstein's relativity, whereby the velocity of light is always c in the vacuum-space and as a consequence of the transformation of coordinates is also c measured in any reference system in this transformation, so without receiving the inertial velocity V_{ESL} of the emitting source of the Earth-Sun-Laser system. But, surprisingly the result was that the laser light came out of the diagonal laser showing that the light received the inertial velocity $V_{ESL} \approx 350$ km/s of the Earth-Sun-Laser system. So this experiment showed that the laser light has the speed c' with respect to cosmic microwave radiation or vacuum-space-ether, greater than c in relation to the Earth-Sun-Laser system, that is, $c' = (c + V_{ESL})$ in the direction of the motion of the Earth-Sun-Laser System and diagonally $c'_{diagonal} = (c^2 + V^2_{ESL})^{1/2}$. With this inertial velocity of light, the Michelson-Morley experiment is fully explained, since round-trip and transverse times are the same. So, as shown in Figure 48, for the distance of 3 m from the laser to the target, in the Earth-Sun-Laser-Target system with the velocity $V_{ESLT} = 350$ km/s, running 3.5 mm in vacuum-space, by the theory of special relativity the contraction of the target in the vacuum-space would be of $(L - L_0)$, in which $L_0 = 3.5$ mm and $L = L_0 [(1 - (V_{ESLT}^2/c^2)]^{1/2} = 3.499997616$ mm, so $(L - L_0) = 0,000002386$mm. That is, an insufficient contraction for the laser light coming out at 90 ° reaches the center of the lattice that has shifted from 3.5 mm in vacuum-space-ether along with the Earth-Sun-Laser-Target. Therefore, it is the velocity of the source that drags the light on the diagonal and not the contraction of space that would cause less wavelength of light in the direction of the motion and the null result of the Michelson-Morley experiment. And in the direction of the motion of the Earth-Sun-Laser-Target system, the velocity of the electromagnetic radiation c_v is variable according to the velocity of the source when correlated with the CERN experiment of 1964 and the laser light experiment. Figure 57.

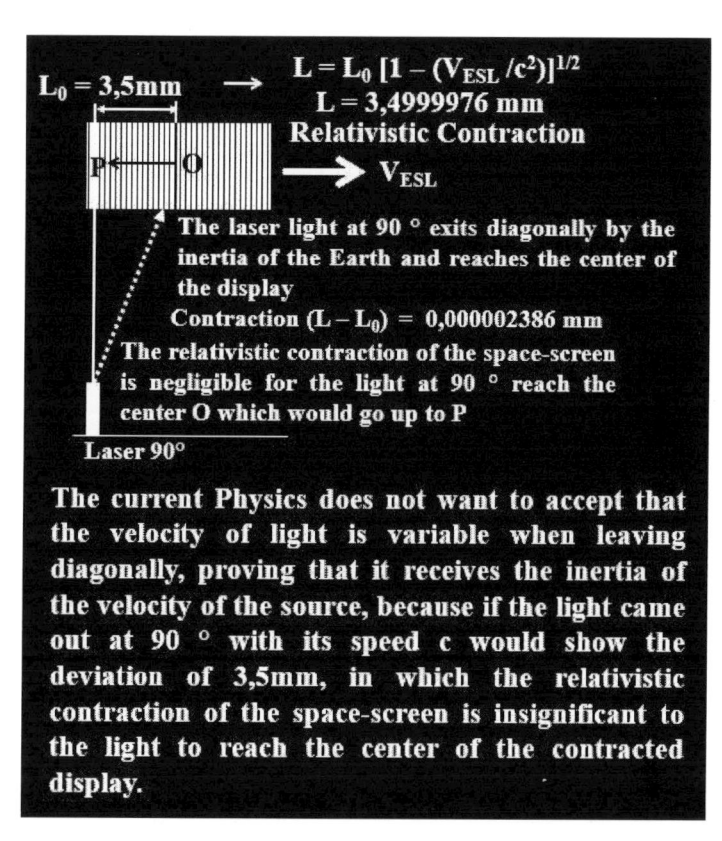

$L_0 = 3,5mm$ → $L = L_0 [1 - (V_{ESL}/c^2)]^{1/2}$
$L = 3,4999976$ mm
Relativistic Contraction

→ V_{ESL}

The laser light at 90° exits diagonally by the inertia of the Earth and reaches the center of the display
Contraction $(L - L_0) = 0,000002386$ mm
The relativistic contraction of the space-screen is negligible for the light at 90° reach the center O which would go up to P

Laser 90°

The current Physics does not want to accept that the velocity of light is variable when leaving diagonally, proving that it receives the inertia of the velocity of the source, because if the light came out at 90° with its speed c would show the deviation of 3,5mm, in which the relativistic contraction of the space-screen is insignificant to the light to reach the center of the contracted display.

Fig. 57 – The referees of the Scientific Journals of the pre-established Physics do not want to accept experiments that show the variable velocity of light in relation to the emitting source of light. By relativity, the light of a laser at 90° should come out perpendicular to the motion of the source at speed c.

New Laser Experiment Proposition – In 2014, it was presented in Physics Essays by J. Drodzynski a proposition for an experiment of light in spacecraft, in which it was also predicted that the beam of light would come out at 90° of the ship's horizontal. According to this work: *"In the motion of the ship with a constant velocity u, a laser source (S) is a direct parallel to the y-axis. Since the laser pulse cannot be influenced by the motion of the source, that is, by the motion of the ship's coordinate system, the laser pulse must move to down and up in the extension of the same straight line.*

Contrary to the one recommended by the special theory of relativity, the observer on the ship will inform that the pulse is not returning to the starting point S. The positions of the light source at the time of exit and return of the pulse are given by S and S', respectively, and can be determined experimentally by the value d". Figure 58.

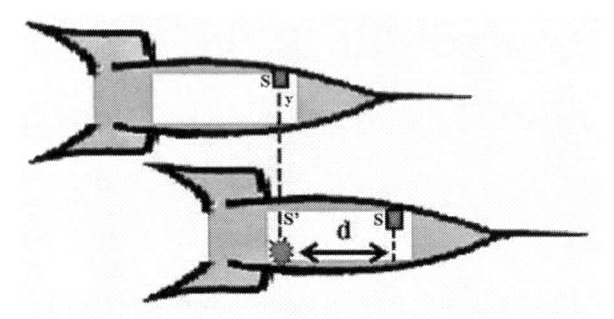

Fig. 58 – As presented in Physics Essays by J. Drodzynski on 2014 one S laser source is directly parallel to the y-axis within the nave. Since by theory of relativity the laser light cannot be influenced by the motion of the source, that is, by the motion of the ship's coordinate system, the laser light should move below and above in the extension of the same straight line and should be obtained a displacement *d*, as Reference.

However, the laser light experiment presented in 2002 in Physics Essays had shown that the light would come out diagonally, so the displacement *d* would be zero, also contrary to the theory of relativity in which the laser light would be independent of the motion of the source and that it should exit perpendicularly to the motion of the ship. So it is evident the inconsistency of the theory of relativity, because in the two experiments, with results opposed, they are contrary to this theory.

Stellar Light Aberration – But why was the aberration of starlight, by the inclination of the telescope, in which the light from the stars is not dragged away by the Earth's source with velocity V_{ES}, while the laser light is dragged diagonally by the same source V_{ES}, when it should exit the laser at 90° by relativity to keep c constant? The aberration of starlight, for example, for the Dragon constellation as Bradley mesured only the angular shift around the Sun. Because, according to this laser experiment, the Dragon constellation also moves in the vacuum-space-ether with a velocity similar to from Earth-Sun to the constellation of Leo, and the emission of light by Dragon would present the inertia of its velocity, then being seen as entering near the perpendicular to the Earth. Therefore, the light from the Dragon would be dragged by the inertial velocity of the emitting Dragon source and entering the distant source of the Earth without being dragged. And in the laser, the light is dragged into the emitting source of the Earth-Laser without being dragged into a distant source, because the inertial velocity is the one originated in the emitting source. However, for other stars, different values were observed for stellar aberration, or the length of the telescope should be adjusted to obtain the angular deviation obtained for the Dragon constellation. Figure 59.

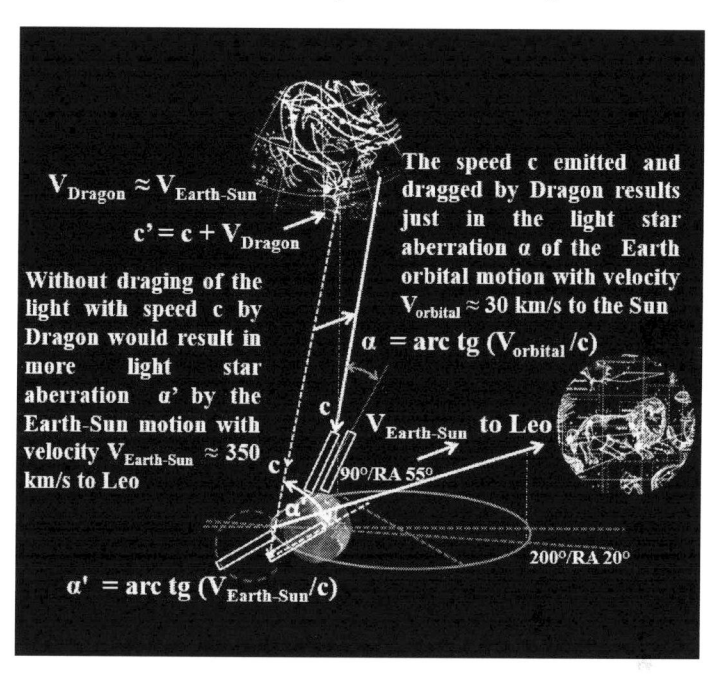

Fig. 59 – The light emitted by the Dragon's constellation is also dragged by Dragon velocity in relation to cosmic microwave radiation or vacuum-space-ether. Adapted from clipart of *Editora Europa* and drawings of constellations of *Astronomia-Nova Carta Celeste* from Linneu Hoffmann, Latt-Mayer Artes Gráficas-AGGS Indústrias Gráficas, Rio de Janeiro, 1973.

Velocity on Relationship to the Cosmic Microwave Radiation of Background or Vacuum-Space-Ether – For the laser light experiment was admitted to the Earth-Sun-Laser system the velocity of 350 km/s in relation to the cosmic microwave radiation or the vacuum-space-ether in which the velocity of the constellation of Leo in relation this reference is $V_{Leo} = 0$. However, Leo could be at any other velocity. It has been observed that the planets revolve around them, as well as solar systems and galaxies, but the entire universe is static in relation to cosmic microwave radiation or vacuum-space-ether. So, admitting an starting point of velocity = 0 for the constellation of Leo, could present a reference of c' speed of light in relation to cosmic microwave radiation or vacuum-space-ether. To c' = c_v + V_{Source} , in which c_v is the variable velocity of electromagnetic radiation as a function of velocity V_{Source} from the emitting source of the Earth-Sun-Laser System, V_{ESL} = 350 km/s in relation to the constellation of Leo, is obtained c' = c + V_{ESL} , where c = 300,000 km/s in relation to source of V_{ESL} velocity , so c' = 300,000 km/s + 350 km/s = 350,000 km/s.

Constant Speed of Electromagnetic Radiation in the Cosmic Microwave Radiation or Vacuum-Space-Ether – By uniting the laser light experiments of 1995, 2002 and CERN of 1964, the smaller the velocity of the emitting source, the higher the velocity of light c_v measured by the observer at the source. Then, for the Leo-Earth-Sun-Laser-meson $\pi°$-gamma-rays System, whether motionless in the cosmic microwave radiation or vacuum-space-ether, the speed of laser light or gamma rays measured by the observer in the Earth source would be c' = $c_{(ESL=0)}$ = 300,350 km/s. With the velocity of the V_0 source = 0, the laser light shows the highest velocity in relation to cosmic microwave radiation or vacuum-space-ether, i.e. $c_{(V=0)}$ = c'.

With velocity V_{ESL} = 350 km/s of the Earth-Sun-Laser system in relation to cosmic microwave or vacuum-space-ether, the observer on Earth measures the speed of light $c_{(ESL=350)}$ = 300,000 km/s, and speed c' in relation this reference is 300,350 km/s. And if the Earth-Sun-Laser System and the mesons $\pi°$ were with 300,350 km/s, the observer on Earth-Sun-Laser would measure the gamma rays or the laser light with velocity $c_{(ESL=300,350)}$ = zero, and c' = 300,350 km/s. Figure 60.

The Pre-Established Physics and the Quantum Impulse

constant light velocity c' relative to space-vacuum

$V_0 = 0$

light velocity $c_{V=0}$ relative to Earth source velocity if $V_0 = 0$

the velocity c of the light measured in the source is smaller than larger the velocity $V_{Earth\text{-}Sun\text{-}Laser}$ of the source

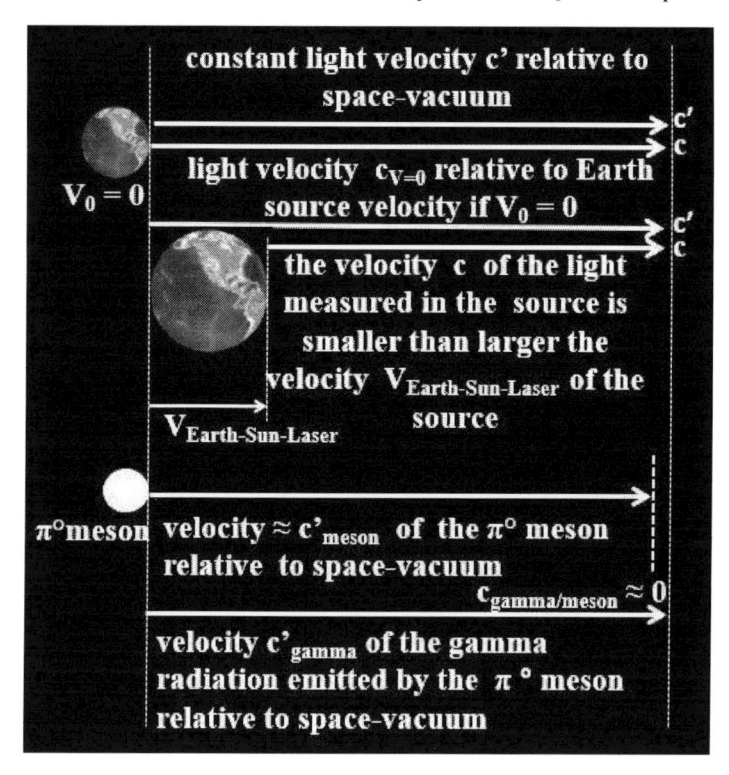

$V_{Earth\text{-}Sun\text{-}Laser}$

$\pi°$meson **velocity ≈ c'$_{meson}$ of the $\pi°$ meson relative to space-vacuum**

$c_{gamma/meson} \approx 0$

velocity c'$_{gamma}$ of the gamma radiation emitted by the $\pi°$ meson relative to space-vacuum

Fig. 60 – By the CERN experiment of 1964 and of 1995 and 2002 laser light, the smaller the source emitting's velocity, the higher the c_v velocity of light measured by the observer at the source. As the velocity of the mesons $\pi°$ was similar to that of the gamma rays, then the velocity of the gamma rays was zero in relation to the mesons $\pi°$, that is, different from speed of light c by the theory of relativity. Earth clipart from *Editora Europa*.

Of course, there are by 1995, 2002 and CERN laser light experiments of 1964 one constant speed c' of electromagnetic radiation in relation to cosmic microwave radiation or vacuum-space-ether propagating electromagnetic waves.

And that it was admitted as constant speed c by the electromagnetic wave theory, theories of the ether and relativity of Einstein, even considered with the speed of light c constant in relation to the vacuum-space-ether in all directions of the Michelson-Morley experiment by the wave theory.

According to Pauli, all the laws of nature should be derived from a simple axiom and the credit he gave to Einstein whereby it was necessary to assume that the speed of light c was independent of the motion of the source, with the value c constant in the vacuum-space. By the theory of relativity any ray of light moves in a possible stationary system or any reference system with a certain speed c, whether it is the ray emitted by the eventual stationary system or by a moving source. And to always keep this measure of speed of light c in all reference systems, there would be space contraction and time dilation in all systems of reference by Einstein's theory of relativity.

However, the electromagnetic wave theory, sound, seismic and maritime waves also present their independent speeds of the source speed. But, these 1995 and 2002 laser light experiments and the CERN experiment of 1964 showed that there is a constant speed c' independent of the velocity of the source in relation to the cosmic microwave radiation or vacuum-space-ether, but the speed c_v of the electromagnetic radiation measured by the observer in relation to the source depends on this velocity of the source, and the smaller the velocity of the emitting source, the higher the velocity of light c_v measured by the observer with respect to that source. No space contraction or dilation time, because is the path of light showing its actual velocity, proven by diagonal light path laser on the transverse. By these experiments of laser light of 1995, 2002 and correlated to the CERN of 1964, the elemental impulse in relation to the cosmic microwave radiation or vacuum-space-ether in the direction of the motion of the source has the speed $c' = c_v + V_{ESL}$, and to $V_0 = 0$, $(c_0 + V_0) = c_0$. Figure 61.

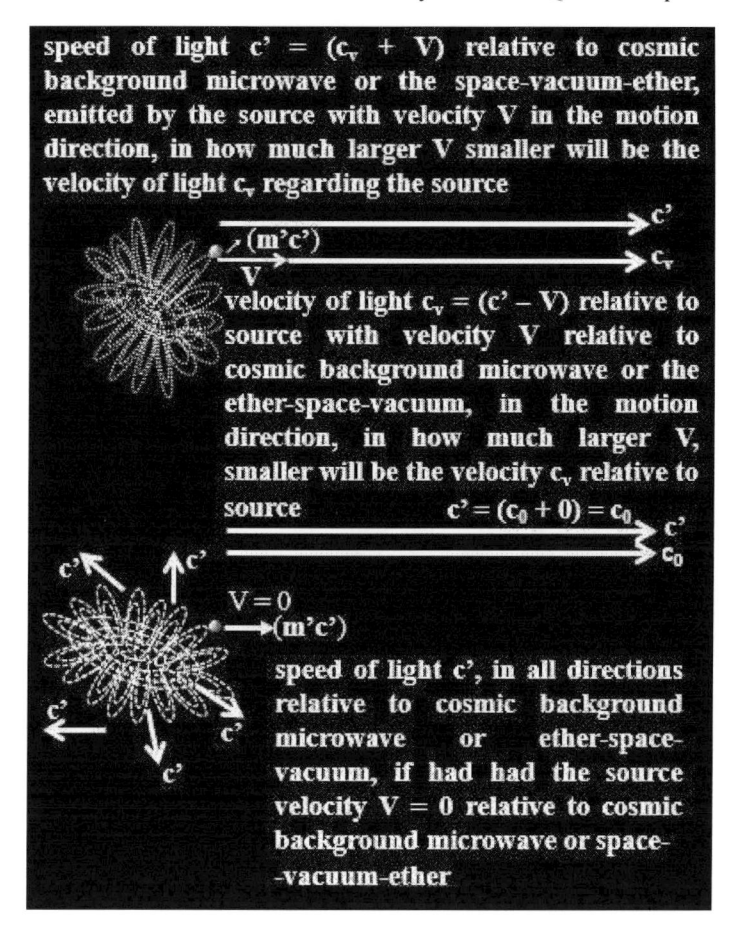

> **speed of light c' = (c_r + V) relative to cosmic background microwave or the space-vacuum-ether, emitted by the source with velocity V in the motion direction, in how much larger V smaller will be the velocity of light c_r regarding the source**
>
> $\longrightarrow$ c'
> (m'c')
> V $\longrightarrow$ c_r
>
> **velocity of light c_r = (c' – V) relative to source with velocity V relative to cosmic background microwave or the ether-space-vacuum, in the motion direction, in how much larger V, smaller will be the velocity c_r relative to source**
>
> c' = (c_0 + 0) = c_0
> $\longrightarrow$ c'
> $\longrightarrow$ c_0
>
> V = 0
> $\longrightarrow$ (m'c')
>
> **speed of light c', in all directions relative to cosmic background microwave or ether-space-vacuum, if had had the source velocity V = 0 relative to cosmic background microwave or space-vacuum-ether**

FIg. 61 – The figure shows the light emitted with the velocity c' by source with velocity V and V_0 immovable in relation to cosmic microwave radiation or vacuum-space-ether, and the velocity of light c_v emitted in relation to the emitting source.

The impulse m'c' which forms the light m'c' x c' = m'c'^{2} comes out of the laser in the direction of the motion of the source, at velocity V_{ESL}, with the velocity [c' = (c + V_{ESL})], relative to the vacuum-space-ether or cosmic microwave radiation, with the energy [m'c' x c'] = [m'c' x (c + V_{ESL})] = [(m'c' x c) + (m'c' x V_{ESL})], preserving the conservation of momentum and energy. And the laser experiment showed that on the laser output at 90° in relation to the source motion, the light impulse m'c' comes out of the laser diagonally, composing vector velocity with c and V_{ESL}, with velocity c'$_{diagonal}$ = $(c^2 + V_{ESL}^2)^{1/2}$ in relation to cosmic microwave radiation or vacuum-space-ether, where the m'c' impulse is dragged with velocity V_{ESL}. Figure 62.

J. P. Hansen

V_{ESL} = Earth-Sun-Laser relative to space-vacuum-ether in the direction to the constellation Leo
c' = speed of light relative to space-vacuum-ether
c = speed of light relative to V_{ESL}

$$c'_{transversal} = (c^2 + V_{ESL}^2)^{1/2}$$

In the experiment laser, in the exit of the laser light in 90°, the elementary impulses m'c' which form the electron runninged through the course in the diagonal, dragged by the electrons of the source Earth-Sun-Laser system with the velocity V_{ESL} in the direction to the constellation Leo, resulting in the velocity $c'_{diagonal} = (c^2 + V_{ESL}^2)^{1/2}$, and in the direction to Leo: $c' = (c + V_{ESL})$ relative to space-vacuum-ether.

For the source velocity V: $c' = (c_v + V)$ relative to space-vacuum-ether, where the velocity of light c_v is variable conform the source velocity V.

velocity of light c_v relative to source of velocity V

source velocity V relative to the space-vacuum-ether

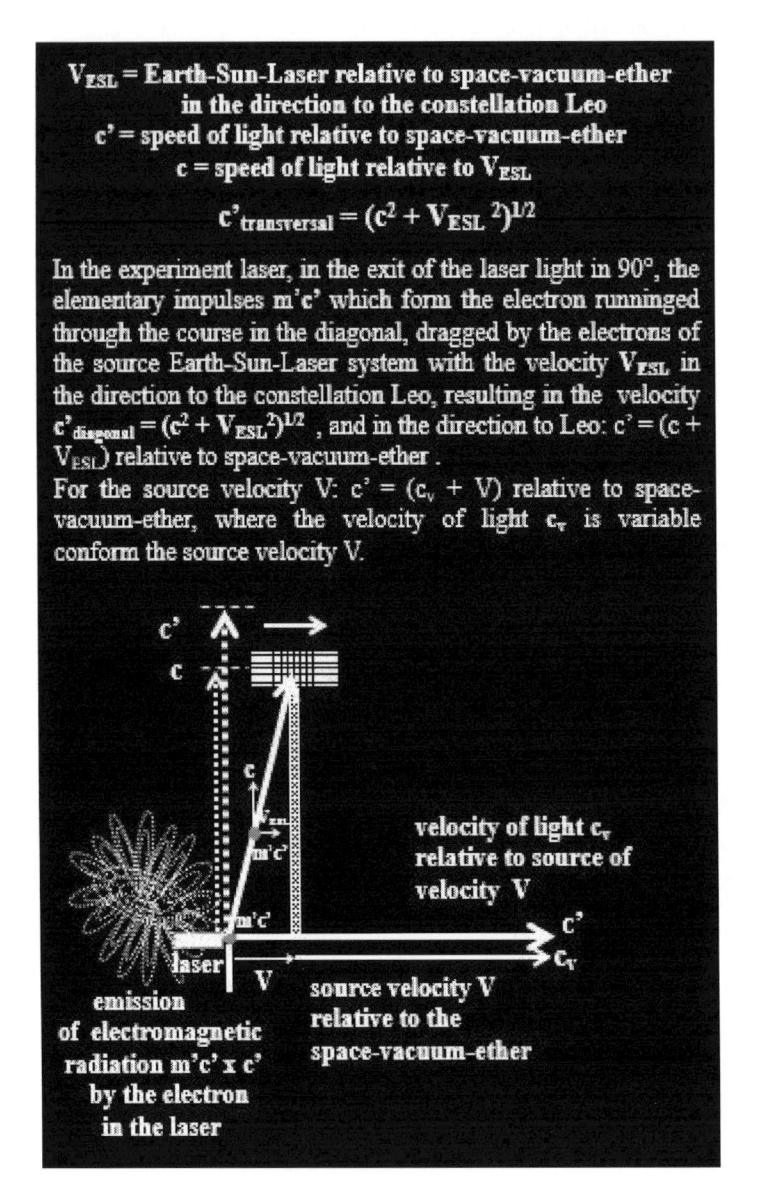

emission of electromagnetic radiation m'c' x c' by the electron in the laser

Fig. 62 – The figure shows the light that comes out on the diagonal of the laser at 90° of the motion of the source to Leo, dragged by the velocity of the source.

Einstein and the imposed Physics, despite considering a constant speed of light c in relation to vacuum-space, did not have the perception that the light (quantum impulse) has a constant velocity $c' = (c + V_{source})$ in relation to vacuum-space-ether, and that when being carried by the source is that it results in the measure of velocity c in relation to the source (Earth-Solar system). In the case of lower source velocities there is less mass and insaturation of impulses in the particles and in their fields, then the quantum impulse being emitted with higher velocity relative to the source, but maintaining constant the velocity of light c' in relation to the vacuum-space-ether. While at high velocities of the source there is a saturation of quantum impulses in the particles and in their fields, as if clinging, and then the velocity of light being lower compared to that source. In the CERN experiment of 1964 and the laser light in 1995 and 2002, it was proven that the velocity of the gamma rays, relative to the emitting source of the meson with close velocity of light, was almost equal to the velocity of the meson, that is, almost zero in relation to the meson. The source carries the light, so that in relation to the vacuum-space-ether the velocity of light c' is always constant, but the higher the velocity of the source the lower the velocity of light relative to that source. Figure 63.

If $V_0 = 0$, $c = c_0 = 300.350$ km/s, in relation to space-vacuum-ether, determined in the direction of the constellation of Leo

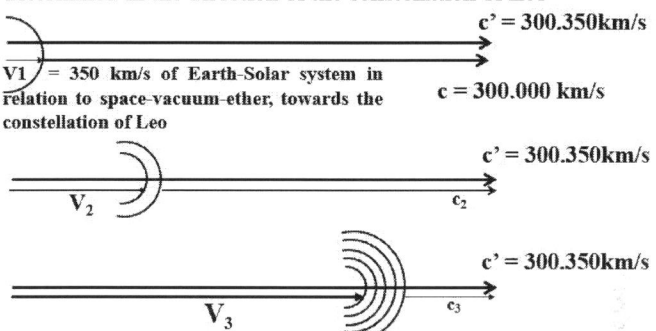

$c' = 300.350$ km/s

V_1 = 350 km/s of Earth-Solar system in relation to space-vacuum-ether, towards the constellation of Leo

$c = 300.000$ km/s

$c' = 300.350$ km/s

V_2, c_2

$c' = 300.350$ km/s

V_3, c_3

The higher the velocity of the emitting source (which always drags the particle) in relation to the space-vacuum-ether, the greater is the mass and tension in the particle field that emits the quantum radiation, resulting in less velocity of the radiation relative to that source.

Fig. 63 – The source at rest or in velocity carries the light (quantum impulse). And when the light is emitted by the source, the lower the velocity of the source the greater is the velocity of light in relation to the source,

And the laser experiment proved irrefutably, for the laser at 90° from the direction to the constellation of Leo, that the light comes out diagonally, proving also irrefutably that the light is carried by the inertia of the source with the laser, that is, the velocity of light is variable in relation to the vacuum-space-ether or cosmic microwave radiation, which is contrary to the theory of relativity. That is, the light is carried by the particles, proving irrefutably that

J. P. Hansen

the particles are formed of quantum impulse of the immortal creation of Ederlyi. Figure 64.

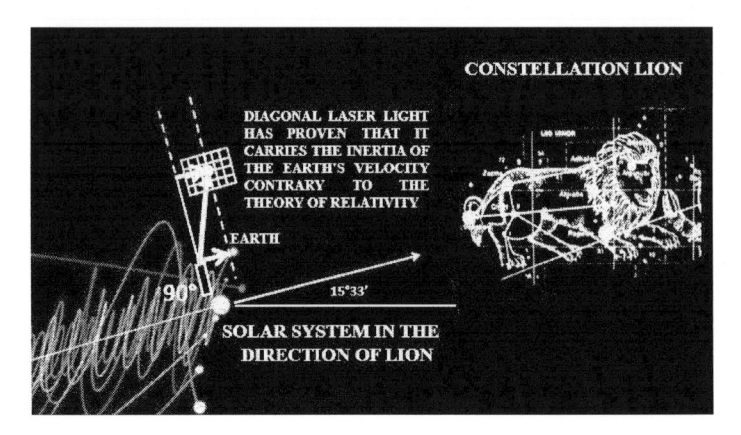

Fig. 64 – In the laser experiment placed in 90° of the direction to the constellation of Leo showed that the light comes out diagonally, therefore showing that is variable the velocity of light which is contrary to the theory of relativity, proving that the light is carried by the inertia of the source, that is, the light is carried by the particles, proving irrefutably that the particles are formed of quantum impulse of the immortal creation of Ederlyi. Adapted from drawings www.pinterest.com, Google Images and *Astronomia-Nova Carta Celeste from Linneu Hoffmann, Latt-Mayer Artes Gráficas-AGGS Indústrias Gráficas,* Rio de Janeiro, 1973.

Kilometer Ruler in the Earth-Sun-Laser Reference System in Cosmic Microwave Radiation or Vacuum-Space-Ether in the Direction of the Constellation of Leo – Figure 65 shows a visualization in the cosmic microwave radiation or vacuum-space-ether of a kilometer ruler over this laser experiment in which for the observer, in the reference system of the source with velocity V_{ESL} = 350 km/s in direction to constellation Leo relative to the vacuum-space-ether or cosmic microwave radiation, it measures the velocity of light at 90° to the motion of the source with c = $[(A_1A_2)/1\,seg] \approx$ 300,000 km/s. But, for an observer in the reference system of cosmic microwave radiation or vacuum-space-ether, the velocity of light is measured in diagonal as $c'_{diagonal}$ = $(c^2 + V^2_{ESL})^{1/2} \approx$ 300,000.2042 km/s with V_{ESL} = 350 km/s. In the direction of the motion, the observer in the reference system of the source measures the speed of light c = $[(A_1A_3)/1\,sec] \approx$ 300,000 km/s, and in the reference system of cosmic microwave radiation or vacuum-space-ether the observer measures c' = $[(P_1P_6)/1\,sec] \approx$ 300,350 km/s. By relativity, in the two reference systems is measured the speed of light c in which in the moving frame there would be contraction and time dilation to keep c constant, but this laser experiment showed that the velocity of light is different in the two reference systems.

Kilometers Ruler in the Earth-Sun-Laser System in the Ether-Space-Vacuum or Cosmic Microwave Radiation in the Direction of Lion

$c'_{diagonal} = (c^2 + V_{ESL}^2)^{1/2} = 300.000,2042$ km/s

P_3 — 300×10^3 km

P_4

A_2 **Source Reference System**

speed of light $c = \underline{A_1 A_2}/1s$ = 300 000 km/s, perpendicular to the motion of the source velocity V_{ESL} source, measured by the observer at the source

speed of light $c = (c' - V) = \underline{A_1 A_3}/1s = 300\,000$ km/s, in the direction of the source motion with velocity V_{ESL}, measured by observer at the source

200×10^3

100×10^3

P_1 P_2 A_1 100×10^3 200×10^3 300×10^3 km A_3 P_6

$V_{ESL} = 350$ km/s

$c' = 300\,350$ km/s

speed of light c' relative to vacuum-space-ether or cosmic microwave radiation

V_{ESL} velocity of the Earth-Sun-Laser relative to vacuum-space-ether or cosmic microwave radiation

Vacuum-Space-Ether or Cosmic Microwave Radiation Reference System

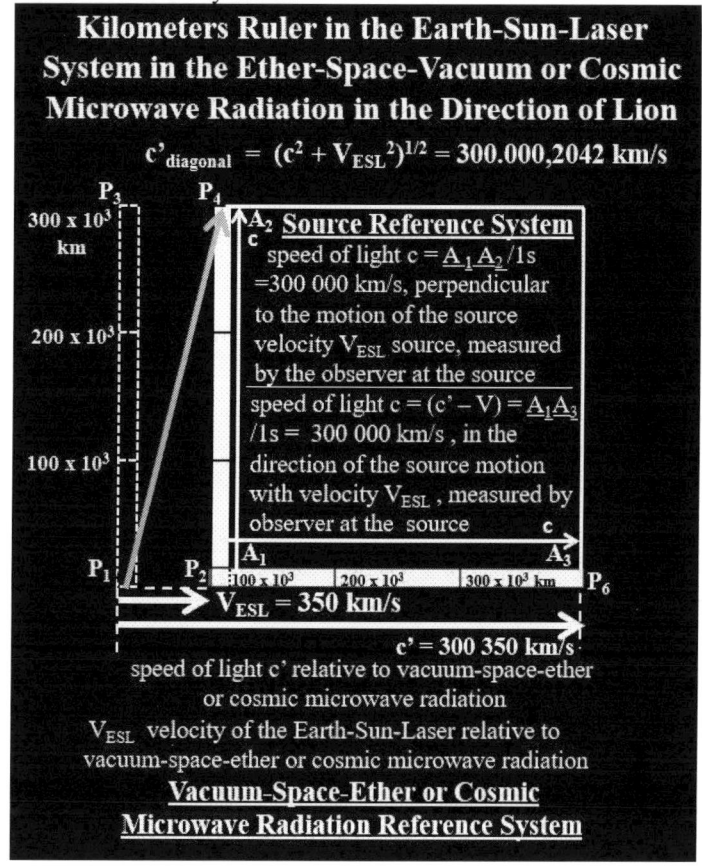

Fig. 65 – The speed of light measurements are considered as the constant c in all reference systems through space contraction and dilation time, but by the laser light experiment the observer at the source measures the velocity of light c in relation to the source, $c'_{diagonal}$ in the diagonal and c' in the source motion direction in relation to the cosmic microwave radiation or vacuum-space-ether.

This laser experiment showed that the light comes out of the laser diagonally P_1P_4 with velocity $c'_{diagonal} = (c^2 + V_{ESL}^2)^{1/2} = 300,000.2042$ km/s in relation to cosmic microwave radiation or vacuum-space-ether, carried by the emitting source with velocity $V_{ESL} = 350$ km/s in relation to cosmic microwave radiation or vacuum-space-ether in the direction of the constellation of Leo. And for the observer in the reference system of the emitting source

Earth-Sun-Laser measures the speed of light as $A_1A_2/s = 300{,}000$ km/s.

And in the direction of the motion of the emitting Earth-Sun-Laser with 350 km/s, by the laser experiment the light travels in the cosmic microwave radiation or vacuum-space-ether the distance of $P_1P_6 = 300{,}350$ km in 1 second, and being c' = (c + V_{ESL}), is measured by the observer in the reference of the emitting Earth-Sun-Laser source, of velocity 350 km/with respect to cosmic microwave radiation or vacuum-space-ether in the direction of Leo, the speed of light c = 300,000 km/s running the distance A_1A_3.

The Figure 66 shows that, as the laser experiment, the speed of light c, measured by the observer in the reference system of the moving earth, in a interferometer, is obtained by the distance of the arm of the interferometer by the time t, D/t. However, the actual measure of the speed of light is the diagonal OA' = D' in time t, that is, D'/t = c_d'. So the speed of light c is not measured because there is space contraction and dilation time, but solely by the diagonal path obtained by the velocity of Earth that moves in space in direction to Leo.

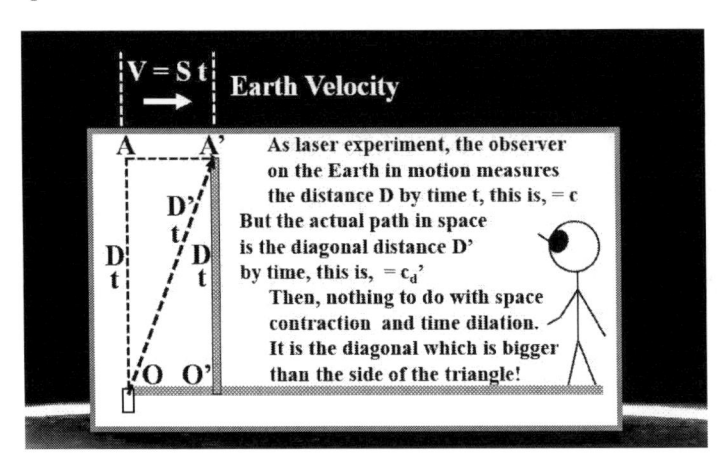

Fig. 66 – As the laser experiment, the speed of light c is obtained by the observer, on Earth that is in motion, by the distance D of the interferometer at time t. However, the distance travelled by the light is the diagonal D' in the time t. So, there's nothing to do with space contraction and time dilation. It's the diagonal that's bigger than the side of the triangle! Clipart from *Editora Europa*.

Particle-Wave, Electromagnetic Wave and the Elementary Quantum of Energy – In the oscillation of electrical charges as the electron is emitted the quantum, following the wave equation of that oscillation, in which the axis of propagation is represented the emission frequency and or wavelength, and on the vertical axis the component of the oscillation. And the maximum oscillation, the amplitude, is increased in quadratic function of the wave energy. Also, the amplitude could be matched to all the energy of the waves, on an oscilloscope, because the frequency is independent of the excursions of the oscillator.

In 1907, Planck had asserted that the significance of Einstein's quantum of action or quantum light was only in the emitter and radiation absorber, because in the space-vacuum was the electromagnetic wave described by Maxwell's equations. And for Einstein in 1909 the quanta were punctual that moved independently with energy $h\mathcal{V}$ with the speed of light c. In 1911 stated that the quanta should have a provisional concept until it is reconcilable with the wave theory.

And in 1926, as Einstein the waves were to indicate the path to the quanta punctual $h\mathcal{V}$ which were propagated with the speed of electromagnetic radiation c. And in 1928, Bohr presented the principle of complementarity of light as a spread now as a wave, now as a particle.

However, Planck's quantum of action is covered by the physical reality of the oscillator, which for each acceleration must be an electromagnetic emission according to the Maxwell equation, that is, an elementary quantum m'c x c = h' while the quantum $h\mathcal{V}$ it is a compact of h' quantities for each emission. And according to Einstein, if as a whole quantum **(h$\mathcal{V}$)** is propagated with speed c, it would be mc^2 energy x c = mc^3 no physical significance!

But, the oscillator emitting the quanta elementary h' results in a wave, in which m'c is that it has the speed c that travels with speed c, resulting in the energy m'c x c = h'.

By the principle of complementarity presented by Bohr, it shows only terms of connection, intending to explain a fact, but without achieving the physical reality of the elementary quantum of electromagnetic radiation.

By the theory of quantum impulse with speed c in relation of Earth source, it is the elementary quantum m'c^2 = h' that forms the quantum $h\mathcal{V} = |\mathcal{V}|$ h', in which h' is emitted as oscillating which results in a wave in the propagates by vacuum-space-ether. The light has polarization plans and also vibrating in all directions emitted by the electron, resulting in the impulses of light that form the wave. Figure 67.

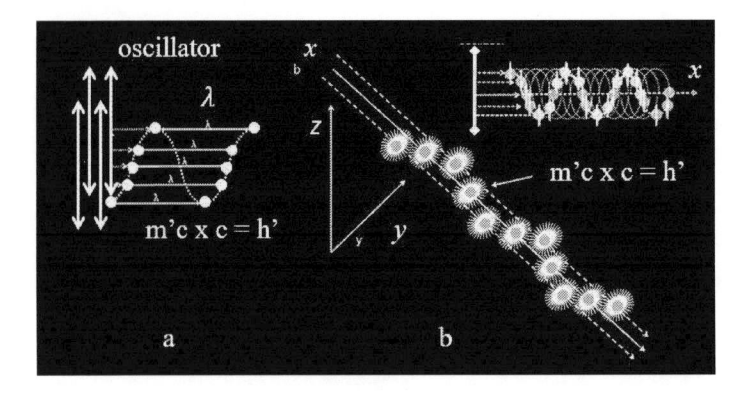

Fig. 67 – (a) The elementary quantum h' is emitted as oscillating. (b) And resulting in a wave in the propagation by space-vacuum-ether.

By quantum elementary impulse theory, the impulses m'c with speed c, m'c x c = h' are emitted with spacing $\lambda_\mathcal{V}$ on the frequency $\mathcal{V}$ of h'. In the case of waves in water, they are formed in a complex way by the applied energy, as climatic phenomena, through the motion of the units of the molecules of the hydrogen and oxygen atoms that makes up the water. In the case of the electromagnetic wave, it is also complex the way the h' units, which form the radiation and the electromagnetic field, are propagated and resulting in the oscillation of the electromagnetic field formed of h' in the frequency $\mathcal{V}$ or quantity $|\mathcal{V}|$. Figure 68 it shows the violet radiation by Planck's theory of radiation and Einstein and the theory of elementary impulse.

GREEN QUANTUM $h\mathcal{V}_{green}$
Green Quantum Energy = $\mathcal{E}_{green} = h\mathcal{V}_{green}$
h = Planck's constant = 6.62×10^{-27} (erg s)
  "action" without more physical meaning
$\mathcal{V}_{green}$ = frequency = 5.6×10^{14} vibrations/s
$\mathcal{E}_{green} = h\mathcal{V}_{green} = 3.7 \times 10^{-13}$ erg

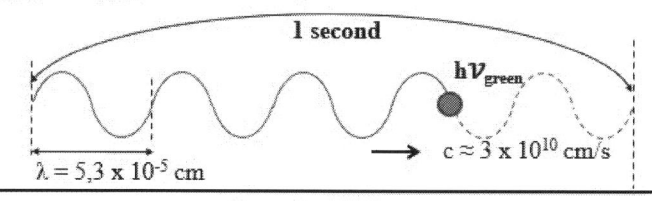

```
                        1 second
                                    hV_green
    λ = 5,3 x 10^-5 cm           c ≈ 3 x 10^10 cm/s
```

GREEN QUANTUM $h'|\mathcal{V}_{green}|$

$h' = m' \times c \times c$
                    1 second
$\mathcal{E}_{green} = h'|\mathcal{V}green|$
● h'   ● h'   ● h'   ● h'   ● h'
$\lambda = 4.9 \times 10^{-5}$ cm $c \approx 3 \times 10^{10}$ cm/s

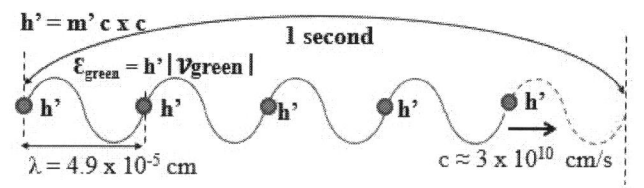

$\mathcal{E}_{green}$ (erg)/s = [h (erg.s) x $\mathcal{V}_{green}$] / s = h'(erg) | $\mathcal{V}_{green}$ |
h' = elementar quantum, this is, for $\mathcal{V} = 1 : \mathcal{E}_{(\mathcal{V}=1)} = \varepsilon = h'$
h' (erg) | $\mathcal{V}_{green}$ | = | $\mathcal{V}_{green}$ | m' c x c = m' c²
h' = 6.62×10^{-27} erg ; $c \approx 3 \times 10^{10}$ cm/s ; m' = 0.73×10^{-47} g
$\mathcal{V}_{green}$ = frequency = 5.6×10^{14} h'/s
$\mathcal{E}_{green}$ = h' x | $\mathcal{V}_{green}$| = 6.62×10^{-27} erg x 5.6×10^{14} = 3.7×10^{-13} erg

Fig. 68 – By quantum mechanics and Einstein's relativity would be emitted a single quantum of frequency $\mathcal{V}$, that is, a single emission of the oscillator! However, by electrodynamics the accelerated electric charge emits electromagnetic radiation, so for each acceleration there must be a quantum emission, which is the quantum h', the elementary quantum corresponding to each oscillation.

In the harmonic oscillator there is a force $F = k \times d$ of restoration proportional to the displacement d where k is a constant balancing of the coordinates $\emptyset$ (x, y, z) in time t: $F = m \times a = m[d^2\emptyset(t)/dt^2] = m[\omega^2\emptyset(t)]$, where ω = angular frequency.
Therefore, the oscillator that emits the energy of the electromagnetic radiation is accelerated in each vibration, and should emit an energy for each vibration. In fact, for two types of quantum $\mathcal{E} = h\mathcal{V}$:
$\mathcal{E}_{(\mathcal{V}=1)} = h\mathcal{V}_{(\mathcal{V}=1)}$ in which $\mathcal{V} = 1$ cycle/sec, so $\mathcal{E}_{(\mathcal{V}=1)} = \varepsilon = |h| \times |1|$ [erg]
$\mathcal{E}_{(\mathcal{V}=1000)} = h\mathcal{V}_{(\mathcal{V}=1000)}$ in which $\mathcal{V} = 1000$ cycles/sec. In this case what should occur in the first cycle of quantum $\mathcal{E} = h\mathcal{V}_{(\mathcal{V}=1000)}$?

In the first cycle the energy would be 1/1000 of the total energy of the compact quantum $\mathcal{E}_{(\nu=1000)} = (h\mathcal{V}_{(\nu=1000)})$, and the time of the first cycle would be 0.001 s.

$(\mathcal{E}_{(\nu=1000)}/1000)$ [erg] = $\{(h\mathcal{V}_{(\nu=1000)})/1000\}$ [erg] = $\{h$ [erg s]$/1000\} \times \{1\text{cycle}/0.001$ [s]$\}$

That is, the constant h appears divided into 1000 parts!
Then it is obtained:

$(\mathcal{E}_{(\nu=1000)}/1000)$ [erg] = h [erg] $|\mathcal{V}_{1\text{ cycle}}|$ = h [erg] $\times$ [1] = [1] h [erg], which is exactly the energy of the quantum of 1 cycle frequency:

$\mathcal{E}_{(\nu=1)} = \varepsilon = h\mathcal{V}_{(\nu=1)} = |h|$ [erg] $\times |1| = |h|$ [erg], where $\mathcal{V}_{(\nu=1)}$ = 1 cycle/s

And by the theory of elementary quantum impulse, $|h|$ [erg] = h' [erg]

ε [erg]/1 [s] = h [erg s]/[s] $\times \mathcal{V}_{(\nu=1)}$ = h' [erg] $\times$ [1 cycle/s]
So, ε [erg] = $|1|$ h' [erg]
And to $\mathcal{E}_{(\nu=1000)} = h\mathcal{V}_{(\nu=1000)}$,

$\mathcal{E}_{(\nu=1000)}$ = h [erg s] $\times 1000$ [cycle/s] = 1000 h' [erg]

That is, to form this quantum $\mathcal{E}_{(\nu=1000)} = h\mathcal{V}_{(\nu=1000)}$ 1000 cycles/sec, 1000 emissions of h', i.e. 1000 ε. And in the electromagnetic emission oscillator the acceleration occurs at each oscillation, so the quantum must be emitted at each oscillation.

In 1972 Hobson presented the quantum h_e equal to h', with quantum and astronomical applications.

In 1974, Csiky Jr. had presented the fundamental aspect of the quantum harmonic oscillator for each oscillation, based on the acceleration force of the oscillator similar to the pendulum and the restoration force.

And between 2000 and 2010 Worsley presented the quantum of quintessence of the harmonic oscillator E = nh where n represents the number of the quanta of Planck's quintessence contained within a quantum per unit of time, which is the quantum h' of the quantum elementary impulse of Erdelyi and Hobson. And in 2011 A. Nikolov also observed that the constant h of Planck should be a constant of energy of an oscillation, and named h = E_1 = E_0 = h* (h with asterisk) as power unit, for an oscillation, which is the elementary quantum h' of Erdelyi.

CXXX. SPIN, HELIX, PARTICLE CHARGE AND QUANTUM IMPULSE THEORY – **1928 to 1969**

By quantum mechanics, fundamental particles have dimensions as a point that rotates only mathematically on themselves, that is, the spin that can rotate to the right and to the left and be positioned as up or down. In the direction of the particle motion and as the right hand is defined as helicity + 1 and in the direction of the motion and the left hand is defined as helicity – 1. But, as the observer exceeds the particle motion, there is a right-hand or left-hand reversal. However, for particles considered without mass, such as the photon, the observer could not exceed speed of light, so the helicity is permanent in this case. But, the chirality hand-right or hand-left of a particle is absolute, without depending on the reference system. However, most physicists refer to positive helicity particles as right hand and negative helicity particles as left-handers, which create conflict with the notion of chirality-hand-left and chirality-right-hand. Then, the current Physics shows indications of the concept of spin and helicity are in divergence between their followers.

According to Griffiths, *"The helicity of massless particles is not an artifact of the observer's reference, but a fixed and fundamental property"*.

Thus, with zero mass and divergences in the current Physics, the origin of the spin remains unknown by quantum mechanics.

By the elementary impulse the chirality is a fundamental property for all particles that have spin, as it is the actual intrinsic helicity of the spin, regardless of the observer that determines the particle or antiparticle, therefore its charge, and explains the attraction and repulsion between charges. By the theory of elementary impulse the particles with negative charge have a single helicity and the positive charge particles have the opposite helicity. But, particles can present a position in space as up and down, but with a same helicity or chirality. By erroneous concepts of current Physics, incoherent conventions are created in an attempt to justify these concepts, while by the greatest physical conceptualization of quantum impulse results in precise understanding. Figure 69.

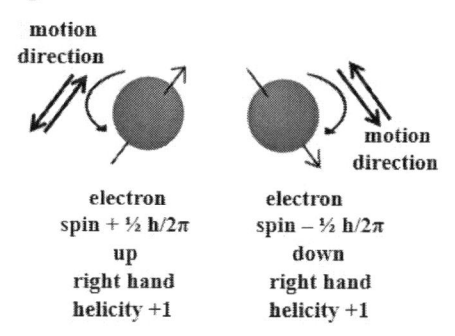

electron
spin $+ \frac{1}{2} h/2\pi$
up
right hand
helicity +1

electron
spin $- \frac{1}{2} h/2\pi$
down
right hand
helicity +1

Fig. 69 – Position of charge in space as up and down, and helicity + 1 which can be defined by the right-hand electron that follows different directions of motion in space. By the theory of quantum impulse the helicity is independent of the direction of the motion, because it is the particle that always has the helicity in one of the senses (clockwise or counterclockwise) for any reference.

By quantum mechanics, the fundamental particles are without dimensions, so they should be without rotating physically around them. Because the neutrino was close to the speed of light, the observer could not overtake it and see it moving away with helicity or chirality contrary from the point of view of this observer. So this is proof that there is only a single intrinsic helicity or chirality for each particle, whether the particle is moving with the speed of light, at a lower velocity or at rest.

According to the theory of elementary impulse, all particles with spin present only one intrinsic helicity. In the case of electrons there is a single intrinsic helicity, but with two different positions in the atom. At point P an observer sees a spin as clockwise and the other counterclockwise, but the intrinsic helicity is the same in both electrons. The electron has an intrinsic helicity and the positron has the other intrinsic helicity. In the case the electron has the intrinsic helicity of right hand-clockwise and the positron has the intrinsic helicity left counterclockwise. Figure 70.

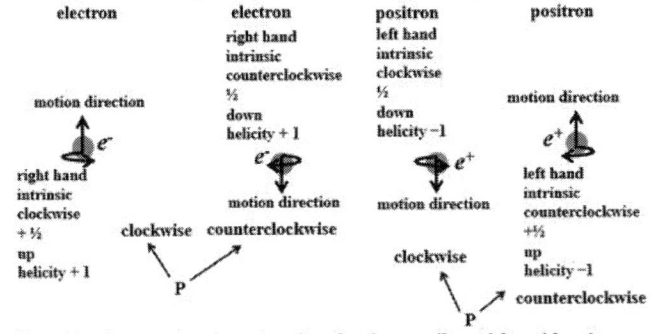

According to quantum elementary impulse theory, all particles with spin present one intrinsic helicity or chirality. For electrons and positrons there is an intrinsic helicity, but with two different positions in the atom. At point P an observer sees a spin as a clockwise and the other spin as a counterclockwise, but the intrinsic helicity is the same. The electron has an intrinsic helicity and the positron has the other intrinsic helicity. In this case, the electron has the intrinsic helicity of right-hand clockwise and the positron has the intrinsic helicity of left-hand counterclockwise.

Fig. 70 – Intrinsic helicity of the electron and the positron.

The different helicities of the particles is the cause of the attraction between negative and positive charges. The resulting vectors and the force lines of the field, of spins and helicities of the electron and positron crossing between them causing the attraction. Figure 71.

The Pre-Established Physics and the Quantum Impulse

By the theory of quantum impulse the different helicities of particles is the cause of the attraction between negative and positive charges. The resulting vectors and the force lines of the electromagnetic field of the electron, spins and helicity, of the electron and positron, formed of m'c cross between them causing the attraction.

Fig. 71 – Attraction between charges of different intrinsic helicities.

The same helicity causes the repulsion between electrons. The vectors and force lines of the field resulting from the spins and helicyties of the electrons converge in parallel and repel each other. Two clockwise hand-right electrons in opposite directions repels less than in the same direction, so when in opposite directions there is a convergency of the power lines. Figure 72.

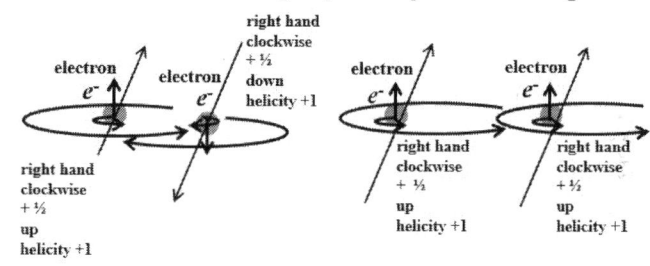

The same helicity causes the repulsion between electrons. The vectors and force lines resulting from the spins and helicities, which form the electrons, convert parallels and repel each other. Two-way right hand-clockwise in opposite directions repel less than in the same direction. When in opposite directions there is some convergence of the force lines of m'c. In the same direction of the electron movement there is repulsion between lines of force of m'c that form the electrons.

Fig. 72 – Repulsion between charges of intrinsic helicities equals.

In the decay of the antipion π in rest that has spin 0 in the muon (μ) and antineutrino: $\pi \rightarrow \mu + \nu_\mu$, the muon and the antineutrino go in opposite directions and their spins must be aligned in opposite directions. As the antineutrino is always hand-right and the helicity of the antimuon is hand-left, by the elementary impulse the muon can only have chirality hand-right.

In fact, in the decay of the immovable pion of spin 0 in antimuon and neutrino of the muon: $\pi^+ \to \mu + v_\mu$ the antimuon as it is always left hand, the neutrino is hand-left.

In the decay of the antipion π in rest spin 0 in electron and electron antineutrino: $\pi \to \mu + v_\mu$ the antineutrino is always a right hand. As by the theory of the elementary impulse the positron is always hand-left, then the electron should only be hand-right.

And in the decay of the pion π° in rest spin 0 in photons γ: $\pi^\circ \to \gamma + \gamma$. The antiparticle γ it's its own private γ. And are hand-left or hand-right helix representing circular or left-right polarization states.

By the elementary impulse protons are formed of positrons-electrons with helicity opposite the electron-positron, and produce in shock of particles, electrons, positrons, neutrinos and gamma rays, as well as electron-positron produce gamma rays, that is, formed of m'c x c.

By the elementary impulse particles of positive and negative charges, i.e. particles and antiparticles, are coupled in the same form, being that the helix to the right or to the left, or positive or negative, is that it results in one or another particle-antiparticle.

In 1932, Heisenberger observed the independence of charges in the nuclei and which maintained a symmetry which resulted in spins of similar properties, i.e. isotropic spins. In 1936, Cassen and Condon observed the independence of charges in the interaction proton-proton, neutron-neutron and neutron-proton. And in 1936, Bartlett presented a scattering experience proton-proton, neutron-neutron and neutron-proton, in which it had detected change of position of the spins between them, maintaining a symmetry that resulted in spins of similar properties, ie isotropic spins. The direction and position of lines of down-up attraction and the intrinsic helicity fundamentally define the particles.

CXXXI. ELECTRON STRUCTURE AND QUANTUM IMPULSE
1897 to 2005

On the electron and the particles is analyzed in 2005 according to P. Lancini and R. Mignani the current state of quantum Physics: *"All properties reported to interactions are well described in the structure of quantum electrodynamics. However, the values of the mass and electric charge of the electron considered in quantum electrodynamics are not intrinsic, but result from its interaction with the vacuum. As matter of fact, there is no model presently of the electron capable for a broad breadth for its behavior and characteristics in a convincing direction. The above facts are well-known generally accepted as a consequence of the inability to describe the classic electron or quantum mechanics. This leads to a contradictory view that the electron is observed as an electrically charged particle as a point, but endowed with an unknown internal structure responsible for its spin and magnetic momentum. This situation, already summarized by Fermi in 1932, was redeclared by A. O. Barut in 1990 in the following terms: "If a rotating particle, of spin, is not punctual, nor a three-dimensional solid. What can it be? What is the structure that may appear to be tested with electromagnetic fields as a point charge, as far as wave and spin properties are concerned to display a dimension of the order of Compton wavelength?".* And in 2005 as M. H. MacGrecgor: *"The electron still remains an enigmatic object".*

As Lancini and Mignani is presented that the origin of the spin of the electron is caused by the helical motion of the electron, but with 3 dimensions of time and it is proposed that the electron and lepton be without mass, moving in 6 dimensions of space-time, and complements that 6 dimensions is perhaps something more than a mere mathematical hypothesis. Other electron models present massless vacuum vortex, with space-time dimensions appearing and disappearing and rotations of the particle as sphere. Also, another electron model is presented as consisting of "electric wires" and "magnetic wires" that forks out of these "electric wires" at the boundary of the wrapper, ("shell"), of the electron or membrane. This membrane would involve the proton that would consist of a "tiny dandelion" of "wires" convert to a tiny sphere that would form heavier atoms and molecules. And these "wires" would be connected to all the atoms of the universe that would be tangled all over the middle through a signal of torque, that is, like light. Therefore, the photon of energy would only travel by the tension of the wire to the atom connected in another at the end. This proposition was that the matter would consist of a simple "wire" of closed ring in ample space. The friction generated at all points along the surface of the electron where the "electric and magnetic wires" branched would be measured as charge, that is, a physical interaction. So, this model of electron and interconnected atoms by "wires" moving in the empty space!

By this model in which the light would be a wave torque generated by a double "wire" would be explained:

-Why the light behaves like a particle and as a wave?

-Why the laws of Maxwell and Faraday, by which the electric and magnetic fields, that is,

the wires are 90° between them and the induction occurs?

-Why the light is sine?

-Why $c = \nu\lambda$ it's a constant?

-The physical interpretation of the amplitude by the vertical torque.

-Why the light travels so quickly?

-Why the light travels rectilinear even during refraction and diffraction?

-Entanglement.

-Simultaneous clockwise and counterclockwise.

-Why the electron has wave, particle "shell" and properties as cloud.

-Why the electron does not spiral into the nucleus, that is, its stability?

-The mystic leap of the quantum, of Bohr, for the atom would be a tiny heart perpetually
  pumping torque waves, with a whole number of wavelengths.

-Why the atom emits and absorbs light in discrete waves packets?

-Why the light has constant speed respectively of the motion of the source?

-The effect of P. L. Mössbauer of indentation in the emission

-Why the light travels like a wave, and comes out and comes as a particle?

-The explanation of the Mach principle.

So the time model of the electron would be discarded for immense reasons, among them: quantum mechanics has to present a physical interpretation of positive and negative; the field of the electron and particles is without mass, a concept, it is not a physical object. Although Einstein considers that a variation of magnetic field results in emission of radiation, i.e. energy, assigning to the field an energy deposit, and stating that energy has mass, that mass represents energy, and that there is no essential distinction any between mass and energy, for motionless radiation, then the rest photon becomes without mass, for by relativity $m_{photon} = m_0/[(1 - (v^2/c^2)]^{1/2}$ and for $v = c$, $m_0 = m_{photon} \times 0 = 0$. Therefore, a contradiction in stating which matter is where the energy concentration is large and the field where the concentration is small, in which mass is energy and energy is mass. And worse, when Feynman writes in his lectures in 1964 that: *"Physicists have no idea what energy is"*. So, quantum mechanics would be unable to define energy levels; also, the electron point model of which is a cloud in the orbit that interacts, without defining what that cloud would be.

And in 2006 as F. Caruso: *"Over 100 years of discovery of the electron is still unknown its structure and also the origin of its mass and its charge"*.

However, this issue had already been answered in 1964 by Erdelyi to Dirac and Scientific American, in that by quantum impulse the electron has the structure of a helix containing the negative and positive charge, in which one is manifested by the direction of the left or right helix motion. And its mass is formed of m' with the speed c in relation to the Earth source or c' in relation to the cosmic microwave radiation or vacuum-space-ether.

And by the elementary impulse the electron consists of m'c x c = h' which explains the particle-wave, the electric and magnetic field, the spin, the constants c, m', h' and the charge e^2 of the electron-positron, the absolute helicity or chirality of the positive and negative particles, the absorption and emission of h' by the energy levels that form the electron field, the principle of h' interaction Mach with all the particle fields, the atomic recoil of emission of photons and the speed c' of electromagnetic radiation that is always constant in relation to cosmic microwave radiation or vacuum-space-ether, while the velocity of light c_v is variable relative to the velocity of the source, according to the experiments of CERN of 1964 and of the laser light of 1995 and 2002.

CXXXII. RADIATION BRAKING ("BREMSSTRAHLUNG"), CHERENKOV RADIATION AND THE PHOTON RADIATION OF QUANTUM IMPULSE THEORY – 1933 to 1969

In 1933, Heitler and Sauter observed the emission of photons by electrons scattered in a field of attraction and nuclear repulsion, that is, the radiation of braking or "bremsstrahlung". and Heitler and Bethe determined the theoretical aspect of this radiation in which the electron at the momentum p_0 and energy $E_{energy(0)}$ passing through a field of a nucleus or atom when presenting an acceleration a with a momentum p_a and an energy $E_{energy(a)}$ in resulting $[E_{energy(a)} - E_{energy(0)}] = h\mathcal{V}$, in which there was an electron interaction with the energy field that produced the electron acceleration, that is the radiation field, and the electron interaction with the atomic field, then causing radiation emission $h\mathcal{V}$. And in the calculations between momentums and energies was obtained the shock section $\delta = Z^2 r_0^2 / 137$, where Z is the number of atomic mass and r_0 the atomic radius.

And in 1934, Cherenkov and also Heitler, Bethe and Racah and after in 1937 Tamm and Frank observed that electrons that had a speed in a medium greater than the speed of light in this medium resulted in variation of the speed and consequent emission of light, according to the theory Maxwell's electromagnetics, in which the variation of electric current resulted in emission of electromagnetic field with the speed of light.

That is, if it were like the beta decay in which the electron is formed only in the instant of the emission of the atomic nucleus according to E. Fermi, the light would be created only at the instant of the emission in the radiation of Cherenkov, despite being issued continuously!

The light that comes out of the electron is still evidence that the electron is decomposing into light, that is, that the electron is formed of light, that is, of m'c x c = h'.

But by the weak electromagnetic interaction of the current quantum and relativistic physics, the light emitted by the electron is only formed when it is outside the electron which emits the light. Or, before the emission of light by the electron the light would not exist in the electron, although the electron has mass, but without being able to emit the mass of the electron! In the radiation the photon has the mass m of the momentum mc in the quantum $h\nu = mc^2$, but by the quantum mechanics and theory de relativity the rest photon is without mass. Because in the equation $m = m_0/\sqrt{1-(v^2/c^2)}$ the resting mass of the photon m_0 must be zero for that $m \neq \infty$ when $v = c$, that is, when the photon is with speed c of the radiation. So the electron could not have mass of the photon of light, because the photon at rest would have to have null mass by the theory of relativity! And the electron could not be formed of light, because the interpretation by relativity of the equation of mass increase is that speed c is absolute for all reference systems, with space contraction and time dilation.

However, the laser experiment showed that velocity c_v is variable relative to the velocity of the source, in which the absolute speed is c' relative to the cosmic microwave radiation or vacuum-space-ether without space contraction or time dilation. The interpretation with physical significance is that the elementary photon m'c x c = h' increases the mass to a close limit of c, then $v \neq c$ in the equation deduced by relativity and the theory of elementary impulse: $m = m_0/\sqrt{1-(v^2/c^2)}$.

By fundamental aspects of Planck's postulate to physical concepts with their real meanings, the mass is provided by the energy h' = m'c x c that forms the Planck radiation of energy $\varepsilon = h\nu = h'|\nu| = m'c^2|\nu|$.

According to the theory of elementary impulse, the elementary quantum mass and its speed of light c relative to the Earth source or c' relative to the vacuum-space-ether are the first constants of nature. And Planck's constant h and the charge are the consequent of these two. Quantum mass is a complex magnitude that is perceived through light, by all other electromagnetic radiation, by the electrical, magnetic and gravitational fields. So the position of the current Physics comes to qualify space-time and abstract spaces without considering fundamental physical principles, by not realizing that this fundamental constant elementary of mass m' is causing the increase in mass and the constitution of the particles. That is, m'c x c = h', in [erg] in the CGS, contrary to the constant h [erg x second] without greater physical significance, the elementary energy h' forming the particle configuration and when injected resulting in the increase of speed of the particles, contrary to the increase of mass by the coordinates of the speed of one reference to another! Even, Dirac considered decades after its electron equation that the symmetry of space-time, that is, the theory of relativity should be replaced.

CXXXIII. NUCLEAR FORCE, ABSTRACT SPACES, STATISTICS, ELECTROMAGNETIC AND GRAVITATIONAL FIELD AND QUANTUM IMPULSE THEORY – 1500 to 2003

Heisenberg, Wigner, Condon, Breit, Feeberg and Cassen in 1936 had devised theory for forces between proton-proton, proton-neutron and neutron-neutron, with mathematics of abstract spaces. And in 1946 Wigner, and in 1947 Wigner and Eisenbud presented an extensive mathematical formalism about the nuclei. That is, space apart, separate, intangible. And also quantum mechanics introducing the imaginary or complexe numbers appeared in the years 1500, as in $(number)^2 = -1$ and $(number) = \sqrt{-1}$, and $\sqrt{-4}$ it is no -2, neither $+2$, because -2^2 and $+2^2$ they are both of the same value $+4$. In the years 1500 this number of $\sqrt{-1}$ came with the resolution of equations of 2° and 3°, by N. Fontana (Tartaglia), S. Ferro, A. M. Fior, and G. Cardano which commented as: *"As subtle as useless"*. And in 1591, almost 50 years later, also A. Viète also disregarded these numbers with $\sqrt{-1}$; around of 1630, R. Descartes named numbers originated from $\sqrt{-1}$ as imaginary; and around 1750, L. Euler named $i^2 = -1$ and $i = \sqrt{-1}$, of imaginary; later these numbers with i were also called complexes.

But, power of connection, symmetry, nuclear wave function, nuclear power exchanges, forces between nuclei, force of interaction of attraction and repulsion, forces of both positions and coordinates of spin of the interaction nuclei involve, by the theory of quantum impulse, real force m'c/t with energy n m'c x c = nh'.

As Infeld and Einstein, in 1938, quantum Physics was not intended to describe individual objects in space and their changes in time, but the probability that the individual object had an indicated property, and there were no laws showing changes in objects individually in time. *"If a probability wave was needed in thirty dimensions to describe entangled ten particles, a probability wave would be required with an infinite number of dimensions to describe the entangled field. Quantum Physics has moved away from the mechanical concept and an indentation seems unlikely and it is based on the concept of matter and field, in which it also moves away from reducing everything to the concept of the field".*

For Einstein, quantum mechanics was incomplete, unfinished.

Also, Feymann on 1963 commented on the field: *"A field is some physical quantity which takes on different values at different points in space. The best direction is to use the abstract field template. That he is abstract is pitiful but necessary"*!

According to Griffiths, on 1999: *"What exactly is an electric field? But, I encourage you to think of the field as a "real" entity, filling all the space in the vicinity of any electric charge"*!

And as Veltman in 2003: *"We have concluded that electromagnetic fields are made of photons. But they have different properties from those photons of light. The photons of light are on the mass of "shell", that is, apparent, and those of electric and magnetic are not"* ! But, "shell" mass, apparent, is the relativistic mass that would be created "phantasmagorically" by speed!

However, by the elementary impulse, the fields, particles and electromagnetic radiation are formed of m'c x c = h', so the photon has mass m', which is the cause of the speed increase.

The spin being an intrinsic electric current formed, the origin of the magnet, generates a magnetic field, that is, the magnetic charge and the magnetic field. And this rotating motion of its own in displacement also generates an electric field. Although Einstein shows that an electric field can be observed as magnetic or otherwise, and the electron spin is the origin of the magnet with its charge and magnetic field, it does not explain how the motion electric charge causes the magnetic field. The transformations of relativity reference systems only exchange one field for the other.

Why does the electric charge in motion cause the magnetic field?

The motion electric charge causes the magnetic field, as the electron in motion has the rotating velocity component, the spin that is the elementary magnet, in which the n (m'c/t) or energy h' energy lines of that component form the nh' magnetic field.

By the theory of elementary impulse, the electron at rest has its charge and lines of its electric field constituted of m'c x c = h', and in motion has its velocity of translation and turning, forming a helicoid. So the moving electric charge has a spin of m'c x c = h' which is the magnetic field, the spin. And that spin of m'c x c next to a coil, by drag, interacts with its lines of the electric field causing motion of the electron and its electric field in that coil. Thus, these interactions of the h' formed field lines explain that the motion electric charge causes the magnetic field.

Why does the coil or magnet motion causes electromagnetic induction, i.e. the electrical current in the other coil?

According to Faraday, the motion of a coil or a magnet, that is, the motion of charges forming electric current, generated electric current in the other coil, by the action of lines of forces of the electric current, in which varied their electric fields and magnetic. By the theory of quantum impulse the charges with its electric and magnetic fields, i.e., nh', through the motion of the coil or the magnet increases the mass of the electron. And it causes its motion which is the electric current with its formed n (m'c/t) lines-force or nh' lines-energy that form the electric field of the electron, this field that exists always at rest or in translation and when rotating forming the component of the magnetic field.

The F_G gravitational force, of attraction between an electron of m_e mass and a positron of m_p mass is calculated by Newton's Law: $F_G = K_G\, m_e\, m_p/d^2$, where K_G is the gravitational constant and d the distance between these masses. In CGS is obtained: $F_G = 6.6... \times 10^{-8} \times (9.1... \times 10^{-28})^2/d^2$. and the $F_{Electric}$ electric force of attraction between an electron e^- and a positron e^+ follows Coulomb's law: $F_{Electric} = K_e\, e^- e^+/d^2$, where K_e is the electric constant in vacuum = 1 in the electromagnetic CGS, is obtained: $F_{Electric} = (1.6... \times 10^{-20})^2/d^2$. And the relationship between $(F_{Electric}/F_G = 4.68... \times 10^{20}$, that is, the electromagnetic force is intensely larger than the gravitational. But, the current Physics has come to explain to this immense difference between the two forces, in which string theory considers that gravitational force acts up to infinity, in other dimensions and in other universes, and only one residue acts in this universe. However, the electrical-magnetic charge is also considered acting up to infinity. Another explanation is that the mass of an electric charge, like that of the electron is negligible, because gravity is noticeable to large masses, but continuing without explaining why the electric charge of attraction is much higher than its gravitational mass. And yet, this explanation of string theory is admitted as the most plausible by today's Physics!

However, by the theory of quantum impulse, the electric-magnetic attraction involves the charge-helix in its entirety of quantum mass m' that forms the force m' c/time, in which m'c x c = m'c² = h' which is the elementary energy quantum of 1 cycle per second. That is, the electric-magnetic charge and the gravitational mass are manifestations of quantum force. As seen, the electric charge of an electron-positron has its mass $m_{e-p} = 1.2... \; 10^{20}$ m', like a helix when at motion, and in rotating results in the magnetic charge, the spin. In gravitational attraction, it is the exchange, interaction, between a lesser m' quantity of the particles that constitute the bodies. In the case of a clip, a minimal interaction of m' with the Earth occurs. And with a block of clips the interaction is proportionately greater of m', so there is similar acceleration in both cases. Figure 73.

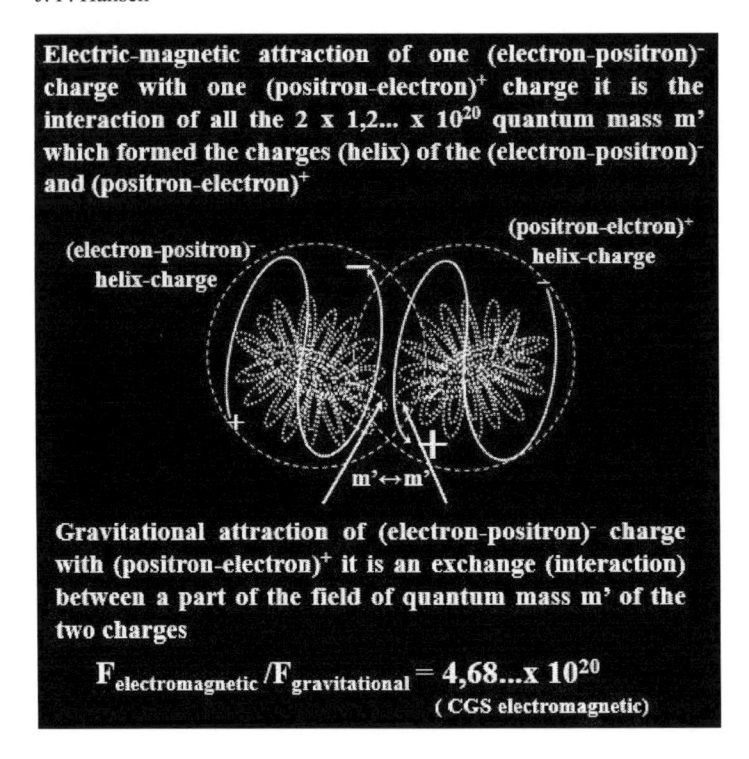

> **Electric-magnetic attraction of one (electron-positron)⁻ charge with one (positron-electron)⁺ charge it is the interaction of all the 2 x 1,2... x 10²⁰ quantum mass m' which formed the charges (helix) of the (electron-positron)⁻ and (positron-electron)⁺**
>
> (electron-positron)⁻ helix-charge   (positron-elctron)⁺ helix-charge
>
> m'↔m'
>
> **Gravitational attraction of (electron-positron)⁻ charge with (positron-electron)⁺ it is an exchange (interaction) between a part of the field of quantum mass m' of the two charges**
>
> $$F_{electromagnetic}/F_{gravitational} = 4,68...x\ 10^{20}$$
> (CGS electromagnetic)

Fig. 73 – Electric-magnetic and gravitational attraction, in which the attraction or repulsion electric-magnetic involves the charge-helix, and the gravitational is the interaction of part of the elemental quantum mass m'.

CXXXIV. CASIMIR-POLDER-LIFSHITZ EFFECT AND THE QUANTUM IMPULSE – 1948 to 2008

Generalized for other materials by Lifshitz in 1956, the Casimir effect in 1948 that showed attraction in the vacuum between plates had been verified many times and was called the force of Casimir-Polder-Lifshitz.

In 1957 and 1958 F. C. Neto, N. P. Neto and in 1958, B. F. Svaiter and N. F. Svaiter presented the Casimir-Polder-Lifshitz effect for space-time relativistic.

In 1958, M. J. Sparnaay using two metallic aluminium and chromium mirrors and a spring scale proved the appearance of this force of Casimir-Polder-Lifshitz.

On 1991 Caruso, R. De Paola and N. F. Svaiter they analyzed the zero-point energy of non-mass scalar fields in the presence of borders.

And in 1997 S. K. Lamoreaux, using copper and gold plates connected to a torsion pendulum has obtained the force of Casimir-Polder-Lifshitz that produced a twist on the pendulum, and observed that the gravitational force produced by the masses of the plates it was equivalent to 1,000 times the force of Casimir-Polder-Lifshitz, thus compensating to obtain that force.

In 1999, Caruso, Paola and Svaiter they analyzed the zero-point energy of non-mass scalar fields in D dimensions.

In 2001 also, M. Bordag, G. T. Gillies and V. M. Mostepanenko; U. Mohideen and A. Roy presented new developments of this effect.

Also, this effect could be applied in fields of other particles such as protons, mesons, quarks, gluons, in spherical surfaces, cylindrical, string theory in 1998 to 2003 by M. V. Cougo-Pinto, C. Farina, A. C. Tort, A. Tam, F. C. Santos, D. T. Alves, F. A. Barone, and in 2002 by A. Lambrecht and U. Leonhardt and T. Philbin.

In 2002, O. Kenneth, I. Klich, A. Mann and M. Revzen, observed that materials of certain permissiveness and permeability, or with a certain configuration, could present an effect Casimir-Lifshitz repulsive.

And in 2008 J. N. Munday, F. Passo and V. A. Parsegian showed that these forces of Casimir-Polder-Lifshitz they could be repulsive with quantum levitation in fluids.

In the Casimir-Polder-Lifshitz effect, quantum fluctuations create intermolecular forces that fill macroscopic bodies. In molecular deselections of a few nanometers or less, these interactions are from Van der Waals forces. However, as recognized in the Casimir-Polder-Lifshitz effect at greater distances and between condensed means macroscopic they reveal effects associated with the finite speed of light. But by the theory of elementary impulse the Casimir-Polder-Lifshitz effect is the result of the interaction of the electromagnetic field of the particles, and even Casimir observed that the effect is identical when seen by the forces of Van der Walls. So, the m'c/t force field is acting on that effect.

CXXXV. PLASMA AND QUANTUM IMPULSE – 1949 to 1969

In 1949, Fermi presented a mechanism in which the ion-formed plasma and its electromagnetic field presented a displacement that approached or drifted away from the cosmic rays causing acceleration or deceleration of these rays.

It is the quantum force of impulse $f_{m'c} = m'c/t$, which forms the plasma particle and its field, which is injected into the particles of the cosmic rays causing the acceleration of these particles of the cosmic rays by approaching the plasma, or in the deceleration of the particles of these rays when $f_{m'c} = m'c/t$ leaves the particles of the cosmic rays during the plasma clearance. Figure 74.

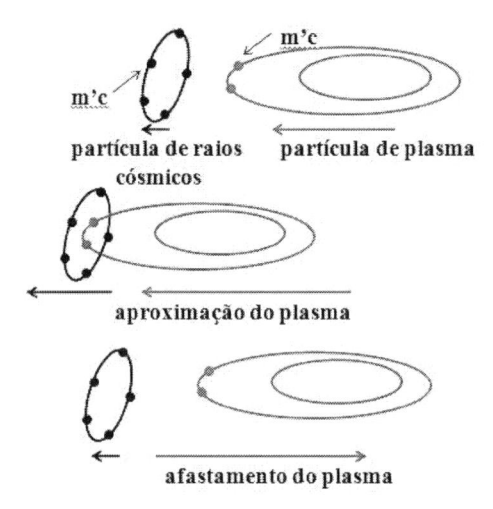

partícula de raios cósmicos   partícula de plasma

aproximação do plasma

afastamento do plasma

Fig. 74 – The plasma approach injects more elementary impulse, m'c, into the particles of cosmic rays causing their acceleration. When the plasma comes out there is the deceleration of the particles of the cosmic rays.

CXXXVI. GLUE WITHOUT MASS, GLUONS AND QUANTUM IMPULSE

In 1968 the SLAC and CERN observed constituents of the proton with the expected charges and spins, the partons of the proton or quarks, with their masses and velocities, that is, their momentums. But the sum of these momentums and masses were inferior to the mass and momentum of the proton. The masses of the quarks were smaller than 1/3 of the proton mass and all the momentums of the quarks were 1/2 of the proton. Neutral particles have also been observed, this is without charge, along with quarks and that should effect the interaction between quarks, just as the massless photons interact with the electron of the particles. Then it was concluded that the quarks were made of gluons that would be without mass similar to the photons without mass, with time to compensate the momentums of the quarks.

And were observed, experimentally in 1968 by the SLAC, three points of higher densities in the proton, which were attributed to the theory of quarks.

However, the quarks by quantum impulse theory would be formed from the quantum impulses m'c and the elementary photon h'.

But for the quark, the momentum is calculated by the product of the mass and the velocity of the quark, $m_{quark} \times v_{quark}$ and the result is also the momentum of gluon without mass! However, as momentum is mass x velocity, and if the mass is zero the momentum is zero! By the elementary impulse the whole particle is formed of m', the elementary mass, so they all have mass.

By quantum impulse theory, protons are formed of positron-electron with helicity opposite the electron-positron, seen in Figure 70, and produce in shock of particles, electrons, positrons, neutrinos and gamma rays. And the electrons and positrons produce gamma rays that are formed of photos m'c x c = h' by quantum impulse theory.

In the case of the proton with mass 1,837 times that of the electron and equal charge, by the elementary impulse are the links of 1,837 helix of positron-electron forming a ring, which would show polarity. But, the helix of this twisting ring come to present the proton-antiproton polarity similar to the electron-positron and positron-electron model. And there are three interlaces of the ring forming three main points of energy density. Figure 75.

Fig. 75 – Schematic of the proton formed of electron-positron helix with three energy concentrations in the structure.

By string theory the proton model is also considered as interlaced. Figure 76.

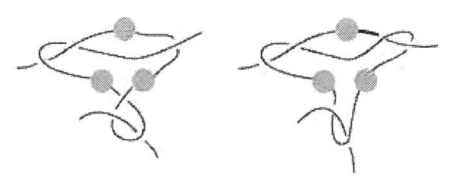

Fig. 76 – The proton model by string theory shows interlacing between three strings that form the proton. Schiller, C., *Creative Commons Attribuition*, www.motionmountain.net, Munich, Germany and Brussels, Belgium, 1990-2014.

CXXXVII. ENTANGLEMENT, MOZART'S SYMPHONY AND QUANTUM IMPULSE – 1926 to 2016

J. Bell in 1964 had observed that probability measurements in the classical statistic presented smaller and more regular values than by the probability of the wave function, then would allow to obtain what the particle behavior would be.

And in 1982 A. Aspect, P. Grangier, G. Roger and J. Dalibard showed experiments with photons of conjugated spins, that is, entagled by laser and interferometer for the degree of interlacing, that were emitted with wavelength in the visible spectrum of green and red, and that in fact the particles were entangled, because they instantly altered the opposite conditions between them. The two photons would remain at speed c, but it would be the direction of a quantum wave of probability that would be instantly modified as the current Physics conceives, to preserve Einstein's theory of relativity.

By quantum impulse theory, conjugated photons are in equilibrium with an elastic medium of submass, a penetrating ether, in a field formed of h', in the cosmic microwave radiation or vacuum-space-ether, where the photon h' = m'c x c is propagated, and by directing a conjugated photon to a directed it results in an instability of the middle and or field, in which the other photon to stabilize the middle and or the field goes in the other direction. Or, the middle and field of the photon conjugate would behave as a rigid body in which when changing the end to one direction, the other end would go in the opposite direction in which the other photon is. A rigid body of matter like iron, is only shifted by applying a force to overcome its inertia. And for a quantum field, a minimal force would already act for the instantaneous shift to a distant point.

In 1994, G. Nimtz presented that the electromagnetic wave when entering a spacing, which is a barrier, between two crystals had its speed increased by 4.3 of the speed of light or electromagnetic radiation and transmitting the Mozart Symphony which was heard from the other side of the crystal set. As Nimtz, the electromagnetic wave transmitted the signal to virtual photons in the barrier in which these photons in the barrier presented the speed of 4.3 greater than the speed of electromagnetic waves in the vacuum-space.

But by the theory of quantum impulse the medium and or electromagnetic and gravitational field between the two crystals, that is, the barrier of the field between the crystals would have the behavior as a rigid body, thus transmitting instantaneously the input signal in the spacing between the crystals for entry into the second crystal. The free electron can receive m'c x c that increases its mass, as occurs in the particle accelerators through the electromagnetic field formed of m'c x c that passes to the electron or the other particles as the proton for acceleration.

However, in the crystal barrier, the existing field is in equilibrium and the increased mass per injection of m'c x c is extremely limited, different from the electron or proton that are free in the particle accelerators.

For the current Physics the entanglement is considered a powerful mystery.

In 2015, an experiment in which two crystals of diamonds containing electrons considered conjugated, that is, with pairs of opposite orientations, in which the diamonds were placed at a distance of 1,200 meters between them, in the measurements carried out in each of them in a same instant they always presented opposite results of the directions between them. It was as if turning the first diamond to the north, with the electron also indicating the north direction, the other fixed diamond was indicating the direction south. And when the first diamond was turned south with the electron also indicating the south, the other fixed diamond presented the measurement of the electron in the north direction. That is, it is as if there is instant communication, "ghostly", between the two electrons. But outside of the mathematical abstraction of interpreting the Schrödinger equation as probability, which considered it a real wave, no ghostly event would occur! In 1926-1927 L. De Boglie and in 1952 Bohm presented the quantum potential that connected all particles with all the space, But without explaining what was this quantum potential energy. However, according to relational mechanics, the inertia is explained by the distance action, which can be related to the ether and or m'c x c impulse field. So, it's this field of quantum energy h' or subphotons in the ether that causes quantum entanglement between atoms, particles and photons. Everything is interconnected by this field! Figure 77.

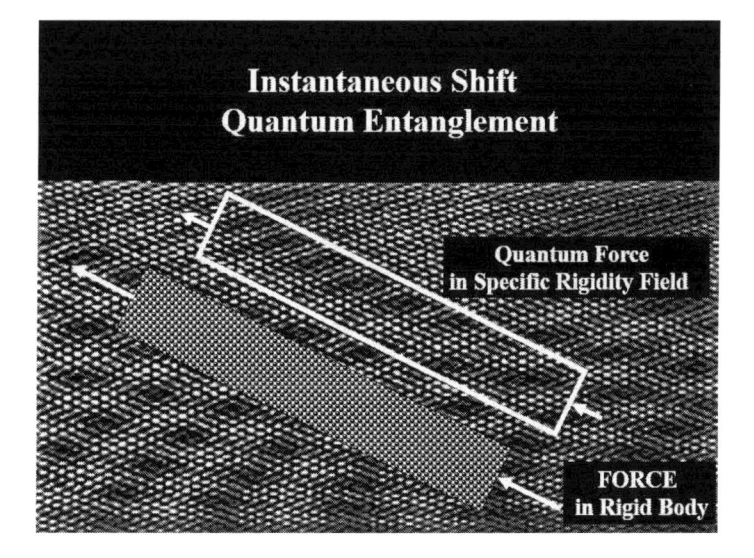

Fig. 77 – The quantum field formed of m'c x c in the vacuum-space-ether has a constitution that presents a specific rigidity which transmits the instantaneous quantum force m'c/time applied, which explains the entanglement. Adapted from www.pinterest.com.

CXXXVIII. HIGGS FIELD AND QUANTUM IMPULSE THEORY
1964 to 2012

Despite the Higgs mechanism built in 1964 by Higgs, Brout, Englert Guralnik, Hagen and Kibble, giving mass to particles by the field created by mathematical artifices, in 1985 Feynman noted that: *"The masses of the particles, m, it is used of these numbers in all theories, but we do not understand it, what they are or where they come from. This is a serious problem"*.

And about field, again mentioning Griffiths, on 1999: *"What exactly is an electric field? But, I encourage you to think of the field as a "real" entity, filling up all the space in the vicinity of any electric charge"*. And Veltman on 2003: *"We have concluded that electromagnetic fields are made of photons. But they have different properties from those photons of light. The photons of light are on the mass of "shell", apparent, and those of electric and magnetic are not"*. But, mass "shell", apparent, is the relativistic mass that would be created "phantasmagorically" by speed.

However, by the elementary impulse, the fields, particles and electromagnetic radiation are formed of m'c x c = h', so the photon has mass m', which is the cause of the speed increase.

Then, the increase in particle mass is originated by the electromagnetic fields, resulting in the acceleration of these particles and the formation of others after collision between them, as occurs in the particle accelerators. The increase in mass causes speed increase, contrary to Einstein's relativistic theory. All the mass of the collision should form h' clusters, in every range of the histogram of energy, which actually occurs.

And, with the collision mass of 100 Gev/c^2, what could be expected to get in the range of that energy histogram of particle accelerators? A much larger concentration of mass should be expected in the nearest range of the largest concentration of injected mass, i.e. n m' of n m'c^2 = n h'. And in fact in the range of 70 to 90 Gev is where the bulk of mass concentration occurs. So, some peaks obtained from 80.4 Gev, from the W boson, is a consequence of the mass injection, of n m'c^2 = n h', in the quantity next to the supplied power of 100 Gev. And these peaks occurring around ten times in over 1 million of collisions at CERN in 1983. Thus, in addition to the peaks found for the particle masses in the range of 0.5 Mev to 9.4 Gev, peaks could appear in the larger range despite the extreme compression of the distances between h', or wavelength, by the high frequency of h' in the particles that result in increased mass by the particle accelerator. And in the collision occurs an extreme instability for particle formation, in which jets should appear and form the particles of greater stability, which actually occurs.

The quark masses of, 5, 10, 200, 1,300, 4,500, 175,000 Mev are seen as varying in extreme discontrol and really intriguing, as mentioned by Veltman.

But, there are many peaks to form particles, by these conditions where the accelerated particle is in the collision between them. So this variation is predictable and orderly! Figure 78.

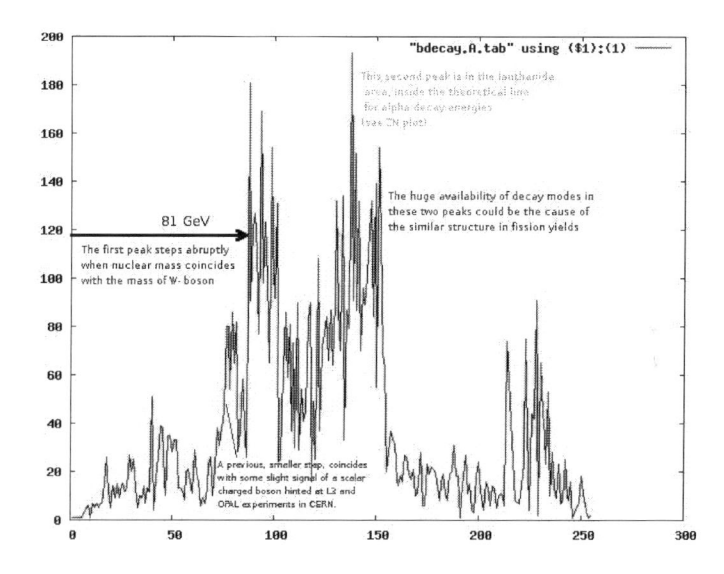

Fig. 78 – Formation of peaks throughout the energy histogram, presented by CERN as stated in Reference.

In the case of Higgs particles, as observed in 2012, a peak with sigma 5 occurred between 120 and 550 Gev, and closer to the smallest value, ie 125 Gev! However, peaks have also occurred between 125 and 550 Gev, with sigma around 1. That is, all of these values are within some exponential in particle formation. In this way, the formation of peaks would be inevitable when the Higgs particles were sought.

For, in every range of Gev there are energy of n h' forming particles within the most stable conditions of them in the smaller bands of energy, and in the larger ranges of energy the various minor peaks of intensities come to indicate that the particles would have only some tendency of formation by the exponential condition. Figure 79.

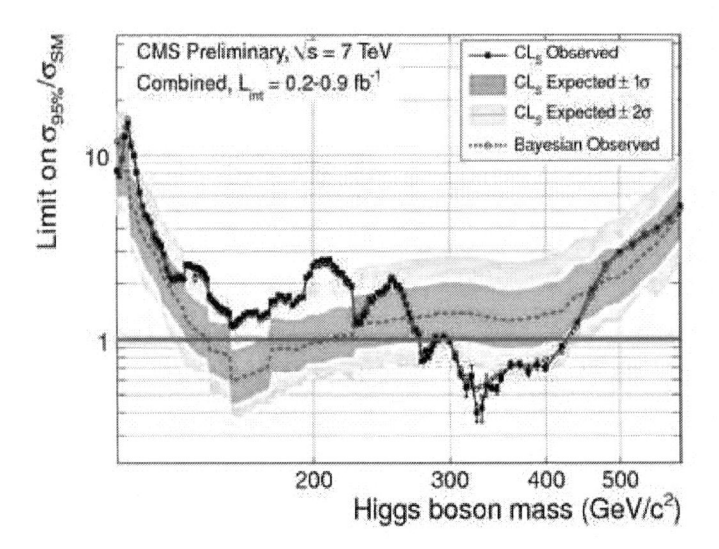

Fig. 79 – Histogram of energy and particle formation peaks throughout the energy range provided, as observed in CERN's work, which is listed in References, when the search for the Higgs boson, as Reference.

By the elementary quantum impulse, the field is formed by the elementary energies h' = m'c x c of the physical significance of Planck's postulate and Einstein in which the mass m' that in the conditions of energy density form the particles. Therefore, in addition to the H particles of the Higgs field, with the Higgs boson, the elementary mass m' forms other particles in the accelerators as placed in the histogram of energies, Figures 78 and 79. There are, beyond the H boson, countless other particles with their energies, unstable, all formed by m' obtained from the quantum theory of Planck and the quantum of light of Einstein, by the theory of elementary quantum impulse.

Particle configuration occurs for some energy clusters, such as electron, positron, proton, neutron, neutrino, mesons, quarks and other particles. And particle accelerators, despite the intense concentration of energy by the impact of electrons, positrons, protons and other particles, which are accelerated almost to the speed of light, in which some particles are only formed for approximately one million impacts, presents an almost zero probability of its formation.

By general relativity the mass curves the space, however a hypothetical mini-space of Riemann would present a background setting to retain the elementary impulses m'c with c speed, i.e. the elementary energy m'c x c = h' forming the particles. It would be the configured space that would rehave the mass, contrary to the general relativity in which the mass shapes the space.

PARTE 9 – THE UNIVERSE OF THE MOTION STARS

CXXXIX. MOTION STARS – 13.7 BILLION YEAR to 2014

"Fixed" Star Motion – After the vision of the universe of Antiquity and Earth Systems, Sun, planets and stars, in 1718 E. Halley had observed that the stars of Aldebran, Arcturus and Sirius had changed position in the vault celestial view of Greece, that is, the stars they'd look like they were fixed. There should be a precession of the motion of the Earth and or the stars that would move.

Variable Stars – In the year 185 was seen in China's firmament, a superbright star visitor, and in 1054 also seen in China an explosion in the sky originating a superstar that shone for 23 days, including during the day, in the constellation of Taurus, and then again in 1181 another superbright star in China, and in 1572 T. Digges and Brahe and other astronomers in China and Korea observed the appearance of a supershiny new star, and another superbright star in 1604. Between 1769 and 1784, E. Pigott had observed in China stars that varied in brightness, as the stars named new ones that were actually stars that almost faded presented in an instant an intense brightness. And other types of stars that varied in brightness, such as Algol also named the Flashing Demon in the constellation of Perseus and Eta of the constellation of Eagle. Between 1782 to 1784, J. Goodricke noted that the star Algol presented a variation of brightness in the period of 68h50', with a symmetry in the curve of brightness and period, and deduced that the variation of brightness was resulting from one star orbiting the other, a binary star, and that one of the stars would have a lower brightness that passed in front of the highest brightness causing its eclipse. And Pigott and Goodricke observed that the stars Eta of the Eagle and Delta of the constellation of Cepheu, presented a type of serrated curve of the variation of brightness and period, different from the binary star. The maximum brightness was obtained around a day and decreased around four days to Delta and six days to Eta. And with the discovery in 1839 of the image printing by light in a metal plate chemical it was possible to standardize the brightness of the stars and to determine their variation better.

Three-Dimensional Distance – Around 1815, W. Herschel admitted to an initial estimate of three-dimensional distance, that the stars emitted a similar intensity of light in which the brightness decreased with the square of the distance, and having as reference Sirius, the brightest star of the night firmament. But as the distance to Sirius was unknown, only multiples of that distance would be calculated for a three-dimensional map of the stars. However, he observed that there was a milky band of stars around the Earth, in the form of a disc with a diameter of 1,000 distances to Sirius and thickness of 100 distances to Sirius, and which was subsequently called the Milky Way, "Milk Path" which in Greek is "Gala"or Galaxy. So, if the distance to Sirius was determined, the distances of the stars and the Milky Way would be obtained by trigonometry.

Stellar Parallax – In 1838, F. Bessel through lens technology, telescope and equipment, presented the first measurement of stellar parallax, observing the constellation Swan, in Star 61 Cygni, with the value of 0.3136" or 0.000087°, corresponding to a distance of $\approx$ 10.4 light years that is the distance travelled from 9.460530×10^{12} kilometers at the speed of 300,000 km/s in one year. And in Centauri for the Alpha Star, in 1839 T. J. Henderson found the value of 1" and distance of $\approx$ 4 light-years. And in 1840, the. Struve measured the stellar parallax of Vega, the Alpha Lyra, in 0.2613" with the distance of $\approx$ 12.5 light years. These distances were from stars closest to Earth, but these measurements were astonishing to point a distance far beyond what was expected. And it also resulted in obtaining an initial estimate for the distances of the Milky Way, obtaining 10,000 light-years in diameter and 1,000 light-years thick.

Chemistry of the Universe – Between 1860 and 1862, W. Huggins and W. A. Miller applied the discovered spectroscopy in 1752 to observe 50 stars, concluding that they had a chemical composition like that of the Sun, containing most of the elements on Earth: hydrogen, sodium, magnesium, calcium, iron, oxygen and other chemical elements, contrary to the hypothesis that they were formed by some kind of unknown substance of the Earth, thus obtaining a similar chemistry throughout the universe.

Radial Star Speed by Doppler-Fizeau Effect – Doppler, in 1842, it had shown that the sound wave emitted by a source moving in the direction of the observer was perceived more acute than that emitted by a source that was moving with the observer, and the sound of a source that moved away from the observer resulted the most deep. And similarly it would occur to the luminous waves. The wave when moving toward the observer had decreased wavelength and as it moved away it presented the increased wavelength. And Fizeau, in 1848, observed that the variation of the wavelength of the light from the stars could be used to measure relative speeds in the same line of distances. In 1868, Huggins noticed that the speed of a beam of light from the star Sirius to Earth, the radial speed, presented a deviation of the spectrum line of 47 km/s, with $\Delta \lambda/\lambda_0 = v/c$, v = star radial speed, c = speed of light, λ_0 = spectrum wavelength, $\Delta\lambda = \lambda$ wavelength shift of λ_0, the Doppler-Fizeau effect.

Curved Space Universe – Riemann, in 1854, had presented the curvature geometry of space, in which the flat geometry was a particular case of infinite radius curvature. and Clifford, in 1870, conceived the motion of masses as originated by the variation of curvature of space. It was the unification of Physics with Geometry. And in 1915, Einstein presented the principle of equivalence, in which the acceleration of a reference system was perceived to an observer in this reference as an equivalent action of gravity. Gravitation would be a particular case of acceleration originating from mass of inertia. And in the curved space-time, forces would be replaced by the variation of the curved spatial coordinates. And so, Einstein presented the theory of general relativity in which the mass would exert distortion in space-time, resulting in acceleration of the bodies of the universe in which there would be the collapse of the system. After, Einstein modified his theory by introducing a constant in the cosmology of the curvature of space, by the attraction between the bodies of the cosmological system, that is, a repulsion to nullify the gravitational collapse. In this gravitational collapse would occur the singularity, that is, a place of concentrated energy density. So, Einstein with the cosmological constant obtained that the singularity would be avoided and the universe would remain eternal. In 1916, K. Schwarzschild calculated the dimension of this singularity of the space-time of general relativity, and presented the boundary of a sphere where light was accumulated and time and space were nullified, while physical variables resulted in infinities.

Nebulae – C. Messier, in 1781, presented a cloud-type luminosity catalogue, called a nebula. And Herschel observed that it could be formed of gases and clumps of other chemical elements, along with a star and situated in the Milky Way. But as Kant many nebulae were formed of millions of stars and they were beyond the Milky Way, in which he commented: *"Eternity is only enough to encompass the manifestations of a supreme Being if combined with the infinity of space"*. Between 1845 to 1850, W. Parsons-Lord Rosse noted that nebulae were made up of stars with spiral structure. But any measurement of parallax from any star in the nebula showed indetermination by the light rays of the surrounding matter. In 1920, H. Shapley presented that nebulae were clouds of gas and dust observed only on the outside of the Milky Way disk and that they were nurseries of stars and planets, so the nebulae would belong to the Milky Way. It had been observed that a new star presented a small brightness that increased in a minimum of time, and in 1885 a new presented a brightness that had risen to 1/10 of the brightness of the entire Andromeda Nebula, and that it would be feasible if Andromeda were constituted by a few stars. However, for Shapley, it would be impossible for a star to shine more than billions of stars in Andromeda and be far from the Milky Way. But, H. Curtis presented that the nebulas were also distant from the Milky Way, but the light of the nebulae at the disk level of the Milky Way was blocked by the stars, gases and interstellar dust in the vicinity of that disk. And the new star of 1885 in Andromeda was the only new one among all new ones that presented an intense glow, while the other new ones were extremely of less brightness, ie, this supershiny new star of 1885 was similar to those previously observed in 185, 1054, 1181, 1572e 1604, and said: *"We, the microscopic inhabitants of a small satellite of one among the millions of suns that form our galaxy, can look beyond its limits and see other similar galaxies, with tens of thousands of light-years in diameter, formed by also millions of other suns"*.

Displacement Meter with Variable Stars – In 1908, H. S. Leavitt observed that the stars that varied in its brightness, like the Delta star of Cepheu, named cepheid stars, that the brightest had a longer period of time to change the brightness, and more significant was that for equal periods of change of brightness the farther stars were less luminous. So it would be possible to measure distances to the stars, compared to the cepheid of shorter displacement of the Earth obtained by stellar parallax, with the distant cepheids of the same period of change of brightness and less luminosity. So based on parallax measurements of a cepheid from the closest to the Earth by Shapley and E. Hertzsprung, the distances to the stars would be more real measurements of the universe.

Nebula Escape – V. Slipher, in 1912, had measured the Doppler shift to blue, from the Andromeda Nebula, equivalent to a escape speed of 300 km/s, considered as most high, because Andromeda was admitted to be in the Milky Way. And to ascertain whether it was correct observed the nebula at the time, of Sombrero, obtaining a detour to the red with speed of 1,000 miles per sec! And in 1917 he observed that 21 out of 25 nebulae had been displaced from Earth and in 1922, 40 nebulae, 36 moved away from Earth and as E. P. Hubble, the four that approached were the closest to the Earth.

Inference from the Beginning of the Universe – As there was the escape of nebulae, there was an inference that had previously been closer to the Earth, and to form the Earth would need igneous matter, and that matter from somewhere, from a dense region that would be the beginning of the universe. In 1922, Friedmann presented the universe as a function of the cosmological constant admitted as zero, which was the initial theory of general relativity without that constant. With an admitted dense region obtained in the universe, if there was a greater density of matter there would be a greater gravitational attraction and the consequent collapse of the universe. With low density, the universe would continue this expansion until the end by entropy thermodynamics. In the case of a density between the two previous ones, adjusted with maximum precision, the excessive expansion would be stopped and the universe would always be in modification.

Galaxy Escape – Between 1923 and 1924 Hubble was able to observe a cepheid in the NGC6822 nebula and with the brightness period and the pattern of brightness and distances of Leavitt to the cepheids and parallax of the referential cepheid by Shapley and Hertzsprung detected that it was much farther from the measurements performed, calculating at 700,000 light-years. And after, another 25 cepheids those observed in M33, estimating distances around 850,000 light-years, in which later measurements were recalculated to 3 million light-years. And following get a cepheid in M1, the Andromeda nebula, with 31 days glow period, obtained a distance around 900,000 light-years from Earth. As the Milky Way had been estimated at 10,000 light-years and after 100,000 light-years, it was distant from the Milky Way. And because it was very far away, the brilliance of Andromeda would indicate that it should contain millions of stars, so it would be a galaxy. With other observations of nebulae, most of them were classified as galaxies while other nebulae in the Milky Way were smaller and formed from clouds of gases and stardust. So the Doppler deviations that were assigned as escape from nebulae, obtained by Slipher, were escape of galaxies.

Primordial Matter, the Primordial "Atom"- As there was the escape of galaxies, there was an inference that was closest to the Earth, and to form the Earth would need igneous matter, and that matter from somewhere, from a dense region that would be the beginning of the universe. In 1927, G. E. Lemaître presented a model of the universe resulting from gas pressures with localized matter that would expand. And in 1931 he changed the beginning of the universe to a singularity that contained all matter, and exploding would form the expanding universe. Between 1922 and 1924, C. Wirtz notted a correlation that the fainter the galaxy the higher its redshift, thus it faster it is receding, and assuming that fainter galaxies are more distant, then galaxy velocity increases with distance as long as the distances of the galaxies were correct.

Proof of the Creation of the Universe and Age of the Universe
– In 1929, Hubble observed by spectroscopy that there was a shift to the red galaxies, and obtained that the velocity v_{galaxy} of remoteness of the galaxies in the universe was proportional to their distance d_{galaxy} and the relationship v_{galaxy}/d_{galaxy} later named H_{Hubble} = Hubble constant. So it is inferred that this expansion to retreat would be an extremely dense region, and that it would originate the stars, planets and all matter of the universe. The constant initially calculated by Hubble was 558 km/s by 30.9 x 10^{17} km = 1 megaparsec, so to traverse that distance from the beginning of the universe would be the time t = 30.9 x 10^{17} km/558 km/s = 1.8 billion years. This relationship of Hubble's constant showed that all the galaxies as they retreated in time converged exactly to a point of extreme density, the origin of the universe, as a proof of when the universe was created. However, it would be necessary to measure more distant galaxies to obtain a different rate of expansion, that is, another value for the Hubble constant. For Hubble, the galaxies expanded through space, that is, the space was continuous and immobile, and the motion was of the expanding galaxies and some on approach. Figures 80 and 81.

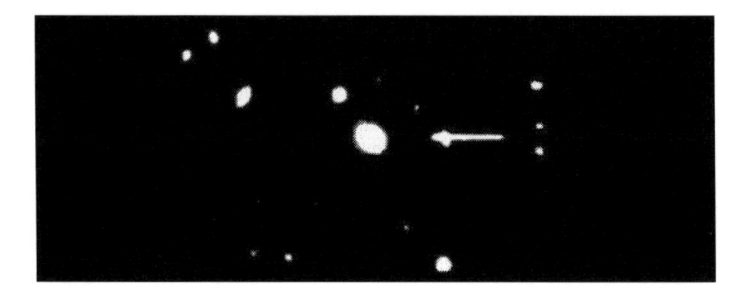

Fig. 80 – Galaxy photography in motion in space signaled by Hubble. The space was admitted continuous and motionless, and it was the galaxy that presented motion. That is, without the space "carrying" the galaxy! Photo as stated in Reference Humason M..

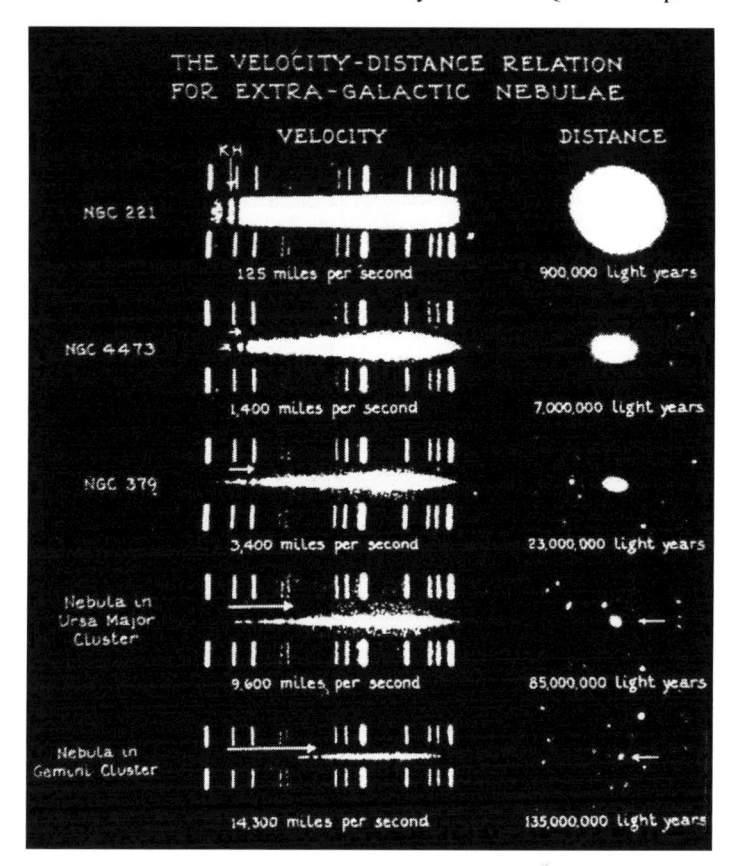

Fig. 81 – Velocity and distance of the galaxies in measurements by Hubble. Photo as stated in Reference Humason M.

In that year of 1929 Rutherford estimated the age of the Earth as less than 3.4 billion years that would be the time of formation of stars. Thus, measurements of distances and or expansion velocities of the galaxies would be smaller than was necessary, so that the time of recoil would encompass the formation of the Earth and the universe.

Inherent Redshift – F. Zwicky In 1929 also admitted that galaxies were fixed or at lower velocities, but by the matter of the galaxies emitting light with a wavelength of diverted to the red would be inherent to the existing conditions in the galaxies for this emission. That wavelength variation being named of inherent redshift, in theory of light "aged" by R. Tolman, or by other mechanisms as gravitational attraction, or still conforming in 1953 by E. Finlay-Freudlich presenting that radiation fields would be responsible for the loss of energy of the photons. However, the calculation of gravity attraction for energy $E = h\nu = (c/\lambda)$ of light presented only a minimum increase of the λ wavelength, that is, a minimal decrease in energy, short of the redshift found by Hubble.

Chemical Elements of the Universe – Through the spectrometric analysis of the stellar bodies and the Earth it was possible to estimate the quantity of chemical elements in the universe, and analyzed that around 90.8% was hydrogen, 9.1% helium, 0.06% oxygen, 0.01% carbon and 0.01% of all other atoms in which gold and platinum were the rarest.

Star Energy – With the discovery of the atomic constituents it was possible to establish the level of stability between the chemical elements with their layers of electrons, radioactivity, isotopes and other properties, which were synthesized in the periodic table. And with the transformation of a minimum mass into energy $E = mc^2$ and quantum mechanics by the tunneling effect, Atkinson and Houtermans, in 1929, presented the hypothesis of the nuclear reactions in the stars that formed their heat and radiation. And predicted that there should be the neutron to keep the two protons in stability, for helium formed by hydrogen that is stable
and that under pressure, volume and temperature conditions there could be its proton-proton fusion and proton-neutron formation, deuterium, and the most stable helium with energy release. And in 1938, Bethe presented the reaction complete for the nuclear reaction of the smaller masses stars: $_1H^1 + {_1}H^1 \rightarrow {_1}D^2 + e^+ +$ neutrino; $_1H^1 + {_1}D^2 \rightarrow {_2}He^3 + \gamma$;
$$_2He^3 + {_2}He^3 \rightarrow {_2}He^4 + {_1}H^1 + {_1}H^1 + 24{,}7 \text{ Mev}.$$
And C. F. Weizsäcker presented nuclear reactions to stars more massive than the Sun:
$$_1H^1 + {_6}C^{12} + {_7}N^{13} + \gamma;\ {_7}N^{13} \rightarrow {_6}H^{13} + e^+ +$$
neutrino;
$$_1H^1 + {_6}C^{13} \rightarrow {_7}N^{14} + \gamma;\ {_1}H^1 + {_7}N^{14} \rightarrow {_8}O^{15} +$$
γ;
$$_8O^{15} \rightarrow {_7}N^{15} + e^+ + \text{neutrino};\ {_1}H^1 + {_7}N^{15} \rightarrow {_6}C^{12}$$
$+ {_2}He^4$

Plastic Balloon that Expands the Nothing! – In 1930, Eddington to defend the theory of general relativity in which space-time would expand without any cosmological constant, which would be necessary if the universe were static, and also to defend the theory of the primordial atom in which the point of density of initial singularity should expand space-time, it presented that the expansion of the galaxies observed by Hubble in relation to space was in reality the space that would expand, while the galaxies would remain fixed. Like a plastic balloon that when being inflated increases away from all points, which are fixed on the surface of the balloon, but the surface is expanding.

So Eddington and the entire Physics community were admitted that the galaxies were fixed, without motion, and it's the space that expands between all the galaxies, resulting in the spacing between the galaxies. However, since the galaxies are in space, it would seem that all the matter that form the galaxies would also be expanding, but the gravitational force within the galaxies would be of incredible precision, in which only the space between the objects within the galaxies is that would expand without increasing any object within the galaxies!

Distant Points Bridge ("Wormhole") – In 1935, Einstein and Rosen presented that the singularity of general relativity calculated by Schwarzschild in 1916 was a double singularity that would bind as a bridge two universes or two distant points of the universe itself, being subsequently named "wormhole".

Dynamic Evolution – In 1948, R. A. Alpher, Bethe, Gamow and R. Herman presented a hypothesis that in an initial universe the temperature would be 10^9 °Kelvin and consisting of radiation without presence of matter, and the singularity of the hypothesis of Lemaître was formed of neutrons that fell in protons and electrons, the plasma, permeated by the light forming a unique and fantastic expanding cloud with the plasma, in which the protons merged with the remaining neutrons that would form nuclei of larger masses, the nucleosynthesis, which in the initial phase would be the formation of helium, lithium, beryllium and boron. The energy field, called the physical entity undetermined, of the electromagnetic light in the vicinity of the field which occupied the space around some electric charge, presented interaction being absorbed and emitted by the nuclei and electrons. And in the decrease of hydrogen for the formation of stable helium atoms and consequent decrease in the quantity of heat formed, this electromagnetic field of light began its propagation through space. Then, with the stable atoms formed, the electromagnetic field of light was only reflected by the electrically stable atoms, that is, by the fields of the positive and negative particles neutralized in the atoms. So it could be detected the emaining of this initial light field of this universe, even with some absorption of that light by electric charges in the universe. With the estimate of $\lambda_i = 0.001$ mm for the initial wavelength of this light, and at the temperature calculated by the thermodynamics of 3,000 °C, with the expansion determined by Hubble in which the deviation to the red occurs, at the present time should be observed a wave of $\lambda_{current} = 1$ mm that would be in the spectrum of microwave, the cosmic microwave radiation. And the prediction that this radiation that was at high temperatures at the beginning of the reactions of the dynamic evolution model, should be in the present at 5 °K.

Stationary Universe, Perpetual Creation and the Big Bang – In 1948, H. Bondi, T. Gold and F. Hoyle presented that as galaxies deviated from the Hubble's law, there would be permanent creation of matter, through a field of space creation, based on the cosmological principle of E. A. Milne which in 1935 presented that the universe maintained uniformity in the course of time. And Hoyle termed the theory of dynamic evolution of the universe ironically as the "Big Bang".

New Variable Stars and the Age of the Universe – Between 1890 to 1902, W. P. Fleming observed that a type of variable star, as RR of the Constellation of Lyra, similar to the cepheids, could be used to measure distances. However, in 1948, W. Baade when searching for this type of star RR in Andromeda, which should show a lesser brightness than the pattern of this star, he noticed that it was observed only in the Milky Way. It would be impossible within the probability that in a large galaxy like Andromeda that kind of star was absent. Baade noted that since 1940 the stars were being grouped into two types: group I – brighter and lower age and group II – less bright and longer age. Then, Baade inferred that the cepheids and the RR could also be grouped into two types. So, what Hubble had compared was between the less-brilliant cepheids of group II of the Milky Way, which would have a greater brightness compared to the one that would observe in Andromeda, with the brightest cepheids stars in group I of Andromeda, that is, if they were compared with the less bright stars of Andromeda Group II would obtain a far greater distance from Andromeda. By correcting the pattern of the cepheids, Baade found that the distance from Andromeda was 2 million light-years. And at that distance, the RR-type stars were extremely low in brightness to be detected. And the fundamental thing was that the time of retreat to travel that greater distance to where the Earth was and to the beginning of the universe was much greater, then calculated for 3.6 million years, compatible with the age of the Earth, but still short of the age of formation of the universe.

Distances from Remote Galaxies – The cepheids variable stars would have imperceptible brightness if observed in remote galaxies. So the estimate for determining distances from remote galaxies was to compare the brightest star of Andromeda to the brightest of the remote galaxies. If the brightest star in a remote galaxy had 1% of the brightness of Andromeda's brightest star, as the brightness diminishes with the square away, the distant star would be 10 times the distance from Andromeda.

But between 1952 and 1954, A. Sandage observed that an intense glow in the distant galaxy was a hydrogen cloud which absorbed the light of stars around increasing its temperature and radiating a brightness greater than any other star. So these stars would look closer to the glare. Thus, Sandage recalculated the distances and found 5.5 billion years for the age of the universe, and subsequently between 10 to 20 billion years, in which the Hubble constant decreased to 180 and in 1980 to 50 to 55.

Stellar Nucleosynthesis – The stars initially came to be classified by type of brightness and dimension and in 1890, Lockyer presented initial hypotheses of its evolution with contraction, temperature variation and radiation emission. After, in 1907, R. Emden presented stellar evolution where contraction, temperature, and heat were related. Between 1917 and 1927, Eddington formulated a hypothesis of gravitational contraction in stars. In 1926, Fowler presented that the stars, when forming helium, lost energy, contracting with increased pressure until the collapse of its structure. In 1931, S. Chandrasekhar presented a relationship between density, gravitational pull, and repulsion in star type. In 1932, Landau presented a relationship between gravitational attraction and atomic repulsion in stars. In 1934, Baade and Zwicky presented the balance between gravitational attraction and atomic repulsion of neutrons in stars with end-formed neutrons. In 1938, E. Öpik presented that in star type of immense mass, the high temperature expanded the outer part of the star. And in synthesis, in the stars there was a balance between the gravitational pull, between the hydrogen atoms and the helium that was formed and the pressure exerted by the temperature in the inner region of the star. Around 1940, Eddington had presented that in the stars there would be fusion of hydrogen forming helium and other atomic nuclei. The spectrometry showed that in the stars were the chemical elements found on Earth. However, the calculation of positive nucleus and negative charge forces of the electrons, the gravitational force and the nuclear force required an energy to allow the fusion of the nuclei in which the temperature would be billions of degrees, well above the observed temperatures of the stars. And also, the nucleus of each chemical element would require a specific energy to occur this fusion, which would imply that there were different types of stars to form each chemical element. Around 1950, Hoyle calculated the nuclear forces, temperatures, pressures, densities, time of reactions for various types of stars, concluding that the chemical elements would be produced according to various types of stars and time of evolution of theirs until its extinction. And also, stellar wastes from its final blast would reach other of them that would come to form stars, several generations of stars, already with larger nuclei, and those stars at the end would explode into atoms of larger numbers of nuclei. So life in the universe would be derived from the extinction of stars that form and provide the chemical elements for the constitution of planets which in special conditions allow the formation of cells of the living organism. But for the nuclear reaction to melt helium into carbon and consequent reactions from carbon to the merging of the other atomic nuclei, helium should react with the lithium and beryllium atoms, but being unstable would prevent the continuity of nuclear reaction. Also, the probability of three helium nuclei colliding in an instant to form carbon was almost nil. And another possibility was that two carbon nuclei formed the beryllium and this one with a nucleus of helium would form carbon, but the beryllium was also extremely unstable. The reaction could occur, but there would be a larger mass of the carbon nucleus formed of helium and beryllium, and the excess of carbon mass to be released as radiation would be greater than the time for the beryllium to decay in helium. However, if the carbon had an additional energy of 7.65 MeV the reaction would be within the lifetime of the beryllium and there would be carbon formation. So, in 1953, W. A. Fowler detected this additional energy carbon.

Extreme Nucleosynthesis and Black Holes – It had appeared in the China firmament, in year 185, a superbright visitor star, and between 1054 and 1937 other stars of this type were observed in which there was variation of the brightness and which would be named supernova in the NGC4157 galaxy. In 1938, Baade and Zwicky presented a description of the brightness periodicity curve of a supernova.

Around 1953, as Hoyle, when the final phase of the nucleosynthesis of larger masses is irreversible, an implosion of greater intensity occurs, releasing an energy similar to 10 billion stars of smaller masses, and these stars emitting a superbrightness, supernova, and presented the most extreme nuclear reactions to the production of major masses. So the stars could evolve into neutron stars that present an initial mass of 10 to 20 times the mass of the Sun to 1.4 to 3 times of the mass of the Sun, dwarfs of initial mass of 8 times the solar mass to 0.6 of the solar mass, in addition to supernova exploding completely and stars with mass greater than 3 times the mass of the Sun and with density twice the dwarf stars, that is, black holes. In black holes gravity presents a greater intensity than the nuclear repulsion of the star. And at the center of the black hole by general relativity there would be a singularity, that is, a point at which the density would be infinite and an opening to a wormhole to another part of the universe or to another universe! Figure 82.

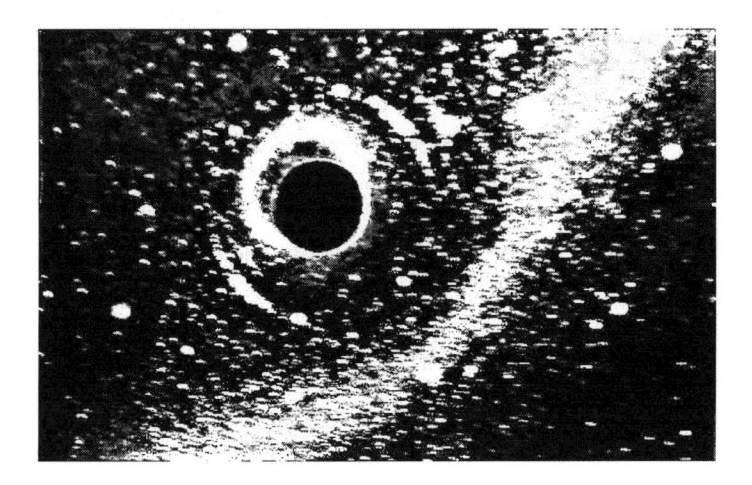

Fig. 82 – Schematic drawing of a black hole, adapted from Google Images.

In 1783, J. Michell had presented that supermassive stellar bodies would rehave the light, so they would be invisible, dark or black bodies. In 1967, M. Kruskal compared the Schwarzschild radius of the singularity to the collapse of a massive star in which Wheeler named this stellar collapse of black hole. In 1971, L. Braes and G. Miley, and also R. M. Hjellming and C. M. Wade observed the star Cygnus X-1, and in 1972 L. Webster, C. T. Bolton and P. Murdin detected the binary star of Cygnus X-1, the HDE 226868 which dragged the matter around the Cignus X-1 to the center of HDE 226868 and that would be a black hole.

In 1974, Hawking by thermodynamics and quantum mechanics observed that a radiation emission would occur around or horizon of the black hole, later named the Hawking's radiation. And quasars would be center of galaxies that would have a black hole that would emit radiation from the matter dragged into the black hole and also light jets, X-rays, gamma rays and radio waves. And also, there would be mini black holes of mass 10^{-13} cm, other black holes of millions of times the mass of the Sun, located in the center of the galaxies and black holes that would be created at the beginning of the Big Bang.

Hawking had presented the black hole theory with electromagnetic radiation emission across the edge, which consisted of any mass falling into the black hole would be transformed into energy, in photons. For Hawking, such energy, such a photon, had the same nature as that which had driven the thought that commanded any action, such as the printing of a word on a computer. And that photon that was printed, it was an energy that could be emitted, because it was added to some atom of the computer. So, according to Hawking, the thought had the same

J. P. Hansen

value as the action, that is, the same energy, the same photon, the information. But as Hawking this information was totally lost, and for any mass when entering a black hole the time would be eternity, for by stopping the time count, time would equal itself to the time of eternity. Without an observer the time would be nonexistent. The time measured by the observer would be a psychological creation, but not real.

However, by the theory of quantum impulse, time, space and mass are determined by the measurements of time, length and mass. In the case of time, a pendulum in space with one cycle per second and the other with 1 billion of cycles per second would record different passages of time, regardless of the time count by an observer. For any fixed point in space, the pendulum presents the motion that can be 1 or 1 billion vibrations, so different times.

But in 2014, Hawking presented that black holes could to be non-existent. Figure 83.

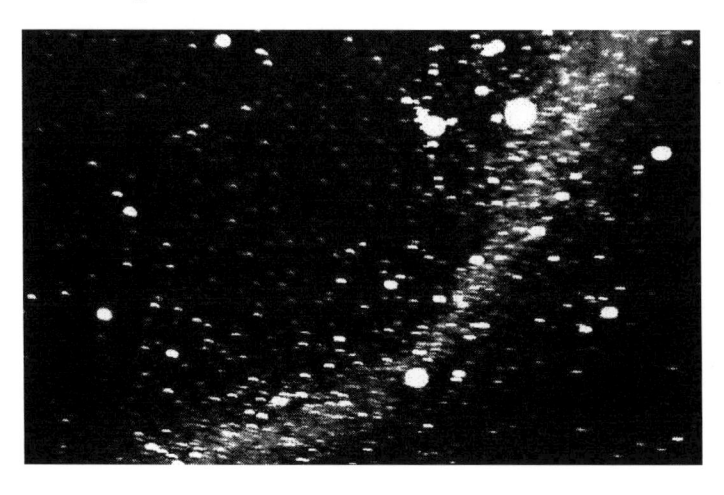

Fig. 83 – As Hawking black holes could even be non-existent. Adapted design of Google Images.

Radioastronomy – Between 1930 and 1933, K. Jansky searched between the radio waves of 1 mm up to 1,000 m the source of radio waves that caused noise in long-distance transmissions. Through metallic receiver, the antenna, which best drives electromagnetic waves like radio waves, it would be possible to measure these radio interference intensities from different directions, and then locate the source of emission of these electromagnetic radio waves. He obtained interference from local and distant climatic conditions, and also a constant noise that reached a peak of intensity at 11pm and 56 minutes, which M. Skellet determined as the sidereal day in which the Earth revolves effectively in relation to the stars of the Universe. Jansky noted that the emission of this radio signal originated in electric disturbances of the central region of the Milky Way. The radioastronomy was set up to observe radio signals from all over the universe. And in 1952, Baade located a radio source that was coming from almost an imperceptible light spot, but it detected by the optical telescope that were galaxies and that the highest brightness was the origin of the emission of radio waves, that is, the radiogalaxy.

Quasars, Redshifts, Intrinsic Shifts, Gravitational Lenses – In 1960, M. Schmidt observed an intense radio source, 3C48, observed in the optic telescope as a small blue star, but containing unidentifiable spectral lines. However, J. G. Bolton presented that these unidentifiable lines were the source of the most high deviation for the red. And M. Schmidt identified it as the hydrogen line with this high shift to the red. And in 1963, M. Schmidt observed another radio source and the optic telescope as a bright point and with a sharp shift to the red, $z = 0.158$, with velocity of 15.8% of the speed of light, that is, 47,400 km/s, therefore at a distance of 2.06 billion light-years. It would look like a nearby star, but it was a distant galaxy of marked brightness and radio source, an almost star galaxy, the quasar. Other quasars were observed until the OQ 172 of $z = 3.53$ was detected, with 91% of the speed of light, i.e. 273,000 km/s, and distance of 11 billion light-years with the Hubble constant = 75 km/s per 3.086×10^{19} km or by 3.26×10^6 light-year. But, the starting from 1966, Arp noted that quasars with a large redshifts were linked by interstellar cloud with galaxies that presented smaller redshifts, such as the association between the NGC3067 galaxy with the 3C232 quasar. And he established that a large, ancient galaxy ejected material that formed smaller and younger companion galaxies around it. And the younger galaxies, in turn, would eject material that would give birth to quasars, and that the younger the object within the group, the higher the shift of the intrinsic red. The so-called Mach principle presented that the inertial mass of the body would be affected by the global distribution of mass in the universe, in which inertia originated from the interaction between the bodies, in which an isolated body in the totally empty space would have its null motion, and that local physical laws were determined by the structure of the universe on a large scale. According to this principle, every particle in the universe derived its inertia from the remainder of the universe's particles. An electron, for example, would have null mass at birth in the universe. So, as time went on, he would get the interaction of the particles from an ever increasing volume of space and consequently his mass would grow in the same proportion. Therefore, if the mass of an electron bouncing from an atomic orbit to a lower level is lower, then the energy of the photon of emitted light is also lower and being diverted to the red. Therefore, it would be the reason why the younger matter has more deviation for the red. Then presented the hypothesis of deviation to intrinsic red. In 1996 Arp using the X-ray Space Telescope, Röntgen Satellit-ROSAT, observed the connection between the Mark205 quasar and the NGC4319 galaxy and others as NGC1097, in addition to the redshift in stars with shifts to the excess red of the supergiants of the clouds of Magellan, which attributed to the shift to the intrinsic red. But, the current Physics attributes the connection of the Arp and the galaxy to noises, imperfections of instruments, and apparent superluminarical motions, that is, the projection of the motion when seen at a certain angle in relation to the target line; the effect of gravitational lenses, that is, the prediction of general relativity in 1936 that a beam of light was diverted as it passed through a massive body, in which the first detection in 1979 by D. Walsh, R. F. Carswell, R. J. Weymann It would be the Q0957 + 561 quasar, of two points emitting identical radio signals separated by a small angle. However, D. Roberts and others identified a difference between the two quasars in radio frequency images, discovering a relativistic jet that emerged from the quasar A, without any equivalence in quasar B, and the distance between the two images, 6 arcseconds, was too large to have been produced by the gravitational effect of galaxy G1, a galaxy identified near quasar B. But, in 1980 P. Young, J. E. Gunn, J. B. Oke, J. A. Westphal and J. Krist discovered that galaxy G1 was part of the galaxy grouping that increased the gravitational deviation and could explain the distance observed between the images. Finally, a team led by M. V. Gorenstein observed relativistic jets essentially identical in small scales from A and B in 1983. The difference between the large radio images is attributed to the special geometry for gravitational lens by the quasar, but not for the whole extent of the emitted jet seen near the image A, and as A. Müller small spectral differences between the quasar A and the B could be explained by different densities of the intergalactic medium in the paths of light, resulting in different extinctions. However, in 2006 new observation showed that the image of the quasar reached the Earth approximately 14 months earlier than the corresponding image B, resulting in a path length difference of 1.1 light.

Although all the scientific instruments developed so far are undefined if quasars are near or far from the galaxies that would appear connected to quasars. But, the James Webb Space Telescope for infrared, with launch predicted in 2020 perhaps you could reveal that distance and end this controversy. Well, if Arp was right about the cause of the redshift, not only was the Big Bang model supposed to be closed, as the distances from the farthest objects in the universe should be reviewed, because they would be wrong by factors from 10 to 100, and their luminosities and masses by factors of up to 10,000. Then there would be a completely different view of extragalactic space.

Missing Mass, Dark Matter, Intrinsic Redshift and Modified Dynamics – Zwicky in 1933 had identified that clusters of galaxies remained at constant distances, although the high velocities of galaxies were insufficient for the attraction exerted among the masses of the galaxies, and in 1937 concluding that there would be a matter invisible, dark, for this attraction between the galaxies. And in 1959, L. Volders demonstrated that the spiral galaxy M33 presented a velocity of rotation greater than that calculated by Newton's gravitational pull. In 1969, V. G. Rubin and W. K. Ford presented that by observing the Andromeda galaxy they obtained that gases and external stars revolved around the center of the galaxy at the same velocity as the most internal gases and stars of the galaxy. That is, it was contrary to the motion of planets around the solar system, where Mercury has the highest velocity. And M. S. Roberts detected by radio astronomy that much further from the gases and stars observed by Rubin and W. K. Ford, which was the same velocity of gases and stars as they had obtained, concluding that there should be missing mass in the galaxy for velocity to be larger than the one predicted by Newton's universal gravitational law. In 1973, J. Ostriker and P. J. Peebles presented that galaxies would only be stable if the missing mass was extremely larger than the estimated mass of the galaxies, by a factor of 10, that is, the observed matter was only 10% of the mass that would be real. In 1978, Rubin and W. K. Ford with observations of ten more galaxies proved that the outer galaxies showed the same velocity as the most inbuilt stars in the galaxy. In 1979, S. M. Faber and J. S. Gallagher presented the missing mass as a massive envelope in the galaxies. and Rubin said: *"No one has ever told us that every matter radiates"*. Therefore, it would be a matter without any emission of light, in which the external light would be fully absorbed, that is, a black body, without presenting irradiation, reflection, refraction or diffraction!

In 1981, M. Milgrom presented an alternative to dark matter, to explain the same phenomenon of the rotation of galaxies, named Modified Newtonian Dynamics-MOND, according to which the gravitational force would be stronger at long distances, when the gravitational acceleration is extremely small. The theory would appear to be consistent with the observations, and contained a relativistic version developed by Milgrom and J. Bekenstein In 1984 called as gravitational scalar vector tensor theory. In 1982, Rubin presented that in the rotation of galaxies, the redshift should be attributed to the galaxy with a lot of undetectable mass, the NGC3067, with almost 16 times greater than the estimated mass of the visible matter. If the universe had a density that would negate the expansion to the gravitational balance between all galaxies, it would be critical mass density. And by the relation of general relativity involving the Hubble constant and Newton's gravitational constant, by Friedmann equations, the critical mass density of the universe is estimated: $\rho = 3H^2/8\pi G$, and for $H = 15$ km/s/10^6 light-years = 50 km/s/3 x 10^7 km and $G = 6.67$ x 10^{-8} cm^{3}g^{-1}s^{-2}, resulting 4.5 x 10^{-30} g/cm^3, that is, like 1 hydrogen atom = 1.66 x 10^{-24} g would be ≈ 3 hydrogen atoms per cubic metre. And the ratio between the average mass density and the critical density called omega, Ω. The estimation of the average density of visible and dark mass was calculated in the similar value of the critical density, i.e., 4.5 x 10^{-30} g/cm^3. So, $\Omega = 1$. With a lot of mass by volume, that is, a lot of density, the universe would be very crowded. With little mass, the expansion would result in freezing and with intermediate density there would be a stable universe. But by general relativity, the mass curves the space resulting in gravity, the universe would be closed as a space globe for greater mass density; and the universe would be opened with space opening by the sides in the case of low mass density; and for an intermediate density it would be a flat universe with the space of minimal curvature or undulations. So, $\Omega > 1$ = closed universe; $\Omega < 1$ = Open universe; $\Omega = 1$ = flat universe.

In 1986, M. J. Geller, J. P. Huchra and V. Lapparent observed large structures or walls of galaxies.

However, despite these large structures, there were also immense gaps, which would present a missing mass, which would require the need for invisible dark matter.

The calculation of astronomers pointed out that 10% of the visible matter would be formed of large structures, clouds of gases, dust, dwarf stars, neutral stars, black holes and massive objects of compact halos of galaxies. And 90% would be of invisible matter that would have been formed in the initial instants of the Big Bang. The interaction with the matter would only be gravitational, without the electromagnetic, that is, without detecting radiation from the spectrum. Dark matter could be formed from other types of atoms without interacting with the formed matter of atoms of protons, neutrons and electrons or other origins.

But, the electron neutrino, the muon and the tau, they could present a total mass around 100% of the invisible matter, if the neutrinos had an energy of 30 electron-volt that corresponds to 30 x 1.6 x 10^{-12} erg/3 x 10^{20} cm^2 s^{-2} = 5.3 x10^{32} g, that is, 0.000003% of the mass of a hydrogen atom. Other invisible materials could be cosmic strings of a dimension, according to A. Vilenkin, Kibble, Y. Zeldovich, N. Turok, D. Spergel and D. Bennett. Also, particles called mass of weak interaction, and another particle of mass < $m_{electron}$ x 10^{-12} g = 9.1091 x 10^{-36} g and infinite virtual life, according to 1975 R. Pecci, H. Quinn, Weinberg and Wilczek, named axion by Wilczek, and which would have been formed also at the beginning of the Big Bang of the disaggregation of quarks to form protons and neutrons and that would be the dominant particle of matter in this phase transition. The interaction with the real matter would be zero. Also, small quarks and small black holes.

In 1986, B. Paczynski presented that dark matter could be observed through gravitational lenses presented by Einstein in 1916 and that in 1917 Zwicky considered that the probability of a galaxy being a lens for dark matter would be almost certain, In which in 1976 a gravitational lens was observed.

In 1988, Arp presented that in the cluster of galaxies COMA which is a cluster of galaxies at a distance of 305 x 10^{21} km or 3.2 x 10^8 light-year of the Earth with about more than 1,000 galaxies identified, with the central region dominated by two giant elliptical galaxies, NGC 4874 and NGC 4889, was an extreme case of proximity galaxies between them with redshifts systematically larger, but caused by a shift to intrinsic red, therefore no discrepancy between the detected matter and the estimated matter, or "missing mass". And in 1988, P. Marmet and G. Reber presented that light would lose energy by colliding with an electron, in which P. Marmet calculated this effect for the Sun and explained a redshift, observed long ago between the edges of the photosphere and the center of the Sun. Also, in 1988, Surveillance and J. C. Pecker presented the redshift by loss of light energy through interaction with virtual particles of quantum vacuum.

But, also in 1988, Arp had observed about the energy-loss mechanism of light by its interaction with the electrons scattered in interstellar space, which in the smallest galactic latitudes where stars were seen through a growing density of gas and dust, of the Milky Way, and until they were almost totally obscured, there has never been demonstrated any increase of reshift for these stars seen through this increasing quantity of material, and indicating their hypothesis of the intrinsic redshift.

In 1989, K. Bibber and L. Rosenberg they searched for the axion by the Axion Dark Matter Experiment-ADMX.

In 1992, E. J. Lerner presented that the gravitational attraction between the rotating galaxies could be explained by the white dwarf stars present in these galaxies, with masses of 10^{11} solar masses and the galactic plasma in ultraviolet detection and X-rays. According to Lerner, this detection was questioned, but dissipated by new observations. The projects Dark Matter-DAMA, and other experiments aiming to detect dark matter presented contradictory results.

In 1983, Sikivie had presented that the axion in an intense magnetic field could decay into a photon that would be detectable. And in 1997 Bibber and L. Rosenberg built an equipment of a resounding cavity of intense magnetization that if the axion were disintegrated into photons it would emit a signal of microwave.

In addition to the ADMX and DAMA, there are other experiments currently being conducted, such as the Cryogenic Dark Matter Search-CDMS, Dark Matter search experiment-

XENON, Cryogenic Rare Event Search with Superconducting Thermometers-CRESST, Directional Recoil Identification From Tracks-DRIFT, Argon Dark Matter-ArDM and Chicago Observatory for Underground Particle Physics-COUPP.

In 2007, A. Riess mentioned that there were 3,000 publications on dark energy, in which the origin of this energy was indicated: holographic, cosmic strings, branes, axion, quintessence or ether inflation, energy coupled to dark matter, ripples universe, other types of ether, oscillations of black holes, tensor-scalar gravity, 5-dimensional space, transformations of empty space of quantum vacuum and other indications.

However, with the neutrino mass and the fields consisting of m'c x c = h' by the theory of quantum impulse in the whole universe, there would be no need for this invisible or "ghostly" matter!

Cosmic Microwave Radiation – Gamow and Alpher in 1948 had presented that in the early universe with the temperature of 10^9 °K should be made entirely by radiation. And between 1948 and 1956, as Gamow, Alpher and Herman, part of this radiation, which would have cooled, would be present in the current universe. And in 1965, A. Penzias and R. Wilson when researching the metallic antenna radio telescope in which electric charges with their fields interacted with the field of electromagnetic waves of radio, causing forces of type F = E x q, (electric field x charge), with the origin of these electromagnetic waves they also detected noises that would originate from electric equipment and other equipments. However, by removing these noises there was another noise that was continuous and coming from all directions. But, R. Dicke, D. Wilkinson and Peebles were building an apparatus in order to detect the cosmic background radiation, predicted by the models of Gamow and themselves, and explained the origin of this noise detected by Penzias and R. Wilson. According to Gamow, Alpher, Herman, and also R. Dicke, D. Wilkinson and Peebles, the temperature of the universe was 3,000 °C to the instant that the primordial light with 0.001 mm wavelength began to spread permeate the neutral atoms by combining of the nuclei with the electrons. And considered by the current Physics that the wavelength of the primordial light was "stretched" along with the own space depending on the theorizing of the space balloon, the primordial light should be at the current time with the wavelength of 1 mm, of microwave, which was the source of noise.

With the discovery of cosmic microwave radiation the model of the universe Big Bang would appear according to the beginning of the universe, however galaxies were formed in discontinuities in space, and different densities of matter should result in different wavelengths this radiation, in that in places of higher densities the gravitational pull should decrease the pulse of radiation emission and consequent lower wavelength, in which the estimated average density of the universe was 1 x 10^{-29} g/m^3. The assumption was that these different wavelengths would come from the time of the start of the Big Bang, so without considering the wavelengths that came from a galaxy of high density and another at the same distance with small density that also it would result in different gravitational attractions. In 1970, E. R. Harrison and also Y. Zeldovich considered that quantum fluctuations could result in structures such as superclusters. That is, quantum fluctuation of probability of quantum vacuum formed of virtual particles from the initial study of Dirac particles. In 1973, E. P. Tryon had presented that the universe could be originated from the quantum vacuum, relating waves like r radius-sphere, bubbles, increasing and decreasing volumes in fluctuations in time t, of type r = a(1 − cosy), where y = 0 to 2π, and t = b(y − sen y), a = $K_G M/E$ and b = $K_G M/E^{3/2}$ are constant, with K_G = Newton's gravitational constant, M = mass, E = energy, related to the space-time of general relativity, temperature, velocity, density and other parameters of quantum mechanics.

In 1974, Hawking had brilliantly calculated that at the edge of the black holes, on the scale of quantum mechanics would occur the fluctuations of space-time, radiation-mass-energy, variations of densities and thermals, in which at a lower peak the gravity would fall allowing radiation emission, later named Hawking's radiation. According to Hawking and Penrose, the space-time singularity would only be a geometric point, but on the small scale of quantum mechanics, space-time fluctuations would occur. And in 1975, Hawking and G. Gibbons applied quantum gravity to the universe in which they would result in thermal fluctuations.

Between 1976 and 1977, G. Smoot, Gorenstein, R. A. Muller and J. A. Tyson sought these different wavelengths, through high-precision aircraft of reconnaissance by a higher altitude of flight with highest terrestrial and spatial visibility. They also sought to determine the motion of the Milky Way in relation to cosmic microwave radiation and to see if the universe revolved around this radiation. If there was the rotation of the universe, there would be variations in temperature in this radiation, even if the expansion of the universe was uniform as a globe or a particular direction was present. Also, this radiation could be detected as equivalent to the absolute space and consequent absolute motion. The Earth revolves around it with 0.46 km/s and around the Sun with 29 km/s, while the solar system has the rotation of 250 km/s around the Milky Way. So they sought to obtain the local rotation of the Milky Way around this radiation of the universe. But, the difference in the celestial temperature found was zero. However, they have obtained a difference in wavelength and between two opposite directions to the flight, that is, they rediscovered by the Doppler effect the motion of the Sun-Earth and the rotation of the solar system around the center of the galaxy of the Milky Way in space, in which a temperature difference was detected between the direction and opposite to the motion, and that the meter when going to meet the waves of this radiation caused the decrease of the wavelength or the spacing resulting in a higher wavelength of this radiation. Thus they obtained ≈ 600 km/s to the galaxy of the Milky Way in the direction of the constellation of Leo and a component of 390 km/s to the Earth-Sun; in 1976, also B. E. Corey and D. Wilkinson obtained the motion of the Earth-Sun with ≈ 270 km/s and D. Muehlner and R. Weiss with 320 km/s; while previously in 1951 and up to 1992, by G. Vaucoulers, W.L. Peters, Rubin, W. K. Ford, N. Thonnard, M. S. Roberts, J. A. Gordon, N. Visvanathan, Sandage, A. Yahil, G. A. Tammann, P Schecter and others, by redshift of galaxies was obtained ≈ 299 to 700 km/s in the direction of Virgo and Orion; in 1974, by Marinov with sprocket wheels in ≈ 303 km/s for the constellation of Leo; from 1983 to 1987, by Silvertooth and S. F. Jacobs with mobile platform displacement with ≈ 378 km/s in the direction of the constellation of Leo; meaning a velocity of ≈ 350 km/s to the component of the Sun-Earth in the direction of the constellation of Leo. In 1992, Smoot with the satellite Cosmic Background Explorer-COBE came to detect a variation of density in the cosmic microwave radiation that was equivalent between the current galaxies and the matter of lower densities, interpreting the hypothesis that these different wavelengths they would come from the time of the start of the Big Bang, in quantum fluctuations, so without considering the wavelengths that came from a galaxy of high density and another with small density, which would also result in different wavelengths. Like this, Hawking considered only the hypothesis of the variation this radiation as originated at the beginning of the Big Bang, even without the origin of this density difference, but stating that it would be the greatest discovery of the century and even the history of science! and S. Singh stated that: *"The term Big Bang implies some kind of explosion, which is not a totally inappropriate analogy, only that the Big Bang was not an explosion in space but space. Similarly, the Big Bang was not an explosion in time but time. Both space and time, froam created at the time of the Big Bang".* However, Narlikar, F. Hoyle and G. Burbidge presented that the cosmic microwave radiation was only the result of the diffusion of starlight. And between 1995 and 2003, Lerner presented that the energy released in the production of helium was enough to explain this radiation, through the high energy electrons that in spiral motions, around lines of magnetic fields, would be emitting and absorbing the radiation in all directions of light released in the production of helium, and used spatial telescope data COBE and Wilkinson Microwave Anisotropy Probe-WMAP to corroborate that theory.

The RMC has been widely investigated through the space telescopes COBE launched in 1989, and WMAP released in 2001, the balloons Ballon Observations of Millimetric Extragalactic Radiation and Geophysics-BOOMERANG released in 1997, 1998 and 2003, and Millimeter Anisotropy Experiment Imaging Array-MAXIMA released on 1998 and 1999, and the terrestrial telescopes Very Small Array-VSA, Degree Angular Scale Interferometer-DASI and Cosmic Backgroud Imager-CBL. And another release was the Satellite for Cosmic Background Radiation Anisotropy-High Frequency Instrument-PLANCK-HFI in 2009 for the purpose of improving cosmic microwave radiation's observations.

Cosmic Plasma – In 1913, K. Birkeland was the first to predict that space would be permeated by plasma. Plasma cosmology presents that the study of the universe, consisting mainly of plasma, it must take into account electromagnetic interactions beyond the only used of gravitational interactions. In 1937, H. Alfvén noted that if the universe was permeated with plasma, then it could be generated a galactic magnetic field. From the work of Alfvén, then, structures on astronomical scales began to be studied from the theoretical framework of plasma Physics. In 1969, A. Peratt simulated the formation of different types of galaxies from the cosmology of plasma and compared its results with those of Arp came to find many similarities. But the current Physics disregarded have magnetic fields on these scales. And from 1992, Lerner came to present models to explain the quasars, the formation of the galaxies and the hierarchy of the universe in galaxies, clusters and superclusters. And its model of quasars was detected according to its prediction by the radio telescope Very Large Array-VSA, but this Physics claiming that the connection between galaxies and plasma physics was disregarded. And as Ostriker would lack evidence for electric and magnetic forces to be important in cosmological scales. Nevertheless, Lerner continued to assert that the evidence existed, for example, the evidence collected by the space probes Pionner and Voyager in the 70, near Jupiter, Saturn and Uranus.

And explained the distribution of chemical elements, in which the cosmic rays generated by the first stars of the galaxy, in collision with the plasma environment, would be able to generate the rare substances that are not created by stellar nucleosynthesis.

"Ad Hoc" Inflationary – In 1981, A. Guth to explain the apparent flatness of the universe presented by Dicke and Peebles in 1969, and the uniformity between the end of a point and the other opposite of the horizon of the universe presented the hypothesis of the inflationary universe in which there would be a modification of the structure of the initial universe of the singularity, forming a momentary vacuum and consequent negative pressure and gravitational repulsion, in which in 10^{-35} seconds the universe would double from space every 10^{-37} seconds, that is, 100 duplications. The apparent flatness of the universe, in which galaxies are observed as more in a plane than in a sphere, because the extremely expanded sphere would appear from the place of observation in the universe as a plane, would be explained by inflation. The similarity between the most opposite and remote regions of the universe would be achieved by inflation by expanding regions that were previously close. In 1981, K. Sato and Sciama observed that quantum fluctuations of inflation could explain a greater variation of density for the formation of galaxies. So the universe would have imperceptible variations of density, but inflation would have enlarged these variations that would give rise to the galaxies.

In 1982, M. Turner, P. Steinhardt, A. Albrecht and A. Lindei in separate studies presented modifications to the time of continuation of inflation and core of initial or bubble formation, of inflation, in which in the transition to form particles and energy, the gravity began to exert attraction and braking the expansion. The inflation of space was treated with the unification of all the forces surrounding the particles: electromagnetic, weak, nuclear and gravitational, in which the unification of these interactions occurred. And involving the phase transition between radiation, energy densities and particle formation. Thus, it would also explain that the hypothetical particle of Dirac of magnetic monopoles at this early stage of the universe would only be found if the Big Bang were of a slow expansion, but with inflation would be found only one monopole in the universe. And inflation would explain the lack of rotation of space, in which the minimum fraction of time expansion of 10^{-35} s would tend to annul a possible rotation of the universe in the phase up to 10^{-40} s from the beginning of the universe.

And then it was presented that inflation would cause the formation of many more initial formation nuclei or bubbles that could constitute other universes, the multiverse.

Therefore, inflation would cover the expansion of space, using the analogy of the air balloon which expands a point on the surface without that point having any proper drive! The inflation of space expansion-nothing-no of the imposed Physics!

Initial Conceptualization of Gravitational Waves – In 1917, Hilbert and in 1923 Eddington had presented on the basis of the general relativity of 1916 of Einstein, studies on gravitational waves and in 1934 Blokhintsev and F. M. Gal'perin named the gravitational quantum that would form gravitational waves as graviton. In 1935 Dirac presented gravitational quantification. This quantization of gravity would be the quantum of space-time propagated in waves, gravitational waves. In 1937 Einstein and Rosen presented cylindrical gravitational waves by general relativity. In 1941, Landau and Lifshitz, and in 1955 V. A. Fock showed gravitational interactions as gravitational waves originating from the quadrupole denomination of Einstein's equation. In 1957, Bondi presented flat gravitational waves by general relativity, and Wheeler and J. Weber analyzed the cylindrical waves of Einstein and Rosen. In 1958, L. Bel presented gravitational quantum studies by general relativity, G. Gaposhkin and P. Kamp presented gravitational waves of binary stars. In 1959, Bondi, Pirani and also I. Robinson presented gravitational waves of the general relativity equation and R. L. Arnowitt presented gravitational waves based on Dirac's gravitational quantum. In 1961 and 1962 R. K. Sachs presented gravitational waves in asymptotic curves plans, and Bondi, M. G. J. Burg and A. W. K. Metzner studied the gravitational waves of isolated systems. In 1963, Schwinger and Feynman presented gravitational quantification by quantum field theory. And in 1964 Weinberg presented the probability of emitting how much gravitational waves by quantum mechanics.

However, while Weinberg presented a probability of gravitational quantum emission, Mesquita in 1965 presented the discovery of this emission of gravitational quantum constituent of gravitational waves. According Mesquita, these waves were made up of elementary energy radiation h' = m'c x c at the frequency of 10^{19} cycles per second, of low-intensity, coming from the atomic nucleus, and forming a gravitational field.

Between 1960 and 1968, J. Weber had presented large aluminium cylinders to detect gravitational waves by resonance. In 1965 V. S. Braginsky discussed how to detect gravitational waves. And in 1968 J. Weber discussed the possibility of the pulsar NP0532, in the nebula of Crab, to be a gravitational wave emitter. In 1969 and 1970 J. Weber announced that he had found experimental evidence of gravitational waves, as he had observed the coincidence of the pulses of these waves in the aluminium cylinders.

Pulsars – In 1967, J. B. Burnell observed a non identified sent signal, in the Crab Nebula of precise periodicity of radio waves, seeming message transmitted by intelligent life in which he named this sign of "little green man".

And in 1968, A. Hewish, Burnell, J. H. D. Pilkington, P. F. Scott and R. A. Collins presented the discovery of a star of the order of the solar mass and radius of the order of 10 km, and which revolved around it itself with a period of ~ 1.337 s. This star received the name of CP 1919 , where CP means Cambridge Pulsar. And in 1973 new 100 pulsars had been discovered. and R. A. Hulse, J. H. Taylor, R. N. Manchester and G. R. Huguenin observed 22 pulsars and Hulse discovered more 40 pulsars between 1973 and 1975. One of them in the constellation of Eagle, he found that it was different from a single star and detected it as a binary pulsar, called 1913 + 16, in which Hulse and J. H. Taylor, found a orbital speed of ~ 300 km/s, with dimension from the orbit of the order of the Sun radius (6.96 x 10^8 m), masses of the binary system of the order of the solar mass (1.99 x 10^{30} kg), and period of 7:45'. And by general relativity, if that binary pulsar system emitted gravitational waves by losing power, the stars of this binary pulsar they would approach each other and consequently there would be a difference in their orbital period. Considering that possibility, J. H. Taylor, L. A. Fowler and P. M. McCulloch began to check whether there was such a difference in the period of the PSR 1913 + 16, and in 1979 presented the first result of this variation, and in 1982 J. H. Taylor and J. M. Weisberg presented the result of 2.40 x 10^{-12} s/s, against a theoretical Einsteiniano value of 2.403 x 10^{-12} s/s, that according to J. H. Taylor was an experiment that presented clear evidence of the existence of gravitational waves. In 2001, S. Kopeikine presented how to effect the detection, through the planet Jupiter, of a gravitational lens to interact with the light emitted by the quasar (JO842 + 1835) that would align with that planet and Earth in 2002, and Kopeikine and E. Formalont used an arrangement of telescopes terrestrial, as the Very Large Baseline Interferometry-VLBI, and through general relativity have found for the gravitational wave speed c-value of the speed of light, which they presented between 2003 to 2007.

And since 2000, teams of search projects by interferometry have begun their search for gravitational waves.

Relational Mechanics in the Universe – Mach between 1868 and 1912 had the realization that the fact that the set of fixed stars did not rotate would coincide with the absolute space. And without mass the absolute space would have no force that could be exerted on the oscillation of the pendulum or any other mass of the universe. Therefore, the mass of the fixed stars is that it would be the real cause acting on the oscillation of the pendulum and all the masses of the universe. Then it would be a force of the interaction between the motion of the planets, like the Earth, and the oscillation of Foucault's pendulum, with the distant masses of the universe. And as Mach the resistance of the mass to the change of motion, the inertia, or to a uniform motion, that is, an inertial motion would have as origin the interaction with all the masses of the universe. That is, the forces of inertia would be originated entirely by the other masses, which would later be named the principle of Mach. In 1846, W. E. Weber had presented a force acting between relative moving electric charges, from the attraction and repulsion of Coulomb and Ampère forces, in which the electric or magnetic force appears related for the first time with the velocity and acceleration .

In 1989, as Assis applied the interaction of Weber's charges with Newton's gravitation in the interaction between gravitational masses identified the kinetic energy with the inertial mass which explained the proportionality or equality of the mass of inertia with the mass gravitational. That is, inertial mass, motion resistance, uniform motion, or inertial motion, was a gravitational interaction with the universe. The masses of inertia and kinetic energy emerged as a direct consequence of the gravitational interaction with the universe.

Thus, in the circular motion as the rotation of the Earth there was a centripetal force for the center of rotation and the real force $F = m_{Earth}\, a$ of gravitational reaction exerted by fixed stars and external galaxies. And this acceleration a of the Earth would occur in relation to the fixed stars and external galaxies, and not in relation to the absolute space. And Assis presented brilliantly the deduction of the actual centrifugal force that would result if the fixed stars and outer galaxies revolved around the Earth.

And also, as Assis, $R_{Universe} = c/H_{Hubble}$, it was the radius of the universe,

And the Doppler shift to the red:

$$\Delta\lambda/\lambda_0 = [(\lambda_{observed} - \lambda_{emitting})/\lambda_{emitting}] = V_{galaxy/Earth}/c = R_{Universe} H_{Hubble}/c$$

Expanding Galaxies – By redshift it resulted in a marked expansion of the galaxies. And galaxies were observed by this redshift with velocities between them greater than that of light, value that led physicists to conclude that it was the space itself that expanded to prevent the speed of light being exceeded, which is the maximum speed of bodies according to Einstein's theory of relativity.

But in the galaxies space would also inflate, but the distances between the stars would remain the same, immutable, for example, the Earth, the Moon and the Sun would remain with the same distances between them, with a fantastic equality between the expansion of space and the attraction gravitational of bodies.

Supernovas – It had appeared in China's firmament in 185, a superbright visitor star and others in 1054, 1181, 1572, 1604, 1769 and 1784. They were stars that varied from brightness, and called new ones, but that were actually stars that almost faded presented in an instant an intense brightness.

In 1934, Zwicky and Baade presented that the nucleus of a star could present nuclear reactions and collapse, based on the recent discovery that had been from the neutron in 1932 by Chadwick. With the implosion of the star would be created a wave of intense vibration expelling the outer layers of the star, in which the nucleus of the stars would be formed by the neutrons, emitting an intense brightness, a supernova (or supernovae).

In 1937, the first supernova was discovered in the NGC4157 galaxy. And in 1938, Baade and Zwicky presented a description of the brightness periodicity curve of a supernova.

In 1940, R. Minkowski observed a supernova that presented hydrogen in its constitution, which would be a larger star than the mass of the sun resulting in implosion of around 65,000 km/s. Another supernova was observed by Zwicky, being similar to the Sun, which at the end it lost the outer layer of hydrogen, contracting to a nucleus, a white dwarf with mass to the diameter of the Earth of intense gravitational field and attracting gases. In 1930, Chandrasekhar presented that this star with 1.4 times the mass of the Sun would explode.

Around 1953, as Hoyle, when the final phase of the nucleosynthesis of larger masses was irreversible, an implosion of greater intensity was occurring, releasing an energy similar to 10 billion stars of smaller masses, and these stars emitting a superbrightness, supernova, and presented the most extreme nuclear reactions to the production of major masses.

The cepheids variables were visible in the nearest galaxies and for distant observations would require another type of higher luminosity pattern, and that would be the search for supernovas studied by Zwicky and Baade in 1934 and discovered in 1937 by Zwicky. There would be two types, I and IIH, both with the explosion. But then it was detected that there was type IbH and type IH was named Ia. In 1979 J. Maza presented an early search for supernovas.

In 1981, Muller, S. Perlmutter and C. Pennypacker began the search for luminous points that occurred in comparing previous dates, by computers, in distant galaxies to detect the supernova, which was found the first by them in 1986 and 3 up to 1987, 500 planned since 1981. For, in addition to taking into account, the dust, gases in the supernova galaxy and the intergalactic distances to the observer, the spectroscopic analysis with the redshift and the photometric to the brightness, the supernova were fast in the period of brightness and rare, and it would be necessary to enlarge the observation covering a set of larger number of galaxies. And to obtain the curves of rise and fall of luminosity, in which the maximum peak is reached in a few days and falling in longer time, which would be the pattern of distance of a supernova, was used the dual-charge device that captured a greater intensity of light. And Muller, Perlmutter, Pennypacker and G. Golhaber in 1992 found a supernova with the largest redshift until then found, 0.45, equivalent to 4.7 billion years.

In 1989, N. Suntzeff also initiated the search for supernovas, and in 1991 Suntzeff, M. Phillips and M. Hamuy found a very bright Ia 1991T supernova and another bright little Ia 1991bg. However, the light curve in the highest brightness rose and descended more elongately , and in the smallest brightness the ascent and descent were narrower. Phillips obtained other similar supernovas, and in 1992 he concluded that these two types of supernova could be used as the standard, because the two types could be compared with nearby supernovas of standard luminosity with the type of the standard curve of each of the two supernovas. On 1994 B. Schmidt, P. Garnavich, R. Kirshner, B. Leibundgut, Phillips, Hamuy, Riess and others, formed another group of search for supernovas, and from 1986 through Suntzeff and others getting around 30 supernovas closer to the Ia type. While in 1994, Muller, Perlmutter and Pennypacker got supernovas farther away in smaller times.

In 1995, Hamuy and also Riess presented the curves of light and distances of supernova to calculate the Hubble constant, finding between 62 and 67 with Hamuy and 67 with Riess, it would correspond to a time of 10 billion years to the universe, which would be less than time of formation of the older stars, as noted in the large structures or walls of galaxies by Geller, Huchra and Lapparent in 1986.

In 1997, Goldhaber noted that the supernova showed by the luminosities and redshift an expansion of the universe in which there was missing mass by up to 40%. And Riess found 36% of missing mass. And by the data of the Hubble Spacial Telescope-HST, P. Nugent and Garnavich observed that the supernova presented between 1.5 to 2 times less bright, i.e. they were farther apart, then less brilliant than expected for the estimated density of the universe of 1. So the constant ω of the universe would be < 1 and the universe should expand forever. And for the space to be with this larger expansion that surpassed the gravitational pull should present a positive value for the cosmological constant lambda, Λ. One consequence was that the current expansion indicated that in the past the velocity was much smaller, so there would be a greater time since the beginning of the Big Bang to the present, that is, a time greater than 15 billion years for the universe and greater than the time for the formation of the stars. And another consequence is that with 30% real and dark mass, plus 70% energy to overcome the gravitational pull, that is, $0.3 + 0.7 = \Omega = 1$, so the universe was as observed, that is, flat.

And in 1998, Turner named the energy that would trigger the acceleration of the empty space of quantum vacuum or false vacuum of dark energy, considered responsible in the model of the Big Bang by cosmic acceleration.

And the supernova, between 1991 and 1998, as Permuter, Ries, B. Schmidt and others, used as standard for measurements of distances between galaxies, were mentioned in 1997 by E. Wright to invalidate the model of "tired light" by not predicting the strong influence of the reshift in the middle superficial brightness of the galaxies and the dilation of the time of supernovas moving away at high velocities, in which the elongated pulse of this type of supernova would be due to the dilation of the decay time of its particles.

But in quantum mechanics by the energy of the zero point space of the quantum vacuum void, the energy of each electromagnetic field fluctuation frequency, in which the sum of the energies that correspond to all frequencies V from 0 to ∞ is $E_{vacuum-quantum} = \sum_v \frac{1}{2}(hV)$, therefore an infinite energy! Then, completely contrary to the energy of 0.7 Ω. Or, by another calculation in which it is obtained by quantum mechanics the zero point energy of the empty space of the quantum vacuum in 10^{120} times greater than 0.7! One of the two theories is far from physical reality! As A. Filippenko there is no theory that is compatible with a rational energy physically of a lower value than 10^{120} times the energy of the vacuum obtained by the supernovae as a measure of distance, and that it would be unacceptable to discard a theory based on which the computer indicates to accept that 2 + 2 = 5!. Or, dark energy would be another epicycle created to mathematically find a solution to the expansion of the universe by supernova!

It would be necessary to prove that the supernova really showed the distances that were obtained. And it was observed brilliantly by Riess that at the beginning of the formation of stars and galaxies, in the initial time of the Big Bang, the expansion of the universe would be minimal, because the density of matter would be immense and gravity would avoid greater expansion. So, in this case, the supernovae would be brighter compared to those observed in the less distant galaxies. For this, Riess should get supernovas from that time that should be in the deep field of the Hubble satellite telescope! And with the observations of R. Gilliland and Phillips in the Hubble telescope and Riess withdrawn from these observations was confirmed in 2001 that the farthest supernovas in the universe were brighter than of other times, confirming a lower initial expansion and the current accelerated universe.

The brilliant work of the supernovas, of Pelmuter, Schmitd, Riess and others between 1991 and 1998, on the accelerated expansion of the universe and confirmation in 2001 by Riess presented the incredible incompatibility between relativity and quantum mechanics, a distance greater than that of the universe with this expansion! The cosmological constant determined in 0.7 by relativity and quantum mechanics with a factor of 10^{120}!

However, it is the theory of quantum impulse that presents h' = m'c x c forming the particle field extending across the universe, with m' = 10^{-47}g, explaining the physical meaning of h, therefore the energy of the zero point duly determined in that h' is the smallest quantity of energy, preventing that it was obtained that incompatible factor of 10^{120} for the energy of the empty space of quantum vacuum.

And in 2007, A. Riess mentioned that there were 3,000 publications on dark energy, in which the origin of this energy was indicated: holographic, cosmic strings, branes, axion, quintessence or ether inflation, energy coupled with dark matter, undulations of universe, other types of ether, black hole oscillations, tensor-scalar gravity, 5-dimensional space, transformations of the empty space of quantum vacuum and other hundreds of assumptions called of "physics"!

CXL. GALACTIC VELOCITY, SPACE AND THE QUANTUM IMPULSE
1929 to 1969

Faraday's electromagnetic field and the equations of this field by Maxwell, in addition to being applied to the particles that constitute the atoms in the formulation of quantum mechanics, were also applied to fields of large extensions in the galaxies.

By quantum impulse theory the frequency is fundamental with its wavelength of electromagnetic radiation m'c x c and when the radiations are emitted from the galaxies are fundamental the variables that influence the frequency deviation and its wavelength. For example, in the case of a galaxy with a velocity of 5,000 km/s of the Earth-Sun System, in the direction of the constellation of Leo, in which the Earth-Sun System has the velocity of 350 km/s, according to the laser light experiment and interpretation of the CERN experiment of 1964 the speed of light c'_{galaxy} emitted by this galaxy and the Earth-Sun System in relation to cosmic microwave radiation or vacuum-space-ether is 300,350 km/s. The speed of light c emitted by the Earth-Sun System in relation to the Earth source is 300,000 km/s, and the speed of light c_{galaxy} in relation to this galaxy is (300,350 km/s – 5,000 km/s – 350 km/s) = 295,000 km/s. And the light emitted against the Earth by the galaxy that moves away from the Earth, in that the galaxy and the Earth are on course for Leo, it has the velocity $c_{galaxy \rightarrow Earth}$ of [(300,350 km/s) – (5,350 km/s of the galaxy's inertial velocity in the direction of Leo) + (350 km/s of the Earth's velocity to the light)] = 295,350 km/s in relation to cosmic microwave or vacuum-space-ether.

For the laser light experiment showed the inertial velocity of light in the direction of the emitting source motion, and the higher the velocity of the source, the lower the velocity of light in relation to the source. In the CERN experiment of 1964, the velocity c_{gamma} of gamma rays emitted by meson μ with velocity $c_{meson} \approx c_{gamma}$ it was ≈ zero in relation to meson μ and 300,350 km/s in relation to cosmic microwave radiation or vacuum-space-ether, with the velocity of 350 km/s of the Earth-Sun System towards the constellation of Leo, widely contrary to Einstein's relativity.

Besides the velocity of radiation by the elementary impulse, it should be considered the energy density received from the galaxy. The radiations are emitted in different directions, and with the distance occurs a lateral shift resulting in a sharp decrease in radiant energy, which would be the real cause of the spectrum redshift. So, the accelerated expansion attributed to the high velocities of the galaxies by the redshift Doppler-Fizeau in which the wavelength is increased, it would be much smaller by the theory of elementary impulse.

In 1966, Arp noted that quasars with a large redshift were linked by interstellar cloud with galaxies that presented smaller redshift, such as the association between the NGC3067 galaxy and the 3C232 quasar. And established that a large, ancient galaxy ejected material that formed smaller and younger companion galaxies around it. And the younger galaxies, in turn, would eject material that would give birth to quasars, and that the younger the object within the group, the higher the intrinsic redshift. In 2020, the James Webb Space Telescope will be able to prove Arp's observations. So the distances of galaxies around 10 billion light-years from Earth by the current redshift Doppler-Fizeau, would be in reality around 300 million light-years.

In 1930, Eddington to defend the theory of general relativity in which space-time would expand without any cosmological constant, which would be necessary if the universe were static, and also to defend the theory of the Big-Bang in which the initial singularity it should expand space-time, it presented that the expansion of galaxies, observed in 1929 by Hubble in relation to space, was the space that expanded, while the galaxies remained fixed, as a plastic balloon that when being inflated increases in distance between all points. The dots were fixed on the surface of the balloon, but the surface would expand. So for Eddington and the entire community of current Physics, the galaxies are fixed, without motion, and it's the space that expands between all galaxies, to result in the remoteness between the galaxies. And in the galaxies, all the celestial bodies would be in incredible gravitational precision, where only the galactic structures were carried and expanded by the vacuum space while among all the materials of the galaxy the space would remain without any spacing between them. That is, only the outer space of the galaxies is that it expands by the imposed Physics! In medieval times there were followers that the planets were moved by celestial creatures, and at the time of this Physics is the empty creature of the space itself that moves the expansion of the galaxies!

However, cosmic inflation presents an inflated "ad hoc" hypothesis as the field responsible for inflation, to explain densities, flatness and uniformity of the horizon, and to try to explain the expansion of space between galaxies and a fantastic gravitational equilibrium within the galaxies without any expansion between the objects of the galaxy, the inflated balloon model of space! While space would expand between galaxies leading all galaxies to expansion, space would expand between the objects of the galaxies and between the atoms, but without taking any photons from within the galaxies along with it!

The painting in the Figure 84 it presents its title as Abstractionism, and it can represent the vacuum-space-ether without any real or virtual particle, ie an abstract space or the "absolute nothing". So, it's this abstract space, this "nothing absolute" that would expand by current theories of this pre-establishead Physics!

Fig. 84 – Painting title: Abstractionism that can represent the abstract space or "absolute nothing" that would expand the galaxies by the current Physics theory of the inflated balloon! Clipart from *Editora Europa*.

To consider space as a physical dimension in which mass presents motion in this space is a fact of nature. But, a space of "nothing absolute" that expands carrying all the mass of the galaxies, while within the galaxy gravity acts with fantastic accuracy at all points that would neutralize the expansion of space would be an incredible mathematical-metaphysics and not Physics!

The Figure 85 it shows the space of "nothing absolute", transporting a spacecraft! That is, the conventional space where the moving ship is being transported by a mathematical-metaphysical space!

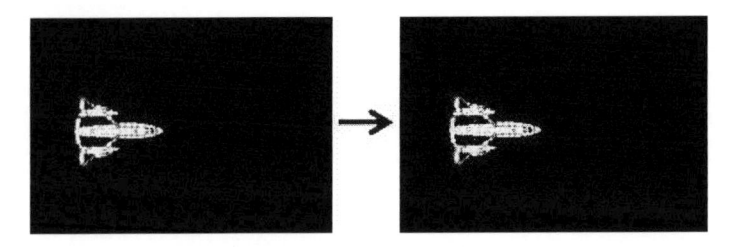

Fig. 85 – The mathematical-metaphysical space of "absolute nothing" transporting by expansion a ship in the conventional space! Ship clipart from *Editora Europa*.

Also, figure 86 it shows a rocket at the edge or curvature of the universe coming from the Big-Bang that emerged from the current Physics of a point, the singularity, being transported through space. But if the spacecraft were in what would be the result of the center of the Big Bang, this artificial space would expand in all directions keeping the spacecraft in the same position, that is, motionless!

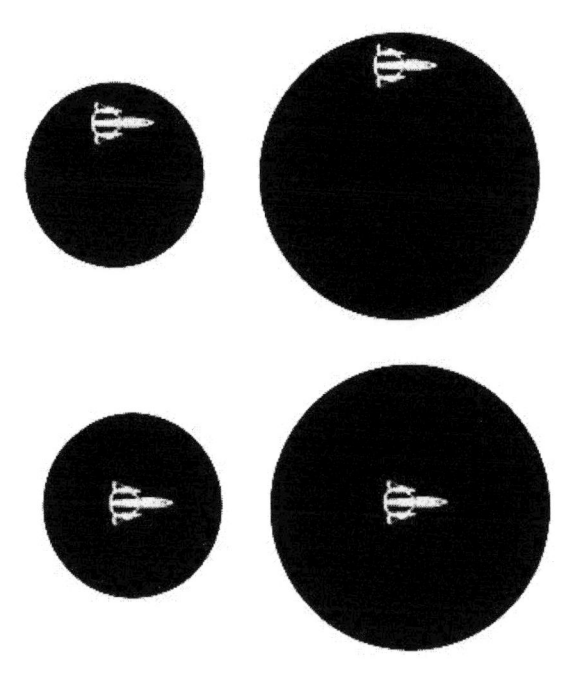

Fig. 86 – The mathematical-metaphysical space expanding and carrying the galaxies and a nave at the edge or curvature of the expanded singularity of the Big-Bang by the current physics. And a ship in the center that would result from the Big Bang, in which the space-vacuum expander and the ship remains motionless by

this artificiality of imposed Physics! Ship clipart from *Editora Europa*.

Newton mentioned about the mediation of gravity without any kind of distance interaction: *"Supposing that one body can act on another at a distance through the vacuum without the mediation of anything, it is to me such a great absurdity, that I believe no being, who had in a philosophical question a competent faculty to think of, could even accept that"*! Thus, in the relational mechanics of Mach, and completed by Assis by deducting the centrifugal force that interacts in the equations between matter and bodies, and the theory of quantum impulse that shows the quanta h' forming the physical fields of gravity and electromagnetic, the Newton's questioning is solved. As for this artificial expansion of the space-vacuum itself, in which removing all real or virtual photons, with or without mass, would remain the "absolute nothing" which would expand by the example presented by the current Physics of the inflated balloon, could be effected a reflection similarly to what Newton questioned: supposing this expansion of the vacuum-space itself, in which removing all the virtual photons or field would remain the "absolute nothing" which would expand is such a great absurdity, that none that had in the matter philosophical a competent college to think of, could even "it fall into that"!

However, gravitational attraction acts only in the presence of mass, and even if by general relativity was the space-time distorted by the galactic mass as a cause of gravitational attraction, this distortion should also be expanding.

For, by general relativity, the radius of the universe is related to time and then Friedmann in 1920 presented the expansion of space-time. That is, space-time would be expanded where there was greater or less mass presence, so the expansion would exist in the presence of greater or lesser distortion of space-time. So, next to the mass of the galaxy the expanding space should reduce the distortion of space-time near the galaxy, by the largest distribution. And it would be an infinite energy if the galaxy continued to distort space-time during the continuity of space-time expansion. And it would never finalize the energy of the galaxy that would always distort the space-time that would be carried by expansion. Einstein had introduced the cosmological constant, in the general relativity theory presented in 1916, to counteract this expansion, which was later detected by Hubble in 1929. But by the absolute space of Newton or Mach relational the expansion is zero.

For Einstein, spaces with his fields are shifted from one to the other, and spaces are superimposed. But, for Newton and Hubble it is the mass that is displaced in the absolute space, and for Mach are the masses and their interactions that are displaced in the relational space. And masses occupy different spaces without overlapping, just as two formed electrons of mass cannot occupy the same level of quantum energy as the Pauli exclusion principle.

Also by the elementary impulse a radiation does not occupy the same space of another, because the impulse is formed of mass m'. By the elementary impulse, space is the place occupied by m'c that to move with velocity c to form the elemental quantum energy h' must go through an elastic medium. And also maintain the principle of conservation of impulse and energy, of extremely minimal density of mass, stationary and widely penetrable, contrary to the ether type material of high resistance almost solid stationary and impenetrable. This ether is contrary to an ether dragged by the bodies only on the surfaces to conform to the aberration of light from the stars, the ether of Stocks with a complex motion in the style of epicycles.

PARTE 10 – UNIVERSE FORCES UNIFICATION

CXLI. UNIFICATION OF ALL PHYSICAL INTERACTIONS – 1909 to 2010

In 1914, G. Nordström had presented a unifying between electromagnetism and gravitation, projecting the 4 dimensions of space-time into a new dimension to apply electromagnetism, that is, a mathematical unification. Nordström himself concluded that it was unification without any physical sense.

And in 1919, O. B. Klein and T. Kaluza presented a curvature in a fifth dimension to cause the force and electromagnetic field, similar to the curvature of space-time with four dimensions that was the cause of gravity by Einstein's relativistic theory. This curvature of the force and electromagnetic field would be within Planck's length, $d_{Planck} = h/cm_{proton} = 1.616... \times 10^{-33}$ cm.

In 1923, Einstein and J. Grommer also presented in 5 dimensions a unifying between electromagnetism and gravitation of general relativity.

And between 1923 to 1955 Einstein sought the unification of all physical fields, obtaining that the equations of the gravitational field responded surprisingly to the electromagnetic field equations. However, concluding that quantum phenomena were absent or only with probability facts in this unification and that the current quantum theory would be improper to unite all physical fields.

Between 1909 and 1917, Einstein considered the photons of light as points that moved independently with energy $h\nu$, and that there should be a development of Physics for a fusion between wave theory and quantum. And in 1948 stated that: *"Planck's discovery of quantum theory has put science to a new goal, which should be* to find a foundation of concepts for all physics. But despite significant gains, this question is still far from a solution". And in 1951 questioned again: "After years of pondering, we lack an approximation of the answer to the question: what are the quantas of light?".

In 1949, Einstein for the unification of the physical fields directed broadly into general relativity for the purpose of moving away from the photons of light that were unfit to achieve this unification. But in 1954 and 1955 stated that the unification of physical fields would be unfeasible with the space-time of general relativity which was unfeasible to explain the atomic structure, electromagnetic radiation and all quantum phenomena. Thus, only significant progress in general relativity and mathematics would be the key to a complete reformulation of quantum mechanics.

In string theory, in which particles are vibrations of lines in different forms, 10 or 11 dimensions occur for the unification of all forces, being six compressed in $1.616... \times 10^{-33}$ cm.

Einstein in three decades sought the unification of Physics with quanta of light and general relativity, in which mathematical abstractions and statistics were formulated in electromagnetism and gravitation, but remaining far from the goal of unifying the fields of Physics. And string theory, with a brilliant development of Mathematics, as Einstein indicated was necessary, came to present a unified Physics with solid mathematical interactions of 11 or n dimensions, but with a physical foundation of hidden dimensions, That is, unification without physical strength!

However, the theory of the elementary impulse of Erdelyi and the gravitational radiations of the Mesquita present the unification of the Physics of all the fields that form the energy of the particles and their electromagnetic and gravitational fields, with m'c x c = $m'c^2 = h'$, $\mathcal{E} = \mathcal{V}h = |\mathcal{V}|$ h' to get the equation of the theory of everything, which synthesizes the equations of everything dreamed by Hawking in 2010. Then, by : Erdelyi's quantum impulse and Mesquita's gravitational radiation, the equation of everything: $F_G \times d = [(K_G m_1 m_2)/d^2] \times d = n\mathcal{V}_G h$, where F_G is the gravitational force in the distance d, m_1 and m_2 masses, n is the quantity of $\mathcal{V}_G h$, $\mathcal{V}_G$ it is the frequency of gravitational radiation in which h/1seg = h' = m'c x c the elemental photon of energy, which gathers the Newton's gravitational K_G constant and the Planck's h constant, with a total physical significance.

CXLII. CREATION IN THE QUANTUM VACUUM: MAKE UP THE LIGHT! 1970 to 2011

In 2011, Physicists from the University of Chalmers in Sweden in publication to Nature presented that they managed to create vacuum light, as the deepest extraction to exist, presenting that the vacuum could be captured light that would be measured. It is an experiment that would confirm a theory of quantum mechanics, in which the vacuum would be a sea of infinites particles that would fluctuate and exist only for a shorter time.

Known as the dynamic Casimir effect, Gerald T. Moore presented in 1970 a study in which virtual photons would become real photons of light if reflected in a mirror that would move near the velocity of light. The mirror at this velocity would provide energy for the quantum fluctuations of the vacuum, and in which mathematically applied the calculations of Dirac's quantum mechanics of particle creation and mathematical supplements to keep the virtual photons in different locations for a while until they received the energy of the mirror motion that would result in real photons.

And an international group of scientists replicated this relativistic reflective effect, providing kinetic energy to virtual particles to materialize. *"Since it is not possible to make a mirror move fast enough, we have developed another method to achieve the same effect. Instead of varying the physical distance to a mirror, we vary the electrical distance in a short electric circuit that acts like a mirror for microwave. Instead of moving something massive we move something that is massless"* said the co-author of the study P. Delsing of Chalmers University of Technology. The researchers used an electronic-quantum component called the superconducting apparatus of quantum interference, which is highly sensitive to magnetic fields. The direction of its magnetic field can be changed billions of times per second to create a vibrant "mirror" above 5% of the velocity of light.

Explained delsing that: *"The result was that photons appeared in pairs from the vacuum, which we could measure in the form of microwave radiation. We were also able to verify that the radiation possessed exactly the same properties that quantum theory claims it should have, when photons appear in pairs that way. "The mirror", that is, the short circuit transferred some of its motion energy to virtual photons so that they could materialize".*

According to quantum mechanics, there are a variety of particles in the vacuum. The photons would appear in the experiment because they wouldn't have mass. According to co-author G. Johansson in the statement: *"Relatively little energy is required to excite them from their virtual state. In principle, it could also create particles from the vacuum, such as electrons or protons, but that would require a lot more energy. But the photons are more possible, because the equivalence of energy and mass, as Einstein implies that the photon being without mass can be stimulated out of his virtual state with relatively little energy".*

As T. Spiller of the Leeds University-United Kingdon: *"But, you're not giving away anything free. The energy comes from the motion. In the vacuum there are fluctuations that if put energy it can return with real photons".*

Other physicists considered that this research was important because it would advance in the basic understanding of the concepts of Physics as fluctuations in the vacuum, which could be related to the dark energy force, which would be linked to the constant acceleration of the expansion of the Universe.

However, the energy supply for the oscillation of the short electric circuit, and the measurement detected as electromagnetic microwave radiation, proved in reality that an increase of energy to the vacuum containing an energy field, which by the theory of quantum impulse is formed of m'c x c = h', it results in the formation of nh', ie $|\mathcal{V}|h' = \mathcal{E} = h\mathcal{V}$ which is the quantum detected as energy of electromagnetic microwave radiation by this Swedish experiment of 2011.

CXLIII. THE GENERAL SCIENCE JOURNAL AND PHYSICS ESSAYS
2015-2016

The Journals *The General Science Journal* and *Physics Essays* publish scientific articles that present different analyses of fundamentals of Physics, which should be an invaluable source of studies by Physics, but which is ignored by the current Physics because there is disagreements with their myths. There are, in these Journals and on the sites in References, hundreds of articles that show theories and experiments that should be observed by own scientific method of Pre-Established Physics which refuses to apply it to cover its outdated fundamentals. On June 4, 2015 (republished the Abstract on March 10, 2016 by copyright in other Journal), the author presented a paper to The General Science Journal on his online website, *"Surprising Result of Inertial Velocity of Photon and Its Consequences in the Physics after 100 Years of the General Relativity and 110 Years of the Special Relativity in 2015"*, which publishes Scientific Papers with scientific responsibility of the authors, therefore without blocking of publication on the part of the current Physics. The article presented that the reality of fields and particles should be made up of elementary photons, which would physically explain all the forces of nature, in which the m' elementary mass of this photon with velocity c (by measurement on Earth) or c' (in relation to the vacuum-space-ether) characteristic of electromagnetic radiation including gravitational electromagnetic radiation, it was the m'c/time, force, forming the elementary energy (m'c /time) x c x time = h', where h' has the value of Planck's constant, but with the dimension of energy. And its consequences pointing to the inconsistency of Einstein's theory of relativity and the current quantum mechanics. So there was little to commemorate the 100 years of general relativity and the 110 of special relativity. Document 29 .

J. P. Hansen

the general science journal

Surprising Result of Inertial Velocity of Photons and its Consequences in the Physics after 100 Years of the General Relativity and 110 Years of the Special Relativity in 2015

June, 2015

J. P. Hansen

Engineer by the Mackenzie University - São Paulo, Brazil
Member of Engineering Institute and Science Progress Brazilian Society-SBPC - Brazil
jph.hansen@gmail.com

Abstract

This work relates consequences of the result of laser experiment presented in 2002 [1] about inertial velocity of photons in cosmic background microwave or space-vacuum, velocity no detected by MINOS-FERMILAB, OPERA-LNGS-ICARUS-CERN and NASA. That experiment involved theory of the elementary quantum energy presented in [1] and extended in this work, with physical meaning interpretation of the M. Planck's postulate for the radiation quantum and of the light quantum of Einstein, and consequences as: the triple correlated postulates explanation of the N. Bohr's postulate that was explained by L. De Broglie's postulate which is explained by the quantum elementary energy which explains with physical meaning the Planck's quantum and Einstein's light quantum; the frequency identification of the gravitational radiations in the electromagnetic spectrum, contrary to the frequencies of these radiations tried there are decades by aluminum and niobium cylinders mechanical resonance, and of other projects by interferometer; the result of that laser experiment is compatible with galaxies acceleration, but incompatible with the expansion of the own space admitted by the special and general relativity, as an inflated plastic balloon; also, the result of that experiment is contrary to the gravitational radiations origin of the cosmic inflation through the cosmic microwave background (cbm), demanded by search teams, by some observations showing corkscrew formation, rotational, polarized of cosmic background microwave and that would be compatible with these gravitational waves originated of cosmic inflation, but that remains distant of nature of that radiation, contrary to the result of that laser experiment; although particle discovery of Higgs by CERN, that laser experiment explains by the elementary momentum or impulse quantum which gives mass to all particles, electromagnetic and to Higgs field, inclusive the demanded particles of super symmetry; the explanation of the energy and dark matter, demanded by the teams of these searches; the application of these foundations of the quantum elementary impulse which are ignored by process of nuclear fusion. Although astronomy brilliant works, of the X-rays observation satellites, gamma-rays, radio-astronomy, microwaves, laser-rays, supernovas, neutrino mass, Higgs particles, particles accelerators and the more advanced mathematical degree in the quantum mechanics, black holes, string theory, the equation of the everything praised by S. W. Hawking, it remains more distant the more advances that present physical and mathematical development.

However, that laser experiment conjugated with an interpretation of the brilliant CERN's experiment of 1964 [2] of gamma-rays emitted by mesons in almost equal velocity of the gamma-rays, it showed that the electromagnetic radiation velocity is variable regarding the source velocity that emits electromagnetic radiation, contrary to the relativity theory, although electromagnetic radiation velocity to be constant regarding the cbm or space-vacuum. And the atoms particles, physical field, the light and all the electromagnetic radiations to are constituted by quantum elementary impulse, with elementary mass which increases the velocity particles and consequent mass increase, in the particles accelerators. And contrary to the mass increase in velocity function, by relativity coordinates transformation of Einstein, which phantasmagoric would increase the mass. And by the elementary impulse theory resulting in equation of everything demanded by Hawking, where ironically this equation is found where he is, in Cambridge, and formulated for who exercised the Mathematical Chair, I. Newton, and now exercised by Hawking, that is the universal gravitation equation of Newton with the introduction in it of the elementary impulse, resulting in: $F_G \times d = [(K_G\ m_1\ m_2)/d^2] \times d = n\ V_G h$, where F_G is the gravitational force in the distance d, with m_1 and m_2 the masses, n is $V_G h$ quantity, V_G is the gravitational radiation frequency where $h/1\ second = h' = m'c \times c$, where m'c is the quantum elementary impulse, considering the velocity c regarding the Earth as a reference velocity, h' is the energy elementary quantum, m' is the elementary mass, in that the light velocity regarding the cosmic microwave background or to the space-vacuum is c', therefore m'c' being the quantum elementary impulse if admitted c' as the reference velocity. And that equation unites the Newton's gravitational K_G constant and the Planck's quantum h constant, with the essential physical meaning, the equation of everything theory praised by Hawking.

Doc. 29 – Paper Abstract on the commemoration of 100 years of general relativity and 110 of special relativity in 2015, presenting that there was little to be celebrated of these two theories of current Physics, publishead in *The General Science Journal*.

Physics Essays

An International Journal Dedicated to Fundamental Questions in Physics

The Elite Journal since 1988

PHYSICS ESSAYS 29, 3 (2016)

Experiments with light that challenge the current theories of relativity, quantum mechanics, and cosmology

J. P. Hansen[a]
Engineering Institute, 1130 Dep. Mario Barros, Curitiba PR-80530-280, Brazil

(Received 22 June 2015; accepted 2 July 2016; published online 6 August 2016)

Abstract: Herein are discussed experiments concerning electromagnetic radiation and velocities, with interpretation different from MINOS-FERMILAB, OPERA-LNGS-ICARUS-CERN, and NASA, relating to Planck's energy quantum and Einstein's light quantum using a correct interpretation of the electric oscillator. It is showed the consequences of contrary results to the theory of relativity and quantum mechanics, such as gravitational radiation frequency, in terms of the search for gravitational waves, the acceleration of galaxies, the constituents of particles, electromagnetic fields, gravitation, the Higgs field, dark energy, and matter, as well as an equation of the everything theory praised by Hawking (S. W. Hawking and L. Mlodinow, *The Grand Design*, New York: Bantam Books, 2010). © 2016 Physics Essays Publication. [http://dx.doi.org/10.4006/0836-1398-29.3.344]

Résumé: Cet article évoque les expériences concernant le rayonnement et les vitesses électromagnétiques avec des interprétations différentes de celles avancées par MINOS-FERMILAB, OPERA-LNGS-ICARUS-CERN et la NASA, en lien avec le quantum d'énergie de Planck et Einstein en utilisant une interprétation correcte de l'oscillateur électrique. Nous montrons les conséquences de résultats contraires à la théorie de la relativité et de la mécanique quantique, notamment la fréquence du rayonnement gravitationnel, pour la recherche d'ondes gravitationnelles, l'accélération des galaxies, les éléments constituant les particules, les champs électromagnétiques, la gravitation, le champ de Higgs, l'énergie et la matière noire, ainsi que une l'équation de la Théorie du tout avancée par Hawking.

Key words: Light Velocity; Michelson-Morley Experiment; Neutrino; Planck's Constant; Quantum Impulse; Quantum Elementary Energy; Relativity Theory; Gravitational Radiation; Fields; High Energy Particles.

a) jph.hansen@gmail.com

Doc. 30 – Paper Abstract on experiments which contradits the quantum relativity and mechanic, publishead in *Physics Essays*.

CXLIV. TEAMS, PROJECTS AND ACRONYMS OF CURRENT PHYSICS
1954 to 2018

High Energy Particles
CERN – Center and Conseil Européean de Recherche Nucléaire - Organization Européene de Recherche Nucléaire - European Organization for Nuclear Research - Laboratoire Européen pour La Physique des Particules - European Laboratory for Particle Physics

FERMILAB – Fermi National Accelerator Laboratory
    Belle II -KEK High Energy Accelarator Research Organization Muon g-2

ICARUS – Imaging Cosmic And Rare Underground Signals

LNGS – Gran Sasso National Laboratories of INFN - Institute Nazionale de Fisica Nucleare
MINOS – Main Injector Neutrino Oscillation Search
NASA – National Aeronautics and Space Adminsitration
OPERA – Oscillation Project Emulsion Racking Apparatus

Gravitational Wave Interferometry Detectors
AIGO – Australian International Gravitational Observatory
ALLEGRO – Antenna Louisiane Gravitational Radiation Observatory
ASTRO – Astrodynamical Space Test Relativity Optical (after 2020)
ASTROD-GW – Astrodynamical Space Test Relativity Optical Gravitational Wave (after 2020)
ASTROD-SUPER – Super Astrodynamical Space Test Relativity Optical (after 2020)
AURIGA – Antenna Ultracriogenica Risonante per l'Inlagine Gravitazionale Astronomica
BBO – Big Bang Observer (after 2020)
CLIO – Cryogenic Laser Interferometer Observatory
DECIGO – Hertz Interferometer Gravitational Wave Observatory (after 2020)
ET-EINSTEIN-TELESCOPE – Einstein Telescope Gravitational Interferometer (after 2020)
EPTA – European Pulsar Timing Array
EXPLORER-CERN – Explorer Cern
GEO – Gravitational Wave Interferomenter Observatory
GEO-HF – Gravitational Wave Interferomenter Observatory High Frequency
INDIGO – Indian Initiative in Gravitational-wave Observations
LCGT-KAGRA – Large Cryogenic Gravitational-Wave Telescope–Kamioka Gravity and Gravitational Waves
LIGO – Laser Interferometer Gravitational Wave
LISA – Laser Interferometer Space Antenna (after 2020)
MINIGRAIL – Gravitational Radiation Antenna In Leiden
NANOGrav – North American Nanohertz Observatory for Gravitational Waves
NASA – National Aeronautics and Space Administration
NAUTILUS – Nautilus Gravitational Waves
NICMOS – Infrared Camera and Multi-Object Spectometer
NIOBIUM-BAR – Niobium Bar Gravitational Wave
PPTA – Parkes Pulsar Timing Array
SHENBERG – Mario Shenberg Gravitational Wave
TAMA – Tsubono And Masaki Ando
TOBA – Torsion Bar
VIRGO – Virgo Interféromètre Gravitation Onde – Interferometer Radiation Gravitational Observatory

Supernovas
BASS – Berkeley Automatic Supernova Search
CSP – Carnegie Supernova Project
DTM – Department Terrestrial Magnetism
HIGHT-z – Hight-z Supernova Search
HST – Hubble Spacial Telescope (ultraviolet-partial-visible lght-infrared partial)
JWST – James Webb Space Telescope (2020)
LBL – Lawrence Berkeley National Laboratory
LOSC – Lick Observatory Supernova Census
NASA – National Aeronautics and Space Administration
NSF – Nearby Supernova Factory
SCP – Supernova Cosmology Project
SAP – Supernova Acceleration Probe
SDSS-II – Sloan Digital Sky Survey-II
SLS – Supernova Legacy Survey

Wave Polarization and Anisotropy
ACBAR – Arcminute Cosmology Balometer Array Receiver
AMIBA – Array Microwave Background Anisotropy
ASTROD – Astrodynamical Space Test of Relativity Optical Devices
ATCA – Australian Telescope Centre for Astrophysics
AVLB – Australian Very Large Baselina Array
BAR–SPORT – Ballon Radiometers for Sky Polarization Observations
BBO – Big Bang Observer (after 2020)
BICEP – Background Imaging of Cosmic Extragalatic
BOOMERANG-B2K – Ballon Observations of Millimetric Extragalactic Radiation and Geophysics
CBI – Cosmic Background Imager
CAPMAP – Cosmic Anisotropy Polarization Maper
CARA – Center for Astrophysical Research in Antarctica
COBE – Cosmic Background Explorer
COBRAS – Cosmic Background Radiation Anisotropy Satellite
COMPASS – Cosmic Microwave Polarization at Small Scale
DASI – Degree Angular Scale Interferometer
EPIC – Einstein Polarization Interferometer for Cosmology
JWST – James Webb Space Telescope (2020)
MAP – Microwave Anisotropy Probe
MAXIMA – Millimeter Anisotropy Experiment Imaging Array
MAXIPOL – Maxi Polarization
NASA – National Aeronautics and Space Administration
PIQUE – Polarization Imaging Q/U Parameters E-mode
PLANCK-HFI – Satellite for Cosmic Background Radiation Anisotropy-High Frequency Instrument
POLAR – Polarization

POLARBEAR – Polarization B-mode and E-mode Angular Array
POLATRON – Polarimeter Telescope Radio Observatory North Celestial Pole
QUAD – Q/U Parameters At Dasi
QUEST – Q/U Parameters Extragalactic Sub-mm Telescope
QUIET – Q/U Parameters Imaging ExperimenT
SAMBA – Satellite for Measurement of Background Anisotropies
SPT – South Pole Telescope
VLA – Very Large Array Mexico
VSA – Very Small Array
WMAP – Wilkinson Microwave Anisotropy Probe

Neutrino
AMANDA – Antarctic Muon And Neutrino Detector Array (ended).
ANITA – Antarctic Impulsive Transient Antenna
ARA – Askaryan Radio Array
ARIANNA – Antarctic Ross Ice-Shelf ANtenna Neutrino Array
ANTARES – Astronomy with a Neutrino Telescope and Abyss environmental RESearch project
BDUNT – Baikal Deep Underwater Neutrino Telescope
BNO – Baksan Neutrino Observatory
BOREXINO – Boronsolar neutrino experiment
CDHS – CERN Dortmund Heidelberg Saclay Neutrino experiment (ended)
CHOOZ – Chooz neutrino oscillation experiment (ended)
CNGS – CERN Neutrinos to Gran Sasso
CUORE – Cryogenic Underground Observatory for Rare Events at LNGS-Laboratori Nazionali del Gran Sasso
DAYSA BAY – Daya Bay Reactor Neutrino Experiment
DUNE – Detector Underground Neutrino
DC – Double Chooz neutrino oscillation experiment
DONUT – Direct Observation of the NU Tau, at FERMILAB (ended)
DUMAND – Deep Underwater Muon And Neutrino Project (ended)
EXO – Enriched Xenon Observatory
GALLEX – Gallium Experiment at LNGS – Laboratori Nazionali del Gran Sasso (ended)
GARGAMELLE – Gargamelle Giant bubble chamber neutrino detector at CERN (ended)
GERDA – GERmanium Detector Array
GNO – Gallium Neutrino Observatory
HALO – Helium And Lead Observatory
HOMESTAKE – Homestake Gold Mine collet and count neutrinos experiment (ended)
HYPER-KAMIOKA – Hyper-Kamioka Observatory Institute for Cosmic Research

ICARUS – Imaging Cosmic And Rare Underground Signals at LNGS-Laboratori Nazionali del Gran Sasso
ICECUBE – IceCube Neutrino Observatory
INO – India-based Neutrino Observatory
IMB – Irvine Michigan Brookhaven detector (ended)
JEM-EUSO – Japanese Module Extreme Universe Space Observatory
JUNU – Jangmen Underground Neutrino Observatory
KAMIOKA – Kamioka Observatory, Institute for Cosmic Ray Research (ended)
KamLAND – Kamioka Liquid Scintillator Antineutrino Detector
KARMEN – KArlsruhe Rutherford Medium Energy Neutrino experiment (ended)
KATRIN – Karlsruhe Tritium Neutrino Experiment
K2K – KEK to Kamioka Experiment (ended)
KGF – Kolar Gold Fields (ended)
KM3NeT – Cubic Kilometre Neutrino Telescope
LAGUNA – Large Apparatus studying Grand Unification and Neutrino Astrophysics
LENA – Low Energy Neutrino
LNGS – Laboratori Nazionali del Gran Sasso in INFN
LVD – Large Volume Detector in LNGS Laboratori Nazionali del Gran Sasso
LSND – Liquid Scintillator Neutrino Detector (ended)
MACRO – Monopole, Astrophysics and Cosmic Ray Observatory at LNGS – Laboratori Nazionali del Gran Sasso (ended)
MAJORANA – Majorana project neutrinoless double-beta decay
MARIACHI – Mixed Apparatus for Radar Investigation of Cosmic-rays of High Ionization
MINERVA – Main Injector Experiment for Eletron Neutrino ν-Appearance
MINOS – Main Injector Neutrino Oscillation Search (ended)
MiniBooNE – Booster Neutrino Experiment
NARC – Neutrino Array Radio Calibration
NEMO – Neutrino Ettore Majorana Observatory (ended)
NESTOR – Neutrino Extended Submarine Telescope with Oceanographic Research Project
NEUTRINO FACTORY – Neutrino Factory fairley focused beam
NEVOD – Nevod Neutrino Water Detector in Russian
NOvA – Neutrinos at the Main Injector Electron Neutrino ν-Appearance
NuMI – Neutrinos at the Main Injector, or NuMI, at Fermilab
OPERA – Oscillation Project with Emulsion-Racking Apparatus at CERN
RENO – Reactor Experiment for Neutrino Oscillations
RICE – Radio Ice Cerenkov Experiment
SAGE – Soviet American Gallium Experiment (ended)

SciBooNE – SciBar Booster Neutrino Experiment SciBooNE at FERMILAB (ended)
SNO – Sudbury Neutrino Observatory
SPS – Super Proton Synchrotron Heavy Ion and Neutrino Experiment at CERN
SUPERNEMO – Neutrino Ettore Majorana Observatory
SOUDA2 – Soudan Mine Proton Decay and Neutrino (ended)
SUPER-KAMIOKA – Super-Kamioka Neutrino Detection Experiment - Super-K - SK
SNEWS – SuperNova Early Warning System
T2K – Tokai to Kamioka
UNO – Underground Nucleon Decay and Neutrino Observatory
WITCH – Weak Interation Trap for Charged particles

Dark Matter
ADEPT – Advanced Dark Energy Physics Telescope
ADMX – Axion Dark Matter Experiment
AMS – Alpha Magnetic Detector
ArDM – Argon Dark Matter
CDMS – Cryogenic Dark Matter Search
CRESST – Cryogenic Rare Event Search with Superconducting Thermometers
DAMA – Dark Matter
DARKSide- Dark Matter Side
DDDM – Double Disk Dark Matter
DEAP – DarkMatter Experiment Argon Pulse
DRIFT – Directional Recoil Identification From Tracks
EDELWEISS – Expérience pour Detecter Les Wimps En Site Souterrain
GAPS – General Antiparticle Spectrometer (2019)
HESS – High Energy Stereoscopic System
JWST – James Webb Space Telescope (2020)
LNGS – Gran Sasso National Laboratories of the INFN – Instituto Nazionale de Fisica Nucleare – Wimps
LNL – AXION – Livermore National Laboratory
LUX – Large Underground Xenon Detector
NASA – National Aeronautics and Space Adminstration
PAMELA – Payload for Antimatter Exploration and Light-Nuclei Astrophysics
SQUIDs – Superconducting Quantum Interference Devices
R-γ CTA – Cherenkov Telescope Array (2022)
RAL – WIMPS – Rutherford-Appleton Laboratory – Weakly Interacting Massive Particle
STANFORD – WIMPs – Stanford Wimps
VERITAS – Very Energetic Radiation Imaging Telescope Array System
XENON – Xenon Dark Matter Search Experiment
ZEPLIN – Zoned Proportional Scintillation in Liquid Noble Gases

Dark Energy
SPT – South Pole Telescope
DES – Dark Energy Survey

Nuclear Fusion, Gamma Rays and Others
ADITYA
AGILE – Astrorivelatore Gamma ad Immagini Leggero (ISA)
ALCATOR – Alcator-C-Mod
ALPHA-g
ARGUS – Argus (ended)
ARIEL – Atmosferic Remote-Sensing Infrared Exoplanet Large-Survey
ASDEX Upgrate
AATERIX-IV-PALS – Prague Asterix Laser System
ATF
BELLA – Berkeley Laboratory Laser Accelerator
BeppoSAX – Beppo Occhialini Satellite per Astronomia a Raggi X
CAT
CDX-U (ended)
CGRO – Compton Gamma Ray Observatory
CHEOPS – Characterizing Exoplanet Satellite
CHS
COMPASS
CONSTELLATION-X
CXO – Chandra X Ray Observatory/AXAT – Advanced X Ray Astronomical Facility
CT-6
CYCLOPS (ended)
DIII-D
DEMO – DEMOnstration Power Plant
DPG – Dense Plasma Focus
EAST
E-ELT – European-Extremely Large Telescope
ESA – European Space Agency
ESO – European South Observatory
EXTRAPT-T2R
ETE
FRT
FINESSE – Fast Infrared Exoplanet Spectroscopy Survey Explorer
FRT
FTU
GAIA – Global Astrometric Interferometer for Astrophysics
GEKKO-XII
FGST – Fermi Gamma-Ray Space Telescope (GLAST – Gamma-Ray Large Area Space Telescope)

GOLEM
GROND-ESA-Gamma Ray Burst Optical/Near-Infrared Detector
H-1NF
HAWC – High Altitude Water Cherenkov Gamma-Ray
  Observatory
HEGRA – High Energy Gamma Ray Astronomy
HETE – High Energy Transiente Explorer (X-Rays and Gamma)
HELIOTRON – HeliotronJ
HIPER – High Power Laser Energy Research Facility
HL-1(M)
HL-2A
HSX
HT-6(B,M)
HT-7(U)
IFMIF – International Fusion Materials Irradiation Facility
IGNITOR
INTEGRAL – INTErnational Gamma Ray Astrophysics
  Laboratory/ (ESA)
IRFM
IRIS – Interface Region Imaging Spectrograph Ultraviolet
IR-T1
ISKRA
ISTTOK
ISA – Italian Space Agency
ISS – International Space Station
ITER – International Thermonuclear Experimental Reactor
JANUS – Janus (ended)
JAXA – Japan Aerospace Exploration Agency
JET – Joint European Torus
JT-60 (ended)
KAGRA (2024)
KEPLER – (photometry)
KSTAR
KTM
LDX
LHD
LINGYUN – Lingyun
LLE – Laboratory for Laser Energetics
LMJ – Laser Mégajoule
LONGPATH – LongPath (ended)
LST – Large Size Telescope (2022)
LT-1
LIT-7(U)(ended)
LULI2000
MAGIC – Major Atmospkeric Gamma-ray Imaging Cherenkov
  Telescope
MAGLIF – Magnetized Liner Inertial Fusion
MAST

MCX
MFTF (ended)
MST – Medium Size Telescope (2022)
MTF
NCFI – National Cold Fusion Intitute (ended 1991)
NCSX
NIF – National Ignition Facility
Nike
Nova (ended)
OMEGA
PACER – Powel Cavity Explosion Repetitive
Pan-STARRS – Panoramic Survey Telescope and Rapid Response System
PBX-M
PEGASUS – Pegasus
PLATO – PLAnetary Transits and Oscillations-Spacecraft Telescope (2026)
PNF – Polywell Nuclear Fusion
QPS
QUEST
R-γ CTA – Cherenkov Telescope Array (2022)
RFX
ROCKOON – Rocket-Baloon (Raios-X e Gama)
ROSAT – Röntgen Satellit (Raios-X), (ended)
RTP
RXTE – Rossi X-Ray Timing Explorer
SCT – Schwarzschild-Couder Telescope
SG-I
SG-II
SG-III
SG-IV
SHENGUANG-II – Shenguang-II High Power Laser
SHIVA – Shiva (ended)
SPITZER-SIRTF-Space Infrared Telescope Facility
SSPX
SST-1 – Space Surveillance Telescope (2022)
START (ended)
STOR-M
SUNIST
Suzako/ASTRO-E11/ASTRO-E2 (JAXA), (X-Ray)
    Swift Satellite Gamma-Ray Burst Mission, (Ultraviolet and Gamma Ray)
T-3
T-4
T-10
T-15
TCV
TESS – Transit Exploration Exoplanets Satellite

TEXT
TEXTOR
TFR (ended)
TFTR (ended)
TJ-I
TJ-II
TJ-IU
TORESUPRA – Toroidal Superconducting Magnetic
TPE-RX
TU-Heliac
TWR – Traveling Wave Reactor
UCLA-ET
UST-1
UST-2
URAGAN –Uragan-1
URAGAN –Uragan-2(M)
URUGAN –Uragan-3(M)
VERITAS – Very Energetic Radiation Imaging Telescope Array System
VULCAN – Vulcan
W7-AS –W7-X – WENDELSTEIN-7-X – Wendelstein Toroid Superconducting Magnetic Coils
WEGA
WISE – Wide Field Infrared Survey Explorer
XMM-Newton/X-Ray Multi-Mirror
ZETA – Zero Energy Termonuclear Assembly (ended)
ZMACHIN – Zmachin
TELESCOPES AND OTHERS

CXLV. THE GREATNESS OF THE UNIVERSE – **2018**

With the improvement of the equipment by the teams observing the universe was calculated the number of galaxies between 100 to 200 billion that can be seen, with 200 to 400 billion stars in each galaxy. But, the estimated total quantity of galaxies is 2 trillion in the universe.

CXLVI. THE FORCE OF THE UNIVERSE – **1952 to 2013**

In 2013 C. W. Lucas Jr. came to publish the theory of the force of the universe in which, based on the work of Mach and of Assis, presented that the residual vibration of particles of charges such as protons and electrons resulted in the attraction of the bodies, the gravity. And in that residual vibration was the emission of electromagnetic waves in which the frequency was calculated in 10^{10} cycles/second, that is, microwave that would explain the origin of the cosmic background radiation. The Mesquita calculation featured 10^{19} cycles/second, coming from the atomic nucleus and could be the origin of this vibration of the charge of the proton and the electron, because by the theory of the elementary impulse the particles are formed by the elemental photon h' = m'c x c, in which fluxes of elementary force m'c/t by constitutes the gravitational radiation as Mesquita theory.

CXLVII. GRAVITATIONAL WAVE DETECTION – 2014 to 2016

In 2014 it was announced by current Physics the discovery of gravitational waves from cosmic inflation, through observations of polarity of cosmic microwave radiations. However, after a few weeks it was verified that they were just noises that showed this polarity. In 2016 it was announced the discovery of gravitational waves from the fusion of two black holes that occurred 1.3 billion years ago.

Gravitational Wave Detection in September 14, 2015 14 at 09:50:45 UTC – In February 1916 was announced by the team of the LIGO-Laser Interferometer Gravitational Wave and VIRGO-Virgo Interferometer Radiation Gravitation Observatory, the detection of gravitational waves at the frequency of 3.5 x 10 to 2.5 x 10^2 hertz originated from space-time oscillations, space and time, which were propagated from two binary bodies of high energy concentration, black holes, at a distance of 410 Mpc, and after 1.3 billion years reaching Earth on September 14, 2015, published immediately in one of the Journals of current Physics, Physical Review Letters of February 2016.

Gravitational Field-Space-Time Distortion and Gravitational Wave – It had been verified as Newton that the mass contained its own gravitational field, observed by the attraction between masses that varied in the direct ratio of the masses and inverse squared of the distance. And by Einstein's theory of general relativity, this gravitational field was termed the distortion of space-time, resulting in the acceleration of masses added to this gravitational field of space-time curved, distorted by the presence of mass. Where by general relativity, space without considering the actual particles or existence of zero mass that are created and extinct in a briefest time by quantum mechanics, is the own "nothing". But, considered by general relativity as a physical entity that curls, flexes, bends, twists, distends, expands and widens! And the general relativity considered by the current Physics, according to the Physicist C. Rovelli of: *"The theory of extreme beauty and the new eyes to see the universe"*! The variation of these distortions of the gravitational field or of these curved coordinates of space-time are the so-called gravitational waves, and propagated in space-time. Although space is considered to contain quantum fluctuation, that is, oscillations involving how much energy $\mathcal{E} = h\nu$ without mass of rest that can be materialized forming particles, it is admitted that the own space and time fluctuate by the variation of distortions, forming these gravitational waves! By general relativity, these waves carry the energy of the mass motion in acceleration, that is, energy that results in the variation of the coordinates of space-time involves these masses, and also named gravitational radiations quantum without mass of the space-time. And according to general relativity, the gravitational field without mass constituted by the space-time curved, which to varying by action of the matter, it was predicted to be propagated at the speed of light. Figure 87 shows representation of the space-time gravitational field as designed by T. Creighton, based on ultra deep field image of the Hubble Spacial Telescope of the National Aeronautics and Space Administration-NASA, European Space Agency-ESA, S. Beckwith Space Telescope Science Institute-STScI and The Hubble Ultra Deep Field-HUDF Team on the cover of the book Gravitational-Wave Physics and Astronomy by J. D. E. Creighton and W. G. Anderson, members of the LIGO.

Fig. 87 – Book cover Gravitational-Wave Physics and Astronomy by J. D. E. Creighton and W. G. Anderson, as Reference, members of the LIGO, Wiley-VCH Verlag GmbH & Co. KGa, by T. Creighton, showing representation of the gravitational field by the general theory of relativity which treats only as a distortion of space-time, macroscopically, without considering the field constituted of quanta with mass of rest.

Macroscopic Frequency of Space-Time Oscillating by the Masses of Gravitation – From the geodesic equation with distance differentials, the differential equations of space-time coordinates in curvature and masses of the bodies, by the vectors or tensors of coordinate transformations, is obtained equation of the v velocity of the particle, which in binary system in which d = distance between the two masses, ω = angular velocity = $2\pi/P$, P = period, $v = d\,\omega$, $f_{orbital}$ = orbital frequency, $f_{gravitational\ wave}$ = gravitational wave frequency = $2\,f_{orbital}$, $\omega\pi = 2/P = 2\,f_{orbital}$, and is obtained $f = v^3/\pi GM$ where G = gravitational constant of Newton = $6.6... \times 10^{-8}$ cm^3 g^{-1} s^{-2} and M = masses of the two bodies = $(m_1 + m_2)$. Where this frequency is the of the orbital masses of the binary system swinging resulting in this frequency to space-time, macroscopic, with its velocity, as mentioned in the book of members of the LIGO, J. D. E. Creighton, W. G. Anderson, Gravitational-Wave Physics and Astronom, Wiley-VCH Verlag GmbH & Co. KGaA, Boschstr. Weinheim, Germany, 2011.

Gamma Rays Detection by the Space Satellite Fermi on <u>14 September 2015 at 09:50:45,4 UTC</u> – On the same day, after the detection of the gravitational wave of the black hole binary by turn on **September 14th, 2015 at 09:50:45 UTC**, the Fermi Space Satellite detected a gamma-rays signal 0.4 seconds after gravitational wave detection by the LIGO, that is, **on <u>September 14th, 2015 at 09:50:45,4 UTC</u>** from the same location of the two emissions, as published by NASA, Figure 88.

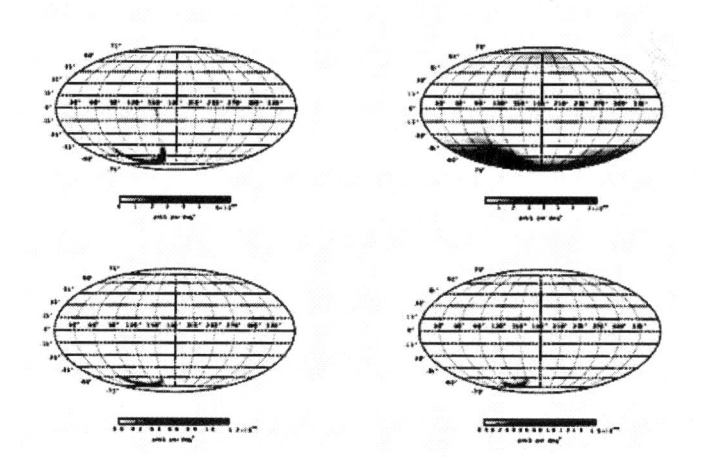

Fig. 88 – The directions of the gamma-rays (right, in black) and the direction of gravitational waves (left, in black) are of extreme proximity as V. Connaughton et al., arXiv:1602.03920 v3 [astro-Ph.HE] 16 Feb 2016, as Reference.

It is clear that the probability of these events occurring almost at the same time and coming from the same direction or location is an impressive coincidence!

But, in the own communication of LIGO it is presented that: *"Emissions of gravitational waves come from unknown sources. By telescopes for visible light, radio waves or gamma-ray detectors, facts have been found that were unexpected and that revolutionized the understanding of the universe. So in gravitational wave emissions we're expecting the unexpected. There is a chance that some systems such as supernovae or gamma-ray emissions can produce gravitational waves, but very little is known about the details of these systems and anticipate the shape that these waves will have. It's hard to say about the origin of these wave".*

Gravitational Waves According to Impulse Quantum Theory of the Correct Interpretation of Planck's and Newton's Constants of Universal Gravitation – As Erdely, between 1952 to 1969, by elementary impulse theory:

$$\text{elementary impulse} = m\text{'}c \text{ [g.cm/s]}$$
$$\text{elementary force} = m\text{'}c / s \text{ [g.cm/s}^2\text{]}$$

So, brilliantly Mesquita, in 1965, introduced the impulse frequency m'c through:

force = $m\text{'}c \cdot \mathcal{V}_{m\text{'}c}$ [g.cm/s .1/s], where $\mathcal{V}_{m\text{'}c}$ is the frequency of m'c

And for elementary energy:

$m\text{'}c \times c \, \mathcal{V}_{h\text{'}} = m\text{'}c^2 \, \mathcal{V}_{h\text{'}} = h\text{'}\mathcal{V}_{h\text{'}}$ [g.cm^2/s .1/s], where $\mathcal{V}_{h\text{'}}$ is the frequency of h'

h' [erg]= m'c x c [erg]= h [erg.s]/1[s] = 6,625...x 10^{-27} [erg] = 6,62 x 10^{-27} [g cm^2/s^2]

h = Planck constant = 6,625... x 10^{-27} [erg.s] = 6,62 x 10^{-27} [g cm^2/s]

In Newton's gravitational law:

$$F_G = m_1 m_2/r^2 \text{ [g}^2/\text{cm}^2\text{] } K_G$$
$$K_G = 6,670... \times 10^{-8} \text{ [cm}^3/\text{s}^2 \text{ g]}$$

And by various determinations is found the value of Newton's constant in 6,627 a 6,625 x 10^{-8} and then the constant h' [erg]= m'c x c [erg]= h [erg.s]/1[s] it can be introduced into Newton's constant without any adjustment.

Introducing h' and energy frequency $\mathcal{V}_{h\text{'}} = \mathcal{V}_G$ in Newton's constant, where $\mathcal{V}_G$ it's the gravitational frequency:

$$K_G \text{ [cm}^3/\text{s}^2 \text{ g]} = h\text{' [g cm}^2/\text{s}^2\text{] } \mathcal{V}_G \text{ [1/s] [cm s /g}^2\text{]}$$
$$K_G = 6,670... \times 10^{-8} \text{ [cm}^3/\text{s}^2\text{g]} = 6,67... \times 10^{-27} \text{ [g cm}^2/\text{s}^2\text{] } 10^{19} \text{ [1/s]}$$
[cm s /g^2]

in which, 10^{19} [1/s] = $\mathcal{V}_G$ [1/s] = frequency of gravitational radiations

λ_G = gravitational wavelength = $c / \mathcal{V}_G$, $\lambda_G = 0{,}2997925 \times 10^{-8}$ cm

K_G [cm³/s² g] = h [g cm²/s²] [s] $\mathcal{V}_G$ [1/s] [cm/g²]
$K_G = 6{,}670... \times 10^{-8}$ [cm³/s² g] = $6{,}67... \times 10^{-27}$ [g cm²/s²] [s] 10^{19} [1/s] [cm /g²]

So:
$F_G = m_1 m_2/r^2$ [g²/cm²] $6{,}67... \times 10^{-27}$ [g cm²/s²] [s] 10^{19} [1/s][cm /g²]

$n = m_1 m_2/r^2$ [g²/cm²] [cm /g²] = n / cm

$F_G = (n/cm)$ h $\mathcal{V}_G$, that is the force between the centre of masses m_1 and m_2

Multiplying both members by cm is obtained the energy:
$\mathcal{E}_G$ = n h $\mathcal{V}_G$, which is the potential for gravitational energy between mass centres m_1 and m_2

$\mathcal{E}_G/s = P_G = n (h/s) \mathcal{V}_G = n h'\mathcal{V}_G$, which is the gravitational power between the mass centres of m_1 and m_2.

So, Newton's gravitational law follows Planck's law $\mathcal{E}$ = h $\mathcal{V}_G$ = (h/s) $\mathcal{V}_G$ x s = h' $\mathcal{V}_G$ x s and
P_G = h' $\mathcal{V}_G$.

The elementary impulse shows that gravitation is caused by the impulse flows m'c x c = m'c² = h' on the frequency of 10^{19} hertz in the atomic nucleus, and a field formed of n' h' involves all the particles, in which the electromagnetic radiation like the light in the vicinity
of the bodies formed by the particles are curved by a field of h' involving that body through the particles of that body.

By the elementary impulse, the electric, magnetic and gravitational field is formed by the elementary impulse m'c x c = h' and the radiation emitted frequently $\mathcal{V}$ of h', ie energy. $\mathcal{E}_G/s$ [erg/s] = h [erg s] $\mathcal{V}$ [1/s] /1s = h' [erg] $\mathcal{V}$ [1/s] and $\mathcal{E}_G$ [erg]= h' $|\mathcal{V}_G|$ [erg].

And what causes the acceleration and increased mass of the bodies and the electrical, magnetic and gravitational field is the mass injected by m'c x c = h', resulting in varying spatial dimensions of the constituent particles of the bodies, which results in the interference of the laser obtained by the LIGO. So, what would be detected by LIGO as frequency between 10^{12} to 10^{-18} hertz of the variation of dimensions of the mirror is the result of the mass variation and motion of the field constituted of radiations m'c x c = h', coming from gravitational radiations of gamma-rays, as they are emitted from atomic nucleus, that reorganize around the electrical, magnetic and gravitational field of the masses in space, and that field can reissue electromagnetic waves at different frequencies. The black holes receive particles and radiations that result in acceleration and mass increase, not the increase of mass by ghostly condensation of speed that would result in mass or acceleration by distortion of the space-time of nothingness of the theory of Einstein's relativity! And the speed of wave propagation is the speed of elementary impulse, that's m'c x c, the speed of light c!

Free Neutron Life – As Communication presented by Mesquita at SBPC in 1967 on the duration of the atom free neutron, as if all its mass was disintegrated by the frequency of gravitational radiation, it would result in incredible probability of being equal to the experimental time!

In fact, Mesquita, applying the concept of energy = force x distance, in Newton's law of gravitation, for a mass M and another unitary in a unitary distance, which represented a potential of the gravitational field, was obtained ≈ 1,807.9 s.

That is, the frequency 10^{19} s^{-1} for decomposition if it were of the entire neutron mass, it is largely between 10.1 to 15 minutes, which is an incredible probability that it is not a simple coincidence, but the reality of the gravitational radiations of 10^{19} s^{-1}.

So the life of the neutron that would be decomposed into gravitational radiation indicated a crucial proof of the existence of these radiations deduced by Mesquita.

Conclusions on Gravitational Waves – By the quantum analysis of the elemental impulse correctly interpreted from the Planck and Newton constants, the electric, magnetic and gravitational field is formed by the quantum impulse m'c x c = h' and the radiations emitted with frequency V of h', and what causes the acceleration and increase of mass of the bodies and of the electric, magnetic and gravitational field is the mass injected in m'c x c = h', resulting in varying spatial dimensions of the constituent particles of the bodies. So, what would be detected by LIGO as frequency between 10^{12} to 10^{-18} hertz of the variation of dimensions of the interferometry is the result of the mass variation and motion of the field constituted of radiations m'c x c = h', coming from gravitational radiations of gamma-rays, as they are emitted from atomic nucleus of massive masses.

The theories of Erdelyi and Mesquita, reported in Physics Essays in March 2002 and September 2016, in General Science Journal on June 4, 2015 and March 10, 2016, the detections by Fermi Space Satellite and the LIGO presentation itself, all of them show that gamma-rays are intrinsically constituting gravitational waves. Not only is coincidence of gamma-rays being related to gravitational waves, but a fact of quantum make up.

The electromagnetic field consisting of all the radiations of the spectrum, including the gravitational, when receiving mass results in variations of its dimensions in space, similar to the variations of particle dimensions by mass increase according to the theory of impulse quantum. Then, there may be the variation of dimensions of the distance between interferometer mirrors and consequent detection of these variations of the electromagnetic-gravitational field. If new detections are reproducible there would be proof of these variations of the electromagnetic-gravitation field. In the case of laser experiment based on the theory of elementary impulse, presented in Physics Essays and General Science Journal, aforementioned, is a reproducible experiment showing that the photon of laser light presents the inertial velocity of the Earth-Sun-Laser System, composing the speed of light c with the velocity of the System Earth-Sun-Laser, ie, leaving diagonally with velocity $c'_{diagonal} = (c^2 + V_{ESL}^2)^{1/2}$. This is, the Earth dragging with its velocity V_{ESL} the light formed of m' of c velocity. Then, showing that the velocity of light c_v is variable in relation to the emitting source, in which the speed c' of light relative to the vacuum-space-ether in the direction of the motion is always constant, that is, $c' = c_v + V_{Source}$, and c_v is the velocity of light <u>variable</u> relative to the velocity of the emitting source, according to this experiment and the CERN of 1964. This laser experiment largely explains the null experiment of the interferometer experiment Michelson-Morley. The Figure 89 shows detail of this laser experiment.

Light coming out diagonally with velocity c'$_{diagonal}$, of laser at 90° of direction motion to Leo, proves that the velocity of light depends on the velocity of the source, contrary to Einstein's theory of relativity

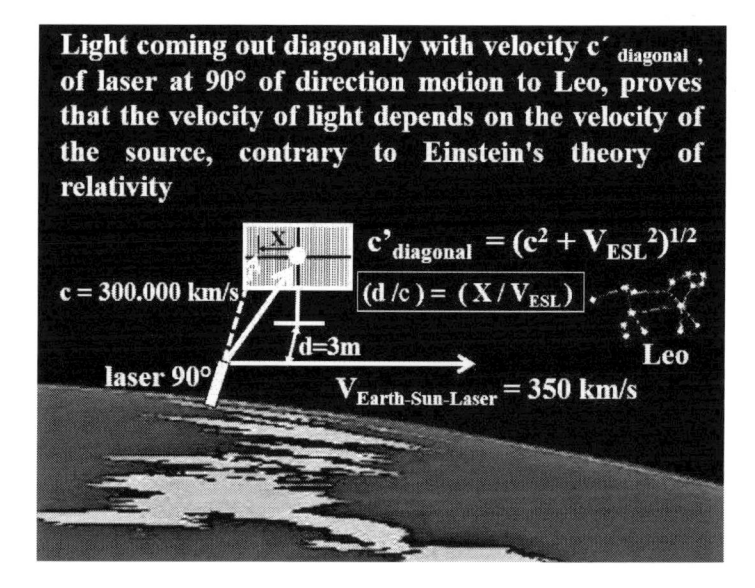

c = 300.000 km/s
$c'_{diagonal} = (c^2 + V_{ESL}^2)^{1/2}$
$(d/c) = (X/V_{ESL})$
d = 3m
laser 90°
$V_{Earth-Sun-Laser} = 350$ km/s
Leo

Fig. 89 – Laser experiment in which the light comes out of the laser diagonally, dragging the formed light formed of m' with speed c with the inertial velocity of the Earth-Sun-Laser in the direction of the constellation of Leo, showing that the velocity of light is variable in relation to the emitting source, contrary to Einstein's theory of relativity, explaining the null result of Michelson-Morley interferometer experiment. Design adapted from Earth photo by NASA. Clipart constellation Leo from *Editora Europe*.

The electric, magnetic and gravitational fields formed from quantum impulse m'c with energy m'c x c = h' = 6.62... x 10^{-27}

[erg] in CGS can have their variable dimensions as an increase in mass and result in propagation of a wave and its detection. However, as Erdelyi and Mesquita by the theory of the elementary impulse the gravitational radiation is in the spectrum of electromagnetic radiations such as electromagnetic wave at the frequency of 10^{19} hertz, contrary to the followed by the current Physics, for example, as mentioned by Physicist Riccardo Sturani, of the Institute of Fundamental Research of South America, of the State University of São Paulo-UNESP, which directed one of the groups of Brazil who collaborated with the LIGO, when he stated in February 2016 that: *"Gravitational radiation is a type of wave that is not in the electromagnetic spectrum"*!

As Riccardo Sturani: *"This gravitational wave detection was something truly unexpected"*. Therefore, only after a few days of detection the equipment would be more likely to detect. And according to the Physicist Odylio Aguiar in the LIGO, Doctor from the University of Louisiana-USA, and responsible for the Detector Mário Schenberg of gravitational waves: *"There is some caution for this discovery to be confirmed. There were previously two false alarms about this detection. One of them by the Background Imaging of Cosmic Extragalactic-BICEP in 2014"*. Then, with the detection of these waves and still considered as unexpected, in addition to previous false alarms, it were necessary to caution as mentioned by the own member of the LIGO, because the current Physics continues to adopt and deyfying as the most beautiful of theory from Physics to Einstein's relativity, and interpreting it as a correct prediction of a part in 100 billion, however contrary to the experiments, interpretations that disapprove of it and forged data from other experiments that would approve general and special relativity. In the case of the gravitational radiation of the Mesquita the correct prediction was of one part in 1 sextillion! This is, the frequence of 10^{19} hertz it was the found by Mesquita for the gravitational radiation, in the spectrum of 1 sextillion of frequencies, and it was the that decays the neutron in its lifetime!

Shortly after the publication of article NASA's Fermi Gamma-Ray Space Satellite of February 16, 2016 which recorded a faint and brief burst of high energy from the same region in the sky, the ESA's INTEGRAL Satellite of X-Rays and Gamma published on 30 March 2016 in the Astrophysical Journal Letters by V. Savchenko and Others of the *François Arago Centre* in France no record was detected from the high energy emission associated with detection by the LIGO! Only a one-millionth gamma-ray emission was detected that was recorded by the LIGO. The detection by the team FERMI Satellite indicated that the fusion of the two black holes of stellar mass could cast gamma-rays together with gravitational waves. But according to ESA, the INTEGRAL Satellite did not record this high energy of gamma-rays, like that of the two black holes or that would be of any other physical phenomenon in the universe, and that by them should not be reported to the detection of gravitational waves!

It is admitted by current Physics that gamma-rays from astrophysics sources are emitted when stars enter collisions or when stellar explosions occur, and accompanied by intense emission of light, considered by NASA as the most luminous and mysterious of the universe. Emissions occur up to two seconds in dense and smaller stars, and emissions over two seconds in extremely massive stars. The astrophysicists have said that the astrophysical gamma-ray emissions short-lived would be created when a pair of superdense stars in a binary system it had spiraled until collide. And that while this event occurs the system would emit gravitational energy that would create waves in space-time. The energy dissipated by the waves would with which the two stars approach even more. In the final milliseconds, before the explosion, the two stars would merge into a deadly spiral expelling highly radioactive material. This stuff would warm up and expand, emitting a jet of light. The emission of gamma-rays would occur by high-temperature gases and high-energy electron jets transferring momentum to infrared photons, then resulting in gamma-ray photons, and by the current Physics not would occur emitting gamma-ray jets of proton and neutron in stellar explosions.

Due to the fact that it was expected there would be no production of any kind of light from the fusion of two black holes, for all the gas that would be the responsible by emitting this radiation already must have been emitted in a long time before the final fusion of the black holes, some scientists of current Physics believe that this detection is a coincidence and nothing would have to do with the gravitational wave event. And also, the current Astronomy theorized that particles of unimaginable dark matter would be the cause of gamma-ray emission, according to publications in their "formal" Journals.

However, gravity is in the mass and if nuclear fusion occurs in the stellar explosions, a part of the mass can be transformed into light and gravitational radiation. Thus, gamma-ray liberation would also be by the atomic nucleus with its fields around the particles and not just by the electrons transferring momentum to the infrared photons, which in reality the quantum impulse of Erdelyi would be energies $h' = m'c \times c$ that they would be added to the infrared rays increasing their quantity, that is, their frequency to follow as gamma-rays.

The analyses by the Fermi satellite team concluded that existed only 0.2% of probability that this joint emission of waves and gamma-rays were a coincidence. And it would be a great landmark in Astronomy and Astrophysics, for it was expected that the fusion of two black holes was without the production of any kind of light.

And the team of researchers working with the Fermi Space Satellite considered this discovery how fantastic. Because if gamma emission was from this event it would be critical to define where gravitational waves came from, that is, gravitational radiation within the electromagnetic spectrum according to the theory elaborated by the Mesquita.

But according to ESA's INTEGRAL team stating that the detection of NASA's Fermi Satellite team should not be reported to detection of gravitational waves or any other physical phenomenon in the universe, the author questioned ESA that if no physical phenomena of the universe could cause this detection by the Fermi Satellite, then it would be reported to what? Document 30.

To: markus.bauer@esa.int – 16, 2016 – Ref: Integral/Fermi your Abstract

Dear, Dr. Markus Bauer,
Congratulations by your Abstract of *INTEGRAL SETS LIMITS ON GAMMA RAYS FROM MERGING BLACK HOLES AT MARCH 30 2016*

You say that:
However, if this gamma-ray flare had had a cosmic origin, either linked to the LIGO gravitational wave source or to any other astrophysical phenomenon in the Universe, it should have been detected by Integral as well. The absence of any such detection by both instruments on Integral suggests that the measurement from Fermi could be unrelated to the gravitational wave detection. But, would that Fermi telescope's detection be related that?

Thank you by your attention,
Sincerely,

José Paulo Hansen
Engineer by the Mackenzie University, São Paulo, Brazil
jph.hansen@gmail.com

Doc. 31 – Email from the author to ESA on how to explain his claim that the Fermi Satellite could not have detected high energy from any gamma emission phenomenon in conjunction with the gravitational radiations obtained by the LIGO, as the INTEGRAL Satellite of the ESA has achieved nothing about high-energy gamma-rays in this event of the LIGO gravitational wave.

However, the ESA preferred not to express itself. For, by following what the theory of general relativity prescribes, as impossibility of light or any other radiation being emitted by black holes, and the project running the LISA Space Satellite by the ESA of gravitational waves similar to those recommended by LIGO, it was really to be expected by no manifestation about any other electromagnetic radiation to have been dected that would set off the pre-establishead.

Previously, the Satellite Compton Gamma Ray Observatory-CGRO had detected gamma-ray explosions indicating it was a phenomenon occurring throughout the universe, coming from stars like the Sun and those exploding, and also from unknown sources, which evidence that results as gravitational radiation discovered by a Mesquita.

In March 2016 in General Science Journal and in Physics Essays Vol. 29. No. 3, 2016 I came to present about what detections are these gravitational waves of space-time-energy undefined by imposed Physics and the quantum reality of the elementary photon of energy h' = m'c x c of quantum impulse theory of Erdelyi and gravitational radiations de Mesquita.

And in August 2017 burst of gamma-rays was detected of neutron star merger by NASA's Fermi Gamma-ray Space Telescope, and shortly after LIGO reported detecting gravitational Waves that arrived 1,7 seconds before the Fermi burst. It was detected the gamma-rays and presented as *Doomed Neutron Blast of Light and Gravitational Waves*, by the emitting visible, ultraviolet, infrared, radio, X-rays, gamma-rays and gravitational wave, as shown in October 2018 by NASA's Goddard Space Flight Center Caltech, MIT, LIGO Lab and ESA. This time INTEGRAL-ESA detected gamma-rays, but by Science Data Center INTEGRAL-ESA continued to disregard the Mesquita gravitational raditions and Erdelyi quantum impulse, despite the previously received information. They presented this new gamma-ray detection as: *"It is an innovative breakthrough, in which we realized that we were detecting something historical with the detection of Fermi gamma-rays and that of an inteferometer of LIGO almost at the same time and then by INTEGRAL. The coincidence with the gamma- rays and a LIGO detector led to the new observation of Ligo that found the second interferometer and that of VIRGO that then consisted of detection"*! The innovative discovery is that of Erdelyi in 1952 and Mesquita in 1965, and this also presented at the International Congress of Committe on Space Research-COSPAR in 1974, to the information that ESA received by email in 2016.

The gravitational energy field m'c' x c' of the neutron stars, originating from the atomic nucleus at the frequency of 10^{19} s^{-1} of m'c' x c', is propagated by the gravitational fields of the galaxies in the vacuum-space-ether causing an increase of energy in the interferometer and consequent dimensional variation of the particles according to the quantum impulse theory, and the resulting oscillation recorded in the interferometer. The propagation of the field is accompanied by these gravitational radiations and other electromagnetic radiation, which circumvent the fields of the galactic medium with the speed of light c' always constant in all media in the vacuum-space-ether, which results in its detection between 1 and 2 seconds after propagation at the velocity of the light c' of the intergalactic gravitational field. In the case of specific conditions of distance and field control electromagnetic-gravitational is obtained a field similar to a rigid body in which can be obtained instantaneous transmission as the entanglement, however in the intergalactic field is by the speed of light.

CXLVIII. NASA – FERMI-JAMES WEBB – SATELLITES – **2018 to 2020**

To report on gamma-ray gravitational radiations by the theory of quantum impulse, it was sent on January 18, 2018, Communication to NASA-FERMI, for involving the detection of gamma-rays, gravitational waves and Arp observations about the distance of quasars that could be analyzed in conjunction with the new James Webb Satellite with infrared ray detection, of launch postponed from 2018 to 2020. For if the quasars are closer to the Milky Way, they would indicate a anomalous deviation to the red. Or, if farther away, the bright light that is received would be gamma radiation, and the radio wave received would be the radiation of light emitted.

PARTE 11 – ADVANCED CONNECTIONS

CXLIX. BASIC CONCEPTS "ADVANCED" – SPACE, TIME, MASS
1949 to 2015

By the basic concepts that Physics follows, space has the standard length of 1 metre; the time, by the fraction of the annual interval around the Sun, of 1 second that is determined precisely by the quantity of 9,19... x 10^9 of the vibrations between two energy levels of the cesium atom; the mass with the standard of 1 kilogram. And by the theory of special relativity, space presents distortions, such as curving and or contraction or dilation; the time that can dilate, this is, the decrease of the vibration rate, or accelerate that is the increase of the rhythm of vibration; and the mass in which it is existing with the continuous field of space-time energy in the universe. However, quantum mechanics is the quantization of energy for particles and fields. For the purpose that Einstein's theory of relativity remained the maximum exponent of current Physics, this came to present the quantization of space, in which there would be no smaller dimension than the Planck's quantum length of 10^{-33} cm; and the quantization of the time when the time would be nonexistent when smaller than around 10^{-44}s. In the decade 90, for space-quantized as a point, a knot making circles. After, the elimination of space and time, in which the variables of an event were only related, without any space or time. And the particles that were considered as dots would be condensed into a massive quantum-particle! Then, the current Physics seeking to create a quantum gravity that unites the two antagonistic theories: relativity and quantum mechanics.

However, the fact that quantum h contains a quantity of energy from the electromagnetic radiation means it occupies a space. Thus, energy levels in space, such as an ether formed of subphoton or elementary submass less than m' of the quantum impulse, would be occupying a space according to their densities. Thus, the quantum h occupying dimension is in the space-continuous and without mathematical artifice of knots and loops for space-quantized or space and time eliminated from the universe. And space and time-quantized is only referring to the energy contained in the Planck's h quantum related to the Newton's gravitational constant and the velocity c of light, and a smaller energy like an ether would have other values. However, this incompatibility between relativity and quantum mechanics occurs by the erroneous concepts of curved space, time dilation, space contraction, massless photons of rest, quantum of constant h with inadequate physical significance, particle-points, mass that increase "ghostly", fields without mass, vagueness in definitions of charges, of fields, of waves and other abstract bases. The explanation that successive divisions find a final form, because a point would be reached and the sum of points would continue to be a point, is of primitive discernment, because the point is introduced precisely to the successive divisions.

The matter to be divided reaches the atom. And proton and other particles are formed by positron-electron helix and a divided positron-electron shows to be formed from the quantum impulse as deduction of quantum impulse theory and the experiments of particle annihilation in gamma rays. And that Mesquita entering m'c in Newton's gravitational constant obtained the impulse frequency of the gravitational radiations between X-rays and gamma-rays originating in particles from the atomic nucleus. However, space and time can be divided, because they are not points, they are infinite.

So, the current Physics, without any rationality, seeks the extinction of space and time with the theory of quantum gravity to keep the dogma of relativity and its creator worshipped as the "God of current Physics". But, space-vacuum-absolute-without mass, this is the total emptiness is an infinite entity of nature, to receive the matter. And without the matter, in this space-nothing, there is a potential, in which if the matter were, a range of vibration or motion of matter would be recorded. This potential, is the time that exists, infinite, eternal, whether with matter in vibration zero or 10^n. For even in an absolute rest of matter, that potential would be flowing between two events that could occur. Even if there was an indetermination of what is an event before and after, as presented by K. Godel, in 1949, by denying the existence of time, this potential of time would continue flowing between any interval before-after or after-before.

But what about the mass quantification? What have you done? The mass of the photon at rest was quantized with nullity: zero mass! Also the field was made up of zero mass! So it is the greatest extinction of the species of nature! The foundations annihilated by the pre-establishead Physics! There is only expanded relationship of words and abstract mathematics, without space, time and mass! Worse still, the current Physics went on to impose in 2000 the guideline of what should be taught around the world: as "dictated" by L. Smolin: *"The basic structure of this theory of quantum gravity will be performed up to 2010 or maximum of 2015. And it will remain fixed. At the end of the 21st century this theory will be taught to secondary school students around the world"*! And in 2014, it is already more offensive to those who do not follow the established, because ignorance is for those who do not follow, according to C. Rovelli: *"It is to look farther, go farther, follow virtue and knowledge. The world revealed by this quantum gravity is a new world, coherent in its simple and limpid beauty. It is a world that does not exist in space and does not develop in time. A world made only of a set of space-time, called the quantum field, which in interactions results in gravitational and electromagnetic field, space, time, particles, waves and light. To look beyond ignorance, to understand what was established by this quantum theory of gravity without time and without space. In that below a minimum scale of the interactions of these how many field there is nothing. Is the truest way"*!

This quantum theory of gravity, of quantum of space-knot-circles, or without space and without time, and of quantum formed from zero mass to the photons and field, is literally a theory of elusive "limpid" terror, because it is really indescribable to express this kind of horror! This is the "advanced" Physics of the present and for future millennia, that *"Doctors of Science Laws and Key"* impose to be obeyed?

CL. BIOPHYSICS, FREE WILL AND DIVINITIES – 1943 to 2018

While the extinction of space, time, mass of photons and fields is sought by the current Physics of abstractions, the quantum impulse with equations from experiments and physical formulations of the greatest scientists of Physics, such as Newton, Faraday, Maxwell, Hertz, Rutherford, Thomson, Planck, it came to obtain the form of the atomic particles, such as the electron, positron, proton, neutron, in which the basic constitution is of a helix formed of quanta elementary $m'c \times c = h'$. And in the annihilation of particles, resulting in the emission of gamma rays, which calculated by the theory of quantum impulses results to the gravity in frequency of 10^{19} s^{-1}. And there is a convergence for the shape of the DNA with its helix form, in which the particles by quantum impulse are formed also by helix, for an advanced towards Biophysics. According to Penrose, in 1987, experiments with electrodes installed in participants showed that when a question was made in which the decision should be yes or no, it was observed that while analyzing the issue there was already a sign of the brain, with the answer that would be after communicated by the participant. So, free will, which involves "mind-conscience-soul-spirit-environment", has been questioned. According to R. Penrose, the "mind-conscience-soul-spirit" to be explained needs a depth of quantum knowledge.

And subphotons could reveal the interconnection of "neurons-synapses-mind- consciousness-soul-spirit". So, in this case, free will is just the result of interaction between the "environment-neurons-synapses-mind-consciousness-soul-spirit". That is, free will it would is only a synonym for all this interaction that decides the actions of the human being. That is, for each being has their own interconnected variables that results in the conduct of each. Then, it would be two ways of treating free will: as a synonymous of biology-environment-mind or accept a pure illusion of freedom of choice.

But, the absence of the free will can be observed in the goddesses of kings and agents of power, in the synthesis that J. P. Sartre mentioned in 1943 by *"The Being and the Nothingness"*: *"Man is fundamentally the desire to be God"*. Newton had uttered God as the creator of the infinite space, time and matter of the universe. And the attempt to eliminate space, time, and mass of photons and fields, reducing to geometry as circles or to nothing, to maintain the theory of relativity, and also the goddessing of quantum fluctuation is to seek to impose basic fundamentals of nature according to dogmas, and as the maximum authority of a god, by current Physics.

CLI. PLANET OF THE APES – **2018**

The Engineer and Writer P. Boulle published the book "La Planéte des Singes" (Planet of the Apes) in 1963 which resulted in film by F. J. Schaffner in 1968. But, at 2018 in a Science Congress on the Planet of the Apes, the scientist Ape I thus manifested: *"For humans, dark matter-energy is one of the greatest constructs of their minds. Their god Einstein equates energy with mass, then his followers teach students different unity equality, when they do mass x acceleration x distance = mass! They seek clarity in the absolute dark"*! Figure 90.

J. P. Hansen

Fig. 90 – In the Congress of Sciences on the Planet of the Apes, the Physicist, Ph. D., Ape I, presented the lecture *"Dark Abstractions that Illuminate the Human Mind"* in which you talk about abstract and "fantastic properties" of dark matter-energy, which act in the absolute darkness of the universe and their minds, annihilating the understanding of the phenomena of nature! Google images.

In another lecture the Physicist Ph. D., Ape II, presented *"Extinction of space, time and matter"*, in which he said: *"Humans have managed to turn everything into nothing, proving that entropy is all level to zero, and even your thoughts"*! Figure 91.

Figure 91 – The scientist Ape II, in the Congress of Sciences, said that humans proved even more and "admirably" that entropy levels to nullity whatever it is, such as matter, time, space and their minds! www.pinterest.com.

The Scientist Ape III commented in his lecture "*Destruction by Humans*" that annihilated their planet with battles, destruction of the environment, omission of birth control, and the basic foundations of science, reducing to nothing, time, space and matter with zero mass for photons and particle fields, in current Physics! Figures 92 and 93.

Fig. 92 – Scientist Ape III presenting the destruction of the basic foundations of Physics by the humans! www.pinterest.com.

Figure 93 – In the Congress of Simian Sciences, it was mentioned by the scientist Ape III the destruction of the planet of humans by wars, environment, and basic foundations for current Physics! Google images.

He was present at the Science Congress on the Planet of the Apes the astronaut Ape IV who showed necessity of new foundations of Physics for quantum theory against the current theories of relativity and quantum mechanics that are obsolete beyond antagonistic. Figure 94.

Fig. 94 – The astronaut Ape IV who presented paper on experiments on necessity of new fundamentals of quantum theory against obsolete ideas of relativity and quantum mechanics.

The scientist Ape IV commented that, by experiment of 1995-2002, the laser light when diagonally exiting the laser placed at 90 ° in the direction of the motion of the source, proved that the light was carried by the inertial velocity of the source, this is, the light formed by the quantum impulse mc in which c is the velocity of the elementary mass m' in relation to the emitting source measured on Earth, or m'c' in which c' is the always constant velocity in relation to the vacuum-space-ether, specific of the elementary quantum impulse and shown by the CERN experiment of 1964 in which the speed of light in the space-vacuum is always constant and independent of the speed of the source. By quantum impulse, the higher the velocity of the emitting source, the lower the velocity of light compared to the source. The emitting source at rest has little internal tension of the quantum impulse field h' in the particles, while at high speed the source has a saturation of h' which "holds" more light in the source, that is, the light coming out of the source with less velocity relative to the source, but at the same speed in relation to the vacuum-space-ether which is the specific speed of the elementary impulse. Figure 95.

The Pre-Established Physics and the Quantum Impulse

Nave-length 300 units emits light into a spatial coordinate P_1 to the target at the end of the ship that is in a space coordinate P_3, with the ship at velocity $v = 100/1s$. When the light hits the target in the space coordinate P_3 the astronaut measures the velocity of light as $200/1s$, as the target 300 is in P3'. And when reaching the target in P3' the light is measured on the ship as $300/1s$. By the laser experiment it was proven that the light exits diagonally from the laser at 90°, so it is carried by the source and the velocity of light is c variable relative to the source velocity, without contraction and or dilation of time contrary to the relativity.

$c' = (c + v)$ it is always constant in relation to vacuum-space, but c is variable relative to the velocity of the source

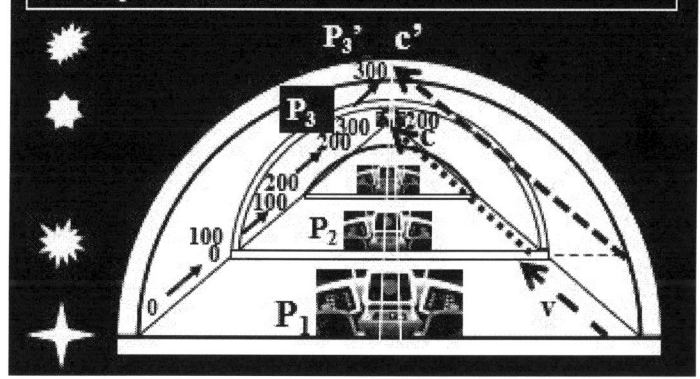

Fig. 95 – As laser experiment the light is carries by the source, and so the speed c' of the light is always constant in relation to the vacuum-space-ether, but the velocity c in relation to the velocity of the emitting source is always variable according to the velocity of the source.

CLII. OTHER CONNECTIONS – 1943 to 2018

"Librorum Prohibitorum"

By the current national security, of absolute UFO coverings sighted in the conflict of World War II, to prevent scientific and technological knowledge by other countries, mankind would be even without the use of the wheel, for its ddiscovery would be covered in the name of that security! The mastery and reverse engineering of UFOs Physics and technology is understood as national security, but the negative that UFOs are real is in reality to cover the absolute power of their own interests. Thus, impeding the study of the knowledge of the universe to mankind. During and after the World War II the sightings, worldwide, of UFOs that were different from: balloons "piloted by dolls", planet Venus, aircraft and military helicopters, hallucinations and other invented explanations, showed the ignorance of current Physics and the existence of a parallel civilian-military power to try the mastery of UFO Physics and technology, looking for time to see what was "that". Therefore, the armed forces of countries that follow this advanced nazi-fascist doctrine, of UFO negative, were discredited for concealing the right to information and scientific knowledge.

In 1969, Physicist Condon of the University of Colorado in the United States, one of the main edifyers of today's quantum Physics, presented a scientific result on Unidentified Atmospheric Phenomena-UAP or Unidentifiable Flying Objects-UFO, at the request of the United States Air Force, in which despite reports of military pilots who observed these objects perform curves maneuvers of almost 90°, was considered by Condon as not be able to add anything to the current Physics! This is, he could not explain these facts for his Physics!

Thus, it should not cause strangeness what astrophysicist J. Vallée and the astronomer J. A. Hynek reported in the decades of 70 and 80, that UFOs could be of other dimensions of matter-mind-thought-psyche-soul-entity-semi-biological and linked to the parallel realities of multiverse, because the formal Physics was inexplicable.

But, in 1976 in France, President Giscard d'Estaing himself In a special state television program presented that UFOs existed and that they would be extraterrestrial origin. Showed dozens of photos and filmes of UFOs flying over under the country and founded a formal entity of uphological research, the "Groupment d'Études des Phénomènes Aeriens", because there were no scientific explanations by current Science.

In 1977-1978 it occurred in northern Brazil, around 100 km from the headquarters of the Regional Air Command in Belém, capital of the state of Pará, and the Naval District of the Navy, an intensified action of luminous objects that were firing luminous rays causing burns, removing blood and leaving needle marks, plus a series of clinical symptoms such as paralysis and tremors in the stricken people, with fatal cases of death from these wounds. The phenomenon was known by the name of sucking, and widely reported by the press, radio and television in Brazil. The population sought to go to other locations, but they were also targets of these attacks. Local doctors considered these burns very serious, because they were not hallucinations! However, because there is no scientific explanation, by the current Physics, these villages panicked. So, in 1977 and 1978 came the largest military intervention already performed in Brazil to investigate and observe the appearance of UFOs in Brazil, the operation "Prato" (Plate). But, by intervention of theS United States the armed forces of Brazil nothing has come to reveal in the name of national security!

In 1986, during an intense wave of UFOs that lasted several days followed culminated with what ficou known as the oficial night of UFOs in Brazil, then Minister of Aeronautical Brigadier Octávio Moreira Lima admitted that the radars of the Integrated Air Defense and Air Traffic Control Center of the Aeronautical Command, from Brasília and Anápolis in the Midwest; Belo Horizonte, Rio de Janeiro; and São Paulo, in the Southeast, detected more than 20 unidentified objects with about 100 m in diameter each, congestion the main air traffic lines in the country. They were intense lights in motion from 250 to 1,500 km per hour, one of them with 15 times the speed of sound reported by the pilot, with radar echoes, in which military aircraft "Mirage" and "F-5" were triggered and followed these lights up to fifteen minutes, and the contacts for 3 hours followed, in which it varied from white to green, orange and red. However, the Minister Brigadeiro Moreira Lima had no scientific explanations for these unknown objects. But, the current Physics at the time had ridiculously explained how to be meteour! One of the two pilots that were in this air traffic, on the aircraft "Xingu EMB-121A1-MBZ", that night of the UFOs, was the aeronautical Engineer Ozires Silva, formed by the most renowned University of Engineering of Brazil, "Instituto Tecnológico da Aeronautica", founder and President of the Brazilian Aeronautical Company-EMBRAER from 1970 to 1986, President of Brazilian Petroleum-Petrobrás in 1986-1987, Minister of Infrastructure of Brazil in 1990-1991 and President of Viação Aero Riograndense-VARIG from 2000 to 2002, reported that: *"As contact with the Air Traffic Control Center of Brasilia I was informed that there were points that were being detected by radar, and that were not registered as regular flights within that area. At the height of 600 meters, we saw luminous spots, of reddish-orange color, with very intense brightness. We tried to get close to the lights, which they erased and lit in different places, between 10 to 15 seconds. We observed very fast variations in speed. The lights had real presences, they were primary targets on the radar, concrete targets. There was a very luminous celestial body that looked like a common chandelier, except for the size, more elongated and at that moment I asked permission to fly in the direction of that object. I was with my co-pilot, only two on this flight, and we turned the bow of the "Xingu" plane in the direction of the object, and each time we approached more "that", which kept more or less stopped. It had orange color, which could be explained even by the pollution of São Paulo, which makes the stars also orange. But the fact is that the radar of Brasilia had detected this object, and celestial stars do not appear on the radar! We were keeping in touch with the control and, at that time, with the control of São Paulo as well, and we went towards the object. But as we approached, it was disappearing, until it disappeared completely, and then we returned to the airport of São José dos Campos. When we were to land in San Jose, the head of local control warned me about another object, now in the direction of Rio de Janeiro and quite visible on the radar. Again I asked for guidance to approach him and I followed him towards you. As we approached, we noticed that he was at a much lower altitude than ours. At that moment, it was a rather elongated body, perhaps the color of a common fluorescent lamp. I circulated it several times with the plane, looking down. It was an elongated and quite clear object, but I could not lower the "Xingu". It was night and the altitude I was flying was already the least for that area, which is quite mountainous. But he went on permanently down there. By my vision of aviator, and with more than 40 years of aviation I have seen similar objects, but always with some explanation, but in that case there was".* He reported that divided the task of piloting the aircraft that night with the Alcir Pereira which commented: *"I've been a pilot for many years and I've never seen anything like it. It was too agile, impossible for a human to be inside. It was a very strong glow".* The conclusion of the Aeronautics was that the phenomena were solid, and reflected a certain form of intelligence The ability to track and maintain distance from observers.

It is evident that in addition to machine, there is a composition of the kind matter-mind! According to Penrose, in 1989, the mind has a composition that would be linked to quantum theory. And quantum entanglement shows exactly the interconnection between fields and particles.

And, in 1993, the Physicist B. Maccabee presented in the Bulletin of American Physical Society, United States, that one of these unidentifiable objects had left a strong magnetic field near the trees from which he levitated for several minutes, so with visible difference within a current Physics of what is different a physical phenomenon.

In 1994, Belgium amid a fantastic uphological wave that has assoded the country that year and following, especially involving observations of mysterious flying triangles, the former Minister of Defence of Belgium admitted that the matter was serious and that the Air Force would search it formally. In 1997, 20 years after Operatio unidentified Prato, Colonel Uyrangê Hollanda, who was the head of this operation, declared that there had occurred in northern Brazil the reality of flying saucers and other types of ships, but was surprised that when more intensified the apparitions of these ships and more proximity of the military team equipped with films and photos, received orders to suppress this operation!

And in 1997 the Physicist P. Sturrock of Stanford University in the United States, and dozens of other physicists presented a report in which they considered these objects unidentifiable as unknown atmospheric phenomena, including detection by radar being attributed to this cause. And in the case of the luminous objects of Hessdalen in Norway, detected by physical instruments, they would also be atmospheric phenomena unknown by the current Physics.

When an aircraft lights its lighthouse it is the light of an object, but when another unknown object emits high luminosity light it is defined as atmospheric phenomenon! The current Physics with this quantum mechanics of Condon and others, and Einstein's relativity, can do little to explain a wrought different atmospheric phenomena. So, by that impossibility, the current Physics remains far from analyzing these unidentifiable objects. Alleging only unknown phenomena, which will remain unknown until it is considered a thorough Physics in real energy fields, of photons with mass that form the particles and increase to their masses and not by ghostly condensation of the speed of the current quantum Physics of Condon, and of his followers and Einstein's theory of relativity! So by a new Physics it would be possible to present new possibilities for the knowledge and control of the gravitational and electromagnetic field, which appear with these objects defined as unidentifiable!

However, the military forces of countries, such as the United States in which Condon by current Physics presented this report that denies any new science and technology facts by Unidentifiable Flying Objects, imposes absolute secrecy in the name of national security. And thus impeding any other conclusion that Condon would make. For, in the case of existing evidence of accidents with these unidentifiable objects the reverse-engineering recovery of these unidentifiable objects would put the country that obtained it in the absolute leadership of science and technology and a political power unreachable. Also, current science receives amounts governmental monies, and thus conditioned to deny any study on these Unidentifiable Objects. However, with this position of cover-up, these military forces, these governments and the current

Physics prevent humanity from having the official knowledge of the reality of evident other civilizations of the universe. In the name of national security, they take the knowledge of all these evidences only for a restricted organization of absolute power, for the control of science, technology and political and military power, treating humanity as a flock of sheep that must be excluded from the evident wonder of reality that occurs in the universe, like the existence of intelligent extraterrestrial life.

But, two exponents of the Pentagon, intelligence and ethics far above the mediocrity of this obscure power, as the General Arthur Trudeau e Lt. Col. Phillip J. Corso, U.S. Army, Member of the National Council of President Eisenhower's, Head of the Development and Research Department, they have come to reveal in 1997 that the Department of Research and Development of the Pentagon, along with selected companies of technology, applied reverse engineering in crashed ship in 1947, in which on this ship there was technology such as the current silicon chips, electronic circuit, laser, fiber optics and fibers high strength. As Corso: *"We've been given orders to deliver the information that came in from Roswell. Inside the Pentagon many were trying to get this information but had never achieved it for their greatest scientists and agents. Artifacts recovered from the Roswell spacecraft brought the integrated circuit chips, fiber optics, lasers, stealth technology"*. Thus, companies such as Bell T. Laboratories, IBM, Monsanto, Dow, General Electric, Hughes and others, were triggered from 1947 and already in 1948 physicist W. Shockley invented together with Bell T. Laboratories the prototype of the chip, the "*silicon wafer*"as mentioned by Corso of the precipitated ship. Also, Corso revealing that even after 50 years of this fall in 1947 continued without resolution the Physics and technology that triggered UFOs. In his reports: *"There were no conventional technological explanations for the way the Roswell craft's propulsion system operated. There were no atomic engines, no rockets, no jets, nor any propeller driven form of thrust. The craft was able to displace gravity through the propagation of magnetic wave, controlled by shifting the magnetic poles around the craft so as to control, or vector, not a propulsion system but the repulsion force of like charges. Once they realized this, engineers at our country's primary defense contractors raced among themselves to figure out how the craft could retain its electric capacity and how the pilots who navigated it could live within the energy field of a wave. The air force discovered that the entire vehicle functioned just like a giant capacitor. In other words, the craft itself stored the energy necessary to propagate the magnetic wave that elevated it, allowed it to achieve escape velocity from the Earth's gravity, and enabled it to achieve speeds of over seven thousand miles per hour. The pilots weren't affected by the tremendous g-forces that build up in the acceleration of conventional aircraft because to aliens inside, it was as if gravity was being folded around the outside of the wave that enveloped the craft. Maybe it was like traveling inside the eye of a hurricane! I reported to General Trudeau that the secret to this system could be found in the single-piece skin-tight coveralls spun around the creatures. The lengthwise atomic alignment of the strange fabric was a clue to me that somehow the pilots became part of the electrical storage and generation of the craft itself. They didn't just pilot or navigate the vehicle; they became part of the electrical circuitry of the vehicle, vectoring it in a way similar to the way you order a voluntary muscle to move. The vehicle was simply an extension of their own bodies because it was tied into their neurological systems in ways that even today we are just beginning to utilize. So the creatures were able to survive extended periods living inside a high energy wave by becoming the primary circuit in the control of the wave. They were protected by their suits, which enclosed them head to feet, but their suits enabled them to become one with the vehicle, literally part of the wave. In 1947 this was a technology so new to us that it was as frightening as it was frustrating. If we could only develop the power source necessary to generate a consistently well defined magnetic wave around a vehicle, we could harness a technology which would have surpassed all forms of rocket and jet propulsion. It's a process we're still trying to master today, fifty years after the craft fell into our possession. The lengthwise alignment of the fibers in the suit also prompted the medical analysts to suggest that the suit might have been capable of protecting the wearer against the low energy cosmic rays that would routinely bombard any craft during a space journey. Where the possibility of some evidence about the workings of the alien brains did exist was in what I referred to in my reports as the "headbands". Among the artifacts we retrieved were devices that looked something like headbands but had neither adornment nor decoration of any kind. Embedded by some very advanced kind of vulcanizing process into a form of flexible plastic were what we now know to have been electric conductors or sensors, similar to the conductors on an electroencephalograph or polygraph. This band was fitted around the part of the alien cranium just above the ears where the skull began to expand to accommodate the large brain. At the time, the field reports from the crash and the subsequent analysis at Wright Field indicated that the engineers at the Air Materiel Command thought these might be communication devices, like the throat mikes our pilots wore during World War II. But, as I would find out when I evaluated the device and sent it into the market for reverse-engineering, this was a throat mike only in a way that a primitive stylus can be considered the forerunner of the color laser-imaging printer. Suffice it to say that in the few hours the material was at Walker Field in Roswell, more than one officer at the 509th gingerly slipped this thing over his head and tried to figure out what it did. At first it did nothing. There were no buttons, no switches, no wires, nothing that could even be considered to have been a control panel. So no one knew how to turn it on or off. Moreover, the band was not really adjustable, though it had enough elasticity to have been one size fits all for the creatures whose skulls were large enough to accommodate them. However, the reports I read stated, the few officers whose heads were just large enough to have made contact with the full array of conductors got the shocks of their lives. In their descriptions of the headband, these officers reported everything from a low tingling sensation inside their heads to a searing headache and a brief array of either dancing or exploding colors on the insides of their eyelids as they rotated the device around their head and brought the sensors into contact with different parts of their skull. These eyewitness reports suggested to me that the sensors stimulated different parts of the brain while at the same time exchanged information with the brain. Again, using the analogy of an EEC, these devices were a very sophisticated mechanism for translating the electric impulses inside the creatures' brains into specific commands. Perhaps these headband devices comprised the pilot interface of the ship's navigational and propulsion system combined with a long range communications device. At first I didn't know, but it was only when we began development of the long brain wave research project toward the end of my tenure at the Pentagon that I realized just what we had and how it might be developed. It took a long time to harvest this technology, but fifty years after Roswell, versions of these devices eventually became a component of the navigational control system for some of the army's most sophisticated helicopters and will soon be on the American consumer electronics market as user input devices for personal computer games. But experiments in replicated alien craft continued to be carried on through the years as engineers tried to adapt the propulsion and navigation systems to our level of technology. This continues to this very day almost in plain sight for people with security clearance who are taken to where the vehicles are kept".*

And without any support from the armed forces, for the defense of nations in the environment, educational system and the development of science and technology, and with the conclusion of formal Physics that UFOs are fantasies and forbidden "Librorum Prohibitorum" studies, by the "real" Physics of them, society itself is the one who seeks knowledge of the real UFO facts.

In 2018, the Pentagon, the United States Armed Force Centrals, with more than 3,000 military and civilian support teams and 20,000, came to release for the first time since its founding in 1943 tree videos about UFO sightings occurring in 2014, filmed by American military aircraft, without being able to explain its identification, certainly were not "balloons with dolls piloting and spying"!

Then in 2018, more than 70 years after the fall of unknown aircraft revealed by Corso, the Physics and technology of the UFO propulsion system remains far from being dominated. When searching for explanations of UFOs, based on the foundations of current Physics with more than 100 years of antagonistic theories of relativity and current quantum, and looking for further abastrations to be combined, womenfolks new real knowledge could arise. So, new fundamentals must be contrary to fields without mass, abstractions mathematical spaces and times quantified or distorted, contracted or elongated. Thus, fields of gravitation, electromagnetic and subquantum, and subphotons, should be observed as real physical meanings, with their masses, energies, moments and quantum forces.

After astronauts entered Earth's orbit, many of them showed different perceptions, reopening a wide variety of facts whose causes or mechanisms cannot be explained by the current stage of scientific knowledge, and that are assigned to forces of mind-consciousness and admitted as pseudoscientific by current Physics. Like this, hallucinogenic experiences, extra sensory perception or precognition, perceived ghosts and spirits, remote viewing, telepathy, near death experiences, out of body experiences, non-human intelligence and Unidentified Aerial or Flying Objects-UFOs, are occurrences where only a few beings exhibit these properties in intermittent conditions or on very special conditions. And researchers observed that these facts could be one unique phenomenon of the mind-consciousness be the key to its understanding. In 2018, in *Beyond UFOS*, of R. Hernandez, R. Schild and J. Klimo, as the Astronomer Ph.D., R. Schild of Harvard-Smithsonian Center: *"Believe that the scientific Physics, Astrophysics and Parapsychology academic community needs to take note and instead of dismissing these phenomena, these academics need to embrace it"*. Conforme And as the Physicist Ph.D., C. Swanson from Princeton University: *"It has been a century since Physics pioneers discovered that consciousness affects quantum mechanics. Until recently, the effects of consciousness have been largely overlooked by mainstream science, hoping they can be treated as a perturbation. But in many cases, these phenomena occur that are so far beyond conventional Physics that at last science will have to address the long needed unification. It appears there is a very real higher dimensional universe that we are a part of, and these phenomena make it clear whether we are ready to see it or not"*. According to scientist D. Radin, Ph.D.: *"These experiences, which often arrive spontaneously "out of the blue," remind us that today's scientific understanding of reality, and especially the role of consciousness in the physical world, are seriously incomplete"*. As D. H. Powell,

M.D.: "*A theory of everything needs to be able to explain all phenomena, including these reports*". As Klaus Heinemann, Ph.D. Former Physics Professor of Research in Materials Science, Stanford University: "*Are thousands of observations that cannot be explained with conventional science. It is a quantum leap in the direction of legitimizing these experiences*".

And by Rey Hernandez, J.D., M.C.P., by Cornell University: "*Physics does not explain the dynamics within the black hole, what is the multiverse and the perception of quantum physics, but theirs are all accepted. However The current physics should not wait any longer to researching these phenomena which are parts of the multiverse and what form the Cosmo*".

When Hawking mentioned that information-thought was energy, and that Penrose considered the mind-consciousness that should be explained by quantum Physics, the current Physics silences broadly, froze, for it was a field of pseudoscience! However, by quantum impulse m'c x c is energy, and subquantum that is the quantum of frequency 1 cycle per second and subphotons that form the ether, also the mind-consciousness could be formed of types of subphotons. And reveal these phenomena of advanced Physics-Biology-Technology as advocated by J. Vallée.

CLIII. ANTIGRAVITY AND GRAVITATIONAL-ELECTRIC-MAGNETIC FIELDS ENERGY – <u>1600 to 2018</u>

Galileo around 1600 and Newton in 1686, dealt with the fall of the heavy bodies, "gravis", in which according to Galileo different masses presented the same acceleration of free fall. And as Newton, gravity occurs by the attraction of masses, in which the greater the mass of a body, the greater the force of attraction between them. In case of free fall on earth: $F = M_{Earth} \times M_{Body} \times K_G/(r + d)^2$, and $F = M_{Body} \times g_d$, where F is the gravitational force between the mass M_{Earth} of the Earth and body mass M_{Body}, K_G the gravitation constant, r the radius of the Earth, d the height of the fall of the body, g_d The acceleration of gravity at height d, where $g_d \sim g$ = acceleration of gravity on the Earth's surface. So the gravitational mass of the Earth exerts a force $F = M_{Body} \times g$ over the body M_{Body}, while for a body mass m_{Body} the gravitational mass of the Earth exerts a lesser force $F = m_{Body} \times g$. In other words, according to the mass of the bodies, the forces that the gravitational mass of the Earth intereage with these masses are different. And it proves that a body of larger mass would have the tendency to remain at rest, this is has more inertia, while a smaller mass body a lesser force already puts it in motion.

Faraday, between 1831 and 1845, had observed that a magnet fixed on a copper disc, in rotating motion or at rest, kept the magnetic field fixed along with the disc that was in the same position. This fact he interpreted that the material of the magnet would seem to induce this field of outer space. And Faraday also concluded that electricity, magnetism, optics and gravity would be intertwined. And between 1855 and 1873, Maxwell and in 1887 Hertz, observed that the speed of electromagnetic disturbances was equal to the speed of light, i.e. the disturbance of an electromagnetic field was attached to the optics.

Then, at the end of the years 1800, N. Tesla and other researchers initiated new electromechanical experiments using different types of equipment, such as magnets and coil rotations, electric oscillers and magnetic flywheels. And in the decades of 1900 to 1940, new researchers observed that objects with electro-magnetic-mechanical action resulted in a decrease in the weight of these objects.

Einstein, in 1916, had presented that by the fact of an equal acceleration in a body of greater inertia (mass) and another of lower inertia (mass) be obtained by proportional forces: $F_1 = M_1 \times a$ and $F_2 = m_2 \times a$, concluded that they were equivalent to the masses in the fall by gravity in which acceleration was also equal. So, by the general relativity the fall by gravity was just an acceleration of mass of inertia caused by deformation of the space-time originated by a mass-energy. Therefore, without any action of gravitational interaction between masses, being direct action of mass-energy deforming space-time. However, without conceating how there was such deformation.

And in the decades from 1950, more researchers had experiments using in addition to electromechanical equipment, also high-voltage electrical current, plasmas, pendulums and gyroscopes. From Faraday's experiments, it had been developed on an industrial scale the electric current induction generator, two-piece, the stator formed of coils, fixed on the outside, and the magnets rotor rotating in the inner part forming the electric current in the coils. But, the Faraday experiment of rotating disc, homopolar, in which the electric current travels the conductor between the center of the magnets disk and the edge of this disc, remained unused for the greatest ease in the time of construction of the stator-reactor technology. However, with the homopolar generator it was observed a formation of electric current that is often higher than by the stator-rotor. According to Engineer P. Tewari, who was Project Director for the largest nuclear power plant in India, he had commented in the 1980 that the lack of this experimental development resulted in an incredible stagnation of the technology of a relevant energy yield. They were obtained with these new equipment also weight reduction and even levitation of metals, plastics and wood, metal ruptures, melting wood in metals, with production of electric current above the input energy.

W. H. Bostick in 1956 presented theory on condensed structure of plasma and magnetic field, and between 1970 and 1980 other researchers observed the cluster positive and negative ions. But in 1980, K. R. Shoulders published an experiment showing that he had obtained a cluster of electrons without other ions in the vacuum, and that an intense quantity of electrical energy occurred. However, upon receiving this work Feynman rejected this and other works of the decade of 1970. But in 1986 Feynman informed Shoulders about this electron cluster. Feynman expressed that he sought to understand how 10^{10} electrons could be kept as a cluster, a ball, in a vacuum without ions, but that he admitted it was a proven physical fact.

A. Overhauses in 1953 had observed the spin of the electron around atomic nuclei of crystalline network in different directions could receive energy by microwave and resulting in a nuclear dinamic orientation. And in that year of 1953, T. R. Curver and C.P. Slichter obtained this alignment when using microwaves. F. Alzofon in 1960 had presented that the matter was condensed electromagnetic energy and that energy was dispersed matter, this is, energy field. Also, stating that the gravitational field was an energy field. And in 1981 Alzofon published that a nucleus receiving an orientation, by the dynamic nuclear orientation theory, would result in a lesser interaction of the gravity between the outside of the material and the Earth's gravity. And in 1994, an experiment was performed showing that when a material was aligned with a magnetic field resulted in obtaining a lower value for the gravitational mass of this material. But, as D. Alzofon in the 1980s, a Nobel Prize in Physics stated that mass reduction would go against the energy conservation law of thermodynamics, because it would be excluding 70% of the kinetic energy of the fall of this mass, so it would be a "dark theory". Ironically, the dark mass and energy were not "darks" by the formal Physics! However, the mass of this body remained the same, with one part aligned and the other as it was, but both with potential energy for the fall of the body. Therefore, the quantum interaction greater or lesser, between the gravitational field and that of the body, is that it resulted in the quantity of kinetic energy, remaining a part as potential energy with energy conservation. So, current Physics doesn't understand the that is energy, and still considers it to be dark to a new principle presented.

Although some academics of formal Physics have observed the authenticity of significant results in high yield of electric current and weight reduction, the theories this Physics were distant from explaining these experimental results. By quantum mechanics there should be a minimum energy value at a point in absolute zero by the quantum theory wave function, as a potential that resulted in half a quantum ($h\mathcal{V}/2$) at one point. But, the quantum mechanics itself existed only as much as positive integers. However, for temperatures above absolute zero it would be foreseeable to exist some other energy potential in the vacuum-space, which in 1948 came Casimir to obtain an attraction of plates, and subsequently by other repulsion, as originated by this quantum energy of the vacuum. And to explain decay, interactions and particle formation, quantum mechanics presented as being by creation and annihilation in the vacuum-space formed of fields without mass and reduced times of 10^{-15} to 10^{-24} seconds, that is, particles called virtual. But, the energy level is indeterminate to explain an intense energy of these electromechanical experiments. To explain this intense energy in the results of these experiments, its authors, Physicists, Engineers, Technicians and Theorists proposed that the motion of these equipments induces the reception of vacuum energy-space-ether or energy from zero point, torsion fields or electromagnetic-gravitational field. Negative gravity was mentioned by gravitational potential, negative mass, in the vacuum-space-ether. The vacuum-space-ether energy was also emphasized to balance the quantum levels of electrons with the positive nucleus in quantum theory.

A example of how there is potential free energy to produce much more energy than the a provided is the case of gravity, in which a body at rest at a height, when falling produces an energy when reaching the fall much higher than necessary to hold the body at that time. On a bridge of height d the energy to shift a block of mass m = 1000 g horizontally to the edge of the bridge is: E_{Block} = m x a x l, a = 1 cm/s^2 and l = 1 cm, where the force of gravity f = ~ mg is annulled by the normal force of the bridge over the block of ~ mg. However, in the fall of 100 meters of this block, the energy in the base is $E_{Block\ Base}$ = (m x g x d) + (m x a x l) = 10.8 x 10^7 [erg] while the input energy is E_{Block} = 10^3 [erg], namely the output power ~ 10,000 times greater than the input power. Figure 96.

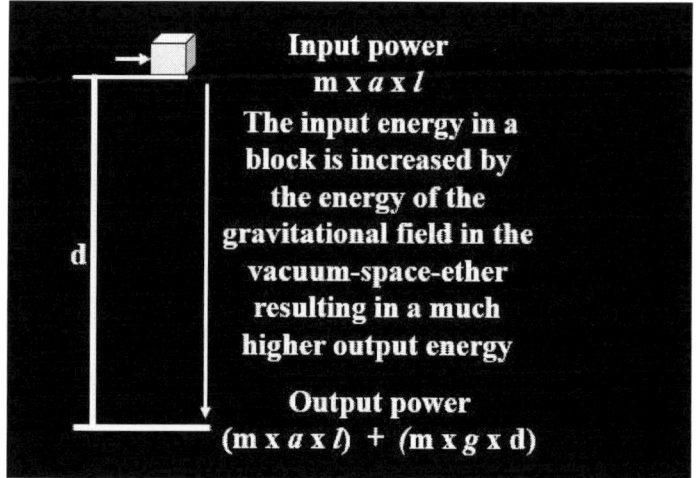

Input power
m x *a* x *l*

The input energy in a block is increased by the energy of the gravitational field in the vacuum-space-ether resulting in a much higher output energy

Output power
(m x *a* x *l*) + (m x g x d)

Fig. 96 – A block receiving an energy less than the final energy produced by the action of gravitational energy.

By quantum impulse, gravitational attraction is precisely the reorganization (interaction) between the particle field of the material, i.e. quantum impulse m'c that would be free, with the field of the other mass. As in the case of gravitational fall, in which the smaller is mass of the material, the smaller is the interaction of m'c with the mass of the Earth. Figure 97.

Mass of the Earth interacts with a body mass: greater gravitational force the lower the alignment of the quantum impulse

greater gravitational force	lower gravitational force
$F_1 = n_1$ m'c/t	$F_2 = n_2$ m'c/t
lower alignment	higher alignment
greater mass of interaction	smallest mass of interaction

Earth Gravitational Field

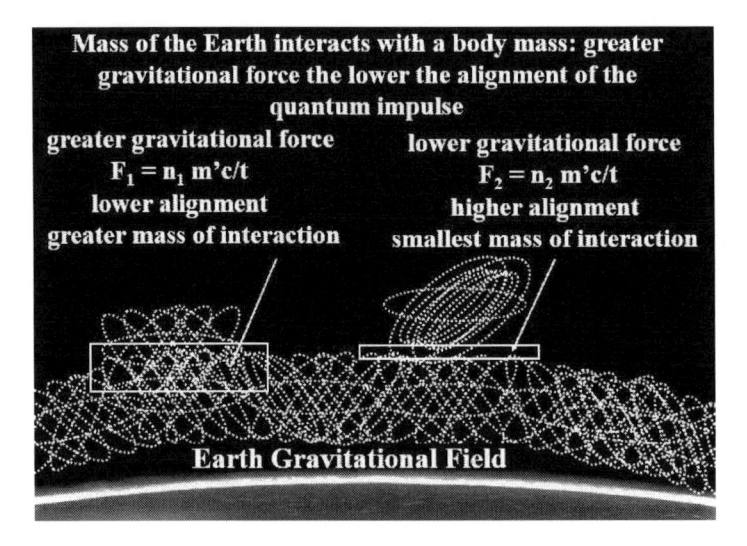

Fig. 97 – By the experiment of 1994 by Physicist F. Alzofon showed that the application of microwave in material results in alignment of gravitational mass more concentrated in the center of the particle, in which the attraction between gravitational masses is occurring more externally, which decreases the gravitational mass of the particles and therefore of the material. By quantum impulse the less free the quantum impulses, the lower the gravitational force.

From World War II when sightings by aircraft pilots, UFOs with maneuvers of almost 90 degrees and speeds far above the existing planes, the armed forces of the dominating nations initiated an intense performance of national security, with cover-up, intimidation, seizure of equipment and censorship of publications, involving new scientific and technological developments directly linked to the propulsion of these UFOs. Yes, whoever had this displacement technology would have the best strategic system of their goals. Thus, as an example, the Physicist Bruce De Palma, graduated at the Massachusetts Institute Technologie and Harvard, built an electromagnetic-gravitational machine with gyroscopes between 1971 and 1978, resulting in intense greater electrical energy, in which the Physicist Dr. Edward Purcell, one of the most eminent experimental Physicists from Harvard, he attested its validity. However, in 1980, the Astronaut Engineer Edgard Mitchell, Apollo 14, 1971, demanded that machine and development stay with NASA, and nothing of his experiment was sent abroad. Also, his patent was veted by the United States. These experiments would allow a minimum cost for electric energy to produce by hydrogen electrolysis, which would be the ideal fuel for automobiles and other vehicles, and dozens of applications reducing the consumption of electricity, oil and coal, which conflicts with the interests of economic groups, preventing this new technology of homopolar and derivative rotary disc. And with the increase of the world population por omission of groups and governments continues this growth exploiting the maximum energy resource, with the destruction of the environment, the devastation of forests with its fauna, flora and climate, for the maintenance of the owners of the world power that control technology and scientific development. So, the formal Physics following only abstract mathematics was stagnofied in its fundamental concepts, preventing new ideas from being disclosed to pressure the reformulation of natural resources to maintain a balance of the environment and other scientific developments.

By quantum impulse theory, m'c' x c' form all the particles and their fields, and in the case of an energy received by gyroscope for its rotation, can result in the interaction of the field m'c' x c' of the particles of the gyroscope with fields of m'c' x c' particles in the vacuum-space-ether, coming from all masses and charges of the universe that form gravitational-electromagnetic fields of m'c' x c', which results in greater output energy than the input energy. It is similar when gamma rays in force field results in electron-positron particle creation. Figure 98.

J. P. Hansen

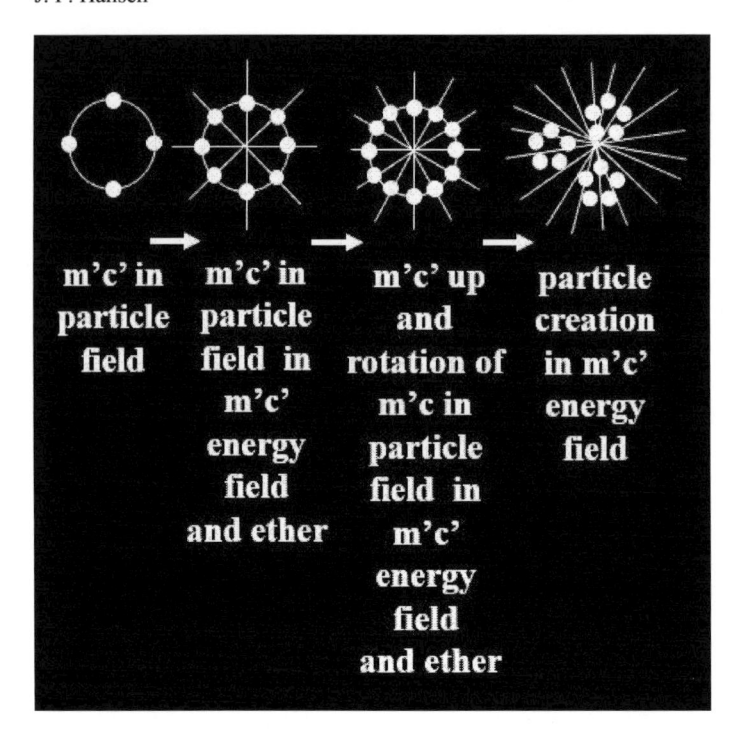

| m'c' in particle field | m'c' in particle field in m'c' energy field and ether | m'c' up and rotation of m'c in particle field in m'c' energy field and ether | particle creation in m'c' energy field |

Fig. 98 – The field of particles receiving an input energy generates in the field-particle rotation more particles of the energy field.

On 2018, CERN and ALPHA-G research the effect of gravity on the antimatter atom, that is, if the atom in the case the anti-hydrogen with positron around the negative charge antiproton core. By quantum impulse theory the proton and antiproton are formed from around 1,800 electrical charges of (electron-positron)$^+$ for the proton or (positron-electron)$^-$ for the antiproton, that is, according to the direction of the right or left helix, that is, positive or negative, of the impulses m'c' x c' that forms the helicoid-particle, which is the charge. The helices of different signal charges are attracted. However, in the case of gravity, the attraction involves the field of m'c'x c' which completes the charge also formed of m'c' x c'. It is only a part of the charge-field mass of the particle that enters into interaction with the gravitational field of the other mass. And this gravitational interaction only exchanges the m'c' x c' impulses independently of the helix orientation. Yes, this gravitational exchange covers both the core charge field and the atom's orbital charge. So, in the case of Earth's gravitational field, its orbital particle with negative helix interacts with the atom's negative helix orbital particle, and the core of the positive helix particle of the Earth field interacts with the nucleus of the atom's positive helix particle . And with the antiparticle atom, the negative helix particle of the Earth's gravitational field interacts with the negative helix of the antiparticle nucleus and the positive helix core particle of the terrestrial gravitational field interacts with the helix of orbital antiparticle of the atom. Figure 99.

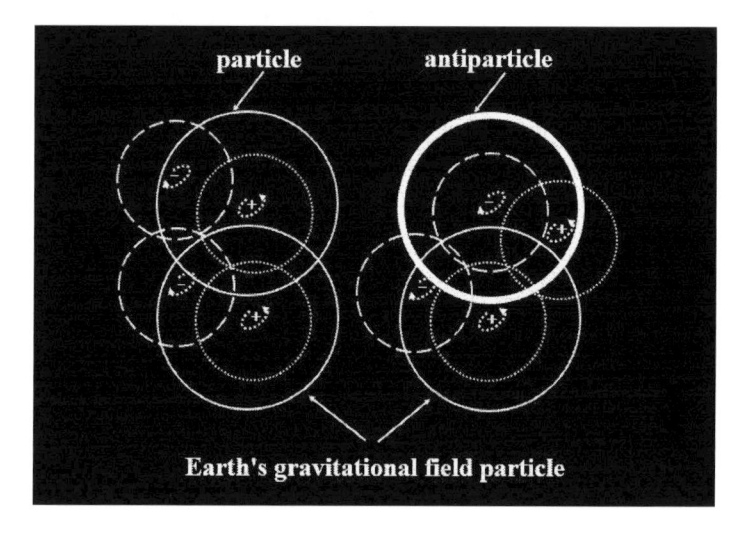

Fig. 99 – Particle and antiparticle interact with the particles of the Earth's gravitational field.

These experiments that involve obtaining higher output power than the input shows Faraday's confirmed prediction on that electricity, magnetism, optics and gravity are interconnected. And by the theory of quantum impulse, the force of interaction $F = n \times m'c'/t$ is that involves all the particles, their charges and all their electric-magnetic-gravitational fields.

PARTE 12 – EPILOGUE

Michelson and Morley, as all scientists at the time admitted that the light had velocity c in relation to the fixed observer on Earth and the exit of the interferometer in motion in space. In the case, as the Earth-Interferometer moved in space with velocity V_{TI}, in the direction of the motion, the speed of light obtained by the observer next to the interferometer to reach a mirror on the opposite side of the light output would then be the speed c minus the velocity of the Earth-Interferometer that was moving away, "fleeing" from the light, ie $(c - V_{Earth})$. And on the turn, when reflecting in the mirror, the velocity that would be obtained by the observer next to the interferometer would be the speed c plus the velocity of the Earth-Interferometer that went against the light, ie,

$(c + V_{EI})$. It is evident that in this case the round-trip times would be different, because the distance traveled D between the exit and the mirror would be traveled with the time of going $t_{going} = D/(c - V_{EI})$ and on the back $t_{back} = D/(c + V_{EI})$. Maxwell had obtained that light was an electromagnetic wave which to if propagating was necessary the ether, as well as for the sound it was the air, maritime waves by the sea, and the earth for propagate seismic waves. As the Michelson-Morley experiment that admitted light with speed c in the ether was null, Einstein and the current Physics concluded that the light was propagated in an electromagnetic field without mass and without ether. However, the laser experiment showed that the light comes out of the interferometer with the velocity $(c + V_{EI})$ in the direction of the constellation of Leo and the 90° it leaves diagonally in relation to cosmic microwave radiation or vacuum-space-ether, thus carried with the inertial velocity of the Sun-Laser-Earth, and hence the null result of Michelson-Morley experiment

And irrefutably this laser experiment proves that light is carried by the emitting source, this is, by the particles constituted of the quantum impulse of the immortal creation of the theory of Erdelyi.

For the going time and of back time of speed of light c in relation to the source Earth, it would be, $t_{going} = D/(c + V_{inertia\ of\ EI} - V_{fleeing\ from\ EI}) = D/(c + V_{EI} - V_{EI})$ and on the back $t_{back} = D/(c - V_{inertia\ of\ EI} + V_{meeting\ to\ EI}) = D/(c - V_{EI} + V_{EI})$, and the time on the diagonal of $t_{diagonal} = D'_{diagonal}/c'_{diagonal} = t_{going} = t_{back}$, that is, equal times. And anyway, to calibrate the zero interference, the mirrors were adjusted, hoping that in the rotation of the interferometer the interference would occur. However, as Janossi, in the rotation of the equipment, the light paths keep the same time difference. Therefore, the Michelson-Morley experiment is explained without contraction of space or time dilation, and the equation of the particle motion by the theory of elementary impulse shows that with mass increase by m'c x c occurs modification in particle dimensions resulting in variations of time for these transformations as in the half life of the mesons μ at different velocities.

The first principle of Einstein's theory of relativity states that in all uniform motion reference systems, that is, inertial, the laws of nature are equal. And the second principle of this theory it states that electromagnetic radiation like light in the empty space is propagated with speed c to any observer in all inertia coordinate systems, regardless of the velocity of the emitting source. But, the CERN experiment of 1964 showed that the meson π° with velocity almost c emits gamma rays with speed c, both in relation to the laboratory. Then, the meson π° and gamma rays are almost together, that is, the velocity of gamma rays in relation to the meson π° depends on the velocity of the meson π°, although the speed of gamma rays in relation to the cosmic microwave or vacuum-space-ether is independent of the velocity of the meson $\pi°$, according to the experiments of CERN and laser light.

By the laser light radiation experiment it was detected that the photon acquires a variable inertial velocity according to the velocity of the source, in which the lower the velocity of the source the greater the velocity of the radiation relative to the emitting source. Therefore, in relation to the observer the velocity of light is variable, this is, c_v, and the relativity that considers the speed c as independent of the observer is contrary with this laser experiment and the correct interpretation from the CERN experiment of 1964.

The referential transformation of Voigt-Larmor-Lorentz applied by Einstein is effected to c constant independent of the velocity of the source. However, by the experiment with laser light the velocity c measured by the observer is variable according to the velocity of the source, in which $c' = (c + V)$ is the constant speed in relation to the cosmic microwave radiation or vacuum-space-ether, and the laser installed at 90° resulted that the speed c of the laser compose with the source velocity, leaving the laser light diagonally, proving that it depends on the source.

Relativity hits when it conceives light with a constant speed relative to vacuum-space. But, also the wave theory admits that constant speed. However, the theory of relativity misses when it maintains that constant speed of light relative to any velocity of the emitting source.

But, Einstein and the pre-establishead Physics, despite considering a constant speed of light in relation to vacuum-space, they did not have the perception that light (quantum impulse) has a constant speed (c') in relation to or space-vacuum-ether or cosmic microwave radiation, and that when the light being carried by the source is that it results the measure of speed c in relation to the Earth-Solar System source. In the case of lower source velocities there is less mass and insaturation of impulses in the particles and in their fields, then the quantum impulse being emitted with higher velocity relative to the source, but maintaining the c' velocity in relation to the vacuum-space-ether or cosmic microwave radiation. While at high velocities of the source there is a saturation of quantum impulses in the particles and in their fields, as if clinging, and then the velocity of light c_v being lower compared to that source. The source carries the light, so that in relation to the vacuum-space-ether or cosmic microwave radiation the speed of light c' is always constant, but the light lowering its velocity c_v relative to the source the higher the velocity of that source. In the CERN experiment of 1964 and the laser light in 1995-2002, it was proven that the velocity of the gamma rays, relative to the emitting source of the meson with close velocity of light, was almost equal to the velocity of the meson, that is, almost zero in relation to the meson.

However, the current Physics follows relativity which extends the constant speed of light c, in relation to vacuum-space, to all the velocities of light sources, contracting the space and or dilating the time. Also, without the perception that in the laser experiment at 90 °, the laser light comes out diagonally, carrying the inertia of the light and varying the velocity of light c_v which is contrary to the theory of relativity. There is nothing to do with space contraction and time dilation, because it is the path in the direction of motion or in diagonal that explains the speed of light c in relation to emitting source and in the vacuum-space-ether. And in the CERN experiment of 1964, the relativity interpreting that gamma rays do not carry the inertial velocity of the meson, without the perception that the meson carries the gamma radiation with velocity similar to gamma radiation which is exactly the always constant velocity c' in relation to the vacuum-space defined by the own relativity by the speed c.

Hubble, the discoverer of galaxy expansion has never admitted expansion as being from space itself, but Eddington and imposed Physics to following the bends of relativity have created an space, which carries the expansion of the galaxies, which remain without motion, as in the case of the inflated balloon in which a spot on the surface remains in the same spot while the balloon inflates! By the theory of general relativity the space is curved the more the matter is increased. But for a particle, in quantum mechanics, it's a minimal mass that could bend space. For example, it is the field between magnets that cause attraction or repulsion, by curving the power lines of that field, not the space that remains the same!

According to Jammer: *"In quantum mechanics, nuclear Physics and field theory, force is essentially a device for calculus, a functional relationship. While in special relativity, force is dependent on systems of referrals where mass is function of speed, in general relativity to force of gravitation is space-time curvature geometry. As a consequence, it could not lead to a review of force for a concept anything more than a mere mathematical relationship".*

By the theory of elementary impulse m'c, the energy $\varepsilon = |v|h'$ = m'c x c = m'c², where h' is in [erg] in the CGS contrary to h [erg x second] without greater physical significance, is obtained simply from the classical physical concepts and the physical significance of Planck's quantum theory and the quantum light of Einstein. And it's this elementary photon h' that constitutes all the elementary particles, so the total energy of the particle E_{total}= nh' = m'c², which constitute all atoms. The evidence of the mass m' is striking as a result of accelerations and collisions in the particle accelerators, because what gives mass to the detected particles, formed peaks and observed energy bands is the resulting mass of the electromagnetic field that accelerate protons and electrons, increasing their masses and consequent velocity. And the fundamental proof that the rest photon has mass m' is that the photon by the equation of the mass increase of Voigt-Larmor-Lorentz and relativity (and quantum impulse with v ≠ c) : m_{photon} = $m_{0photon}$/ $\sqrt{1-(v^2/c^2)}$ with speed c ≈ 300,000 km/s its mass would be infinite, $m_{photon} = \infty$, but it has its rest photon mass m' and at the speed c of the electromagnetic radiation in relation to the emitting source or c' in relation to the space-vacuum-ether. This equation is deduced by the theory of quantum impulse, but with velocity v at the limit of reaching the speed of light c, in which particles when receiving energy m'c x c = h' results in the increase of mass and not "frighteningly" by relativity in which the mass increases "phantasmagorically" by increased speed.

By the theory of elementary impulse is the phenomenon of configuration that is responsible for the shape of the particles of the physical nature of the universe. In which it occurs by thermodynamics an increase in the order in the system in the physical systems and RNA and DNA in biological systems, through a partial entropy of configuration that decreases, but having an increase of dissipation energy, in which the entropy of the total system is increased.

And by different aspects of research is observed the elementary photon h' of nature towards what was predicted in 2005 by H. Terezawa in which phases of the development of Physics would be:
-1900-1925-Atomic Physics
-1925-1950-Nuclear Physics
-1950-1975-Hadronic Physics
-1975-2000-Quark Physics-Leptons
-2000-2025 –Subquark Physics.

And in Subquark Physics the constitution of fractional charges be according to the theory of the elementary photon or elementary energy h'.

When in a nuclear reaction, as occurs in the fission of uranium in the atomic bomb or in the fusion of hydrogen deuterium in helium in the hydrogen bomb and in the stars generating all kinds of atoms, a fraction of the mass of the surrounding atoms in the reaction is emitted as energy $\mathcal{E}$ = h'$|v|$ = $|v|$ m'c x c = $|v|$ m'c². For, by the concept of energy = mass x (speed of light c)² = force x distance = (m'c/t) x (|c|1cm) = m'c², where m' is the mass of the elementary photon, c is the speed of light for the propagation of the momentum m'c of radiation relative to the Earth as a referential velocity and h' the photon of elementary energy that corresponds to the frequency of 1 cycle/s, which forms the particles as the electron, positron, proton, meson neutron, bosons, neutrino and all the others.

The control of nuclear fusion remained for the development of technology, in which by the successive failures that in 30 years would be obtained this control, the current Physics decided to make the expression "30 years from now" as a motto of joke, as if this Physics had nothing to do with this project! However, it is the current Physics that wants to not know what happens within the nuclear reactors. For example, the neutron by acquiring greater energy before that decay in its proton, electron and antineutrino, has its mass increased and its dimensions diminished by the shortest distance between impulses m'c x c, by the particle equation of the theory of quantum impulse, then leaving more of the so-called bottle of Fermi, inside the reactor. In 2018, new research technology reduces the control of nuclear fusion to 15 years. But with the foundations of new ideas, this technology could be achieved within an established period and not indefinable. According to the Physicist and Historian T. S. Kuhn, evolutionary leaps of scientific knowledge can be obtained by changing paradigms.

In the current quantum mechanics of the standard model, the energy field is formed by motions of mathematical abstractions of massless photons, requiring another scalar mass field, the Higgs field, to provide mass in particle formation. In 1963, Feynman presented that: *"The field is a physical amount of functions abstract mathematics of position and time"*! In 1979, physicists presented that: *"We don't know what charge is"* ! In 1985, despite the Higgs field being presented in 1964, to give mass to the particles, by abstraction and mathematical artifice, Feynman commented that: *"Mass we don't understand what they are or where they came from. This is a serious problem"*! At the end of the twentieth century in 1999, Griffiths uttered: *"I encourage you to think of the field as a real entity"*! But what entity? In the new millennium, in 2003, Veltman came to say what was magnetic field: *"It's made of photons different from the apparent photons of light"*! But what would these different photons be?

But, by the theory of elementary impulse is m'c = h' the elemental photon that forms the electromagnetic, gravitational energy field and the Higgs field. And the quantum impulse also explains, through the positive-negative helix dipole that forms the high charge squared, as the charge of the electron e^2, in which the current Physics ceased to speak on this question of Dirac in 1963, about the charge appearing squared in the charge equation. This question was answered in 1964, by Erdelyi who created the theory of quantum impulse, directly to Dirac and the Scientific American.

For the reality of physical facts and meanings, concepts must be followed, such as mass, space, momentum, impulse, force and energy. And the best understanding about the atom and its particles is at the base of all scientific and technological development.

Between 397 to 400 as St. Augustine, the universe would have been created by an absolute God who lives in a dimension of eternity, and who was the creator of the universe. Therefore, time would created only when the creation of the universe, and also of someone to evade the question of what was before the creation of the universe would tell whom inquire about what was before this divine creation, that God was setting hell to be sent to those who did this kind of questioning. And as Hawking, with a permanent initial space-time in unsingular dimensions, the universe could exist eternally, and its laws would be the consequence of this eternal initial structure. So by Hawking, it would be unnecessary for a creation by an absolute God of an already eternal initial structure, and this initial structure as Hawking would be a matter of Theology and Metaphysics, not of Physics.

But Theology itself and Metaphysics are contradicted by Physics, because as an example, wars, destructions, slavery, misery from the beginning of life showed an insufficiency of this Almighty Creator who would have created Heaven (Infinite Space), Eternity (Time) and Matter. So, by Physics, space-time-matter or just coordinate points, with their laws of the universe were self-created, because it would unnecessary or non-existent any Almighty Creator! The current Physics prefers to know only the materialism of the physical phenomena and follows only concepts and their relationships that form the laws of the universe. However, an All-Powerful Creator, is an intentional premise, because the universe is motion in all directions, shocks, explosions, that is, constructions and destructions! So what would exist is light and darkness, that is, God and Satan, both sides of the universe or multiverse! The Creator would have presented the ideal design, but political dissent in the heavenly dimension, Satan, would introduced his diabolical ideology of knowing without need of any Creator! And emerged by the formal Physics the self-creation of the universe without any Creator! For the Absolute Nothing without space, time, photons real or virtual there is one total darkness: Eternal Darkness! But, something was created by the Nothing Absolute, a sense of existence against the Eternal Darkness! And the reason why there would be something rather than nothing! And it was created from the dimension of the Nothing Absolute the eternal-space-time-ether-mass-matter with life-mind-spirit-soul-thoughts-ideas. In 1988, Engineer General Officer A. M. M. Uchôa, Professor of Rational Mechanics of Military Academy of Agulhas Negras in Brazil, concluded by the analysis of the tetrade "tetraktys" de Pitágoras, na qual no tetraedo estaria o espaço da fonte e a raiz da eterna natureza, that the existence of the universe came as a cause by Absolute Nothing which is the Divine Absolute! The mass in the universe is a complex constant and is the fundamental factor of existing something. Thus, the elementary m' mass of the elementary photon theory constitutes the m'c impulse and the elementary photon of energy m'c x c = h' in a ether, contrary to the massless photons of the physical field of today's Physics. In the vacuum-space that allows the count of time, is placed the ether and the light that with its mass and motion shapes the existence of the particles and their fields of force that form matter, the universe! In biblical theological metaphysics, the beginning was space and time ("sky") and the disformed matter ("Earth"), and then the light (coming-emitting from the matter)!

And the current Physics created chimeras such as mass and dark energy, mass particles formed from fields without mass, theory of long-time relativity and deformed space of continuous field incompatible with quantum theory, high-abstraction mathematical quantum mechanics, theory of strings of n dimensions and no mass, then getting the formal Physics on the side "zero mass material", in which neither "Le Diable" expected this!

As R. Pietro: *"When a theory becomes more important than experimental facts, all these experimental results are manipulated to fit the theory. If they don't fit they are discarded. That's not how you do Science. But that's how it's done".*

However, with experimental facts should be lacking arguments, so that scientific facts could be preserved against these discards effected with the purpose of maintaining superficial theories, of only relationship of words, far from the physical meaning and inconsistent. Are discards practiced with the justification that the depth of the physical phenomenon should be explained by mathematical abstractions, but stagnating development of the real meaning of the bases of the Physics, as with the current quantum mechanics of principles and relations of words and Einstein's relativity of general terms independent of the quantum constitution, and with attempts at unification in which one moves away from the other.

According to Einstein himself in 1922: *"For real progress, a new general principle is needed".*

In 1947, the creator of the quantum in 1905, Planck, said: *"The quantum of action has a much more fundamental meaning than it had come to imagine".* That is, he had the perception that there was something much more than the action, that there was little or no physical significance. And then it would be explained by quantum impulse with quantum energy of ample physical significance.

And what Planck revealed in 1947, Einstein completed in 1948 in which: *"Quantum's discovery by Planck has put Science to a new goal: to find a conceptual basis for all Physics, and despite marked gains, the problem is still far from a satisfactory solution".* And in 1950 Einstein claimed: *"I thought a hundred times more about quantum problems than the theory of relativity".* And in 1951 it completed: *"All these years 50 years of pondering has not approximated me to the answer to the question: what are the quanta of light?".*

In 1952 Erdelyi presented the explanation of Planck's constant h with the theory of quantum impulse, yet the pre-establishead Physics prevented the publication in its specialized Journals of Physics, controlled by the Editors of this Physics of relativity and quantum mechanics that were followed without admitting any objection.

In 1953, Schrödinger who had always considered his electron equation as a real wave, attributed to the story to be fundamental to the origin of knowledge of the existing issues: *"History is fundamental Science because all human knowledge loses its scientific character when men forget the conditions that originated them, the questions that came to be challenged and the functions they were created for".*

In 1963, Dirac, Nobel Prize for quantum mechanics, commented on a medium that transmitted the light: *"You must introduce something corresponding to the ether. A new image that conforms to our present ideas of quantum theory".* However, between 1952 to 1954 Erdelyi explained the theory of quantum impulse in which the impulse m'c in vacuum-space needed a medium, an ether, for it to move with the c' velocity relative to that ether, or the velocity c in the specific case of the Earth source.

Also, in 1963, M. K. Hubbert wrote in the Science Magazine the article We Are Retrogressing in Science? *"The books repeat affirmations of specialists, reviewers and "authorities", without taking the trouble to prove or analyze them critically. In 5 of the best books in Physics, all exemplify the calculation of Earth's mass by a non-valid method! Manuals ignore the theory and history of the unit systems and dimensional equations developed by Fourier, Gauss and Maxwell, which result in incredible atrocities that then appear in the treaties as basic truths, and the foundations of structure of Physics are being lost of sight!".*

The Physicist Weisskopf, of current Physics, commented on 1968 that: *"The meson muon is unstable and transforms into an electron and two neutrinos, thus indicating a possible structure of electrons and neutrinos, and that must be related to the nature of the electric charge and the meaning of the charge unit, of fundamental importance in nature".*

And according to F. J. Dyson, on 1968: *"A student early in the studies of quantum mechanics does not understand it, and after more than six months still does not understand. Finally, after some time come to understand that there is nothing to understand"*!

In 1976, Segré, Nobel Prize in Physics, thus manifested: *"Particle Physics is the only area where there are the greatest number of unresolved intellectual problems. Some of them are almost a hundred years old and the tendency of physicists is to forget them because there is no indication of a solution. A trend of modern art to abstraction resembles that it seems to predominate in the development of current Physics. One of the abstractions is the expansion of space-nothing of current Physics! I do not believe that Galileo, Newton and Einstein were the last of their kind, for the philosophy of Physics changes over time and there is every reason to think that they will continue to change".*

In 1978, Lederman, Nobel laureate, described: *"The History of Science suggests that the proliferation of physical particles constitutes tangible proof of whether these entities are not elementary. And the quarks that form particles are seen as the last limit of knowledge, that it would be to hide the last secret forever, however the own electron that was once observed connected to the atom came to be observed in a free state. So the quarks would also be released, and there would be simpler structure than quarks?"*.

Almost the entire academic community learns that the theory of relativity has been widely proven in all the items that have been experienced, however it is widely erroneous information because there are flaws in the interpretation of the experimental results obtained, and thus comments the Physicist Jammer, awarded by the Academy of Sciences and Arts of the United States in 1979: *"The fundamentals of the theory of relativity are not a matter of universal consensus and have not been thoroughly tested by the experiments"*. Also, there is different interpretation of non-relativistic theories.

According to the Physicist, of the current Physics, Nobel Prize, G. T. Hooft, in 1980 related that despite the mathematical mechanism of the Higgs that should provide the mass of particles, he questioned: *"What determines the mass of the Higgs particle? And what determines the fundamental unit of the electrical charge?. There is a need for a more general theory and that could not come from the existing standard theory"*. But, the impulse theory explains admirably what determines the Higgs mass and the electrical charge unit.

In 1984, as Gribbin: *"Schrödinger thought he had eliminated quantum jumps between the orbits of the electrons determined in the atom, as if they were vibration modifications. But, the waves in the equation showed abstract, complex forms of vibration in a phase space, with the wave-particle dualism remaining hidden in the mathematical formalism. After 50 years of 1930, the time of the assembly of statistical quantum mechanics, is almost at the same point as to the understanding of it, in which physicists were at that time"*.

The own exponent of the Brazil current Physics, J. L. Lopes, cited in 1985 for his words in 1972, which stated: *"It is not available in Physics that is followed by a quantum theory of gravitation that is satisfactory. It is not even known whether a new type of treatment other than that of Newton-Maxwell-Einstein and quantum mechanics would be required to better understand the* *nuclear phenomenology. Perhaps the notions of space and time, though necessary as the scenario of current knowledge followed by the physical world, no longer apply to domains of extension less than 10^{-13} cm or high energies. What other ideas will replace space and time?"* . And between 1984 and 1987 completed: *"It is possible that we have to make a theory completely with new conceptions about elementary particle physics"*.

According to the Nobel Prize for Physics of the current Physics, Feynman, in 1985 says: *"The number linking the electron between the charge, the Planck constant and the speed of light, that is, the fine structure constant, has been a mystery. There is not in current Physics a reasonable theory of quantum gravity, and the theory of gravity still remains isolated from other theories. So far gravity is not understandable in terms of other phenomena in Physics"*. But by quantum impulse to the equation of the electron deduced by this theory, and gravity is explained by the unification of Planck's constant h within the constant K_G of Newton's gravitation, by quantum impulse, a fantastic deduction, unprecedented, in the History of Science!

In 1987, the Physicist N. Bernardes of the current Physics commented: *"Since the analysis of Bohm, in quantum mechanics, the concept of probability has been criticized, and some observing that if it was eliminated this concept would occur the development of Physics, thus being the concept of probability in the dock"*.

Feynman had pronounced on the double-slit experiment in 1971 that: *"No explanation is successful in relation to the shape of the curve in function of the electron individually as it passes through two slits, so it is concluded that everything is quite mysterious and the more one explains the more mysterious it gets"*! And Selleri in 1986 on this statement Feynman commented that: *"The learning of official quantum mechanics is basically avoiding the major problem"*.

In 1982, according to Pais the Einstein's biographer, presented that the concept of electromagnetic mass celebrated 100 years, originating from Abraham, Lorentz and Poincaré: *"But what was left of these ancient times was the fact that this question is not yet understood"*. However, electromagnetic mass is the quantum radiation m'c impulse which explains the constitution of the particles and the force field of nature.

Selleri in 1986 said: *"The books of current Physics are purely pedagogical, and if it is considered as having a sharp scientific concept they would not be better than a tourist brochure. The consequence is that many physicists learn to use their instruments without actually understanding them"*.

And the first theoretical Physicist in the Brazil, Schenberg, so pronounced between 1983 and 1987: *"Quantum mechanics does not explain Planck's constant h. There should arise some new concept that the current field and particle. The thin structure constant remains a riddle. As Einstein had been explained this constant would shed "new light" on the constant h"*. By the irony of these words, the "new light" is in reality the light m'c x c which constitutes the light of the theory of quantum impulse, that is, it is in the view of the current Physics, but which is not accepted by going contrary to the relativity of the space-time that contracts, expands and curves. And complements that: *"The unification of electromagnetism with the gravitational field was not performed by quantum mechanics, nor by relativity. It is necessary to seek even more to reach the fundamental structures of the particles, being quite obscure that position. The concept of mass is had as related to force by Newton, but there are misunderstood aspects about the mass that seems complex, and are extremely difficult to understand in depth the ideas of Newton on mass and force"*. Also, in 1908 Poincaré had commented that: *"One must find an electromagnetic explanation of all forces, and in particular the force of gravitation"*. In fact, Newton did not have quantum theory, and then the force is by the impulse: m'c/time. And the mass that forms the matter is really complex, because it reveals itself as mass m' with the c' velocity in relation to a medium of submass, an ether in the absolute space. Continued Schenberg: *"There are many dogmas and you must go against the old ideas and there should be all freedom of creation. It is very near or at the end of the period of the history of the development of Physics, in which there must be another period in which the methods will be others. If something essentially original arises in Physics it should be related to the Planck length that relates the constant h, the constant K_G of Newton and the speed of light c, which seems to be the great fundamental constant of Physics"*. It's actually not the length, but the constant h itself explained as energy and being related to K_G getting the gravitational energy, uniting the electromagnetic field of light quantum radiation with the gravity of the universal attraction of Newton's constant K_G.

Beck thus commented on 1987: *"Currently live under the weight of the past and it is not seen that the current scientific generation of recent decades has been innovative"*.

Tiomno in 1987, another Physicist of the current physics revealed: *"In Physics, mathematics advances in matters more and more abstrats"*. However, not recognizing the theory of quantum impulse because it is contrary to Einstein's relativity.

In 2015 were completed 100 years of the theory of general relativity and 110 of special relativity that are followed by the current Physics. But, the theory of relativity with its macroscopic and distant aspect of quantum phenomena maintains the inability to unite quantum mechanics and general relativity. The formal Physics says that the theory of relativity is successfully tested at almost 100%, but in simple examples as experiments in 2002 and 2014, indicated in the figures 71 and 75, the two possible results, that is, the laser light coming out at 90 ° or diagonally from the direction of the motion, they are contrary to the theory of relativity. Yes, or it would be possible to obtain the displacement of the reference system in the space-vacuum within a spacecraft if the light came out at 90 ° in the direction of the motion, which is contrary to relativity, which does not conceive absolute reference in which this displacement of the ship would be obtained, or that laser light coming out diagonally it proved that it carried the inertia velocity of the emitting source in the direction of the motion. That is, these two possible results of the experiments, it proves that the theory of relativity is inconsistent, totally dissociated from quantum physics, without recognizing that physical phenomena are associated with the constituents of physical nature.

The time of the existence of the theory of relativity and the interpretation of quantum mechanics by the current Physics of indetermination; probability and statistics; particles, photons and physical field without mass; and without considering the physical fact that the harmonic oscillator, which emits the electromagnetic radiation, is accelerated in each restoration force emitting the photon at each oscillation, are the causes of the delay in the development of the bases of Physics. This Physics detected the existence of atoms, the particles that constitute the atoms, aces or quarks, but also gluons without mass that would constitute the nucleus of particles, and mathematical photon without mass, than the elementary photon h', the "joker" of mass elementary m' that forms all the particles and the physical field. It was 100 and 110 years in 2015 of the theory of general relativity and of the special, and with the present interpretation of quantum physics there is little to celebrate by the stagnant foundations of today's Physics.

And on the basis of current Physics: there is also the inability to explain the matter and dark energy, for decades without any discovery of the search teams. There is no shortage of gravitational mass in the galaxy that acts on the velocity of stars around the galaxy. They are closed force lines of the gravitational field formed of m'c x c or subphotons that act on these stars. A 30° tilt ramp causes an acceleration of only 50% to that of a free fall. That is, the gravitational mass is the same, but it is the slope that varies the acceleration of the motion. So it's not matter or dark energy that acts on stars in the galaxies, it's the force field of closed lines around the matter. A matter, in which light is not emitted, absorbed, reflected, refracted or difrated, is a non-existent matter! So what works is the picture of current Physics, the total dark! Figure 100.

J. P. Hansen

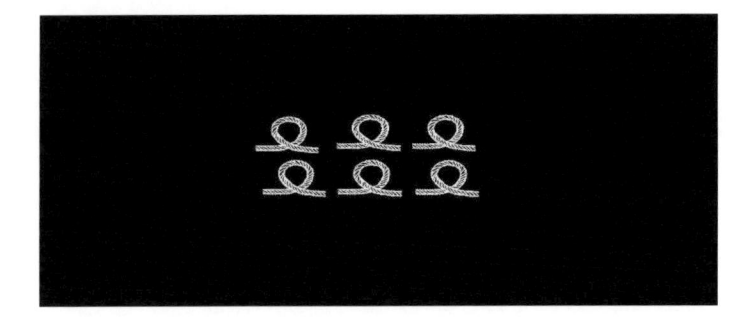

Fig. 100 – Picture of the current Physics: quantization of space-time: nodes and circles, photons and field of zero mass, extinction of space and time, and the total dark of matter and dark energy! According to current Physics, it's their theory of quantum gravity that's named "limpid beauty"! Better to say "limpid darkness"! Google images.

The impossibility of to achieve results that have been sought for decades real origin of the gravitational waves by mechanical resonance or interferometry, without seeking the reality of the constitution of the gravitational wave that is quantum gravity, which is incompatible with the theory of relativity of a continuous space-time antiquantum; to prove the big bang by the gravitational waves of space inflation, that after decades when announced their discovery in 2014, the same year was withdrawn because it's just noises; although the discovery of the Higgs particle continues not to include gravity in the standard model. And current Physics continuing with the exclusion of ether and photons with mass; continues admitting the expansion of the space provided by Eddington to preserve the theory of general relativity, and to be contrary to the expansion of galaxies observed by Hubble in relation to the vacuum-space-ether or cosmic microwave radiation, because the universe as a whole is motionless, without turning, and getting farther from the reality of the expansion of galaxies and distance of quasars; and continuing to get farther from the unification of the physical fields and even concluding which would be an impossibility, trying to justify the dissociation between Einstein's relativity and the current quantum mechanics.

Despite the brilliant work of teams of researchers in astrophysics, X-rays observation satellites, gamma-rays, radio astronomy, microwaves, laser rays, supernovae, neutrino mass, Higgs particles, particle accelerators, mathematics of the most advanced degree in quantum mechanics, black holes and string theory, the equating of the theory of everything advocated by Hawking remains farther as the further advances this physical and mathematical development of the current Physics. However new foundations in Physics will lead to new discoveries, such as the laser light experiment, combined with an interpretation of the brilliant CERN experiment of 1964 gamma-rays emitted by mesons $\pi°$ at almost equal velocity to gamma-rays, showing that the velocity of electromagnetic radiation is variable in relation to the emission rate of the electromagnetic radiation, although the speed of electromagnetic radiation is constant in relation to the vacuum-space-ether or cosmic microwave radiation. And the particles of atoms, physical fields that appear dark matter, dark energy, light and all electromagnetic radiations are formed according to the elementary impulse of elementary mass that increases the mass and consequent increase in velocity in particle accelerators. And contrary to the increase in mass as a function of the velocity by transformation of the coordinates of Einstein's relativity, that "phantasmagorically" would increase the mass. In the last page and last paragraph of the text of Weinberg"s *"The Discovery of Subatomic Particles"*, he states: *"When the poet William Blake needed to summarize all the science in one line, he said: 'The atoms of Democritus, and the particle of light of Newton'*. And Weinberg finishes: *"From Greece of Democritus and Leucippus to the time of Blake and our own, the idea of the fundamental particle has always been emblematic of the profound goal of science: to understand the complexity of nature in simple terms"*.

And it's amazing that when Einstein disagreed with quantum mechanics, denominating it incomplete, he received the sentence of the "Judges of Knowledge" of current Physics: Then Einstein missed, they exclaimed! And later, when experiments that proved the interweaving of particles and photons, they confirmed the verdict. But by quantum impulse, it is the quantum field in the ether that can act as a rigid body and present the instantaneous velocity. But, the current physics maintains the theory of relativity by hitting that the speed of light is constant in space-vacuum, although it is also by wave theory, but it comes to err when it maintains the constant speed of light relative to the emitting source. Therefore, a light launched at 90° of the direction of the motion should maintain its velocity at 90°, or when leaving diagonally it would be composing with the velocity of the source, that is, varying according to the velocity of the source as shown in the laser light experiment. The current Physics follows the inconsistency of relativity created by Einstein, but rejects what more Einstein hit, that only new principles in quantum mechanics could explain the phenomena and achieve the unification of the electromagnetic fields and gravitational. Einstein tried to unify, for decades, these fields with general relativity, but without the new ideas, the new principles, the result was of sharp mathematical abstraction.

And by the theory of elementary impulse resulting in the theory of everything sought by Hawking, or in an equation of everything in which ironically this equation lies where he resided in Cambridge, and formulated by those who exercised the mathematics chair, Newton, and after exerted by Hawking, which is the equation of Newton's universal gravitation with the introduction of it from the frequency of impulse elementary resulting in: $F_G \times d = [(K_G M_1 M_2)/d^2] \times d = n\mathcal{V}_G h$, where F_G is the gravitational force in the distance d, M_1 and M_2 masses, n is the quantity of $\mathcal{V}_G h$, where $\mathcal{V}_G$ it is the electromagnetic frequency of gravitational radiation, h the Planck constant, in which h/1 second = h' = m'c × c is the elementary photon of energy, m' the elementary mass and c the speed of light relative to the Earth as a referential velocity, which gathers the gravitational constant K_G of Newton and the quantum constant h of Planck, with a total physical significance of h' in energy unit, with a complete physical meaning, the equation of everything praised by Hawking.

The current Physics with this quantum mechanics of Condon and others, and Einstein's relativity, little can do to explain unidentifiable atmospheric phenomenons. So by this impossibility, the current Physics remains far from analyzing these unidentifiable phenomena. And claiming only as strangers, that they will remain unknown until it is considered an in-depth Physics in real energy fields, with mass photons that form the particles and increase to their masses and not by "ghostly" condensation of speed of the current quantum Physics of Condon and Einstein's Relativity! Then it would be possible to present new possibilities for the knowledge and control of the gravitational and electromagnetic field, without just by chance of high-confidential reverse engineering could discover the cause of these unidentifiable phenomena, which present intense electromagnetic radiation of broad spectrum, as displayed by all colors of light, by radar detection and magnetic effects.

In the current quantum era, Newton's law of action and reaction is still considered to be a simple mathematical or word relationship, but Newton did not have quantum mechanics. But in current Physics it is treated without the greatest physical significance. In the case of a rocket, it is the force m'c/t released in the reaction that boost the combustion chamber on one side, for example, between the large mass of oxygen-hydrogen and the smallest mass of the rocket that receiving m'c/t presents a greater acceleration than that of the mass of oxygen-hydrogen in the opposite direction. So, who has the force? The ancient force of mathematical relationship, Force = mass x acceleration, must be interpreted by the actual quantum force applied in a time interval, i.e. Force = n x m'c/time. Figure 101.

The Pre-Established Physics and the Quantum Impulse

Fig. 101 – "Who has the Force"? It is who has the real force, the light, the quantum force, that is, the quantum impulse applied in a time interval: $F = n \times m'c / \Delta time$. Google images.

At the beginning of June 2015, CERN presented in its accelerator the highest concentrated energy level in the clash between particles, approaching the energy of the cosmic rays in the upper terrestrial atmosphere. All these shocks result in light radiation, electromagnetic radiation, that is, elementary photons that form and give mass to all particles and also constitutes gravity, by the theory of the elementary photon of energy.

The stubbornness of the reviewers and "zealous" of the current Science continues to apply a broad policing to any deviation to their beliefs, as in the beginning of 2016 in which Chinese authors presented a scientific article, which related the motion of robotic members as the anatomy of structural elements of the human hand, the scientific journal Plos-One who published it in this article they concluded that it was a work of the "Creator" that should be interpreted as "Nature" in the translation of the Chinese to the English effected by the authors. The article mentioned the motion and structure of the anatomy of the hand without citing how it had arisen, therefore it advocated the unreferenced nature of those who had created it. However, as a "Creator" the article was criticized by the "zealous" readers of the current Science and the Editor of the journal informing in the following number its disqualification as a scientific article for mentioning the "Creator".

According to the Physicist Alexandre Zabot, Doctor in Astrophysics and Professor at the University of Santa Catarina-Brazil, in article "Censorship in Science" published in April 2016, considered the position of this Plos-One Journal as pure censorship and intransigence, and that Newton, Galileo and many other scientists referred to the "Creator", God, constantly in his texts, and by chance could be accused of pseudoscientifics?

In 2016 have shown that infrared rays absorbed by certain materials are re-emitted as visible light, which shows the formation of the matter by elemental photon which by the quantity is determining the frequency of the radiation as predicted by quantum impulse. And the discovery of the Fermi Satellite over gamma-rays connected to gravitational waves, like the previous Compton Satellite, in which gamma radiations come from explosive sources of stars and unknown sources, also evince these gamma-rays as gravitational radiation.

For Maddox, Physicist who was Editor of Nature Magazine, came to mention in 1999: *"Only when Science better understand theories gravitation is that one will have a better idea of the real universe"*. And in 2016, gravity remained outside the standard model of particles and the unification of atomic forces. According to the greatest scientists in the History of Physics, such as Rutherford and J. J. Thomson, they advocated that the atomic nucleus would be made up of positive electrons and Newton mentioned in their Optics Questions that: *"Wouldn't the bodies and the light convertible into each other? And the bodies could not owe a large part of their activity to the particles of light that would enter into their composition, because all the bodies emit light? Transmutations follow nature, and why couldn't nature turn the bodies into light and light into bodies?"*. And for another prominent scientist Maxwell, who found that the speed of a wave of the electric charge field and the magnetic field was equal to that of the speed of light, that is, that it was also an electromagnetic wave, admitted that the mass was hidden in the electric and magnetic forces. By quantum impulse the proton is formed by the electron-positron helix of quantum impulse m'c x c and all atomic particles are made up of quantum impulses, including gravity. And in all the forces of nature is present the mass, as Maxwell admitted, then also all kinds of energy being constituted by the elementary energy m'c x c = h'. Figure 102 shows part of the painting "pure beauty" considered one of the most valuable, of the artist Beatriz Milhazes. Although its significance is the extraordinary beauty of shapes and colours (in Google Images), it can be observed as lines forming atomic orbits, dots forming figures and particles with attenuation of its central concentration. And that could be a representation of the matter formed by particles and these by points, which in the case these points would be the quantum impulses forming the particles and their fields around their particles.

Fig. 102 – Part of the painting "Pure Beauty" considered one of the most valuable, of the artist Beatriz Milhazes, with significant beauty of its colours (in Google Images) and figures, could be observed as a form of points, lines and field of quantum impulse. Photo of the author, authorized in exhibition museum Oscar Niemeyer-Curitiba-Brazil.

However, despite these greatest names in Science, leading scientists of the pre-establishead Physics as Steven Weinberg continue to teach that: *"Protons do not contain electrons"*. However, by neurolinguistics the not as in this Weinberg's phrase nothing physically represents, which is the case that in a staircase there is a plaque with the inscription: "Do not go down the stairs running". The passerby who descends the ladder, by neurolinguistics understands the phrase as: "Go down the stairs running"! When for full safety the board should be written: "Lower the ladder slowly".

So what Weinberg should teach is, "Protons contain electrons". That is, they are made of electron-positron which is the particle of the theory of quantum impulse that has the two joint charges, contained of helix of quantum impulses m'c x c.

In a Congress where the data on radar wave velocity was questionable, a congressman asked Weinberg to intervene to show that this data was against Einstein's theory of relativity. However, Weinberg intervened differently: *"We must take care of our Science"*, that is, of the imposed Physics. Where is the scientific method of this current Physics to verify the veracity of the scientific proposition? But, in this way what is being taken care of is the preservation of the delay of scientific bases, by keeping without contesting Einstein's theories of relativity and the current quantum mechanics, in which one completely excludes the other. For, facts and experiments presented in this book show the fallacy of these current theories of the fundamental bases of nature within this current Physics. Members its behave as pygmies when they want to follow their ideas which differ from those presented from the bases of the Physics by the greatest scientists of Mankind, such as Galileo, Descartes, Kepler, Newton, Faraday, Maxwell, Hertz, Rutherford, J. J. Thomson, and many others. And there is only one Physics to be developed and cared with solid foundations, in which everyone should observe all the data and interpretations that the experiments show, different from two Physics where each one would take care of their part.

The current Physics summit follows "religiously" the so-called scientific method in which the theory has to be proven by experiments. However, when there are experiments, such as the laser light presented in scientific journal and commented on in this work, contrary to the theories followed by the current Physics, a reason is issued for the method not to be applied. In this case, it is the requirement of publication in authorized scientific journals, although these journals have received the work. C. Sagan published his horror by the medieval inquisition that "In Name of God" acted against those who disagreed with the beliefs of the "*Doctors of the Laws*", in your book *The World Haunted by Demons* of 1995. However, ironically weaves his subservience to the Pope of the Modern Inquisition, Martin Gardner, who on the cover of book of Gardner brings the motto *In the Name of Science*, with the same insurgency against the new scientific ideas denominated by them of pseudoscientific. Sagan criticizes those who are contrary to his dogmas as Einstein's relativity, denominating of "*A lot of crazy people*" those who are contrary to that theory. In the 2018 edition of Jean de Climont, on the list of all theories and criticisms contrary to current Physics, more than 50% they're doctors, physicists and engineers, reaching 9,000 names. E a lesser number of these scientists who disagree with relativity are considered heretics, in which neither Gardner nor Sagan had sold any indulgence but only the referral to the bonfire! Excellent publications such as *Physics Essays – An International Journal Dedicated to Fundamental Questions in Physics, General Science Journal, John Chappell Natural Philosophy Society, Applied Physics Journal, Journal of Theoretics*, and other publications and websites are treated as publications of works that should be included in the *Index Librorum Prohibitorum*. For more than 2000 years, Christ with new ideas faced the "*Doctors of Laws who Took the Key of Science*". But, it is continued by the Summit of imposed Physics the Crusade Anti New Ideas that defy dogmas and impositions of those who remain with the "*Key of Science*".

The current Physics performs gravitational attraction calculation with energy variations smaller than all existing potential, as analyzed by T. Mees in 2018, and by other researchers.

Then, attempt to extinguish space, time, mass of fields and photons and create dark matter and energy, which does not emit, absorb, reflect, refract and diffracts any radiation. It is to want to undo what Newton uttered that the infinite of space and time, and matter, were created by a single God. And the current Physics by reducing the knot, circles or nothing, seeking to impose the basic fundamentals of nature according to its dogmas, of maximum authority as divinity!

In 2018, recent experiments, such as the photons released into atoms of rubidium, at a temperature near the absolute zero in which the atoms tend to form a single unit of properties as the magnetic field, that is, in resonance, show that the interaction of photons with these atoms incorporates a fraction of the mass of their electrons, forming a microparticle, proving that particles are formed of photons according to the theory of quantum impulse.

In 2018, researchers A. M. Archibald et al., carried out measurements of a distant triple-star in orbits, the PSR J0337+1715 system, to show that the accelerations were equal in bodies of different masses in an energy field of an intense quantity of mass that distorted space-time. As Einstein, in an elevator in space that was being pulled by the top, that is, being accelerated, the observer inside that elevator could have the perception that the elevator was stopped and that it would be a force of gravity that pulled it down in the direction of the ground. And Einstein considered that gravity was just an acceleration, caused by a distortion of space-time through the matter involved in an electromagnetic energy field, which altered the dimensions of that space-time, therefore without action of any force of gravity. And it would be difficult against Einstein! However, Galileo, Newton and many others, had already presented by classical physics that bodies of different free-fall masses were accelerated equally by proportional gravitational force of a much larger mass. In relativity, the mass-energy and null-mass field at rest of space-time cause the curvature of that space-time in a position that results in equal accelerations in that position. But, without explaining what this energy is and how this energy curves the space-empty, that is, the absolute space, the absolutely nothing. The existence of absolute space is an infinite entity of nature, to receive the matter, and there is a potential of the time that exists, infinite, eternal, whether with matter in vibration zero or of 10^n, but neither space nor time it would be distorted. However, if Einstein considered the opposite, that is, the acceleration is that it would cause a real force of gravity? In this case, a quantum force of this field involving matter would interact with another body to cause its acceleration! And ironically, Einstein regarded it as fundamentally the energy field in space-time and the need for greater explanation for what were the quanta of light! And the quantum impulse theory is exactly the explanation for the quanta of light and the energy field in the vacuum-space-ether, in which the interaction of the quantum force m'c/time, of energy m'c x c = h', between the bodies and their gravitational fields, is increased when they are approximate, resulting in the acceleration of the bodies between the masses and their gravitational fields. And a smaller body of mass interacts with a lesser amount of h' of its mass and its field, with a lesser amount of h' of the gravitational field in which it is placed. While the body of larger mass and its field interacts proportionately with a greater amount of h', which results in equal acceleration of the bodies in the gravitational field formed of that energy h'. Figure 103 shows three-star system in an intense gravitational field of quantum impulse, which is the real cause of the same acceleration for different masses, without any curving of the absolute energy-empty-space-time by relativity which is infinitely far from quantum nature!

> **3 bodies of different masses with their gravitational energy fields formed from elementary quantum energy h' = m'c x c, where the greater the mass of the body, the higher the quantum force F = n x m/t interacting, resulting in equal acceleration of gravity for a field gravitational of great intensity and mass**
>
> **m'c x c = h'**
> **gravitational field**

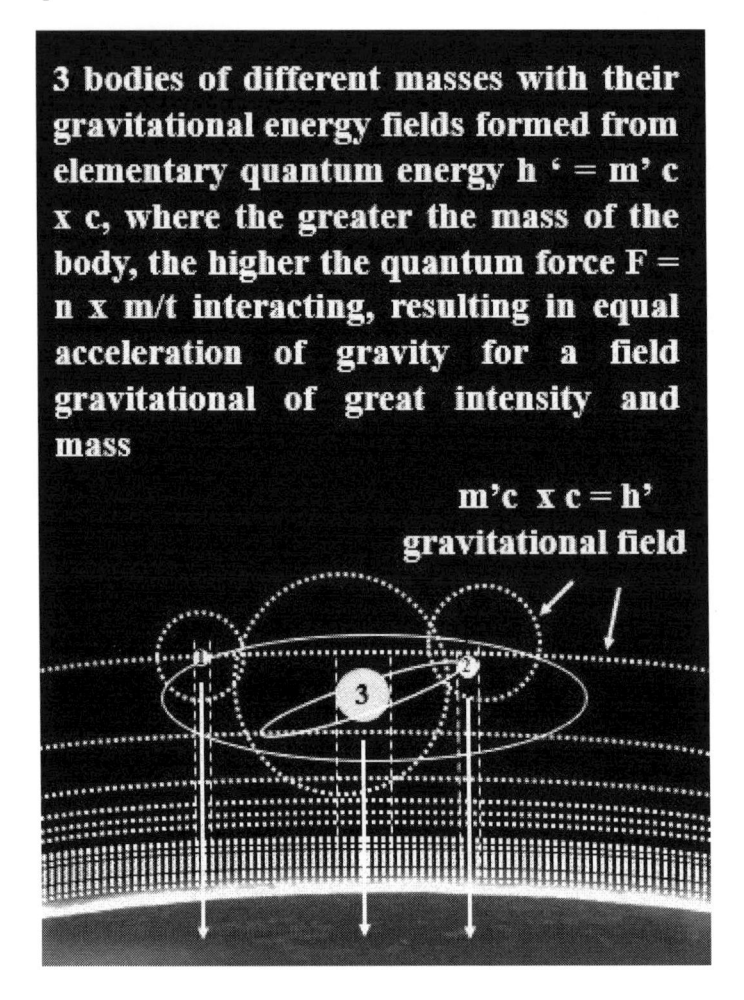

Fig. 103 – Three bodies of different masses fall with equal accelerations in a gravitational field position, through interaction-balance of quantum energy between field masses, while by the theory of relativity these equal accelerations are by the position of the space-time distorted.

Also, in 2018, N. Ashby, T. E. Parker and B. R. Patla, after 14 years of measurements in atomic clocks on Earth, tried to test the theory of relativity as to the equivalence between floating in the distant empty space of mass-gravitational field, caused by an acceleration, and floating in free fall in this acceleration by gravity action. As the result is the same, Einstein considered acceleration as a cause of gravity. And, like any other body in these conditions of floating in this gravitational field or in the empty space by that acceleration received, all these bodies would have the same acceleration in a free fall, without a body causing a different acceleration in the other. So, like a free-fall elevator, they considered the Earth as a floating body in the solar system which would be the elevator shaft. And on Earth, despite variations in the position of planets and other stars in the solar system, with differences in gravitational potential, in those 14 years, these atomic clocks should maintain an equal acceleration. So, no difference in rhythm between them, equal to all the bodies that were inside the elevator in freefall. And in fact, the difference was nil. However, this equivalence of principle of relativity that excludes gravity, replacing it with spatial curving geometry-time, is only a consequence of quantum energy in the interaction of masses-gravitational field that causes attraction-acceleration of gravity resulting in the balance of output-input of quantum energy h', resulting from gravitational quantum force and m'c/time. And the velocity v received of attractions of the solar system it is despicable in a reason v/c of the speed of the light c, because the "ticking" of the clock is significant only for velocity close to c, as the meson that increases its life time with 0, 99c.Figure 104.

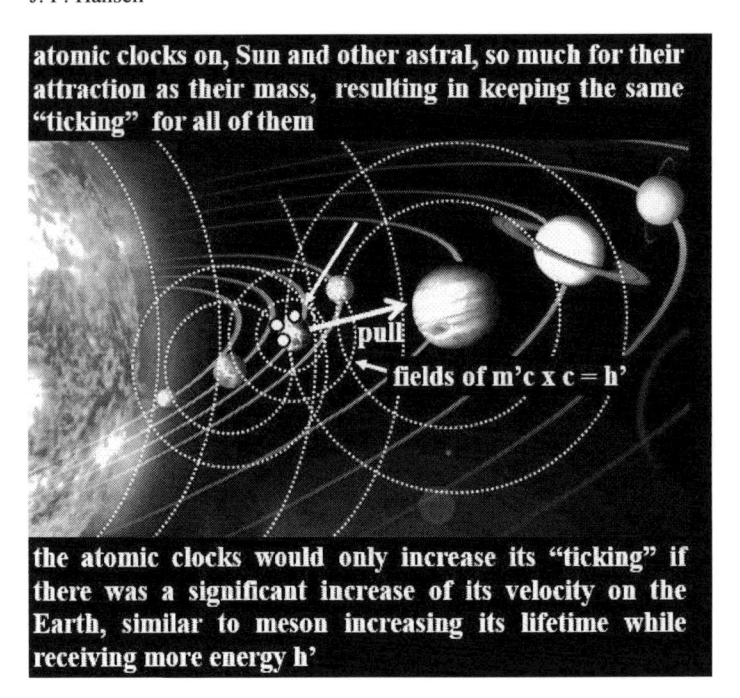

atomic clocks on, Sun and other astral, so much for their attraction as their mass, resulting in keeping the same "ticking" for all of them

pull

fields of m'c x c = h'

the atomic clocks would only increase its "ticking" if there was a significant increase of its velocity on the Earth, similar to meson increasing its lifetime while receiving more energy h'

Fig. 104 – Atomic clocks in "free fall with the Earth in the solar system" show the same acceleration, even receiving attractions of Jupiter and the Sun, because the v velocity received from these attractions is despicable in a rate v/c of the speed of light c, because the "ticking" of the clock is significant only for velocity close to c, like the meson that increases its lifetime with 0.99…x c. Adapted from Google images.

And in another experiment in 2018, by Abuter, R. et al., after 26 years of measurements in the star system named S2 from a highly elliptical orbit around Sagittarius, a supermassive black hole at the center of the Milky Way galaxy, of intense gravitational field, in that by relativity the empty space-time is curved, the velocity of the bodies as S2 is increased at up 7,650 km/s or 3% of the speed of light, and the light emitted has loses energy because of this intense gravity, it was observed the gravitational redshift as the theory of relativity. By relativity, for the light always maintain constant speed c in relation to the source and vacuum-space, the time is dilated and the space contracted. And because the frequency of the light is the inverse of the period, the frequency decreases when the time is dilated. However, by the quantum impulse the gravitational redshift is explained by the increase of mass of the particles and their energy field, with consequent increase in velocity and greater tension of the elementary quantum energy in the particles, as the equation deducted from the electron by quantum impulse theory, which results in emission of less frequency of quantum energy radiations by particles of electric charge. While relativity equals mass-energy, from different units and "stretches the light by pull of curvature", that is, space-time distortion, and being incompatible with quantum theory, the quantum impulse presents the real explanation of nature. The speed of light c' in the vacuum-space-ether is always constant by impulse m'c', or speed c in relation to the emitting source Earth, with the variable velocity of light c_v in relation to the source, and the frequency of the light emitted varying as the energy conditions in which the electric charge particle is situated. But, by relativity the velocity of light c is always constant in relation to the vacuum-space and the frequency of emission of light varies by the energy loss of the quantum hν for a quantum hν' of less frequency, in which to maintain c constant, the time is dilated resulting in this lower frequency ν'. However, what is proved is the inveracity of the constant speed of light c, because by the laser experiment the light comes out in the diagonal of the laser at 90 ° in the direction of the constellation of Leo, and consequently the unreality of time dilation, of space contraction and of space-time distortion. Figure 105.

J. P. Hansen

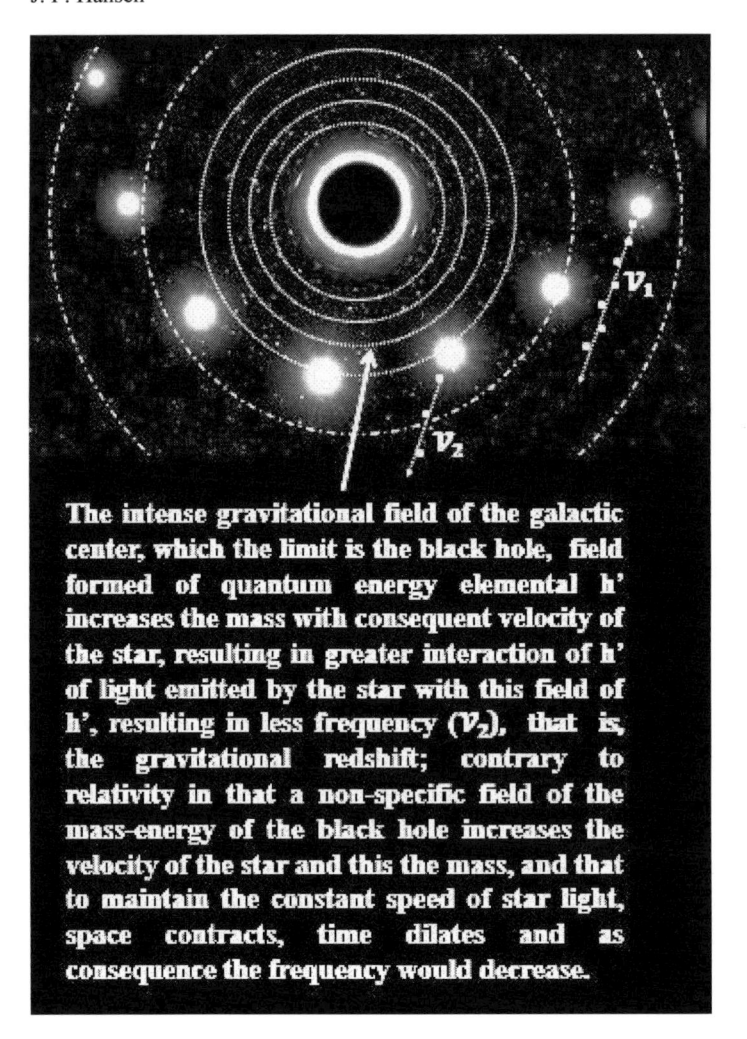

> The intense gravitational field of the galactic center, which the limit is the black hole, field formed of quantum energy elemental h' increases the mass with consequent velocity of the star, resulting in greater interaction of h' of light emitted by the star with this field of h', resulting in less frequency (V_2), that is, the gravitational redshift; contrary to relativity in that a non-specific field of the mass-energy of the black hole increases the velocity of the star and this the mass, and that to maintain the constant speed of star light, space contracts, time dilates and as consequence the frequency would decrease.

Fig. 105 – The gravitational deviation caused by greater interaction of quantum energy between gravitational field and star mass, resulting in the emission of light at a lesser frequency, while the theory of relativity "stretches the light by pull of curvature", that is, by space-time distortion, and being incompatible with quantum theory. Adapted Google images.

Still in 2018, to emphasize even more the need to try to prove the theory of relativistic gravity, although incompatible with quantum mechanics, satellites of incorrect orbits let scientists, P. Dalva et al., and S. Hermann et al., test how clocks change speed in a gravitational field of the Earth. In 2014, two satellites intended for Europe's Galileo network, the equivalent of the United States' GPS network, were placed into orbit incorrectly, causing them to travel around Earth in ellipses rather than circles. That wasn't ideal

The Pre-Established Physics and the Quantum Impulse

for the satellites originally intended navigational use, but scientists realized the wayward satellites were perfect for another purpose: Testing Einstein's theory of gravity, the general theory of relativity. According to general relativity, gravity affects not just space, but also time. The deeper within a gravitational field you are, the slower time passes. So a clock at a higher altitude will tick faster than one closer to Earth's surface, where Earth's gravity is stronger, known as gravitational redshift. As the two misplaced satellites move in their elliptical orbits, their distance from Earth periodically increases and decreases by about 8,500 kilometers. Using the precise atomic clocks on the satellites, the scientists studied how that altitude change affected the flow of time. The clocks sped up and slowed down by tiny fractions of a second as expected, agreeing with the predictions of general relativity within a few thousandths of a percent, the teams report. It is evident that time is related in $t = n/\mathcal{V}$, n = cycles/second and $\mathcal{V}$ = atomic frequency, and for a lower frequency of the atom closest to the gravitational field this time is longer to complete the cycle. However, the time is greater because there is a greater mass of satellite interaction closer to the gravitational field, while the absolute eternal time continues to exist independent of the higher or lower frequency of the atomic clock. A bear in hibernation has his biorithmus diminished, but does not mean the time has been frozen! The previous denomination of 1 second igual equal 1/86400 of the solar day, in case the Earth stopped in the vacuum-space-ether would continue to exist. For, time would continue to spend in the absolute eternal space! Figure 106.

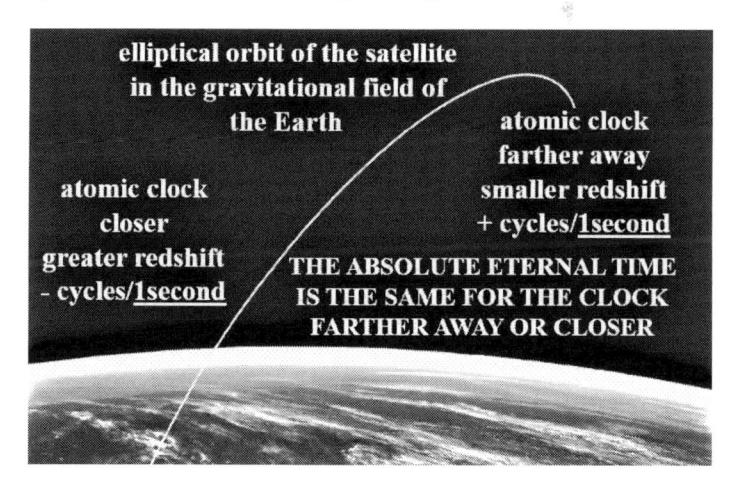

Fig. 106 – An atomic clock tIn a smaller cycle per second when closer to the gravitational field, as there is greater interaction of mass between atoms and the gravitational field, in absolute eternal time continues flowing independently of the frequency being greater or lesser.

J. P. Hansen

In 2018, the Brazilian Post presented a stamp honing the Physicist Lattes, which with Muirhead, Occhialini and Powell, in 1948, discovered the meson π from cosmic rays, which resulted in the Nobel Prize for Physics for Powell. However, Lattes with E. Gardner, also in 1948, planned and determined the detection of the particle accelerator meson π at the University of California at Berkeley. Lattes was the holder of universities in Brazil, the Brazilian Center for Physical Research and the Brazilian Academy of Sciences. Thus, scientists from Brazil recognized Lattes as who would have all the merit to receive a Nobel prize in Physics for the science of Brazil. Figure 107. In 1980 Lattes presented a Communication about an experiment with mercury light in diffraction network, showing that there was variation of the light paths as a function of composing velocities of the rotation and translation motion of the Earth in vacuum-space, which was contrary to the constancy of the speed of light that had no composition whatsoever. Although, Lattes had mentioned that the theory of relativity erroneously considered measurements as equivalent to physical quantities, space and time should be absolute and not contracted or dilated, this variation of paths was attributed to the difference of temperature between day and night that altered the refraction index and consequent light paths. But, when in 1995 I came to present the paper of the laser experiment in the Science Progress Brazilian Society-SBPC, it was proved that the path of light went diagonally when in 90 ° from the direction of the Earth toward the constellation Leo in the vacuum-space-ether, ie , there is in fact composition of the light with the motion of the Earth, and therefore according to Lattes, without contraction of space or time dilation which would be the cause of the constant speed of light.

Fig. 107 – The Brazilian Post in 2018 honored with stamp one of Brazil greatest scientists, the Physicist César Lattes, by the co-discovery of the meson π in the cosmic rays and team leader in the planning and detection of the meson π in particle accelerator in 1948, and that in 1980 Lattes sought to show composition of the speed of light as motion of the Earth in vacuum-space contrary to the theory of relativity, which was evidenced in another type of experiment by the author in 1995 presented in this book.

Recently, admirers of the mathematician-astronomer Tales of Mileto, one of the admitted seven wise men of ancient Greece around 600 BC, came to refer to the flat-Earth admitted by Tales. So, the Pre-Established Physics annexed the flat-Earth with any other divergent theories to include all of them as pseudosciences and conspiratorial! In fact, there is no flat Earth, but there is the Earth of distorted space-nothing, matter and energy dark of obscurantism, mechanical abstract quantum energy, followed by current Physics?

According to W. Smith, Physicist and Ph. D. in Mathematics from the University of Columbia in the United States, in which his studies provided solutions to the problems of re-entry into the atmosphere of spaceships, and who taught at the Massachusetts Institute Technology-MIT , commented in 2014 that the current Physics was moving away from the essence of true scientific knowledge to follow the course of abstractions, writing: *"The essential cannot be discarded and what would be left would only be dissolution and nothing. They sought to build a hyperphysical and what they achieved was a pseudoscience that lost touch with physical reality. In fundamental Physics quantum theory is treated as merely operational, and a quantum particle as an entity in a metaphysical domain. Physics has become scientism, not scientific. And that's why it occurs quantum strangeness, quantum paradox, and Feynman's observation, Nobel laureate of Physics, that no one can understand quantum mechanics. The current Physics annihilated the possibility of authentic knowledge, leaving only a inevitably knowledge disturbed of illusions, a false knowledge that alienates the knowledge of reality. As had been expressed to N. Whitehead the Physics had become a mystical type in praise of an unintelligible universe"*. In fact, it is the scientism of dogmatic belief in the authority of the scientific method, but it does not use it when the result is contrary to its theories, and servility to the summit of the authority of the pre-established Physics. And the theory and or an equation of everything sought by Hawking in current Physics, is noted by Smith as: *"Physics becomes not a theory of everything, as they like to think physicists, but a theory of nothing"*. However, with quantum impulse and gravitational radiation from them, these new foundations of Physics can present a new direction for real scientific knowledge.

The laser experiment showed surprisingly, the light, when leaving diagonally, is carried by the particles of the emitting source, and is in accordance with Lorentz's theory of the moving referential. Then, by the theory of relativity, in which the speed of light is measured constant in relation to the source, by contraction of space and time dilation, it is in fact due to measure different distances traveled by light in the vacuum-space-ether. Although in the theory of relativity, as in the wave theory, the speed of light c is always constant in the vacuum-space and ether, and independent of the velocity of the emitting source, the laser experiment presented that the light also has a velocity $c' = (c_v + V)$ constant in relation to the vacuum-space-ether and independent of the speed V of the source, but the variable speed of light c_v relative to the source. The laser experiment showed that the higher the velocity of the emitting source, the lower the velocity of light in relation to the emitting source.

Newton, one of mankind's greatest scientists, had not in his time the quantum theory, but left a spectacular prediction, in which he emphasized that he used the notion of impulse as the quantity of force for mathematical proportions, and that he would not define in the "Principia" treaty the nature of that impulse that caused force. Pois, he formalized the F force relating resistant m mass, inertial, with the variation of velocity, the *a* acceleration, as F = m*a*, which was due to the variation in the quantity of motion Δmv = I, the impulse, by time, that is, F = I/time. And according to Newton, the principles of the motion being of a very large extent left their causes to be discovered, and claimed that all objects would appear to have been composed of particles. These particles would have a force accompanied by motion laws and would also be governed by active principles of the cohesion of objects. And when he formulated the questions in "Optica.": *"It would not be convertibles bodies and light one on the other?",* and *gravity, electricity and magnetism could not be by impulse, in which particles would act among them "?* When in 2016 it was published, by the LIGO team, the discovery of waves produced by space-time distortion, that is, gravitationL originated from spiraling-shock-melting massive masses stellar, according to general relativity, the Fermi team detected in the same location and instant near the emission of gamma-rays. However, this gamma emission was attributed to the expansion of gases and or jets of electrons. Including, the ESA-INTEGRAL denying that the gamma-rays had any connection with these waves of space-no-time! However, in 2017, it was published by the teams LIGO, Virgo, Fermi and ESA-INTEGRAL that new detections in 1917 of these waves were observed along with gamma-rays, light and a wide region of the electromagnetic spectrum, which lasted for several days, showing the source these gravitational waves visible to telescopes. With the quantum theory, came the discovery by Erdelyi in 1952 of the quantum impulse which forms all particles and them fields, also the quantum force explaining the motion of the bodies. And explains that is the frequency quantum of gamma-rays that form the electron-positron, which is proven by the particle annihilation experiment as the electron-positron. And it's the gamma-rays that are in Newton's gravitational law as presented by Mesquita in 1965. Then, in mass mass shock are the gravitational radiation gamma, light and other radiations, of the electromagnetic spectrum, which are emitted, being these radiation energies that cause the dimensional variation in the interferometer, revealing the source of gravity, and Newton's prediction!

REFERENCES

Abuter, R., et al., Astronomical & Astrophysical, Vol. 615, July 2018.
Aczel, A. D., *Pendulum,* Atria Books, New York, 2003.
Adam, T., et al., *arXiv*:1109.4897V2.2011.
Adams, F. and Lauglin, G., *The Five Ages of the Universe inside the Physics of Eternity*, The Free Press- Simon & Schuster, New York, 1999.
Adamson, P. et al., *arXiv*:0706.0437.2007 e *Physical Review,*. D76, 072005, 2007.
Agafonova, N. et al., *arXiv*: 1102.1882V1. 2011.
Almeida, H., *Landell de Moura-Inventor do Rádio-Televisão e Teletipo*, Editora Record, Rio de Janeiro, 2006.
Alves, G., Caruso, F., Motta, H., Santoro, A., Marques, A., Lattes, C., Lopes, J. L., Quigg, C., Souza, M., Shellard, R., Álvares, B. A., Carvalho, B., Monserrat Filho, J. and Bassalo, J. M. F., *O Mundo das Partículas de Hoje e de Ontem*, Livraria da Física, São Paulo, 2012.
Alzofon, D., *The Top-Ten UFO Riddles*, Middletown, DE, USA, 2018.
Alzofon, F., *Gravity Control with Present Technology*, Middletown, DE, USA, 2018.
Anastopoulos, C., *Particle or Wave*, Princeton University Press, Princeton, 2008.
Antonello, N. M. et al., *arXiv*:1203.3433.2012.
Allais, M., *L'Anisotropie de L'Espace les Donneés de l'Expérience*, Éditions Clément Jugler, Paris, 1997.
Almeida, E. G., *The Elementary Impulse Theory and the Gravity Control*, Private Communication, 2005.
Alvager, T., Farley, F. J. M., Kjellman, J. and Walin, I., *Physics Letters.,* 12, 269,1964.
Anastopoulos, C., *Particle or Wave*, Princeton University Press,

Princeton, 2008.
Anderson, D., *Physics Essays*, 24, 163, 2011.
Andrade, P. N., Revista Pesquisa Naval, Rio de Janeiro, 3, 1, 1990.
Arany-Prado, L. I., *À Luz das Estrelas*, DP&A Editora, Rio de Janeiro, 2006.
Arp, H., *Quasars, Reshifts and Controversies*, Interstellar Media, USA, 1987-1990.
Arp, H., *Seeing Red:Redsfiths, Cosmology and Academic*, Apeiron-C. Roy Keys, Montreal, Quebec,1998-2000.
Archibald, A. M., *et al.*, *Nature* 559: 73-76, 2018.
Artsimovitch, L., *Physique Élémentaire des Plasmas*, Éditions de La Paix, Moscou, Decade of 1960.
Aquino, F. De, *Proca Equations and the Photon Imaginary Mass*, Universidade do Estado do Maranhão, Brasil, 2011.
Ashby, N., Parker, T. E., Patla, B. R., Nature Physics, Vol. 14, pages 822-826, 2018.
Assis, A. K. T., *Mecânica Relacional*, Centro de Lógica, Epistemologia e História da Ciência-Universidade de Campinas-UNICAMP, Campinas, Brasil, 1998.
Assis, A. K. T. and Hernandes, J. A., *The Electric Force of a Current, Apeiron*, C. Roy Keys, Montreal, Quebec, 2007.
Assis, A. K. T., *The Experimental and Historical Foundations of Electricity*, Apeiron-C. Roy Keys, Montreal, Quebec, 2010.
Assis, A. K. T. and Chaib, J. P. M. C., *Eletrodinâmica de Ampère*, Editora da Universidade de Campinas-UNICAMP, Campinas, Brasil, 2011.
Assis, A. K. T., *Relational Mechanics and Implementation of Mach's Principle with Weber's Gravitational Force,* Apeiron-C. Roy Keys, Montreal, Quebec, Montreal, 2014.
Asimov, I., *Chronology of Science & Discovery*, HarperCollins Publishers, New York, 1994.
Auger, P., Born, M., Heisenberg, W. and Schrödinger, E., *Discussione sulla Fisica Moderna*, Torino, Boringhieri,1960, and of Éditions de La Baconnière, Neuchâtel, Switzerland.
Baggott, J., *The Quantum Story A History in 40 Moments*, Oxford University Press, Oxford, 2011.
Baggot, J., *Farewall to Reality*, Pegasus Books, New York, London, 2014.
Baker, J., *Quantum Physics Ideas*, Quercus Publishing, London, 2013.
Baptista, J. P. and Ferracioli, L., *Da Physis à Física*, Editora Universidade Federal do Espírito Santo-EDUFES, Vitória, 2003.
Barrow, J. D., *Theories of Everything-The Quest for Ultimate Explanation*, Oxford University Press, Oxford, 1991.
Barthem, R., *A Luz*, Livraria da Física, São Paulo, Brasil, 2005.
Bassalo, J. M. F. *Nascimentos da Física*, Editora Universitária Federal do Pará-EDUFPA, Belém, Brasil, Vol.1, 1996, Vol. 2,

2000.

Bassalo, J. M. F. and Cattani, M. S. D., *Osciladores Harmônicos Clássicos e Quânticos*, Editora Livraria da Física, São Paulo, 2009.

Beiser, A., *Concepts of Modern Physics*, The McGraw-Hill, New York, 1963-1967.

Bekefi, G. and Barret, A. H., *Electromagnetic Vibrations, Waves, and Radiations*, The Massachusetts Institute of Technology-MIT Press, Massachusetts, 1977.

Bell., M. S., *Lavoisier in the Year One-The Birth of a New Science in an a Age of Revolution*, Atlas Book, L. L. C./W. W. Norton & Company, New York, 2005.

Bergman, D. L., *Common Sense Science*, www.commonsensescience.org

Bergmann, P. G., *The Riddle of Gravitation*, Dover Publications, Mineola, NY, 1992, de Charles Scribner's Sons, New York, 1968.

Berkson, W., *Fields of Force-The Development of a World View from Faraday to Einstein*, Routledge and Kegan Paul, London, 1974.

Berlinski, D., *Newton Gift-How Sir Isaac Newton Unlocked the System of the World*, The Free Press-Simon & Schuster, New York, 2002.

Bilenky, S. M., et al., *Pontecorvo Selected Scientific Works*, Societa Italiana di Fisica, Bologna, 1997.

Biriukov, V. A., Lebedenko, M. M. and Rizhov, A. M., *Instituto Unificado de Investigaciones Nucleares-Scientific Works of the Institute of 1956 until 1961*, Ediciones in Lenguas Extrangeras, Moscou, Decade de 1960.

Bitter, F., Magnets, Doubleday & Company, New York, 1959.

Bohm, D., *Quantum Theory*, Dover Publications, Mineola, NY, 1989, of Prentice-Hall, New Jersey, 1951.

Bohm, D., *Physical Review,* 85, p. 166; 180, 1952.

Bohr, N., Sobre a Constituição de Átomos e Moléculas, Fundação Calouste Gulbenkian, 1989, of Philosophical Magazine, 1913.

Bohr, N., *Atomic Physics and Human Knowledge*, Niels Bohr Archive, Copenhage, 1958.

Born, M., *Einstein's Theory of Relativity*, Dover Publications, Mineola, NY, 1965, of Methuen Company, London, 1924.

Born, M., *Atomic Theory and the Description of Nature*, Cambridge University Press, Cambridge, 1961, and John Wiley and Sons, New York, 1958.

Born, M., Stoyle., R. J. B. e Radcliffe, J. M., *Atomic Physics*, Blackie & Son Ltd.,Glasgow, 1969.

Borovói, A., *Como se Registran Las Particulas*, Editorial Mir, Moscou, 1985, of 1981 of the original Russian.

Boslough, J., *Stephen Hawking's Universe*, Avon Books-The Hearst Corporation, New York, 1989, of *Is the End in Sight for*

Theoretical Physics?, Cambridge University Press, Cambridge, 1980.

Boutroux, E., *Aristote*, Étude d'Histoire de Philosphie, Félix Alcan, Paris, 1925.

Boyer, C. B., *A History of Mathematics*, John Wiley & Sons, New York, 1991.

Braguiski, V. and Manoukine, A., *Mesure de Petites Forces dans les Expériences Physiques*, Editions Mir, Moscou, 1976.

Bricmont, J., *Quantum Sense and Nonsense*, Spring International Publishing AG, Basel, Switzerland, 2017.

Bridgan, P. W., *A Sophisticate's Primer of Relativity*, Dover Publications, Mineola, NY, 1962.

Brockman, J., Matson, K. et All, *How Things Are*, Phoenix-Orion Publishing Group, London, 1997, of William Morrow-HarpenCollins, New York, 1995.

Broglie, L. V. P. R. De, *Comptes Rendues Hebdomadaires des Séances de l'Académie des Sciences de Paris, 183, p. 24; 447, 1926, and Comptes Rendues Hebdomadaires des Séances de l'Académie des Sciences de Paris, 184; 185, p. 273; 380. 49, 1927.*

Broglie, L. V. P. R. De, *Une Nouvelle Théorie de la Lumière, la Mécanique Ondulatoire du Photon,* tome I: *La Lumière dans le Vide,* Hermann, Paris, 1940.

Brown, L. M.; Hoddeson, L. et All, *The Birth of Particle Physics*, Cambridge University Press, Cambridge, 1986.

Boulle, P. F. M. L., *La Planète des Singes*, Édition Julliard, Paris, 1963.

Cahill, R. T. and K. Kitto, K., *Apeiron,* 10,104, 2003.

Cahill, R. T., *Progress in Physics*, 3, 25, 2005.

Cahill, R. T., *Progress in Physics* 3, 60, 2006.

Cahill, R. T.,*Progress in Physics* 4, 73, 2006.

Calder, N., *The Key of the Universe*, Viking Press-Penguin-Randon House, New York, 1977.

Calder, N., *Einstein's Universe*, Viking Press-Penguin-Randon House, New York, 1979.

Cao, G., *Physics Essays*, 24, 381, 2011.

Capria, M. M. et All, *La Construzione dell'Immagine Scientifica del Mondo*, Instituto Italiano per gli Studi Filosofici, Napoli, 1999.

Caruso, F., Oguri, V., Santoro, A., Bassalo, J. M. F., Salmeron, R., Joffily, S., Shellard, R., Motta, H., Barreto, J. L. V., Alves, G., Marques, A., Guzzo, M. M., Natale, A. A., Begalli, M. and Green, D., *Partículas Elementares, 100 Anos de Descobertas*, Editora Livraria de Física, São Paulo, 2012.

Caruso, F. and Oguri, V., *Física Moderna*, Elsevier Editora Ltda. and Editora Campus, Rio de Janeiro, 2006.

Caruso, F., Santoro, A., Lopes, J. L., Predazzi, E., Begalli, M., Bassalo, J. M. F., Shellard, R., Barros, C.,Miranda, M.,

Alvarenga, B., Novello, M., Temer, M. and Pimenta, A., *Do Átomo Grego á Física das Interações Fundamentais*, Livraria da Física, São Paulo, 2012.

Caruso, F., Oguri, V., Santoro, A., Marques, A., Cavalcanti, C. C., Silveira, C., Marques, G. C., Bediaga, I., Neto, J. A. H., Bassalo, J. M. F., Giordano, J., Begalli, M., Pietrocola, M., Siqueira, M., Neto, N. P., Bergliaffa, S. E. P. and Salmeron, R., *O que são Quarks, Glúons, Bósons de Higgs, Buracos Negros e outras coisas estranhas?* Livraria da Física, São Paulo, 2012.

Cherman, A., *Sobre os Ombros de Gigantes*, Jorge Zahar Editor, Rio de Janeiro, 2004.

Cherman, A. and Mendonça, B. R., *Por que as Coisas Caem?*, Jorge Zahar Editor, Rio de Janeiro, 2010.

Chung, Dongwoo, *Cold Fusion: A Study in Scientific Controversy*, Stanford University, Winter 2015, March, 14, 2015.

Climont, J., *The Worldwide List of Alternative Theories and Critics*, Editions d' Assailly, www.editionsassaily.com, 2018.

Close, F., *Cosmic Onion*, CRC Press-Taylor & Francis Group, Boca Raton, FL, 1983.

Close, F., *The Void*, Oxford University Press, Oxford, 2007.

Close, F., *Neutrino*, Oxford University Press, Oxford, 2010.

Close, F., *The Infinity Puzzle*, Basic Books, New York, 2011.

Collins, H. and Pinch, T., *The Golem-What You Should Know About Science*, Cambridge University Press, Cambridge, 1993.

Couderc, P. and F. Perrin, *La Relativité, Presse Universitaire de France*, Paris, 1981, and *Que Sais-Je?* Paris, 1973.

Cramp, L. G., *UFOs and Anti-Gravity*, Adventures Unlimited Press, Kempton, IL, USA, 1996.

Crease, R. P., *The Prism and the Pendulum-The Ten Most Beautiful Experiments in Science*, Random House, New York, 2003.

Creighton, J. D. E. and Anderson, W. G., *Gravitational-Wave Physics and Astronomy*, Wiley-VCH Verlag GmbH&Co, KGaA, Germany, 2011.

Cremo, M. A. and Thompson, R. L., *The Hidden History of the Human Race*, Bhaktivedanta Book Publishing, Alachua, FL, 1999.

Crossley, D., *Physics Essays*, 24, 435, 2011.

Cruz, F. F. S., *Faraday & Maxwell: Luz sobre os Campos*, Odysseus Editora, São Paulo, 2005.

Csiky, G. A., *The Meaning of the Momentum Theory*, Scientific Conference, Instituto de Engenharia, São Paulo, Brasil, 1974.

Corso P. J. and W. J. Birnes, *The Day After Roswell*, Rosewood Woods Productions Inc., Pocket Books-Simon & Schuster Inc., New York, NY, 1997.

Cougo-Pinto, M. V., www-*The Quantum Vacuum and the Casimir Effects*, Universidade Federal do Rio de Janeiro, 2013.

Cougo-Pinto, M. V., Farina, C. and Tort, *A. C., Revista Brasileira de Ensino de Física* 22, p. 122, São Paulo, 2000.

Crease, R. P., *As Grandes Equações*, Jorge Zahar Editor, Rio de Janeiro, 2011, of *The Great Equations*, W. M. Norton & Company, New York, 2008.

Cunha, O. A. R., *Les Sources Théoriques de la Matière*, Escola Politécnica Universidade Federal do Rio de Janeiro, 1963.

Curie, M., *Radioactive Substances*, Dover Publications, Mineola, NY, 2002 of Philosophical Library, New York in 1961, and Recherches sur les Substances Radioactives, Gauthier-Villars, Paris, 1903.

Das, S., *Physics Essays*, 27, 134, 2014.

Davies, P. C. W., Other Worlds, J. M. Dent & Sons, London, 1980.

Davies, P. C. W., *Superforce*, Glenister Gavin-Touchstone Edition- Simon & Shuster, New York, 1885.

Davies, P. C. W. and Brow, J. R. and Others, *A Theory of Everything?*, The Press Syndicate of the Cambridge University, Cambridge, 1988-2000.

Davies, P. C. W. and Brown, J. R., *The Host in the Atom-A Discussion of the Mysteries of Quantum Physics*, Cambridge University Press, Cambridge, 1986.

Davies, P. C. W. , *About Time*, Orion Productions, London, 1995.

Davis Jr., L., Goldhaber, A. S. and Nieto, M. M., *Physical Review Letters*, 35, 1402, 1975.

Deutsch, D., *The Fabric of Reality*, Penguin Books, New York, 1998.

Deutsch, S., *Return of the Ether*, SciTech Publishing, Mendham, NJ, 1999.

Deutsch, S., *Einstein's Greatest Mistake-The Abandonment of the Aether*, iUniverse, Lincol, NJ, New York, 2005.

Dingle, H. and McCrea, W. H., Nature, 177, 782; 178, 680, 1956.

Dingle, H., *Science at the Crossroads*, Martin Brian and O'Keeffe, London, 1972.

Dirac, P. M., Heisenberg, W. and Salam, A., *Unification of Fundamental Forces*, Syndicate of the Press of the University of Cambridge, Cambridge, 1990.

Dirac, P. A. M., *The Physicist's Picture of Nature, Scientific American*, May, 1963.

Dirac, P. A. M., *Lecture on Quantum Mechanics*, Dover Publications, 2001, of Belfer Graduate School of Science, Yeshiva University, New York, 1964.

Ditchburn, R. W., *Light*, Dover Publications, Mineola, NY, 1991, of Interscience Publishers, New York, 1961 and Blackie & Son, Bishopbriggs, Glasgow, 1953.

Dober, V., Kuznetsov and Others, *Philosophical Problems of Elementary-Particle Physics,* Progress Publishers, Moscou, 1968.

Doria, M. M. and Marinho, *Ondas e Bits*, F. C., Livraria da Física, 2006.
Drozdzynski, J., *Physics Essays*, 27, 458, 2014
Duquesne, M., *Matière et Antimatière*, Presses Universitaires de France, Paris, 1958.
Dyson, F. W. and Eddington, A. S., *Philosophical Transactions of the Royal Society* 220, 291, 1920.
Eddington A. S., *The Nature of the Physical World*, The Macmillan Company, New York, 1948.
Edwards, M. R. et al., *Pushing Gravity, Apeiron*-C. Roy Keys, Montreal, Quebec, 2002.
Eisberg, R. and Resnick, R., of *Quantum Physics*, John Wiley, New York, 1974.
Einstein, A., Annalen Physics, 17, 891, 1905.
Einstein, A., *A Teoria da Relatividade Especial e Geral*, Contraponto Editora, 1999, of *Über die Spezielle und die allgemeine Relativitätstheorie,* The Jewish National and University Library, Jerusalem, 1916.
Einstein, A., *Relativity*, Prometheus Books, New York, 1995, of Henry Holt Company, New York, 1920.
Einstein, A., Lorentz, H. A., Weil, H. and Minkowski, H., *The Principle of Relativity*, Dovers Publications, Mineola, NY, 1952, of Methuen and Company, London, 1923.
Einstein, A. and Infeld, L., *The Evolution of Physics*, Touchstone and Colophon-Simon & Shuster Inc.,New York, 1967 of 1938.
Einstein, A., *The Meaning of Relativity*, Princeton University Press, Princeton, 1956.
Einstein, A., Grünbaum, A. Eddington, S. and Other, *Relativity Theory: Its Origins and Impact on Modern Thought*, John Wiley & Sons Inc., New York, 1968.
Einstein, A. and Ono, Y. A., *How I Created the Theory of Relativity, PhysicToday*, August, p. 45, 1982.
Engelhardt, W., *Physics Essays* 27, 586, 2014.
Ensle, H. E., *The Electromagnetic Universe*, Lexington, KY, USA, 2017.
Ensle. H. E., Alien Technology, Lexington, KY, USA, 2017.
Erdelyi, I. L., The Examination and Correct Interpretation of Planck's Constant h, Copyright by Istvan Laszlo Erdelyi, Linotipadora Grafica Ltda, São Paulo, 1952.
Erdelyi, I. L., *Momentum Theory*, I. L. Erdelyi, São Paulo,1954; *Teoria da Impulsão*, Universidade Mackenzie, São Paulo, 1965; Revista Astronáutica-Sociedade Interplanetária Brasileira-Centro de Pesquisa Bradesco, Osasco-São Paulo, 24, 2; 29, 1; 30, 1, 1970; 39, 1, 1971; 12, 7, 1968.
Erdelyi, I. L., *Some New Aspects of Matter, Energy Relationship by the Momentum Theory*, São Paulo, 1954.
Erdelyi, I. L., *Is it Possible to Maintain the Theory of Relativity without a Foundation?*, Mackenzie University, São Paulo,

1954.

Erdelyi, I. L., *What is Electricity?* – Revista Indústria, Abril, 1954 and Sociedade Interplanetária Brasileira-SIB, São Paulo, 1962.

Erdelyi, I. L., *O Triângulo Mágico – O Novo Modelo do Elétron e do Pósitron*, Universidade Mackenzie, São Paulo, 1964.

Euler, L., *Reflexiones sobre El Espacio, la Fuerza y la Materia*, Alianza Editorial, Madrid, 1985, de Pasquier, G., *Léonard et sés Amis*, Librairie Scientifique, J. Hermann, Paris, 1927; Thiele, R., *Leonhard Euler*, BSB.B, Teubner Verlagsgesellschaft, Leipzig, 1982; Marie, M. M., *Histoire des Sciences Mathématiques et Physics*, Paris 1815; tome VIII, Kraus Reprint, tome VIII, Nendeln/Lichtenstein, 1815.

Évora, F. R. R., *Espaço e Tempo*, Centro de Lógica, Epistemologia e História da Ciência-CLE-Universidade de Campinas-UNICAMP, Brasil, 1995.

Falk, D., *Universe on a T-Shirt*, Penguin Books, London, 2002.

Fara, P., of *A Four Thousand Year History,* Oxford University Press,Oxford, 2010.

Faraday, M., *Mr. Faraday's Chemical History of a Candle-The Forces of Matter*, Book Jungle-Kessing Publishing, Whitefish, Montana, of W. Crookes-Chatto and Windaus, London, 1886, of M. Faraday, 1861.

Faria, M. C. B., *Aristóteles*, Editora Moderna, São Paulo, 1994.

Filgueras, C. A. L., *Lavoisier-O Estabelecimento da Química Moderna*, Odisseas Editora, Lexicon, Rio de Janeiro, 2002.

Fischer, A., *Cold Fusion*, Mosaic, Vol. 21, Number 2, Summer 1990, pages 12-23.

Fitzgerald, G. F., *Science*, 13, 390, 1889.

Flamsteed, S., *Discovery*, 3, 66, 1995.

Ford, K. W. *101 Quantum Questions*, Harvard University Press, Cambridge, MA, 2011.

Fox, J. G., *American Journal Physics*, 30, 297, 1962 and 33 , 1, 1965.

Freire Jr., O., Pessoa, O., Bromberg, J. L. and Others, *Teoria Quântica*, Universidade da Paraíba-EDEPB, Campina Grande, Editora da Física, São Paulo, 2011.

Friedman, A., *Zeitschrift für Physik*, 10, 377, 1922.

Foucault, L., *Comptes Rendues de l' Academie des Sciences de Paris*, 3 Février, p. 135, 1851.

Fritzsch, H., *Los Quarks, La Material Prima de Nuestro Universo*, Alianza Editorial, Madrid, 1984, of *Quarks, Urstoff unserer Welt*, R. Piper & Co. Verlag München, 1981.

Fukuda Y. et al (Super-Kamiokande), *Physical Review Letters*, 81, 1562, 1998.

Galiana,TH., *As Ondas Eletromagnéticas*, Editorial Estúdios Cor, Lisboa, 1966.

Galilei, G., Duas Novas Ciências, Nova Stella Editorial-Ched Editorial, of *Discorsi e Dimostrazioni Matematiche intorno a*

Due Nuove Scienzie Attenenti alla Mecanica ed ai Movimenti Locali, Adriano Salani, Editore, Firenze, 1935. Original Edition, Leiden 1638.

Geiger, E. E., *Experimento Michelson-Morley*, Monograph at XXI, and XXIII, Reunião Anual da Sociedade Brasileira para o Progresso da Ciência-SBPC, Brasil, 1969 e 1971.

Gamow, G., *Thirty Years that Shook Physics*, Dovers Publications, New York, 1985, of Doubleday & Company, New York, 1966.

Gapaillard, J. *Et Pourtant, Elle Tourne! Le Mouvement de la Terre*, Éditions du Seuil, Paris, 1993.

Gardner, M., *Fads & Fallacies in the Name of Science*, Dover Publications, Mineola, NY, 1957, of *In the Name of Science*, G. P. Putnam's Sons-Penguin Books, New York, 1952.

Gardner, M., Relativity Simply Explained, Dover Publications, Mineola, NY, 1962.

Gell-Mann, M., *The Quark and the Jaguar*, Alfred A. Knopf, New York and Andre Deutsch, London, 1960.

Gaukroger, S., *Descartes: An Intellectual Biography*, Oxford University Press, New York, 1995.

Ghosh, A., *Origin of Inertia*, Apeiron-C. Roy Keys, Montreal, Quebec, 2000.

Gift, S. J. G., *Physics Essays*, 23, 1, 2010.

Gift, S. J. G., *Physics Essays*, 23, 271, 2010.

Gift, S. J. G., *Physics Essays*, 25, 387, 2012.

Gigov, E., The Laser of Einstein, www.gsjournal.net, 2018.

Gilbert, W., *De Magnete*, Dover Publications, Mineola, NY, 1958 of John Wiley and Sons, New York, 1893.

Ginzburg V., *Sur La Physique et L'Ástrophysique*, Editions Mir, Moscou, 1976.

Giordmaine, J. A., *The Interaction of Light with Light, Scientific American*, April, 1964.

Gladkov, K., *La Energia del Atomo*, Ediciones en Lenguas Extrangeras, Moscou, Decade de 1960.

Goldemberg, J. and Torizuka, V., Physical Review, vol. 129, n°1, pag 312, 1963.

Goldemberg, J. and Torizuka, V., Physical Review, vol. 129, n°6, pag 2580, 1963.

Goldemberg, J., Física Geral e Experimental, Editora Nacional e da Universidade São Paulo, Brasil, 1970.

Good, I. J., *Physics Essays* 12, 190, 1999.

Gouiran, R. and Weidenfeld, G., *Particles and Accelerators*, World University Library/MaGraw-Hill, New York, 1967.

Greene, B., *The Fabric of the Cosmos*, Vintage Books-Random House, New York, 2005 and Alfred A. Knopf-Random House, New York; Random House, Toronto, 2003.

Greene, B., *The Hidden Reality-Parallel Universe and the Deep Laws of the Cosmos.* Alfred A. Knopf-Random House, New

York, 2011.
Greene, B., *The Elegant Universe*, W. W. Norton & Company Ltd., New York, 1999.
Gribbin, J., *White Holes-Cosmic Gushers in the Universe*, John Gribbin Publisher, Delta Printing, 1978; Delta Books, 1977.
Gribbin, J., *Time Warps*, J. and M. Gribbin, 1979.
Grigoryev,V. and Myakishev, *The Forces of Nature*, Mir Publishers, Moscou, 1967.
Griffiths, D., *Introduction to Elementary Particles*, Wiley-VCH Verlag GmbH & Co. KGaA, Weinheim, Germany, 2008.
Guillemin, V., *The Story of Quantum Mechanics*, Dover Publications, Mineola, NY, 2003, of Charles Scribner's Sons, New York, 1968.
Guimarães, A. P., *From Lodestone to Supermagnets, Understanding Magnetic Phenomena*, Wiley-VCH Verlag GmbH & Co. KGaA, Weinheim Germany, 2005.
Guiragossián, Z. G. T., Rothbart, G. B. and Yearian, M. R., *Physics Review Letters*, 34, 335, 1975.
Gunzig, E., Diner, S. et All, *Le Vide Univers du Tout et du Rien*, Revue de L'Université de Bruxelles, Éditions Complexe, Bruxelles, 1998.
Guth, A. H., *The Inflationary Universe*, Addison-Wesley Publishing Company, Boston, 1997.
Hafele, J. C. and Keating, R. E., *Science*, 177, 166, 1972.
Hatch, R., *Escape from Einstein*, Kneat Company, Wilmington, CA, 1992.
Hansen, J. P., *Interpretação Física Racional da Constante h de Planck conduz à Solução Racional do Campo Unificado da Física*, 35ª. Reuniâo Anual da Sociedade Brasileira para o Progresso da Ciência-SBPC, Brasil, 1983.
Hansen, J. P., *Novos Fundamentos Físico-Matemáticos Aplicáveis à Detecção de Radiações Gravitacionais e Consequente Desenvolvimento de Nova Estrutura do Cosmos*, 18ª. Reunião Anual da Sociedade Brasileira de Física-SBF e 36ª. Reunião da Sociedade Brasileira para o Progresso da Ciência-SBPC, Brasil, 1984.
Hansen, J. P.,*Comprovação do Desvio de Radiações por Computador apresenta Prova Fundamental da Teoria de Cálculos Físico-Matemáticos de Campo Unificado Racional da Física*, 37ª. Reuniâo Anual Sociedade Brasileira para o Progresso da Ciência-SBPC, Brasil, 1985.
Hansen, J. P., *Teoria Unificada da Física: Opinião de Dr. Hawking e um Comentário sobre a mesma- Translação/Extrato de Carta a Dr. Hawking*",
41ª. Reunião Anual da Sociedade Brasileira para o Progresso da Ciência-SBPC, Brasil, 1989.
Hansen, J. P., *Experimento com Raios-Laser sobre Determinação de Movimento Absoluto de Sistema Inercial no Espaço*, 2ª.

Reunião Especial da Sociedade Brasileira para o Progresso da Ciência-SBPC, Brasil, 1995.
Hansen, J. P., *Valores Coincidentes de Constantes Físicas e Consequentes Descobertas Fundamentais na Ciência*, 3ª. Reunião Especial da Sociedade Brasileira para o Progresso da Ciência-SBPC, Brasil, 1996.
Hansen, J. P., *Physics Essays,* 15, 61, 2002.
Hansen, J. P., *The General Science Journal, www.gsjournal.net*, March 10, 2016 of June 4-2016.
Hansen, J. P., *The General Science Journal, www.gsjournal.net*, March 10-2016.
Hansen, J. P., *Physics Essays*, 29, 3, 2016.
Harada, M. and Sachs, M., *Physics Essays*, 11, 521, 1998.
Harré, R., *Great Scientific Experiments*, Dover Publications, Mineola, NY, 2002, of Phaidon Press, Oxford, 1981.
Hawking, S. W., *A Brief History of Time*, Bantam Books, New York, 1988.
Hawking, S. W. and Mlodinow, L., *The Grand Design*, Bantam Books, New York, 2010.
Hawking, S. W. and Penrose, R., *The Nature of Space and Time*, Princeton University Press, Princeton, 1996.
Hawking, S. W. and Mlodinov, L., *Brifer History of Time*, Bantan Books, New York, 2005.
Hayden, H. C., Physics Essays, 8, 366, 1995.
Hedman, M., *The Age of Everything*, The University of Chicago Press, Chicago, 2007.
Heidmann, J., *Intelligences Extra-Terrestres*, Éditions Odile Jacob, Paris, 1992.
Heisenberg, H., *The Physical Principles of the Quantum Theory*, Dover Publications, New York, 1949, of University of Chicago Press, Chicago, 1930.
Heisenberg, W., *Physics and Beyond: Encounters and Conversations*, HarperCollins Publishers, New York, 1971.
Heisenberg, W., *Physics and Philosophy The Revolution in Modern Science*, Harper Perennial Modern Thought, 2007 of
Harper Torchbooks-HarperCollins Publishers, New York, 1962.
Heitler, W., *The Quantum Theory of Radiation*, Dover Publications, New York, 1984 of Oxford University Press, Oxford, 1936.
Herbert, N., *Quantum Reality*, Anchor Books-Randon House, New York, 1987.
Hernandez, R., Schil R. and Klimo, J., *Beyond UFOS*, Create Space Independent Publishing Platform, USA, 2018.
Hesse, M. B., *Forces and Fields*, Dover Publications, 2005, of T. Nelson, London, 1961.
Hill, P. R., *Unconventional Flying Objects: A Scientific Analysis*, Hampton Roads Publishing Company, Newburyport, MA, 1995.

Hoag, J. B., *Electron and Nuclear Physics*, D. Van Nostrand Company, New York, 1938-1946, of *Electron Physics*, D. Van Nostrand Company, New York, 1929.
Hobson, V. G., *The Toth-Maatian Review*, 5, 2697, 1987.
Hobson, V. G., *The Toth-Maatian Review*, 5, 2801, 1987.
Hobson, V. G., *The Toth-Maatian Review*, 6, 3181, 1988.
Hobson, V. G., *The Toth-Maatian Review*, 6, 3345, 1988.
Hobson, V. G., *The Toth-Maatian Review*, 7, 3517, 1989.
Hobson, V. G., *The Unified Quantum Field Theory*, National Library of Australia Card No., 1972.
Hoffman, B., *The Strange Story of the Quantum*, Dover Publications, Mineola, NY, 1959, of Harper and Brothers, New York, 1947.
Hoffmann, L., *Astronomia-Nova Carta Celeste*, Iatt-Mayer S. A. Arts Gráficas, AGGS Indústrias Gráficas, Rio de Janeiro, 1973.
Hogan, C. J., *The Little Book of the Big Bang-A Cosmic Primer*, Copernicus Springer-Verlag, New York, 1998.
Holton, G., *Ensayos sobre El Pensamento Científico em La Época de Einstein*, Editorial Universidad, Madrid, 1973.
Horgan, J. *The End of Science*, Addison-Weslag Publisher-Pearson PLC, Boston, London, 1996.
Hoyle, F., *The Intelligent Universe*, Dorling Kindersley Book, London, 1983.
Humason, M., The Realm of the Nebulae, Yale University Press, New Haven, Connecticut, London, 1936, of Minnnesota University, Minneapolis, MN, USA.
Incandela, J., *Status of the CMS SM Higgs Search*, CERN, July 4, 2012.
Ivanov. B., *Contemporary Physics*, Peace Publishers, Moscou, Decade of 1960.
Ivanov. B., *Physique Nouvelle*, Reditions of Mir, Moscou, 1966.
Ives, H. I., Journal Optics Society American, 29, 472, 1939.
Jacob, M., *Au Coeur de La Matière*, Editions Odile Jacob, 2001.
Jaffe, B., Crucibles: *The Story of Chemistry*, Dover Publications, Mineola, NY, 1976, of Simon & Shuster, New York, 1930.
Jaffe, B., *Michelson and the Speed of Light*, Doubleday, New York, 1960.
Jammer, M., *Concepts of Force*, Dover Publications, Mineola, NY, 1999, of Harvard University Press, Cambridge, MA, 1957.
Jammer, M., *Concepts of Mass*, Dover Publications, Mineola, NY, 1997, of Harvard University Press, Cambridge, MA, 1961.
Jammer, M., *Concepts of Space*, Dover Publications, Mineola, NY, 1993, of Harvard University Press, Cambridge, MA, 1969.
Janossi, L.,*Theory of Relativity Based on Physical Reality*, Akadémiai Kiado, Budapest, 1971.

Jastrow, R., *Until the Sun Dies*, W. W. Norton & Company, New York, 1977.

Jauch, J. M., Are *Quanta Real?*, Indiana University Press, Bloomington, IN, 1973.

Javadi, H., *Beyond the Standard Model*, San Bernardino, CA, USA, 2018.

Jeans, J., *Physics and Philosophy*, Dover Publications, Mineola, NY, 1981, of Cambridge University Press, Cambridge and Macmillan Company, London, 1943.

Jefimenko, O. D., *Causality Electromagnetic Induction and Gravitation*, Electret Scientific Company Star City, USA, 2000.

Jefimenko, O. D., *Electromagnetic Retardation and Theory of Relativity*, Electret Scientific Company Star City, USA, 2004.

Kamenov, K. G., *Space, Time and Matter, and the Falsity of Einstein's Theory of Relativity*, Vantage Press, New York, 2000.

Kaplan, I., *Nuclear Physics*, Addison-Wesley Publishing Company, Massachusetts, 1977.

Kassir, R. M., *Physics Essays* 27, 16 2014.

Kean, L., *Ufos: Generals, Pilots, and Government Officials Go on the Record*, Three Rivers Press, New York, 2010.

Keller, A., *The Infancy of Atomic Physics*, Dover Publications, Mineola, NY, 2006, of Claredon Press, New York; Oxford University Press, Oxford, 1983.

Kelly, A., *Challenging Modern Physics-Questioning Einstein's Relativity Theories*, Brown Walker Press, Boca Raton, FL, 2005.

Kennedy, R. J. e Thorndike, E. M., *Physical Review*, 42, 400, 1932.

Kondratyev, V., *The Structure of Atoms and Molecules*, Mir Publishers, Moscou, 1967.

Kostro, L., *Einstein and the Ether*, Apeiron-C. Roy Keys, Montreal, Quebec, 2000.

Kragh, H., *Quantum Generations-A History of Physics in the Twentieth Century*, Princeton University Press, Princeton, 2002.

Krauss, L. M., *Fear of Physics-A Guide for the Perplexed*, Basic Books-HarperCollins Publishers, New York, 1993.

Krauss, L. M., *A Universe from Nothing*, Free Press-Simon Schuster, New York, 2012.

Laszlo, E., *Aux Racines de L'Univers*, Librairie Anthème Fayard, Paris, France, 1992.

LaViolette, P. A., *Secrets of Antigravity Propulsion*, Bear & Company, Rochester, VT, USA, 2008.

Lederman, L. and Teresi, D., *The God Particle*, Houghton Mifflin Company- Mariner Book, Boston, New York, 1993.

Lederman, L. M. and Hill, C. T., *Symmetry and the Beautiful*

Universe, Prometheus Books, Amherst, NY, 2008.

Lefort, M., *Les Radiations Nucléaires*, Presses Universitaires de France, 1964.

Lerner, E. J., *The Big Bang Never Happened*, Vintage Books-Randon House Inc., NewYork, 1992, of Times Book, New York, 1991.

Lesche, B., *Teoria da Relatividade*, Editora da Física, São Paulo, 2005.

Lévy, J., *From Galileo to Lorentz... and Beyond*, Apeiron-C. Roy Keys, Montreal, Quebec, 2003.

Longair, M. S., *The Origens of our Universe*, Cambridge University Press, Cambridge, 1991.

Lorentz, H. A., *The Theory of Electrons*, B. G. Teubner, Leipzig, 1916.

Lorentz, H. A., *Lectures on Theoretical Physics*, Vol.1, *Aether Theories and Aether Models*, MacMillan, London, 1927.

Lucas, C. W. Jr., *The Universe Force*, Mechanicsville, MD, USA, 2013.

Lugo, C., *Los Rayos Simultáneos-Pruebas Experimentales Contraries a La Teoria de la Relatividad*, Editorial Américale, Buenos Aires, 1967.

Luklin, G. B., *Physics Today*, 31(1), 18, 1978.

Lyne W., *Occult Ether Physics*, Creatopia Productions, New Mexico, 1997- 2008.

Maccbee, B., *Bulletin of the American Physical Society*, Vol.38, p. 1041, 1993.

Maddox, J., *What Remains To Be Discovered*, The Free Press-Simon & Shuster, New York, 1998.

Maia, N. B., *O Caminho para a Física Quântica*, Editora Livraria da Física, 2009.

Maillet, H. et al., *Le Laser Principes et Techniques d'Application, Technique et Documentation*, Paris, 1984.

Marinov, S., *Czechoslovakia Journal Physics B*, 24, 965, 1974.

Marinov, S., *General Relativity and Gravitation*, 12, 57, 1980.

Marmet, L., *Red-Shift*, www.marmet.org

Marques, G. C. et al., *Física-Tendências e Perspectivas*, Editora Livraria da Física, São Paulo, 2005.

Martin, B. R., *Particle Physics*, Oneworld Publications, London, 2011.

Martins, J. B., *Teoria da Relatividade-O Caminho de Lorentz, a Revolução de Einstein*, Editora Ciência Moderna, Rio de Janeiro, 2011.

Martins, J. B., *O Spin*, Livraria da Física, São Paulo, 2014.

Martins, R. A., *O Universo*, Editora Livraria de Física, São Paulo, 2012 de Editora Moderna, Curitiba, Brasil, 1994.

Martins, R. A., *A História do Átomo de Demócrito aos Quarks*, Editora Ciência Moderna, Rio de Janeiro, 2002.

Martins, R. A., *Searching for the Ether: Leopold Courvoiser's*

Attemtps to Measure the Absolute Velocity of the Solar System, The International Journal of Scientific History,17, 1, 2011.
Martins, R. A., *Teoria da Relatividade Especial*, Editora da Física, São Paulo, 2011.
Martins, R. A., *Becquerel e a Descoberta da Radioatividade-Uma Análise Crítica*, Editora da Universidade Estadual da Paraíba-EDUEPB, Campina Grande, Brasil, e Livraria da Física, São Paulo, 2012.
Martins, R. A., *O Universo*, Editora Livraria de Física, São Paulo, 2012.
Martins, R. A. and Rosa, P. S., *História da Teoria Quântica: A Dualidade Onda-Partícula de Einstein a De Broglie*, Editora Livraria da Física, São Paulo, 2014.
Martins, R. A., *A Origem Histórica da Relatividade Especial*, Editora da Física, São Paulo, 2015.
Martits, J., *Experimento sobre Movimento da Terra*, Monografia ao Conselho da Sociedade Interplanetária Brasileira-SIB, São Paulo, 1973 e Comunicação Pessoal, 1991.
Maxwell, *A Treatise on Electricity & Magnetism*, Vol. 1 and 2, Dover Publications, Mineola, NY, 1954, of Claredon Press, Oxford, 1891.
Maxwell, J. C., *A Treatise on Electricity & Magnetism,* Dover Publications, New York, Vol 1 and 2, 1954 of Clarendon Press, 1891, 1873.
Maxwell, J. C., *Matter and Motion*, Dover Publications, New York, 1952, of Cambridge Press University, 1877.
Mazur, J., *Zeno's Paradox*, Plume-Penguin Group, New York, 2008, of *The Motion Paradox, Dutton Edition, 2007.*
McCausland, I., *Physics Essays* 18, 530, 2005.
McDowell, All, *Uncommon Knowledge, New Cience of Gravity, Light, the Origin of Life and the Mind of Man*, AuthorHouse, Bloomington, IN, USA, 2010.
McEnvoy, J. P. and Zarate, O., *Quantum Theory*, 2007, Penguin Books, Totem Books, Icon Books, Faber and Faber, Allen & Unwin Pty, USA, Canada, Australia, 1996.
McMahon, C. R., gsjournal.net, 2010.
McMahon, D., *Relativity Demystified*, The McGraw-Hill, New York, 2006.
McMahon, D., *Quantum Field Theory*, The McGraw-Hill, New York, 2008.
Mees, T. De, *General Science Journal, www.gsjournal.net, Extended Analysis of the Erroneous Use of the Virial Theorem for Elliptical- and Disc Galaxies, and for Galaxy Clusters, which leads to Dark Matter*, April, 2, 2018.
Mesquita, P. F., *Significado da Constante Gravitacional de Isaac Newton e consequente descoberta específica de radiações gravitacionais revelam e comprova "data venia" equívoco de Albert Einstein, Max Planck e contemporâneos-"The Meaning*

of the Universal Gravitation Constant of Newton Evinces the Specific Existence of Gravitational Radiations", Universidade São Paulo, 1965; 2º· Congresso Brasileiro de Cartografia e apresentado à Academia Brasileira de Ciências, Rio de Janeiro, 1965; Revista Engenharia-Instituto de Engenharia, São Paulo, 393, 52, 1977.

Mesquita, P. F., *A Duração da Vida do Nêutron Livre Calculada Racionalmente Constitui Novo Teste Crucial da Existência Específica de Radiações Gravitacionais Identificadas no Espectro do Eletromagnetismo*, 19ª. Reunião Anual da Sociedade Brasileira para o Progresso da Ciência-SBPC, Brasil, 1967.

Mesquita, P. F., *Curso Fundamentos de Física*, Escola Politécnica da Universidade São Paulo, 1970.

Mesquita, P. F., *O Postulado de Planck Obedece a Lei de Newton*, Reunião da 22ª. Reunião da SBPC, Brasil, 1970.

Mesquita, P. F., *Processo para determinar a Posição Aparente do Ponto na Esfera Celeste para o qual se encontra dirigido o Movimento Absoluto da Terra*, 23ª. Reunião Anual da Sociedade Brasileira para o Progresso da Ciência-SBPC, Brasil, 1971.

Mesquita, P. F. *A Lei de Coulomb obedece ao Postulado de Planck,* 24ª. Reunião Anual da Sociedade Brasileira para o Progresso da Ciência-SBPC, Brasil, 1972.

Mesquita, P. F., *O Movimento do Periélio Independente da Relatividade Geral*, 26ª. Reunião Anual da Sociedade Brasileira para o Progresso da Ciência-SBPC, Brasil, 1974.

Mesquita, P. F., *"The Physical Meaning Meaning of the Newtonian Constant Reveals a Way for Studing Satellite Dynamics within a Unified Electromagnetic and Gravitational Field"* in the Symposium on Satellite Dynamics do 17º· Plenary Meeting of the Committee on Space Research by the International Council of Scientific Unions-COSPAR, 1974.

Mesquita, P. F., *O Gráviton em CGS e MKS*, Reunião da 29ª. Reunião da SBPC, Brasil, 1977.

Mesquita, P. F., *Da Lei de Newton ao Postulado de Planck em CGS e MKS*, Reunião da 29ª. SBPC, Brasil, 1977.

Mesquita, P. F., *Porquê a Frequência $\nu = 10^{19}$ s^{-1} é Invariante da Gravitação Universal*, 4º. Simpósio de Ensino de Física organizado pela Sociedade Brasileira de Física-SBF, Brasil, 1979.

Mesquita, P. F., *Solução Algébrica para o Diâmetro e Buraco Negro*, 31ª. Reunião Anual da Sociedade para o Progresso da Ciência-SBPC, Brasil, 1979.

Mesquita, P. F., *Buraco Negro – "Black Hole" Que é?*, Revista de Engenharia-Instituto de Engenharia, Número 427 de 1980, São Paulo.

Mesquita, P. F., *Noção da Teoria da Impulsão em Física*

Fundamental, 14ª.Reunião Anual da Sociedade Brasileira da Física-SBF e 32ª. da Sociedade Brasileira para o Progresso da Ciência-SBPC, Brasil, 1980.

Mesquita, P. F., *O Significado Físico da Constante de Loschmidt na Teoria da Impulsão,* 14ª.Reunião Anual da Sociedade Brasileira da Física-SBF e 32ª. da Sociedade Brasileira para o Progresso da Ciência-SBPC, Brasil, 1980.

Mesquita, P. F., *Existência e Unicidade Físico Matemática do Invariante Gravitacional* $\mathcal{V} = 10^{19}\ s^{-1}$, Sociedade Brasileira para o Progresso da Ciência-SBPC, Brasil, 1981.

Mesquita, P. F., *Densidade Média do Universo Físico pela Teoria da Impulsão de Erdelyi*, 34ª. Reunião Anual da Sociedade Brasileira para o Progresso da Ciência-SBPC, Brasil, 1982.

Mesquita, P. F., *Idade do Universo pela Teoria da Impulsão de Erdelyi na Hipótese do "Big Bang"*, 34ª. Reunião Anual da Sociedade Brasileira para o Progresso da Ciência-SBPC, Brasil, 1982.

Mesquita, P. F., *Invariança Físico-Matemática das Cônicas na Teoria da Impulsão*, 35ª. Reunião Anual da Reunião Anual da Sociedade Brasileira para o Progresso da Ciência-SBPC, Brasil, 1983.

Michelson, A. A. and E. W. Morley E. W., *American Journal Science* 33, 377, 1886 and 34, 333, 1887.

Michelson, A. A., *Studies in Optics*, Dover Publications, Mineola, NY, 1995, of The University of Chicago Press, Chicago, 1927.

Mike, J., *The Anatomy of a Flying Saucer*, Lexington, KY, USA, 2017.

Mitchell, W. C., *Bye Bye Big Bang Hello Reality*, Cosmic Sense Books, Carson City, NV-USA, 2002.

Monti, R. A., *Physics Essays*, 9, 238, 1996.

Morales-Riviera, E., W*as Einstein Wrong?*, Trafford Publishing, Victoria, Canada, 2007.

Morris, R., *Achilles in the Quantum Universe-The Definitive History of Infinity*, Henry Holt and Company, New York, 1997.

Mourão, R. R. F., *Buracos Negros-Universo em Colapso*, Editora Vozes, Petrópolis, Brasil, 1986.

Múnera, H. A., *Apeiron*, 4, 80, 1997.

Musser, G., *Spooky Action at a Distance*, Scientific American FSG, New York, 2015.

Natale, A. A., Vieira, C. L. e Outros, *O Universo sem Mistério*, Vieira & Lent, Brasil, 2003.

Navega Neto, J. C., Couto, H. S., Kassar, E., and Tilman, N. V., *Difração do Elétron como Teoria do Momento Elementar*, 28ª. Reunião Anual Sociedade Brasileira para o Progresso da Ciência-SBPC, Brasil, 1976.

Navia, C. E.; Augusto, C. R. A.;Franceschini, D. F., Robba, M. B. and Tsui, K. H., *Progress in Physics*, 1, 53, 2007.

Newman, D., Ford, G. W., Rich, A. and Sweetman, E., *Physical Review Letters,* 40, 1355, 1978.
Newton, I., *Óptica,* Editora Universidade de São Paulo,1996, by A. K. T. Assis, da versão inglesa *Optics,* of Opticks, Benj. Walford Royal Society, and W. Innys, London, 1704, 1717, 1721 e 1730.
Newton, I., *Principia,* Nova Stella Editorial, São Paulo, 1990, by Ricci, T. S. P., Brunet, L. G., and Ghering, S. T. and Célia, M. H. C., of English translation of 1686, 1729, 1934, of *"Philosophiae Naturalis Principia Mathematica"*, S. Pepys, Keg. Soc. Praeses, London, 1686.
Newton, I., *Principia,* Vols. 2 and 3, Editora Universidade de São Paulo-EDUSP, 2008, by Assis, A.K. T. of versão inglesa de 1729 e 1934, University of California Press, de *"Philosophiae Naturalis Principia Mathematica"*, Isaac Newton's *Mathematical Principles of Natural Philosophy, and his System of the World,* S. Pepys, Keg. Soc. Praeses, London, 1686.
Neto, J. B., *Mecânica Newtoniana, Langragiana e Hamiltoniana,* Editora Livraria de Física, São Paulo, 2004.
Nicolson, I., *Gravidade, Buracos Negros e o Universo,* Livraria Francisco Alves Editora, Rio de Janeiro, 1983, of *Gravity, Black Holes and the Universe,* David & Charles Devon, UK, 1981.
Nikolov, A., *Essence of Planck's Constant,* gsjournal.net, 2011.
Nimtz, G. and Haibel, A., *Zero Time Space,* Wiley-VCH Verlag GmbH & Co. KGaA, Weinheim, Germany, 2008.
Novello, M., *Do Big Bang ao Universo Eterno,* Jorge Zahar Editor Ltda., Rio de Janeiro, 2010.
Novello, M., *O que é Cosmologia?,* Jorge Zahar Editor, Rio de Janeiro, 2006.
Nussenzveig, H. M., *Física Básica,* Editora Edgard Blücher, São Paulo, Vol. 1 e Vol. 2 1981, Vol. 3 1997, Vol. 4 1998.
Nussenzveig, H. M., Lobo Carneiro, F. and Rosa, L. P., *300 Anos do "Principia" de Newton,* Núcleo de Publicações-COPPE-Aula Editora, Rio de Janeiro, 1988.
Oliveira, I. S., *Física Moderna,* Editora Livraria de Física, São Paulo, Vol.1 e Vol. 2, 2005.
Oliveira, I. S., *Física Moderna,* Livraria da Física, São Paulo, 2010.
O'Rahilly, A., *Electromagnetic Theory-A Critical Examination of Fundamentals,* Dover Publications, Mineola, New York, 1965.
Pagels, H. R., *Simetria Perfeita,* Gradiva Publicações, Lisboa, 1990, of *Perfect Symmetry,* Simon & Schuster-CBS Corporation, New York, 1985.
Pais, A., *Subtle is the Lord,* Oxford University Press, Oxford,1982.
Palacios, J., *Relatividad,* Espasa-Calpe, Madrid, 1960.
Panek, R., *The 4 Percent Universe-Dark Matter, Dark Energy, and the Race to Discover the Rest of Reality,* Houghton Mifflin

Harcourt, New York, 2011.
Parker, E. N., *Scientific American Review*, 4, 66, 1964.
Pardy, M., *arXiv*:hep-ph/9512544V1 30 Dec,1999.
Pauli, W., *Theory of Relativity*, Dover Publications, Mineola, NY, 1981, of Pergamon Press, 1958.
Pecker, J. C. and Vigier, J. P., 2, 19, Apeiron, 1988.
Pellanda, E. B., *As Ilusórias Teorias da Física do Século 20*, Editora Assessoria Gráfica e Editorial-AGE, Porto Alegre, 2007.
Penrose, R., *The Emperor's New Mind*, Oxford University Press, Oxford, 1989.
Pessoa Jr., Os *Conceitos de Física Quântica*, Editora Livraria de Física, São Paulo, Vol.1 – 2003, Vol.2, 2006.
Pierce, J. R., *Almost All About Waves*, Dover Publications, Mineola, NY, 2006, of Massachusetts Institute of Technology, 1974-1981.
Phipps Jr., T. E., *Old Physics for New,* Apeiron-C. Roy Keys, Montreal, Quebec, 2006.
Pickover, C. A., *Archimedes to Hawking*, Oxford University Press, Oxford, 2008.
Pietro, R., WWW-http://rogeriopietro.wordpress.com
Pikelner, S., *Physics of interstellar Space*, The Foreign Languages Publishing House, Moscow, Decade of 1960.
Pires, A. S. T., *Evolução das Ideias da Física*, Livraria da Física, São Paulo, 2008.
Pires, A. S. T. and Carvalho, R. P., *Por Dentro do Átomo*, Livraria da Física, São Paulo, 2014.
Pla, C., *Velocidad de la Luz y Relatividad*, Espasa-Caple-Argentina, Buenos Aires, Mexico, 1947.
Planck, M., *The Theory of Heat Radiation*, Dover Publication, Mineola, NY, 1991 of P. Blakiston Son & Co., Philadelphia, 1914.
Podgornik, R., www-*50 Years of the L.Lifshitz Theory of Van der Waals Forces*, 2006.
Poincaré, H., *O Valor da Ciência,* Contraponto Editora, Rio de Janeiro, 1995, of *La Valeur de La Science*, Ernest Flammarion Éditeur, Paris, 1908.
Potter P., *Anti-Gravity Propulsion Dynamics*, Adventures Unlimited Press, Kempton, IL, USA, 2016.
Prigogine, I., *O Nascimento do Tempo*, Edições 70., Lisboa, 1990, of *La Nascita Del Tempo*, Edizione Theoria, Roma, Napoli, 1988, Italy, 1993.
Prigogine, I., *Le Leggi del Caos*, Gius. Laterza & Figli, Roma, 1993.
Potter, C., *You Are Here-A Portable History of the Universe,* Hutchinson, UK, Random House Group-Penguin-HarperCollins Publishers, New York, 2009.
Price, D. S., *Science Since Babylon*, Yale University Press, New

Haven, Connecticut, CT, 1960.
Primack, J. R. and Abrams, N. E., *The View from the Center of the Universe*, Riverhead-Penguin, New York, 2006.
Pullman, B., *The Atom in the History of Human Thought*, Oxford University Press, Oxford, 1998.
QiYu Liang, et al., *Observation of three-photon bound states in a quantum nonlinear medium*, Science, February 2018.
Rae, A. I. M., *Quantum Physics*, Oneworld Publications, New York, 2005.
Randall, L., *Knocking on Heaven's Door*, Ecco-HarperCollins Publishers, New York, 2012.
Randall, L., *Dark Matter and The Dinosaurs*, HarperCollinsPublishers, NewYork, 2015.
Randall, L., *Higgs Discovery*, HarperCollins Publishers, New York, 2013.
Rosenfeld, R., *O Cerne da Matéria*, Companhia das Letras-Editora Schwarcz, São Paulo, 2013.
Ranzan, C., *Physics Essays*, 25, 327, 2012.
Ranzan, C., *Physics Essays*, 26, 40, 2013.
Ray, C., *Time, Space and Philosophy*, Routledge-Taylor & Francis Group, Abingdon, England, 1991.
Rasul, M., *Physics Essays*, 23, 293, 2010.
Rees, M., *Just Six Numbers-The Deep Forces that Shape the Universe*, Orion Publishing Group, London, 1999.
Rehead, M. *From Physics to Metaphysics,* Cambridge University Press, Cambridge, 1996.
Resnick, R. , *Introduction to Special Relativity*, John Wiley & Sons, New York, 1968.
Riabov, Y. *Les Mouvements Des Corps Célestes*, Éditions Mir, Moscou, 1967.
Ribeiro, J. E. A., *Lorentz Force*, Universidade São Paulo-USP, 2008.
Ritz, W., *Oeuvres*, Gauthier Villars, Paris, 1911.
Rival M., *Les Grandes Scientifiques, Éditions Du Seuil*, Paris, 1996.
Rocha, J. F. et al., *Origens e Evolução das Ideias da Física*, Editora Universidade da Bahia-EDUB, Salvador, Brasil, 2002.
Roditi, I., *Dicionário Houaiss de Física*, Editora Objetiva, Rio de Janeiro, 2005.
Rodrigues, W. and Tiomno, J., *Foundations Physics*, 15, 945, 1985.
Ron, J. M. S., *El Origen y Desarrollo de la Relatividad*, Alianza Editorial, Madrid, 1983-1985.
Rohden, H., *De Alma para Alma*, Editora Martin Claret Alvorada-São Paulo, 1989.
Ronan, C. A., *The Cambridge Illustrated History of the World's Science*, The Press Syndicate of the University of Cambridge, Cambridge and Newnes Books, London, 1984.

Rosenblum, B. and Kuttner, F., *Quantum Enigma*, Oxford University Press, New York, 2008.

Rothman, T., *Physics From Aristotle to Einstein and Beyond*, A Byron Press Book-Fawcwtt Books, New York, 1995.

Rothman, T., *Everything's Relative*, John Wiley & Sons, New York, 2003.

Rovelli, C., *Sette Brevi Lezione di Fisica,* Adelphi Edizione S.p.A, Milão, 2014.

Rovelli, C., *La realtà non è come appare: la struttura elementare delle cose*, Raffaello Cortina Editore, Milano, Italy, 2014.

Roychoudhuri, C., Kracklauer, A. F., Creath, K. e Outros, *The Nature of Light: What is a Photon?* CRC Press-Taylor & Francis Group, Boca Raton, FL, 2008.

Rutherford, E., *Radio-activity*, Dover Publications, Mineola, NY, 2004, of Cambridge University Press, Cambridge, 1904.

Rydnik, V., *ABC's of Quantum Mechanics*, Peace Publishers, Moscou, Decade of 1960.

Sagan, C., *The Demon-Haunted World*, The Random House Publishing Group-A Ballantine Book, New York, 1996.

Salvetti, A. R., *A História da Luz*, Livraria da Física, São Paulo, 2008.

Santoro, A., *Física Moderna*, Departamento de Publicações Acadêmico Horácio Lane, Universidade Mackenzie, São Paulo, 1975.

Sarg, S., *Field Propulsion by Control of Gravity, Theory and Experiments*, Lexington, KY, USA, 2017.

Sato, M., *Physics Essays*, 23, 405, 2010.

Sato, M. and Sato, H., *Physics Essays*, 24, 467, 2011.

Sauerheber, R. D., *Physics Essays*, 27, 116, 2014.

Schawlow, A., *Advances in Optical Masers*, *Scientific American*, July, 1963.

Schenberg, M., *Pensando a Física*, Nova Stella Editorial, São Paulo, 1984.

Schiafly,R., *How Einstein Ruined Physics*, Dark Buzz, USA, 2011.

Schiller, C., *Motion Mountain*, Creative Commons Attribuition, www.motionmountain.net, Munique, Alemanha e Bruxelas, Bélgica, 1990-2014.

Segré E., *Les Physiciene Classiques et Leurs Découvertes de la Chute des Corps aux Ondes Hertziennes*, Librairie Arthème Fayard, Paris,1987, *From Falling Bodies to Radio Waves*, Freeman and Company, New York, 1984, Dover Publications, Mineola, NY, of *Personaggi e Scoperte nella Fisica Contemporanea*, Mondadori Editore, Milano, 1980-1983.

Seife, C., *Alpha and Omega-The Search for the Beginning and the End of the Universe*, Wendy Wolf-Viking, Penguin, New York, 2003.

Shamos, M. H., *Great Experiments in Physics*, Dover Publicationa, Mineola, NY, 1987, of Holt-Rinehartand, New

York, 1959.

Shapiro, I. I., *Physical Review Letters*, 13, 789, 1964.

Silvertooth, E. W. Jacobs, S. F., *Applied Optics*, Vol. 22, 1274, 1983.

Selleri, F., *El Debate de la Teoría Cuántica*, Alianza Universidad, Madrid, 1896, of *Die debatte um die Quantentheorie*, Germany.

Selleri, F. *Foundations Physics,* 26, 641, 1996.

Selleri, F. et al., *Open Questions in Relativistic Physics*, Apeiron-C. Roy Keys, Montreal, Quebec, 1998.

Sena, L. A., *Units of Physical Quantities and Their Dimensions*, Mir Publishers, Moscou, 1972.

Serres, M. et al., *Élements pour une Histoire des Sciences*, Bordas, Paris, 1989.

Shadowitz, A., *The Electromagnetic Field*, Dover Publications, Mineola, NY, 1988 of McGraw-Hill Book Company, New York, 1975.

Siegfried, T., *The Bit and the Pendulum*, John Wiley & Sons, New York, 2000.

Silk, J., *The Big Bang-The Creation and Evolution of the Universe*, W. F. Freeman and Company, New York, 1980.

Simaan, A. and Fontaine, J., *L'Image du Monde des Babyloniens à Newton*, Adapt Éditions, Paris, 1999.

Simhony, M., *Invitation to the Natural Physics of Matter, Space and Radiation*, World Scientific Publishing, Singapore, New Jersey, London, Hong Kong, 1994.

Simulik, V., Keller, V. J., Terezawa, H., Phipps Jr., T. E. and Others, *What is the Electron?* Apeiron-C. Roy Keys, Montreal, Quebec, Montreal, 2005.

Singh, S., *Big Bang*, Editora Record, Rio de Janeiro, 2006, de *Big Bang Fourth Estate*, HarperCollins Publishers, New York, 2004.

Sitchin, Z., *The 12th Planet*, Bear & Company-Inner Traditions International-Rocherster-VT-USA, 1977.

Smith, W., *Science & Myth: With a Response to Stephen Hawking's The Grand Design*, Angelico Press Publisher, USA, 2012, of *Science & Myth: What We Are Never Told,* Angelico Press, USA, 2010.

Smolin, L., *Three Roads to Quantum Gravity*, Basic Books, New York, 2002.

Smoot, G. F., Gorenstein, M. V. and Muller, R. A., *Physical Review Letters*, 39, 898, 1977.

Smoot, G. F. and Davidson, K., *Dobras no Tempo*, Editora Rocco, 1995, of *Wrinkles in Time*, Little Brown Company, 1993.

Speyer, E., *Six Roads from Newton*, John Wiley & Sons, 1994.

Steinhardt. P. J. and Turok, N., *Endless Universe: Beyond the Big Bang*, Doubleday Publishing Group-Random House, New York, 2007.

Sujak, P., *Einstein's Destruction of Physics*, CreateSpace, North Charleston-SC, 2017.
Susskind, L., *The Black Hole War*, Hachette Book Group, New York, 2008.
Swenson Jr., L. S., *The Ethereal Aether-A History of the Michelson-Morley-Miller Aether Drift Experiments*, University of Texas Printing, Austin, London, 1930.
Szamosi, G., *The Twin Dimensions Inventing Time and Space*, McGraw-Hill, New York, 1986.
Tavares, O. A. P., *Descobrindo o Núcleo Atômico*, Livraria da Física, São Paulo, Centro Brasileiro de Pesquisas Físicas-CBPF, Rio de Janeiro, 2012.
Taylor, J., *Black Holes-The End of the Universe?*, Avon Books, New York, 1978.
Taylor, J. H., Wolszczan, A., Damour, T. and Weisberg, J. M., *Nature*, 355, 132, 1992.
Teller, E. and Latter, A. L., *Our Nuclear Future*, Criterion Books, New York, 1958.
Thibaud, J., *Vie et Transmutations des Atomes*, Éditions Albin Michel, Paris, 1952
Thorne, K. S., *Black Holes & Time Warps*, W. W. Norton & Company, New York, 1993.
Tipler, P.A., *Física*, Editora Guanabara Dois, 1986-1990, of *Physics*, Worth Publishers, New York, 1982.
Toben, B. and Wolf, F. A., *Space-Time and Beyond*, 1975-1982,
E. P. Dutton-New American Library, York, 1975-1982.
Tolchelnikova-Murri, S. A., *Galilean Electrodynamics*, 3b, 72, 1992.
Toomey, D., *The a Journey to the New Time Frontier of Physics Travelers*, W. W. Norton & Company, New York, 2007.
Trémolières, J., *Os Lasers*, Editorial Estúdios Cor, Lisboa, 1966.
Trigg, G. L., *Landmark Experiments in Twentieth Century Physics*, Dover Publications, Mineola, NY, 1995, of Crane-Russak and Company, New York, 1975.
Troup, G., *Masers and Lasers*, Methuen & Co., London, 1969.
Tsau, J., *Discovery of Ether and its Science*, Infinity Publishing, West Conshohocken, PA, 2005.
Tu, L-C, Luo, J. and Gillies, G. T., *Rep. Prog. Phys.*,68, 77, 2005.
Uchôa, A. M. M., *Uma Busca da Verdade*, Editora do Conhecimento, Limeira-SP-Brazil, 1991.
Unzicker, A., *The Higgs Fake*, Alexander Unziker, Middletown-DE, 2017.
Valone, T., Electrogravitics Systems, Integrity Research Institute, Beltsville, MD, USA, 2008.
Vanclair, S., *La Symphonie des Étoiles,* Éditions Albin Michel, Paris, 1997.
Vanco, B., *Revista Astronáutica*, Sociedade Interplanetária Brasileira-Centro de Pesquisa Bradesco, Osasco, São Paulo,

50, 16, 1964.

Videira, A. A. P. and Coelho, R. L., *Física, Mecânica e Filosofia-O Legado de Hertz*, Editora Universidade do Estado do Rio de Janeiro-EDUERJ, 2012, of *Die Prinzipien der Mechanik in neuem Zuzammenhange dargestellt*, J. A. Barth, Leipzig, 1894.

Vigier, J. P., *Apeiron*, 4, 71, 1997.

Veltman, M., *Facts and Mysteries in Elementary Particle Physics*, World Publishing; New Jersey, London, Singapore, Hong Kong, 2003.

Voinoff, I., *Revista Engenharia Mackenzie*, Universidade Mackenzie, São Paulo, pag. 5, Novembro 1942.

Watkeys, C. W., Allin, H. L. et All, *An Orientation in Science*, McGraw Hill Book, New York, 1938.

Weber, R., *Dialogues with Scientists and Sages-The Search for Unity*, Routledge & Kegan Paul, London, 1986.

Weinberg, S., *Gravitation and Cosmology*, Wiley & Sons, New York, 1972.

Weinberg, S., *The Discovery of Subatomic Particles*, Cambridge University Press, Cambridge, 2003, of *Scientific American Library, New York, 1983*.

Weinberg, S., *Dreams of a Final Theory*, Pantheon Books, New York, 1992.

Wesley, J. P., *Progress in Space Time Physics*, edited by J. P. Wesley, Weiherdarmstr. 24, 7712, Blumberg, Germany, 1987, pages 11 a 15.

Wesley, J. P., *Selected Topics in Advanced Fundamental Physics*, Benjamin Wesley Publishers, Blumberg, 1991.

Wennerström, H. e Westlund, P. O., *Physics Essays*, 26, 174, 2013.

Wher, M. R. e Richard Jr., J. A., *Physics of the Atom*, Addison Wesley Publishing, Massachusetts, 1959.

Whittaker, E., *A History of the Theories of Aether and Electricity*, Vol. 1, 1951 and Vol. 2, 1953, Humanities Press, Austrália-Thomas Nelson & Sons, Nashville, TN and Harper Torchbook-HarperCollins, 1960.

Whitehead, A. N., *The Concept of Nature*, Dover Publications, Mineola, NY, 2004, of Cambridge University Press, Cambridge, 1920.

Wilczek, F., *The Lightness of Being, Mass, Ether, and Unification of Forces,* Basic Books-Perseus Books Group, New York, 2008.

Williams, L. P., *Einstein, A., Grünbaum, Eddington, A. S., Relativity Theory-Its Origins and Impact in Modern Thought*, John Wiley & Sons, New York, 1968.

Wilson, R. R. and Littauer, R., *Accelerators Machines of Nuclear Physics*, Doubleday & Company, New York, 1960.

Whitrow, G. J., *Time in History-Views of Time from the Prehistory*

to the Present Day, Oxford University Press, Oxford, 1988.
Will, C. M., *Was Einstein Right?*, Basic Books, New York, 1986.
Woit, P., *Not Even Wrong*, Basic Books-Perseu Books Group, New York, 2007.
Worsley, A., *Physics Essays*, 23, 311, 2010.
WWW – *Applied Physics Research* – www.ccsenet.org
WWW – *Arxiv: arxiv.org of the Cornell University Library, Ithaca, NY*
WWW – *AstronomyToday: astronomytoday.com/cosmology/quantumgrav.html.*
WWW – *Bruce DePalma*: brucedepalma.com
WWW – *Davi Pratt*: davipratt.info/gravity.htm
WWW – *Electrogravityphysics: electrogravityphysics.com*
WWW – *Gyroscopes: gyroscopes.org*
WWW – *Infinite Energy Magazine of New Energy and Technology: infinite-energy.com/images/pdfs/index.pdf*
WWW – *John Chappell Natural Philosophy Society: natural philosophy.org*
WWW – *Journal of Theoretics: journaloftheorectics.com*
WWW – *KeelyNet: keelynet.com*
WWW – *MagnetoSynergie: magnetosynergie.com*
WWW – Physics Essays: physicsessays.org
WWW – *Project Earth Online: projectearth.com*
WWW – *Rex Research* – rexresearch.com
WWW – *Science Daily: sciencedaily.com*
WWW – ScienceHistory: sciencehystory.org/distillation/magazine/the-frontiersman
WWW – *Simple Guide to Free Energy: free-energy-info-tuks.nl*
WWW – *The General Science Journal*: gsjournal.net
WWW – *Wikipedia: pt.wikipedia.org*
WWW – *Zpernergy: zpenergy.com*
York, D., *In Search of Lost Time*, Institute of Physics Publishing, Philadelphia, PA, and Cambridge University Press, Cambridge. 1997.
Yourgrau, P., *A World without Time*, Basic Books, New York, 2005.
Zee, A., *Fearful Symmetry*, Princeton University Press, Princeton, 2007.
Zeilinger, A., *Dance of the Photons,* Farrar, Straus and Giroux, New York, 2010, of *Einsteins Spuk- teleportation und weitere Mysterian der Quantenphysik*, C. Bertelmann, Germany, 2005.
Znidarsic, F., *Energy Cold Fusion & Antigravity*, Lexington, KY, USA, 2017.

THE AUTHOR

Jose Paulo Hansen, Engineer by the School of Engineering at Mackenzie University, Brazil, with specialization in Physics-Chemistry and Technology in France, with dozens of presentations, scientific and technology publications in Brazil and Abroad. Between companies that acted and scientific activities remained more than 20 years in company founded on a Nobel Prize in Physics Gustaf Dalen and by Scientist Karl Von Linde of Company that participated in refining of uranium by gaseous diffusion for the initial production of nuclear fission. He is currently a Consultant in the Energy field. Resides in Curitiba-Brazil.

Printed in Great Britain
by Amazon